BIG IDEAS MATH®

GREEN

Resources by Chapter

- Family and Community Involvement
- Start Thinking! and Warm Up
- Extra Practice (A and B)
- Enrichment and Extension
- Puzzle Time
- School-to-Work
- Graphic Organizers / Study Help
- Financial Literacy
- Technology Connection
- Life Connections
- Stories in History
- Cumulative Practice
- Unit Project with Rubric

BIG IDEAS LEARNING®

Erie, Pennsylvania

ISBN 13: 978-1-60840-039-3
ISBN 10: 1-60840-039-5

456789-VLP-13 12 11

Contents

About the Resources by Chapter

Family and Community Involvement (English, Spanish, and Haitian Creole)

The Family and Community Involvement letters provide a way to quickly communicate to family members how they can help their student with the material of the chapter. They make the mathematics less intimidating and provide suggestions for helping students see mathematical concepts in common activities.

Start Thinking! and Warm Up

Each Start Thinking! and Warm Up includes two options for getting the class started. The Start Thinking! questions provide students with an opportunity to discuss thought-provoking questions and analyze real-world situations. The Warm Up questions review prerequisite skills needed for the lesson.

Extra Practice

The Extra Practice exercises provide additional practice on the key concepts taught in the lesson. There are two levels of practice provided for each lesson: A (basic) and B (average).

Enrichment and Extension

Each Enrichment and Extension extends the lesson and provides a challenging application of the key concepts.

Puzzle Time

Each Puzzle Time provides additional practice in a fun format in which students use their mathematical knowledge to solve a riddle. This format allows students to self-check their work.

School-to-Work

The School-to-Work projects show students how the concepts of the chapter are applied in various careers.

Graphic Organizers / Study Help

The ready-to-use Graphic Organizers assist students with organizing key ideas and concepts of the chapter

About the Resources by Chapter (continued)

Financial Literacy

The Financial Literacy projects help students develop an awareness of good financial practices. Students will consider risks and trade offs as they practice decision making.

Technology Connection

Each Technology Connection provides opportunities for students to explore mathematical concepts using tools such as scientific and graphing calculators, spreadsheets, geometry software, and the Internet.

Life Connections

The Life Connections connect the mathematics of the chapter to real-life situations.

Stories in History

The Stories in History allow students to learn about the development of mathematical ideas in an entertaining and engaging way.

Cumulative Practice

The Cumulative Practice exercises help students maintain skills through ongoing review.

Unit Project with Rubric

The Unit Projects summarize the key concepts of the unit. They require students to investigate a concept, gather and analyze data, and summarize the results. Scoring rubrics are provided.

Chapter 1

Name ______________________________ Date __________

Chapter 1 Expressions and Number Properties

Dear Family,

When you shop at the grocery store and you place your items on the checkout stand, you usually don't ask, "Does it matter if the price of milk is added before the price of bread?" This is because you know that the order you add the items doesn't matter—but you may not realize you are using one of the mathematical properties of addition!

You and your student can have fun relating the day-to-day mathematics you use to other number properties. For example, you can ask your student:

- "Does it matter if the cashier adds three cans of peas one at a time or multiplies the price of one can by three?" Your student may answer "The total is the same either way."
- "If you take three apples from the display and put two back, is that the same as taking two apples from the display and putting three back? How does that make subtraction different?" Your student may answer "Order does matter with subtraction. You can't put back more than you took originally!"
- "Does it matter if you multiply the price of an item you buy by the quantity or if you multiply the quantity by the price?" Your student may answer, "The total price is the same either way. Order doesn't matter when you multiply."
- "Does order matter with division? If you divide the price of a pack of pudding cups by the number of pudding cups will you get the same answer if you divide the number of pudding cups by the price?" Your student may answer, "No, the answers are different, so order does matter with division."

Your student will be studying concepts like these in math class. See if you and your student can find other examples of mathematical properties or formulas in your day-to-day life.

Happy Shopping!

Nombre ____ Fecha ________

Capítulo 1 Expresiones y Propiedades Numéricas

Estimada Familia:

Cuando compra en la tienda de abarrotes y coloca sus productos en el mostrador de compra, generalmente no pregunta: ¿"Importa si el precio de la leche se suma antes del precio del pan"? Esto se debe a que sabe que no importa el orden en que se suman los productos—¡pero quizás no se está dando cuenta de que está usando una de las propiedades matemáticas de la adición!

Usted y su estudiante pueden divertirse relacionando las matemáticas de la vida diaria que usan con otras propiedades numéricas. Por ejemplo, puede preguntarle a su estudiante:

- ¿"Importa si el cajero suma tres latas de arvejas de una en una o si multiplica el precio de una de ellas por tres"? Su estudiante puede responder: "El total es igual de cualquier forma".
- "Si tomas tres manzanas del mostrador y regresas dos, ¿es igual que tomar dos manzanas y regresar tres? ¿Cómo hace esto que la resta sea diferente?". Su estudiante puede responder: "El orden sí importa con la resta. ¡No se puede regresar más de lo que se tomó originalmente"!
- ¿"Importa si multiplicas el precio de un producto que compras por la cantidad o si multiplicas la cantidad por el precio"? Su estudiante puede responder: "El precio total es igual de cualquier forma. El orden no importa cuando se multiplica".
- ¿"Importa el orden con la división? Si divides el precio de una caja de envases de pudín entre el número de envases de pudín, ¿obtienes la misma respuesta que si divides el número de envases de pudín entre el precio"? Su estudiante puede responder: "No, las respuestas son diferentes, así que el orden sí importa con la división".

Su estudiante estará estudiando conceptos como estos en la clase de matemáticas. Vea si su estudiante puede encontrar otros ejemplos de propiedades o fórmulas matemáticas en su vida diaria.

¡Feliz Compra!

Non ______________________________ Dat ____________

Chapít 1 Espresyon ak Pwopriyete Chif

Chè Fanmi:

Lè w'ap achte nan makèt epi ou mete atik ou yo sou kontwa kès la, ou abityèlman pa mande "Èske sa enpòtan si yo ajoute pri lèt la anvan pri pen an?" Se paske ou konnen lòd ou ajoute atik yo pa enpòtan—men ou gen dwa pa reyalize w'ap sèvi ak youn nan pwopriyete matematik adisyon!

Ou menm ak elèv ou a kapab amize nou fè relasyon ant matematik toulejou nou itilize yo ak lòt pwopriyete chif. Paregzanp, ou kapab mande elèv ou a:

- "Èske sa enpòtan si kesye a ajoute twa bwat pwa youn alafwa oubyen si li miltipliye pri yon bwat pa twa?" Elèv ou a gen dwa reponn "Total la men kèlkeswa jan an."
- "Si ou pran twa pòm nan etalaj la epi ou remete de nan plas yo, èske se menm bagay la si ou pran de pòm nan etalaj la epi ou remete twa nan plas yo? Kijan sa rann soustraksyon diferan?" Elèv ou a gen dwa reponn "Lòd gen enpòtans ak soustraksyon. Ou pa ka remete plis pase sa ou te pran nan kòmansman!"
- "Èske sa enpòtan si ou miltipliye pri yon atik ou achte pa kantite a oubyen si ou miltipliye kantite a pa pri a?" Elèv ou a gen dwa reponn, "Pri total la menm kèlkeswa jan an. Lòd pa enpòtan lè w'ap miltipliye."
- "Èske lòd enpòtan ak divizyon? Si ou divize pri yon pake bwat pouding pa kantite bwat pouding nan èske w'ap jwenn menm repons ak si ou divize kantite bwat pouding nan pa pri a?" Elèv ou a gen dwa reponn, "Non, repons yo diferan, donk lòd enpòtan ak divizyon."

Elèv ou a pwal etidye konsèp tankou sa yo nan kou matematik. Gade si ou menm ak elèv ou a kapab jwenn lòt egzanp pwopriyete oswa fòmil matematik nan lavi toulejou nou.

Fè Bon Pwovizyon!

Activity 1.1

Start Thinking!

For use before Activity 1.1

Have you ever read a newspaper article? A good article should answer the five W's: Who? What? When? Where? Why?

How is a newspaper article similar to a word problem you would solve in math class?

Activity 1.1

Warm Up

For use before Activity 1.1

Evaluate the expression.

1. $7 + 64$

2. $139 - 25$

3. $150 - 67$

4. 5×18

5. 52×9

6. $250 \div 5$

Lesson 1.1 Start Thinking!

For use before Lesson 1.1

How would you plan a trip in a car? You want to travel to Washington, D.C. What information might you need to make your plans? What math is involved?

Lesson 1.1 Warm Up

For use before Lesson 1.1

Write and evaluate an expression for the problem.

1. You buy 2 scarves for \$15 each. How much do you spend?
2. It takes you 15 minutes to bike 5 miles. How long does it take you to bike 1 mile?
3. Four friends split a lunch bill for \$24. How much does each person spend?
4. You have traveled 70 miles of a 200-mile trip. How much farther do you have to travel?

Name ______________________________ Date __________

1.1 Practice A

Evaluate the expression when $c = 4$, $d = 6$, and $e = 10$.

1. $c + 9$ **2.** $e - c$ **3.** $7d$

4. $20 \div c$ **5.** $c + d$ **6.** $15 - d$

7. $10 \bullet e$ **8.** $\dfrac{24}{c}$ **9.** $c + e$

10. $16 - e$ **11.** $30d$ **12.** $\dfrac{60}{d}$

Evaluate the expression when $m = 5$ and $n = 8$.

13. $4m - 7$ **14.** $4n - 3m$ **15.** $\dfrac{6m}{n - 3}$

16. Describe and correct the error in evaluating the expression when $m = 10$.

X
$$\begin{aligned} 3m - 6 &= 3 \bullet 10 - 6 \\ &= 3 \bullet 4 \\ &= 12 \end{aligned}$$

17. Twenty-five students go to lunch. Pizza costs \$3 and sandwiches cost \$2. Twelve students buy pizza. What is the total amount of money spent on sandwiches?

18. Each student in your class must do 14 homework problems. There are 23 students in the class. How many problems will your teacher have to grade?

19. A summer job pays \$5 per hour.

a. You work for 12 hours. How much do you make?

b. After working 24 hours, do you have enough money to buy an MP3 player that costs \$100?

20. You can read 52 pages per hour.

a. You read for an hour and a half. How many pages have you read?

b. You must read 130 pages for homework. Can you complete the assignment in two hours? Explain your reasoning.

Name ______________________________ Date __________

1.1 Practice B

Evaluate the expression when $c = 6$, $d = 8$, and $e = 16$.

1. $c + 12$
2. $d + e$
3. $2c + e$
4. $48 \div d$
5. $\frac{d + e}{c}$
6. $3e - c$
7. $4d + 3$
8. $\frac{2e}{d}$
9. $\frac{2d}{e}$
10. Copy and complete the table.

x	5	8	12
$9x - 5$			
$x^2 + 4$			
$2x^2 + 3x$			

Evaluate the expression when $m = 9$ and $n = 12$.

11. $mn - 4 \bullet 2$
12. $3n - 4m$
13. $\frac{m^2}{3} - 7$
14. On a field trip, your class stops to get ice cream cones. Each cone costs \$3. There are 68 students in your class. How much will the ice cream cones cost?
15. You earn money by washing windows at \$14 per house or washing cars at \$5 per car. How many cars would you have to wash to make the same amount for washing windows at 10 houses?
16. Your basketball team made ten 3-pointers, eight 2-point field goals, and twelve 1-point foul shots. Your opponents made eight 3-pointers, twelve 2-point field goals, and eleven 1-point foul shots. Who won? By how many points did they win?
17. There are 46 legs in a zoo exhibit. Every animal has either two or four legs.
    a. List two possible choices for the number of each type of animal in the exhibit.
    b. The number of 2-legged animals is 2 more than the number of 4-legged animals. How many of each are in the exhibit?
    c. The number of legs in the exhibit increases by 4. What might account for this increase?

Name ____ Date __________

1.1 Enrichment and Extension

Evaluating Algebraic Expressions

Evaluate the expression when $x = 4$, $y = 2$, and $z = 5$.

1. $\dfrac{x}{y} + \dfrac{y}{x}$

2. $\dfrac{xyz}{zy}$

3. $(x + y)^2$

4. $(z - y)^3$

5. $\left[(z - x)^2 + 4\right]^2$

6. $\left[x^2 - (z - y)^2\right] \div 7$

7. Using the expression $4x - 7$, what is the smallest whole number value for x that gives a result that is more than 150?

8. Using the expression $10 + 3x$, what is the smallest whole number value for x that gives a result that is more than 400?

Guess My Value Which value of x makes the expressions equal?

9. $3x,\ x + 4$

10. $5x,\ 7x - 6$

You roll two number cubes. The first number cube determines the value of x and the second number cube determines the value of y in the expression $\dfrac{60}{x} - y$.

11. What is the largest value of the expression?

12. What is the smallest value of the expression?

13. Use the expression to complete the table.

14. If you rolled the number cubes 100 times, which expression value do you think you would get most often? Explain.

First Number Cube (columns) / **Second Number Cube** (rows)

	⚀	⚁	⚂	⚃	⚄	⚅
⚀						
⚁						
⚂						
⚃						
⚄						

Name ____________________ Date __________

1.1 Puzzle Time

Why Were King Edward's Soldiers Too Tired To Fight?

Write the letter of each answer in the box containing the exercise number.

Evaluate the expression when $a = 4$, $b = 3$, and $c = 10$.

1. $5 + a$

2. $b \bullet 6$

3. $c - 2$

4. $28 \div a$

5. $13 - b$

6. $c + 4$

7. $3a$

8. $c \div 5$

9. $b + c$

10. $c - a$

11. $a \bullet c$

12. c^2

Evaluate the expression when $a = 12$, $b = 5$, and $c = 2$.

13. $3a + 4$

14. $5c - 6.7$

15. $\frac{a}{3} + 4$

16. $\frac{25}{b} + 7$

17. $c^2 + 7$

18. $a^2 - 24$

19. $a + 7c$

20. $bc + 11.2$

21. $\frac{a}{c} - 2$

22. $ab - 30c$

23. $4a + 3b - 7c$

Answers for Exercises 1–12

N. 13	**E.** 100
S. 8	**L.** 10
A. 9	**P.** 6
K. 2	**S.** 12
T. 7	**H.** 18
G. 40	**E.** 14

Answers for Exercises 13–23

T. 0	**I.** 26
Y. 120	**D.** 8
S. 40	**E.** 21.2
H. 11	**S.** 49
I. 4	**H.** 3.3
E. 12	

4	17	6	18		2	1	15		3	21	12	20	10	5	16	7	23		8	9	19	11	14	22	13

Activity 1.2 Start Thinking!
For use before Activity 1.2

Many common words have mathematical meanings. What operation do you think of when you hear these phrases?

Bigger than Older than Higher than

Think of some other everyday phrases that have mathematical meanings. Try to write a phrase for each operation: addition, subtraction, multiplication, and division.

Activity 1.2 Warm Up
For use before Activity 1.2

Evaluate the expression when $x = 2$ and $y = 6$.

1. $x + 8$
2. $12 - y$
3. $x \bullet y$
4. $x + y$
5. $\frac{y}{x}$
6. $y - x$

Lesson 1.2 Start Thinking!

For use before Lesson 1.2

What does it mean when you hear the sentence "To get to school, Susie walks 5 blocks less than Bill does."?

Can the sentence be written mathematically?

If you knew how many blocks Bill walked to school, how would you figure out how many blocks Susie walked?

Lesson 1.2 Warm Up

For use before Lesson 1.2

Write the phrase as an expression.

1. the sum of 12 and a number t

2. a number h times 8

3. the difference of 7 and a number f

4. 11 decreased by a number w

5. a number n divided by 2

6. twice a number p

Name ______________________________ Date __________

1.2 Practice A

Match the phrase with the expression.

1.	twelve more than a number x	**A.**	$15 - x$
2.	17 less than a number x	**B.**	$\frac{x}{7}$
3.	15 decreased by a number x	**C.**	$14x$
4.	the quotient of a number x and 7	**D.**	$x + 12$
5.	the product of 9 and a number x	**E.**	$x - 17$
6.	a number x multiplied by 14	**F.**	$9x$

Write the phrase as an expression.

7. 6 more than 4

8. 7 less than 15

9. 3 times a number x

10. the quotient of a number m and 4

11. the product of 8 and a number p

12. twice a number x

13. a number y decreased by 10

14. 7 fewer than a number k

15. the total of a number c and 3

16. the difference of 11 and a number d

17. Describe and correct the error made in writing the phrase as an expression.

12 less than a number x

✗ $12 - x$

18. Write a phrase for $x + 8$.

19. Write a phrase for $20 - n$.

20. You eat five slices of bread. Your friend eats two slices fewer than you. Write an expression that describes the number of slices your friend eats.

21. Your uncle is 2 years older than 3 times your age.

a. You are x years old. Write an expression to describe your uncle's age.

b. You are 12 years old. How old is your uncle?

Name____ Date__________

1.2 Practice B

Match the phrase with the expression.

1. fifteen more than a number x | **A.** $20 - x$

2. 14 less than a number x | **B.** $x \div 9$

3. 20 decreased by a number x | **C.** $x - 14$

4. the quotient of a number x and 9 | **D.** $15 + x$

Write the phrase as an expression.

5. 8 more than 5

6. 11 fewer than 24

7. 6 times a number y

8. the quotient of a number n and 7

9. the sum of 8 and a number e

10. twice a number n plus 6

11. a number x decreased by 13

12. 7 less than 3 times a number m

13. the total of a number f and 3

14. the difference of 25 and a number w

15. You have two cats. Each cat has a litter of 6 kittens. Write an expression that describes the total number of cats and kittens you have.

16. The total of your dinner bill plus tip is $16.00. You left a $3.20 tip. Write an expression to describe this situation.

17. In the sequence, 2, 5, 8, 11, …, the next number will be the result of which expression? Explain your choice.

A. $x + 3$ **B.** $x - 3$ **C.** $3x$ **D.** $x \div 3$

18. You are baking cookies.

Recipe
2 cups sugar
$1\frac{1}{2}$ cups flour
2 eggs

a. You make one and one-half batches of cookies. How many eggs have you used?

b. Each batch makes 24 cookies. You make x batches of cookies, but eat 5 cookies as you are baking. Write an expression for the number of cookies that you have.

c. You make 3 batches of cookies to make bags for a sale. You put 5 cookies in each bag. Given $14(5) + 2 = 3(24)$, what do the terms represent?

Name ______________________________ Date __________

1.2 Enrichment and Extension

Math Joke

Fill in the squares with the missing phrases.

1. Fifteen _____ the _____ 15 and 5 is 5.

   First phrase

   Second phrase

2. Thirty _____ the _____ 18 and 2 is 46.

   First phrase

   Second phrase

3. Six _____ the _____ 8 and 3 is 18.

   First phrase

   Second phrase

Rearrange the shaded letters to answer the following question.

What did the zero say to the eight?

Name______________________________ Date__________

1.2 Puzzle Time

What Did the Rubber Band Do When It Enlisted In The Navy?

Write the letter of each answer in the box containing the exercise number.

Write the phrase as an expression.

1. 4 less than 7
2. 3 times a number x
3. 27 divided by 9
4. the total of 7 and 13
5. 5 fewer than 21
6. 18 added to 16
7. 10 subtracted from a number x
8. the product of 2 and 11
9. the quotient of 17 and a number x
10. the difference of a number x and 6
11. 6 increased by a number x
12. the difference of 15 and a number x

Write the phrase as an expression. Then evaluate when $x = 4$ and $y = 24$.

13. twice a number y
14. the sum of a number x and 19
15. a number x multiplied by 14
16. a number y take away 12
17. 5 less than the quotient of a number y and 2
18. the sum of a number x and 5, all divided by 3
19. 8 more than the product of 3 and a number y
20. the quotient of 36 and the difference of a number x and 2

Answers for Exercises 1–12

T. $21 - 5$	**N.** $x - 10$
O. $27 \div 9$	**I.** $17 \div x$
P. $16 + 18$	**T.** $7 - 4$
S. $15 - x$	**A.** $x - 6$
N. $7 + 13$	**E.** $6 + x$
T. $2 \bullet 11$	**D.** $3x$

Answers for Exercises 13–20

I. 7	**T.** 23
P. 18	**N.** 3
E. 48	**A.** 80
T. 12	**O.** 56

17	5		12	18	10	6	20	11	2		1	3		19	16	8	13	7	14	9	15	4

Activity 1.3 Start Thinking!
For use before Activity 1.3

How many different ways can you add the following numbers? Use your methods to add them.

$$13 + 24 + 27 + 6 + 19 + 11$$

Do you get the same answer each time? Why?

Activity 1.3 Warm Up
For use before Activity 1.3

Find the sum.

1. $13 + 4 + 5$

2. $1 + 4 + 7$

3. $2 + 5 + 8$

4. $14 + 9 + 6$

5. $6 + (12 + 4)$

6. $(23 + 8) + 17$

Lesson 1.3 Start Thinking!
For use before Lesson 1.3

Describe a real-life situation that uses the Commutative Property.

You and two friends are at a movie theater. Describe a situation at the movie theater that demonstrates the Associative Property.

Lesson 1.3 Warm Up
For use before Lesson 1.3

Tell which property is illustrated by the statement.

1. $5 \times 12 = 12 \times 5$

2. $1 + (2 + 4) = (1 + 2) + 4$

3. $14 + q = q + 14$

4. $6 \bullet (9 \bullet 1) = (6 \bullet 9) \bullet 1$

5. $2 \bullet 6r = 6r \bullet 2$

6. $4 + (1 + 6) = (1 + 6) + 4$

Name ______________________________ Date __________

1.3 Practice A

Match the expression with the corresponding property.

1. $3 + 5 = 5 + 3$
2. $12 + 0 = 0$
3. $6 \bullet 7 = 7 \bullet 6$
4. $8 \bullet (10 \bullet 7) = (8 \bullet 10) \bullet 7$
5. $17 \bullet 1 = 17$
6. $8 + (7 + 5) = (8 + 7) + 5$

A. Associative Property of Multiplication
B. Multiplication Property of One
C. Commutative Property of Addition
D. Associative Property of Addition
E. Commutative Property of Multiplication
F. Addition Property of Zero

Simplify the expression. Explain each step.

7. $2 + (a + 8)$
8. $(2x) \bullet 5$
9 $7(4y)$
10. $(0 + c) + 12$
11. $(8e) \bullet 1$
12. $7 \bullet x \bullet 5$

Copy and complete the statement using the specified property.

13. Commutative Property of Addition: $a + 7 = \underline{\ ?\ }$
14. Commutative Property of Multiplication: $12 \bullet 5 = \underline{\ ?\ }$
15. Associative Property of Addition: $6 + (9 + 7) = \underline{\ ?\ }$
16. Associative Property of Multiplication: $2 \bullet (5 \bullet 7) = \underline{\ ?\ }$
17. Describe and correct the error made in simplifying the expression.

✗	$54 \bullet 1 = 1$ Multiplication Property of One

18. You and your friend are selling your old CDs. Your friend sells 14 the first day and x the next day. You sell y the first day and six the next day. Write an expression that shows that the same number of CDs was sold both days. Rewrite your answers using the Commutative Property of Addition.

Name____ Date__________

1.3 Practice B

Simplify the expression. Explain each step.

1. $16 + (f + 4)$

2. $(3y) \bullet 9$

3. $5(8y)$

4. $(0 + n) + 15$

5. $(21 \bullet y) \bullet 1$

6. $10 \bullet x \bullet 4$

7. $34 \bullet y \bullet 0$

8. $35 + (p + 5)$

Copy and complete the statement using the specified property.

9. Commutative Property of Addition: $h + 11 = \underline{\ ?\ }$

10. Commutative Property of Multiplication: $12 \bullet k = \underline{\ ?\ }$

11. Associative Property of Addition: $21 + (9 + 8) = \underline{\ ?\ }$

12. Associative Property of Multiplication: $12 \bullet (5 \bullet 4) = \underline{\ ?\ }$

13. Multiplication Property of One: $18 \bullet w \bullet 1 = \underline{\ ?\ }$

14. Addition Property of Zero: $26 + c + 0 = \underline{\ ?\ }$

15. Describe and correct the error made in identifying the property.

X	$(2 \bullet x) \bullet 4 = 2 \bullet (x \bullet 4)$ Commutative Property of Multiplication

16. On a bike trip, you traveled 21 miles on the first day, 19 miles on the second day, and n miles on the third day.

a. Write an expression for the number of miles traveled in three days.

b. Simplify the expression. Explain each step.

17. You practiced your guitar 37 minutes on Monday, t minutes on Wednesday, and 29 minutes on Friday. Write and simplify an expression for the number of minutes you practiced.

Name ______________________________ Date __________

1.3 Enrichment and Extension

Finding Sums

Find the sum of the whole numbers from 1 to 100. You can use properties of addition to find the sum more quickly than adding the numbers one by one. Carl Friedrich Gauss, a famous mathematician from the late 1700s, is believed to have used a method similar to the one below as a child.

Step 1: Write the sum. Write the sum backwards below it.

$$1 + 2 + 3 + 4 + 5 + \cdots + 96 + 97 + 98 + 99 + 100$$
$$100 + 99 + 98 + 97 + 96 + \cdots + 5 + 4 + 3 + 2 + 1$$

Step 2: Circle the first numbers, second numbers, third numbers, and so on.

$$\begin{matrix} 1 & + & 2 & + & 3 & + & 4 & + & 5 & + & \cdots & + & 96 & + & 97 & + & 98 & + & 99 & + & 100 \\ 100 & + & 99 & + & 98 & + & 97 & + & 96 & + & \cdots & + & 5 & + & 4 & + & 3 & + & 2 & + & 1 \end{matrix}$$

Step 3: Identify the number of pairs and the sum of each pair.

There are 100 pairs. The sum of each pair is 101.

Step 4: Find the sum.

The sum of both lines in step 2 is $100 \bullet 101 = 10{,}100$. Because the sum was written twice, divide by 2 to find the desired sum.

So, the sum of the whole numbers from 1 to 100 is $\dfrac{100 \bullet 101}{2} = 5050$.

Find the sum by using the method above to complete the table.

	Number of pairs	Sum of each pair	Sum
1. 1 through 9			
2. 1 through 20			
3. 1 through 50			
4. 1 through 199			
5. 1 through 1000			

6. Let n be a whole number. Write an expression that can be used to find the sum of the whole numbers from 1 through n.

Name____________________ Date__________

1.3 Puzzle Time

Why Was A Book In The Frying Pan On The Stove?

Write the letter of each answer in the box containing the exercise number.

Tell which property is illustrated by the statement.

1. $4 \bullet x = x \bullet 4$

2. $3 + (11 + p) = (3 + 11) + p$

3. $6 \bullet (r \bullet 9) = (6 \bullet r) \bullet 9$

4. $c + 13.4 = 13.4 + c$

5. $(y + 7) + 0 = y + 7$

6. $b \bullet 1 = b$

Simplify the expression.

7. $5 + (4 + x)$

8. $7(3x)$

9. $(0 + x) + 9$

10. $11 \bullet x \bullet 3$

11. $(6x + 5.1) + 4.2$

12. $(5x) \bullet 12$

13. $(17.3 \bullet x) \bullet 1$

14. $x \bullet 0 \bullet 16$

Answers for Exercises 1–6

B. Addition Property of Zero

O. Commutative Property of Addition

A. Multiplication Property of One

K. Associative Property of Addition

T. Commutative Property of Multiplication

O. Associative Property of Multiplication

Answers for Exercises 7–14

C. $60x$ **O.** $x + 9$

K. $21x$ **I.** $33x$

S. $9 + x$ **A.** 0

O. $6x + 9.3$ **W.** $17.3x$

10	1		13	6	7		14		12	3	11	8	5	4	9	2

Activity 1.4 Start Thinking!

For use before Activity 1.4

John says that 16×40 is the same thing as 8×80. Is he correct? How do you know?

Activity 1.4 Warm Up

For use before Activity 1.4

Find the product.

1. 7×13

2. 9×23

3. 14×4

4. 5×41

5. 6×37

6. 18×7

Lesson 1.4 Start Thinking!

For use before Lesson 1.4

How can you relate the Distributive Property to preparing for a birthday party?

Lesson 1.4 Warm Up

For use before Lesson 1.4

Use the Distributive Property and mental math to find the product.

1. 4×29

2. 3×42

3. 7×52

4. 6×19

5. $5(48)$

6. $8(67)$

Name ______________________________ Date __________

1.4 Practice A

Use the Distributive Property to complete the statement.

1. $6(7 + 8) = \underline{\ ?\ } \times \underline{\ ?\ } + 6 \times \underline{\ ?\ }$

2. $\underline{\ ?\ }(8 + 6) = 7 \times \underline{\ ?\ } + \underline{\ ?\ } \times \underline{\ ?\ }$

3. $12(10 + \underline{\ ?\ }) = \underline{\ ?\ } \times 10 + \underline{\ ?\ } \times 6$

Use the Distributive Property and mental math to find the product.

4. 5×23

5. 6×25

6. 4×61

7. 8×42

8. $9(54)$

9. $7(59)$

Use the Distributive Property to simplify the expression.

10. $4(x + 6)$

11. $8(c - 5)$

12. $7(y + 8)$

13. $9(e - 4)$

14. $6(4 + n)$

15. $7(3 + x + 4)$

16. $9(w + 6) + 4$

17. $5(3 + m) - 7$

18. Describe and correct the error in rewriting the expression.

X	$5(x + 9) = x + 45$

19. Each day you do homework for m minutes and watch TV for 30 minutes. Which expression can you use to find how many minutes you do both activities in 5 days? Explain your reasoning.

A. $5m + 30$ **B.** $5(m + 6)$ **C.** $5(m + 30)$ **D.** $m(5 + 30)$

20. The school office uses r reams of white paper and 3 reams of blue paper every day. Use the Distributive Property to write and simplify an expression for how much paper the school office uses in five days.

Name________________________________ Date__________

1.4 Practice B

Use the Distributive Property and mental math to find the product.

1. 6×24

2. 9×27

3. 7×89

4. 8×78

5. 5×72

6. 4×99

Use the Distributive Property to simplify the expression.

7. $8(a + 6)$

8. $7(p - 5)$

9. $10(9 + x)$

10. $12(f - 5)$

11. $6(2 + a + 9)$

12. $4 + 3(x + 5)$

13. $8(t + 5) + 15$

14. $4(y + 11) - 10$

15. Describe and correct the error in rewriting the expression.

✗ $6(x + 7) = 6x + 13$

16. A restaurant uses c cups of flour for baking rolls and 12 cups of flour for baking pies each day. Use the Distributive Property to write and simplify an expression for how much flour the restaurant uses in seven days.

17. You are y years old. Your sister is 3 years older than you. Your uncle is 4 times older than your sister. Write and simplify an expression that represents your uncle's age.

Use the Distributive Property to rewrite the expression as a product.

18. $8x + 40$

19. $36 + 9x$

20. $20 - 2p$

21. $6 + 18y + 24$

22. Add one set of parentheses to the expression $8 \bullet x + 10 + 2x + 4 \bullet x + 8 - 6$ so that it is equivalent to $2(7x + 18)$.

Name ______________________________ Date __________

1.4 Enrichment and Extension

Using the Distributive Property

An electronics company sells two types of headphones. The in-ear model costs $50. The over-the-ear model costs $100.

The table shows how many units of each model Stores A, B, and C want to buy.

	Store A	Store B	Store C
In-ear	128	192	256
Over-the-ear	96	192	160

1. How can the Distributive Property and mental math be used to find the total amount the company receives from the three stores for the in-ear headphone model?

2. How can the Distributive Property and mental math be used to find the total amount the company receives from the three stores for the over-the-ear headphone model?

3. What is the total amount the company receives from the three stores for both models?

4. Suppose each store orders twice as many of the in-ear models as the over-the-ear models. Without doing any math, explain how you know that each store will pay the same amount for each type of model.

5. The steps below show that the Distributive Property

$$a(b + c) = ab + ac$$

can be written as $(b + c)a = ba + ca$. Fill in each blank with a property you know to justify the steps.

$(b + c)a = a(b + c)$ ______ **Property of** ______

$= ab + ac$ ______ **Property**

$= ba + ca$ ______ **Property of** ______

Name________________________________ Date__________

1.4 Puzzle Time

Why Was The Shoelace Told To Stay After School?

Write the letter of each answer in the box containing the exercise number.

Use the Distributive Property and mental math to find the product.

1. 4×22

2. 8×75

3. $11(42)$

4. $6(89)$

5. 17×51

6. 9×28

7. $7(62)$

8. $3(35)$

Use the Distributive Property to simplify the expression.

9. $2(x - 8)$

10. $4(x + 5)$

11. $9(x - 3)$

12. $7(x - 10)$

13. $6(9 + x)$

14. $8(13 + x)$

15. $5(x + 2)$

16. $10(2 + x + 3)$

17. $3(9 + x + 4)$

18. $7(11 + 1 + x)$

19. $4(x + 6) + 9$

20. $11 + 5(x + 3)$

Answers for Exercises 1–8

T. 600 **I.** 462

N. 105 **O.** 867

K. 434 **A.** 534

T. 88 **C.** 252

Answers for Exercises 9–20

N. $5x + 10$ **A.** $6x + 54$

O. $10x + 50$ **T.** $7x + 84$

C. $4x + 20$ **N.** $7x - 70$

W. $3x + 39$ **O.** $4x + 33$

T. $8x + 104$ **Y.** $9x - 27$

U. $5x + 26$ **S.** $2x - 16$

19	12		4	10	6	16	20	8	18		3	1		17	13	9		7	15	5	14	2	11

Activity 1.5 Start Thinking!

For use before Activity 1.5

New carpeting needs to be installed in a bascmcnt with thc floor plan shown. Copy the diagram and divide the figure into rectangles to represent rectangular pieces of carpet.

How can you divide the floor plan into 4 different polygons? 5 different polygons?

Activity 1.5 Warm Up

For use before Activity 1.5

Name the polygon.

1. 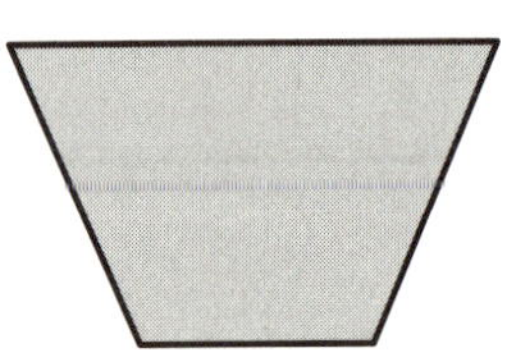

2. 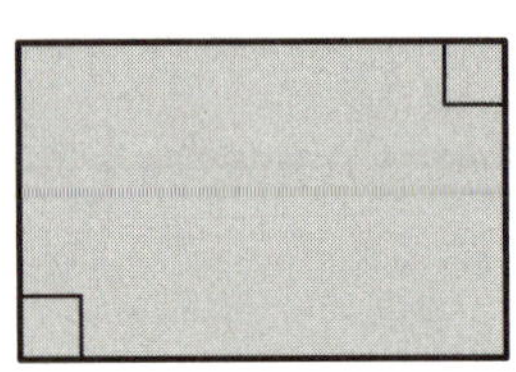

3. 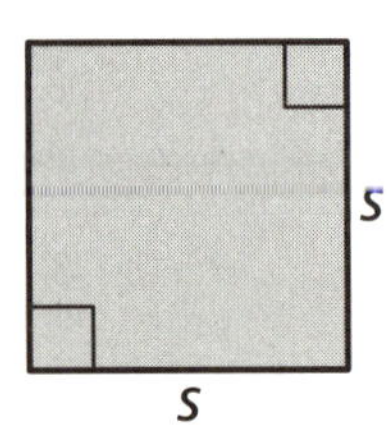

4. 

5. 

6. 

Lesson 1.5 Start Thinking!
For use before Lesson 1.5

You have used formulas for finding areas.

What are some other common formulas that you know?

Lesson 1.5 Warm Up
For use before Lesson 1.5

Use a formula to find the area of the figure.

1.
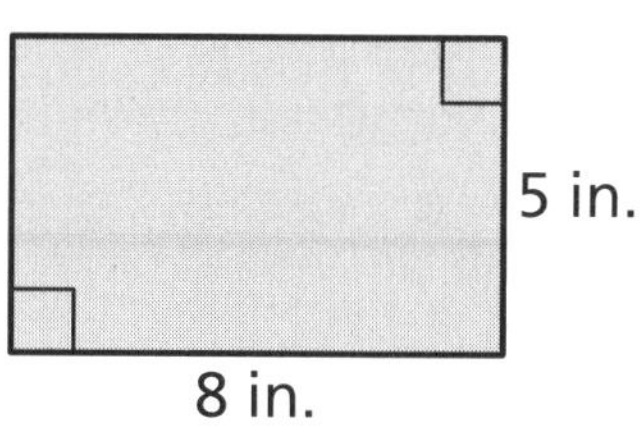

2.
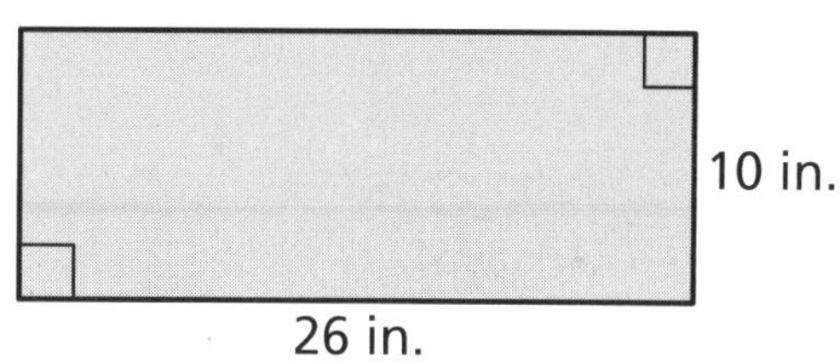

3.
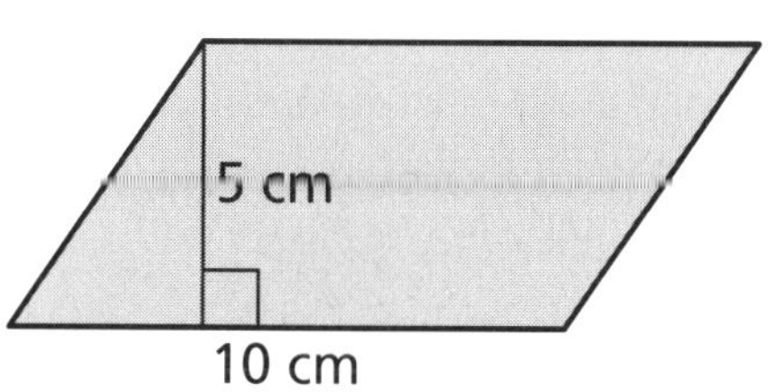

4.
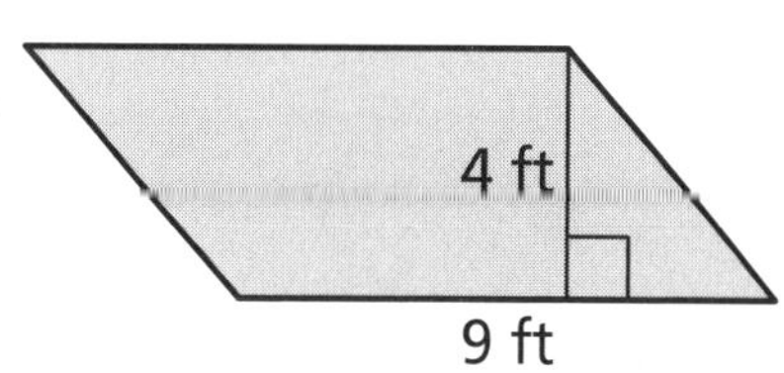

5.
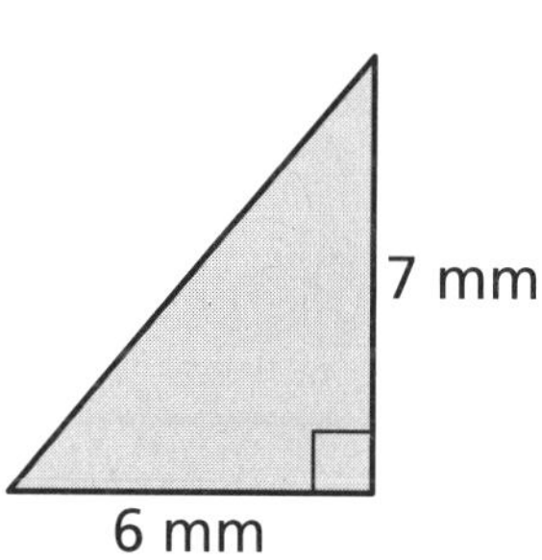

6.
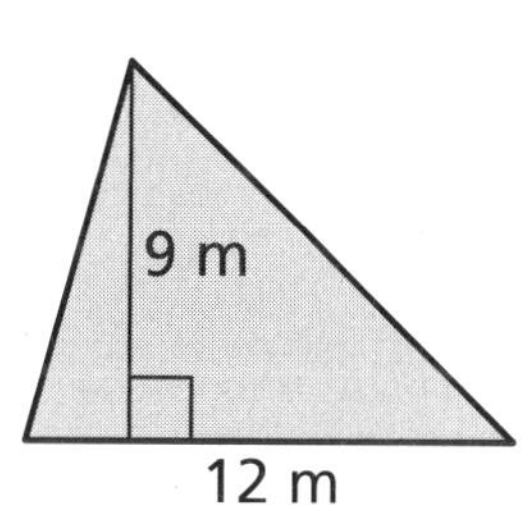

Name ______________________________ Date __________

1.5 Practice A

Evaluate the formula for the given value(s) of the variable(s).

1. $M = 220 - a;\ a = 14$

2. $A = \ell \times w;\ \ell = 12$ in. and $w = 9$ in.

3. $V = \ell wh;\ \ell = 8,\ w = 5$, and $h = 4$

Use a formula to find the area of the figure.

4.

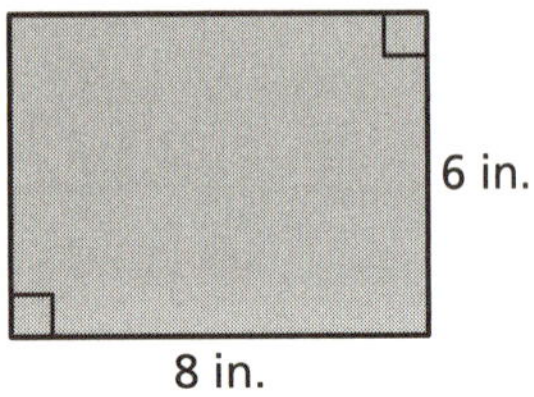

5.

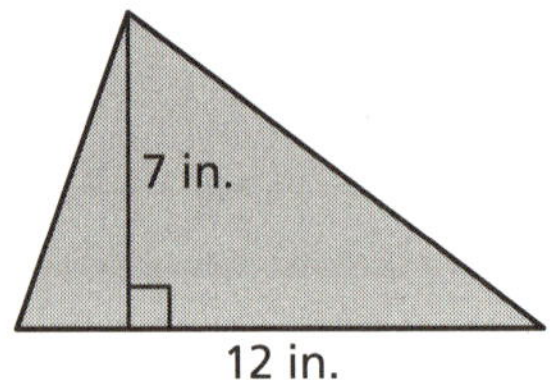

6.

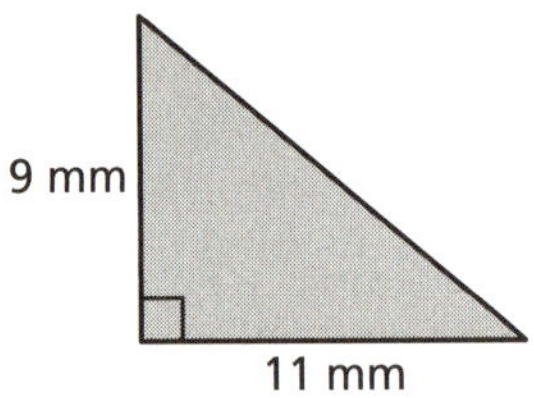

7.

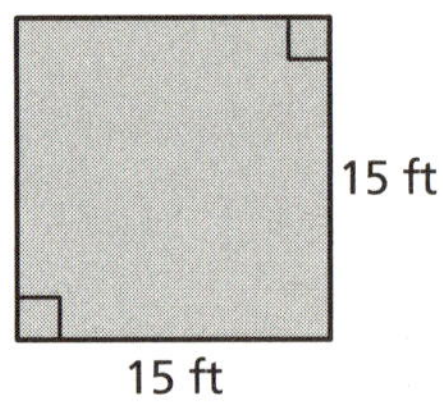

8.

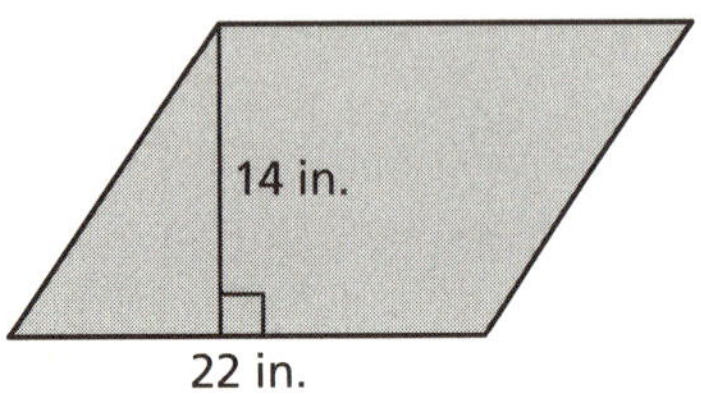

9.

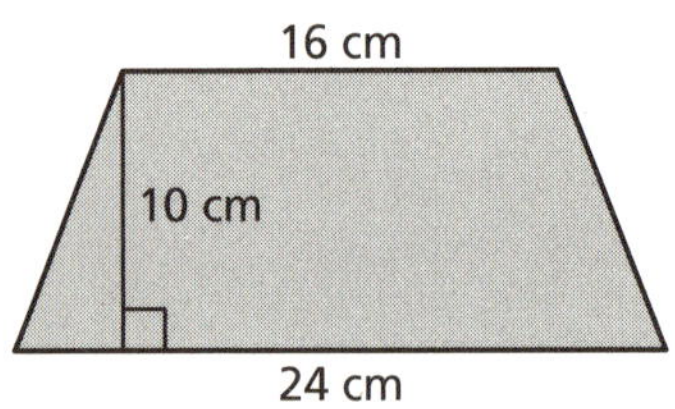

10. The formula $M = 220 - a$ gives a person's maximum heart rate M, where a is the person's age in years. What is the maximum heart rate of a person who is 61 years old?

11. A person's earned run average is the average number of earned runs given up per nine innings. The formula below shows the earned run average, where R is the number of earned runs and I is the number of innings pitched. What is the earned run average of a pitcher who gave up 40 earned runs in 80 innings?

$$\text{Earned Run Average} = \frac{9R}{I}$$

12. The area of a rectangle is 481 square inches. The length is 37 inches. Explain how you can find the width.

Name________________________________ Date__________

1.5 Practice B

Use a formula to find the area of the figure.

1.

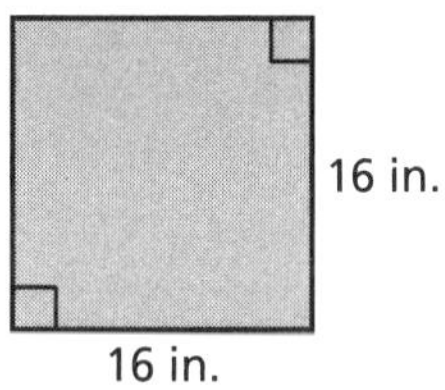

2.

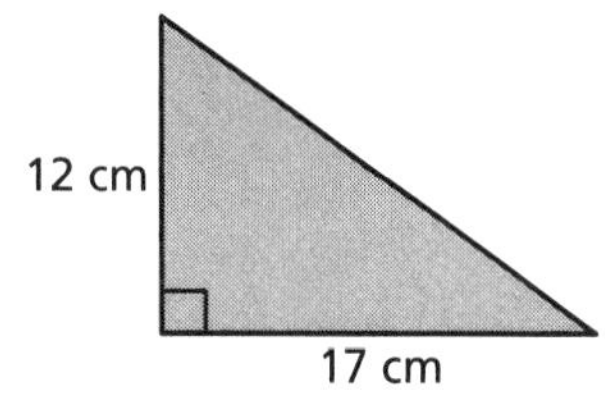

3.

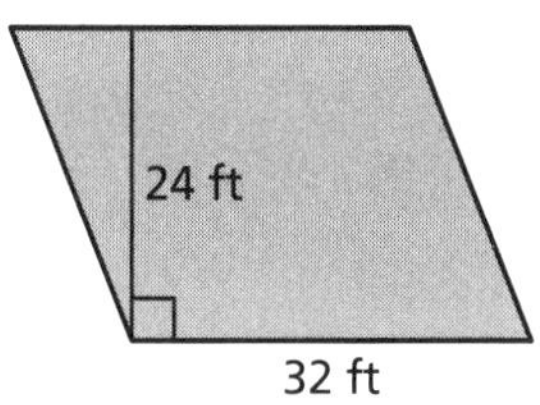

4.

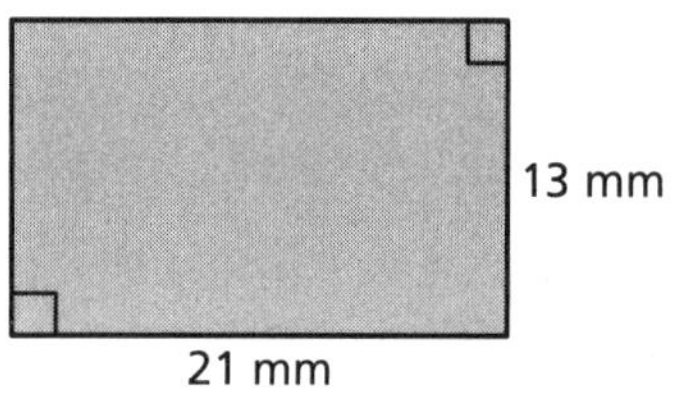

5.

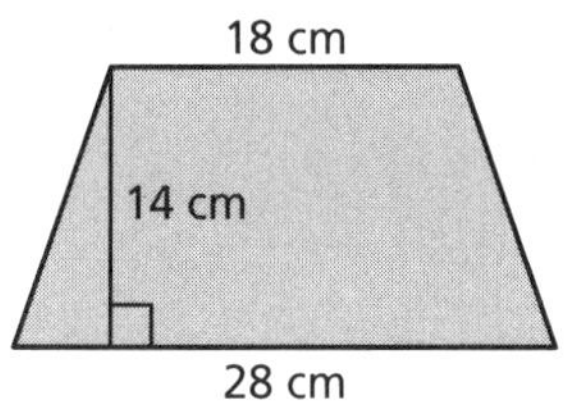

6.

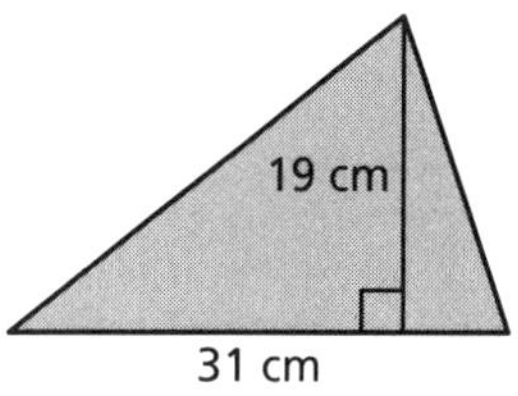

Write a formula for the area of the shaded region in terms of *x*.

7.

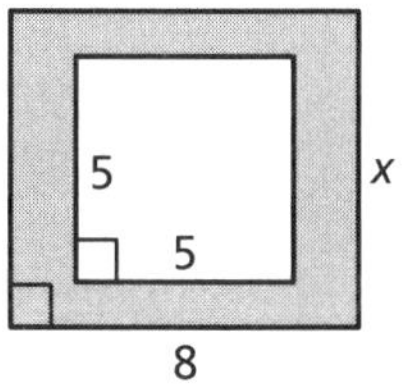

8.

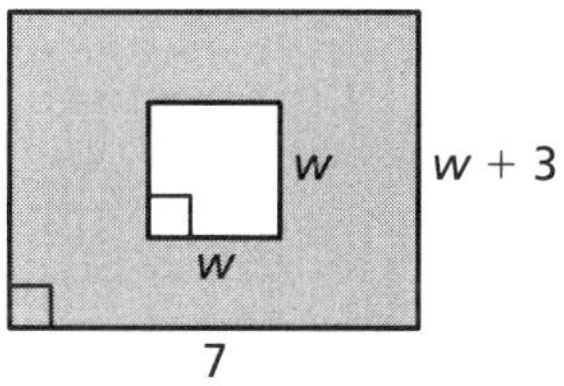

9.

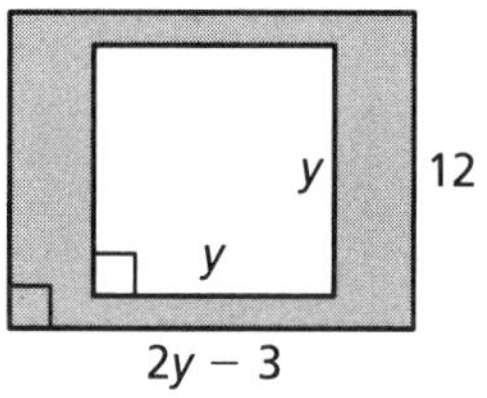

10. The formula $M = 220 - a$ gives a person's maximum heart rate M, where a is the person's age in years. What is the maximum heart rate of a person who is 12 years old?

11. Find three possible sets of dimensions for a parallelogram whose area is 120 square feet.

Name ______________________________ Date ________

1.5 Enrichment and Extension

The Energy of Motion

In science class, you may have learned about *kinetic energy*, which is the energy of motion. Here are some examples of things that have kinetic energy:

- **Your own body when you walk or run**
- **Water falling down a waterfall**
- **A roller coaster going down a hill**

The amount of kinetic energy depends on the *mass* of an object and its speed. To understand the difference between mass and weight, think about your weight on Earth and how it changes if you were on the moon. It is less. But your mass is the same no matter where you are, because the matter that makes up your body does not change.

1. The formula for kinetic energy E is equal to the product of the mass of the object m (in kilograms) and the square of the speed of the object v (in meters per second), divided by 2. Write a formula for kinetic energy.

2. Which of the following objects do you think has the greatest kinetic energy? Why?

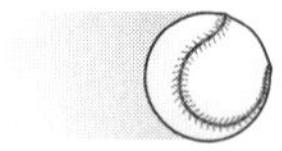

Kinetic energy is measured using a unit called a Joule (J), named after physicist James Prescott Joule.

3. Which has more kinetic energy—a 10-kilogram dog running at a speed of 8 meters per second or an 8-kilogram dog running at a speed of 10 meters per second? Explain.

4. A 1200-kilogram car is traveling at a speed of 22 meters per second. Find the kinetic energy of the car.

5. The speed of the car in Exercise 4 is decreased by one half to 11 meters per second. Did the kinetic energy of the car decrease by one half? Explain.

Name______________________________ Date__________

1.5 Puzzle Time

Where Does A Penguin Store All Of Its Money And Other Valuables?

Write the letter of each answer in the box containing the exercise number.

Use a formula to find the area of the figure.

1.

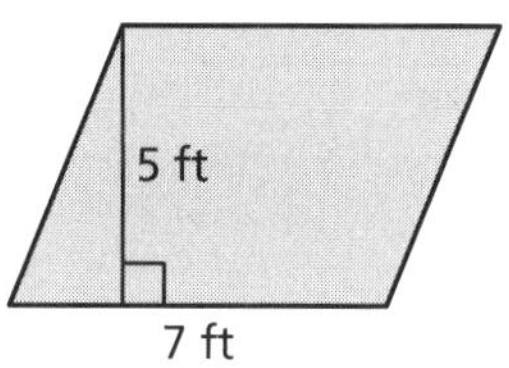

2.

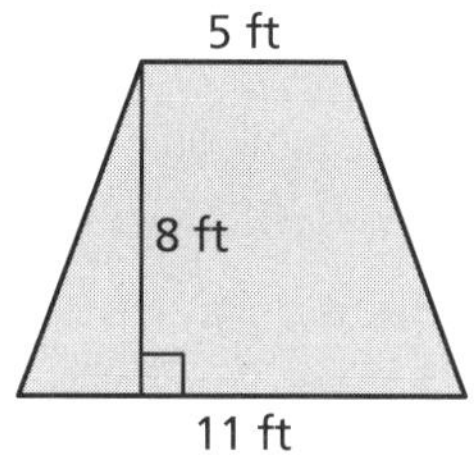

3.

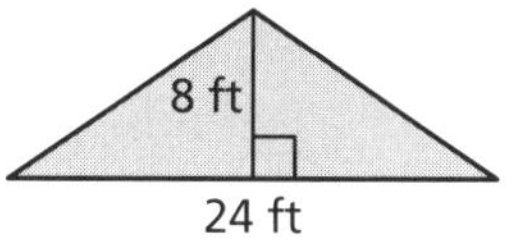

4.

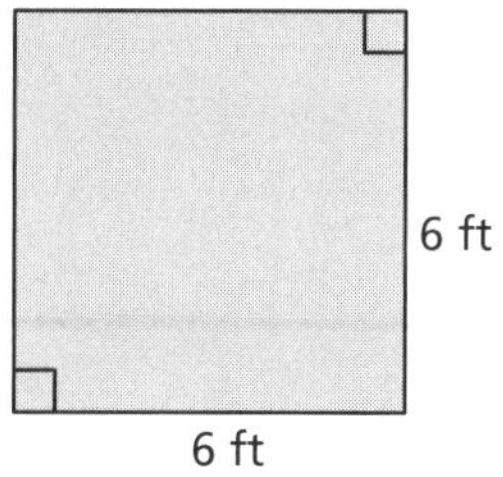

5.

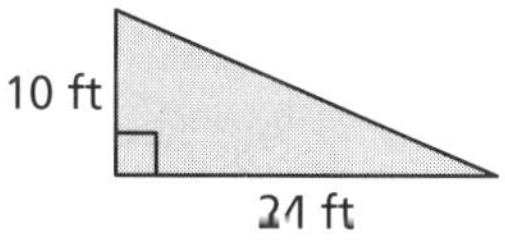

6.

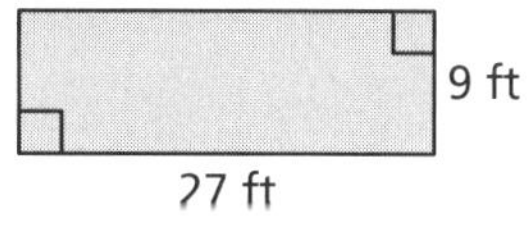

Answers for Exercises 1–6

N.	243 ft^2	**T.**	240 ft^2
A.	35 ft^2	**O.**	96 ft^2
E.	121.5 ft^2	**K.**	64 ft^2
R.	17.5 ft^2	**W.**	36 ft^2
S.	192 ft^2	**M.**	18 ft^2
B.	120 ft^2	**F.**	220 ft^2

Answers for Exercises 7–9

S. $12x - 9$ square units

I. $16 - 5x$ square units

D. $x^2 - 30x - 20$ square units

N. $5x - 16$ square units

C. $9 - 12x$ square units

A. $30x - 20 - x^2$ square units

Find the area of the shaded region in terms of *x*.

7.

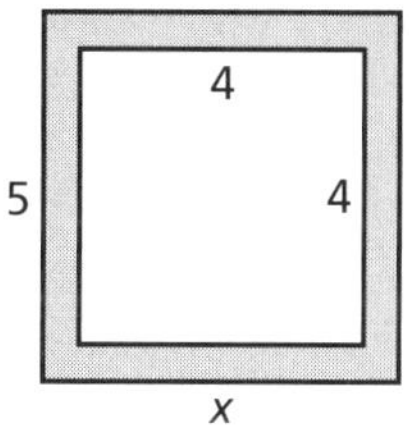

8.

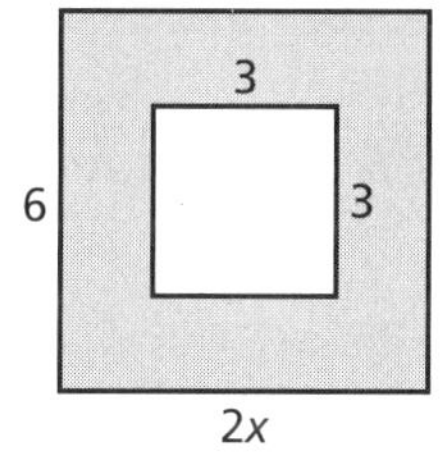

9.

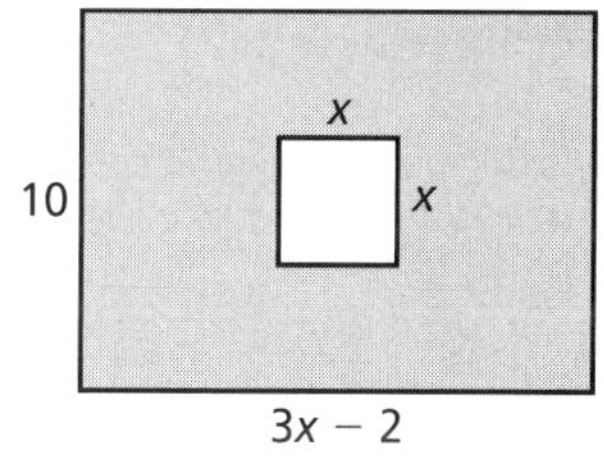

9		8	6	3	4		5	1	7	2

Name ______________________________ Date __________

Chapter 1

School-to-Work

For use after Section 1.3

Making Decisions as a Flooring Contractor

As a flooring contractor, you need to determine the amount to charge for a kitchen tile project. Tile is to be laid everywhere except under the cabinetry.

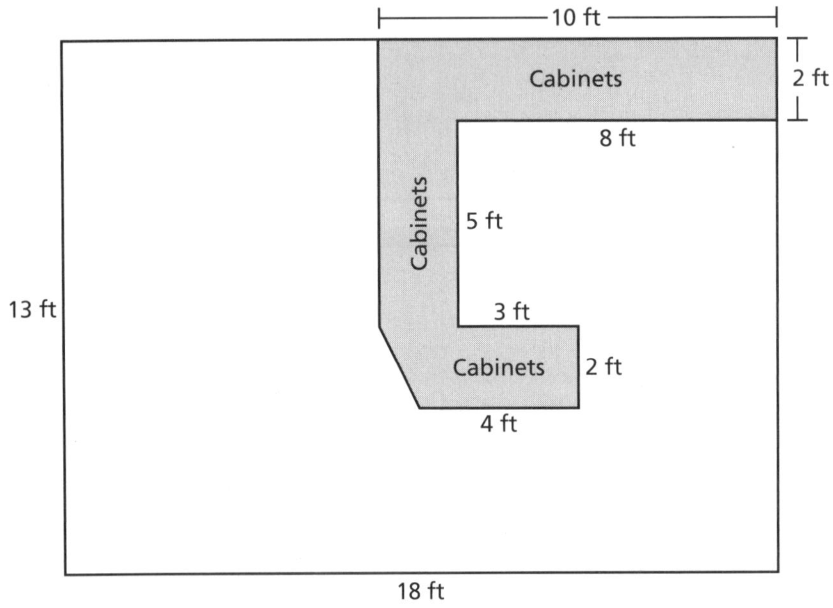

1. Make a plan for finding the area of the floor to be tiled. Include the geometric formulas you will use to find the area. Then find the area of the floor that will be tiled.

2. You estimate that it will take 4 minutes to install each square foot of tile. Find the number of hours it will take you to complete the job.

3. Determine how much you will charge per hour for labor. Remember that your labor costs need to be competitive or you will not get any work.

4. What other costs do you need to consider? Write an expression to represent the cost of your expenses if the customer chooses a tile that costs t dollars per square foot.

5. How much will you charge the owner of the kitchen shown above to install tile that costs \$8 per square foot? Provide a detailed explanation that you can share with the customer.

6. At the last minute, your customer selects a new style of tile that costs \$10 per square foot. How much will you increase the amount you charge the customer?

Name____ Date __________

Chapter 1 Study Help

You can use an **information frame** to help you organize and remember concepts.

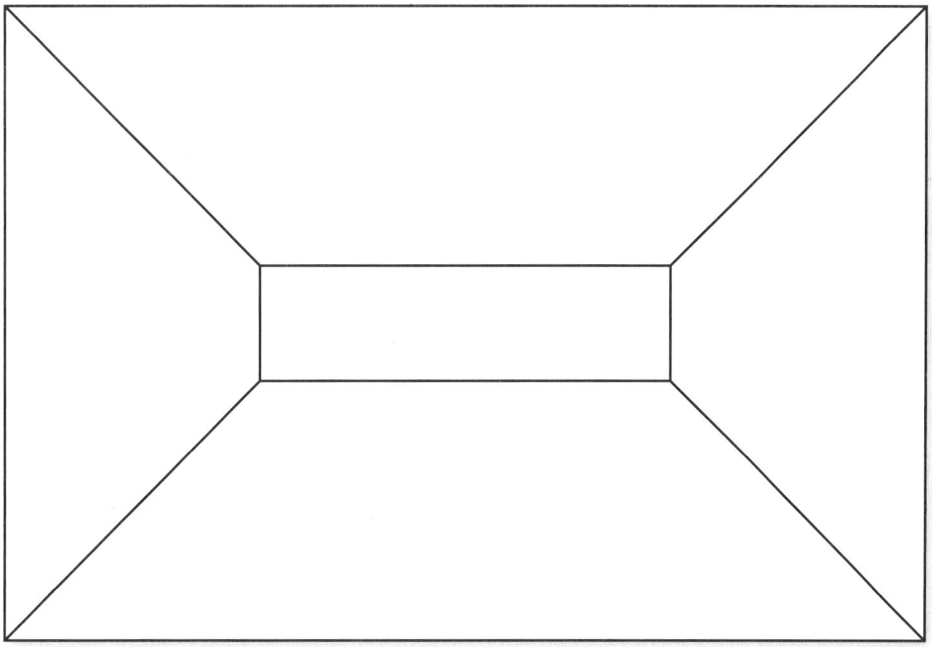

On Your Own

Make an information frame to help you study these topics.

1. Commutative Property of Multiplication
2. Associative Property of Addition
3. Associative Property of Multiplication
4. Addition Property of Zero
5. Multiplication Property of Zero
6. Multiplication Property of One
7. Expressions

After you complete this chapter, make information frames for the following topics.

8. Distributive Property
9. Formulas

Name ______________________________ Date __________

Chapter 1

Financial Literacy

For use after Section 1.5

Planning a Fundraising Event

Your school is raising money for a new sports field. The student council is sponsoring a contest to determine the best plan for raising the money. Each group needs to submit a detailed plan of their fundraising event. The group with the best plan will win the contest.

- Consider some fundraising events you think your class could organize. Decide on a fundraising event.
- What are some ways you can advertise your event? What are good locations for you to advertise? How many potential customers will see the advertisements? Determine how much the different advertising options will cost. Based on your findings, decide on two ways to advertise your event.
- Make a list of advertising materials and supplies you will need to prepare for your event. Research the stores in your area to find the least expensive place to buy your supplies. Based on your research, what is the total cost for advertising materials and supplies?
- Decide on a location. Which location will have the most customers? Do you have to pay for the location? Will you have access to everything you need? What else do you need to consider?
- Decide on the price you will charge for your event. Is there a maximum price that you think you could charge? Explain.
- How many people will need to attend your event so that you raise more than you spend? Do you need to reconsider the price you decided to charge?
- Write an expression for your total income. Write another expression for your total expenses.
- Your profit from the event is profit = income – expenses. Write a formula for your profit. How much money would you raise if you had 50 people participate in your event? 100 people?
- What do you think is a reasonable amount of money to make at your event? Explain.
- Prepare your presentation. Include the price you have decided to charge for the event, why you decided to charge that price, a description of the best location for the event, why you chose the location, a list of the supplies you need and where you will purchase these items, a table showing your income and expenses, and anything else you think will convince the student council to choose your plan.

Name______________________________ Date__________

Chapter 1 Cumulative Practice

Simplify the expression.

1. $4(3 + c)$	**2.** $42 - 6 \bullet 7$	**3.** $9 + (m + 8)$
4. $2(3w)$	**5.** $4 + 2 \bullet 8^2$	**6.** $x \bullet 1 \bullet 3.2$
7. $7 + p + 13$	**8.** $5(y + 10)$	**9.** $18 \div 3 + 3$
10. $23 - 5 - 2$	**11.** $11 \bullet h \bullet 2$	**12.** $2(6 + a) + 14$
13. $9(k + 13)$	**14.** $(g + 2.1) + 6.3$	**15.** $4(3v)$

16. You and five friends go to the zoo. The cost of each ticket is \$15. Use the Distributive Property and mental math to find the product 6×15 for the total cost.

17. The expression $10 + v + 15$ represents the perimeter of a triangle with side lengths 10, 15, and v. Simplify the expression.

Evaluate the expression for the given value(s) of the variable(s).

18. s^2 when $s = 3$	**19.** $\frac{8p}{3}$ when $p = 6$
20. $7 + \frac{18}{c}$ when $c = 3$	**21.** $k^2 - 3w$ when $k = 7$ and $w = 5$
22. $8 - a \div 5$ when $a = 15$	**23.** $m - 7$ when $m = 9$
24. $rc - 2.3$ when $r = 14$ and $c = 20$	**25.** $4y + z - 5$ when $z = 1$ and $y = 6$

26. The expression $5g + 3w$ represents the total weight in pounds of g bags of gerbil food and w bottles of water. What is the total weight of 8 bags of gerbil food and 4 bottles of water?

Write the phrase as an expression.

27. the sum of 11 and a number n	**28.** the difference of 7 and a number y
29. 3 times a number x	**30.** twenty divided by 4
31. 13 fewer than a number k	**32.** the total of a number c and 5
33. the quotient of a number p and 8	**34.** the product of 25 and a number g

35. Your brother is 5 years older than you. Write an expression for your brother's age.

36. Your friend lives six fewer than three times as many miles away from school as you. Write an expression for the number of miles your friend lives from school. How far away does your friend live from school if you live 4 miles away from school?

Chapter 2

Name________________________________ Date __________

Chapter 2 Multiplying and Dividing Fractions

Dear Family,

Many of us have a number of recipes that we enjoy preparing. Perhaps they are old family recipes or simply someone's favorite meal. Some recipes we can prepare without much effort: a quarter cup of butter, a quarter cup of flour, and two cups of milk for a white sauce.

Homemade Macaroni & Cheese

1/4 c butter	2 1/2 c grated cheese
1/4 c flour	1/2 16 oz box of macaroni
2 c milk	salt & pepper

Make the macaroni using instructions on the box. Make a roux of the butter and flour. Add the milk over low heat while stirring until smooth. Add cheese to thickened sauce. Salt & pepper to taste. Stir in macaroni and serve.

However, when we have guests over for dinner, we find these familiar recipes have to be adjusted. Maybe we have to make three times as many servings. Now we need three quarter cups of butter, three quarter cups of flour, and so on.

Sometimes we want to make a smaller portion, such as when part of the family is away. Now we find ourselves using half a quarter cup of butter, half a quarter cup of flour, and so on.

You and your student can discuss strategies for preparing a meal when you have to change the recipe. For example, you might ask your student:

- "We have to use two quarter cups of sugar. Should we use the 1/4 cup measure twice, or use the 1/2 cup measure?" Your student may answer, "The amount will be the same either way."
- "The recipe calls for 2/3 cup of milk, but we only want a half batch. What measuring cup should we use?" Your student may answer, "Use the 1/3 cup measure, but only once."
- "Grandma's brownie muffin recipe makes enough for three dozen muffins, but we only want one dozen. What should we do?" Your student may answer, "We can divide the recipe measures by three."

Sometimes you have to make a judgment call. How do you cut the recipe in half, if it calls for three eggs? Talk with your student about different strategies for changing a recipe like this.

Enjoy your cooking time together!

Nombre ______________________________ Fecha ________

Capítulo 2 Multiplicando y Dividiendo Fracciones

Estimada Familia:

Muchos de nosotros tenemos una serie de recetas que disfrutamos preparar. Tal vez son antiguas recetas de la familia o simplemente la comida favorita de alguien. Algunas recetas las preparamos sin mucho esfuerzo: un cuarto de taza de mantequilla, un cuarto de taza de harina y dos tazas de leche para una salsa blanca. Sin embargo, cuando tenemos invitados a cenar, tenemos que adaptar estas recetas familiares. Quizás tenemos que preparar el triple de porciones. Ahora necesitaremos tres cuartos de taza de mantequilla, tres cuartos de taza de harina, etc.

Macarrones con Queso Preparados en Casa

1/4 taza de mantequilla	2 1/2 tazas de queso rallado
1/4 taza de harina	1/2 caja de macarrones de 16 onzas
2 tazas de leche	sal y pimienta

Prepare los macarrones siguiendo las instrucciones de la caja. Haga una salsa con la mantequilla y la harina. Vierta la leche a fuego lento y revuelva hasta que quede suave. Agregue el queso a la salsa espesa. Ponga sal y pimienta al gusto. Vierta los macarrones, revuelva y sirva.

A veces queremos hacer una porción más pequeña, cuando por ejemplo, parte de la familia no se encuentra en casa. Entonces tendremos que usar medio cuarto de taza de mantequilla, medio cuarto de taza de harina, etc.

Usted y su estudiante pueden conversar acerca de las estrategias que hay que usar para preparar una comida cuando tiene que cambiar la receta. Por ejemplo, puede preguntar a su estudiante:

- "Tenemos que usar dos cuartos de taza de azúcar. ¿Usamos dos veces la medida de 1/4 de taza, o usamos la medida de 1/2 taza?" Su estudiante puede responder: "La cantidad será igual de cualquier manera."
- "La receta indica 2/3 de taza de leche, pero sólo queremos tener la mitad de la hornada. ¿Qué taza de medir debemos usar?" Su estudiante puede responder: "Usemos la medida de 1/3 de taza, pero sólo una vez."
- "La receta de brownie muffins de la abuela rinde tres docenas de muffins, pero sólo queremos una docena. ¿Qué hacemos?" Su estudiante puede responder: "Podemos dividir las medidas de la receta entre tres."

A veces tenemos que tomar nuestra propia decisión. ¿Cómo dividir una receta en la mitad, si ésta incluye tres huevos? Converse con su estudiante acerca de estrategias diferentes para cambiar una receta como ésta.

¡Disfruten su momento de cocina juntos!

Non ____ Dat ___________

Chapít 2 Ann Miltipliye ak Divize Fraksyon

Chè Fanmi:

Anpil nan nou gen yon kantite resèt ke nou renmen prepare. Petèt se ansyen resèt familyal oswa senpleman pla prefere yon moun. Gen resèt nou kapab prepare san anpil efò: yon ka tas bè, yon ka tas farin, ak de tas lèt pou yon sòs blanch.

Makawoni ak Fwomaj Fèt Lakay

1/4 t. bè	2 1/2 t. fwomaj rape
1/4 t. farin	1/2 bwat makawoni 16 ons
2 t. lèt	sèl ak pwav

Swiv enstriksyon ki sou bwat la pou fè makawoni an. Fè farin nan pran you ti koulè nan bè a. Ajoute lèt la sou dife dou pandn w'ap brase jiskaske li vin inifòm. Ajoute fwomaj nan sòs epè a. Mete otan sèl ak pwav ou vie. Ajoute makawoni an pandan w'ap vire li epi sèvi.

Sepandan, lè nou gen envite ki vin lakay nou, nou wè ke resèt sa yo nou fin abitye avèk yo bezwen ajiste. Petèt nou gen pou prepare twa fwa plis pòsyon. Kounyeya nou bezwen twa ka tas bè, twa ka tas farin, ets.

Pafwa nou vle prepare pòsyon ki pi piti, tankou lè yon pati nan fanmi an pa la. Kounyeya nou vin trouve nou ap sèvi ak mwatye yo ka tas bè, mwatye yon ka tas farin, ets.

Ou menm ak elèv ou a kapab diskite sou estrateji pou prepare yon repa lè ou oblije chanje resèt la. Paregzanp, ou ta gen dwa mande elèv ou a:

- "Nou dwe sèvi ak de ka tas sik. Èske nou ta dwe sèvi ak mezi 1/4 tas la de fwa, oubyen sèvi ak mezi 1/2 tas la?" Elèv ou a gen dwa reponn, "Kantite a ap menm kèlkeswa jan an."
- "Resèt la mande 2/3 tas lèt, men nou sèlman vle mwatye yon founen. Ki tas amezire nou ta dwe itilize?" Elèv ou a gen dwa reponn, "Sèvi ak mezi 1/3 tas la, men yon fwa sèlman."
- "Resèt ponmkèt chokola Granmè a ase pou twa douzèn ponmkèt, men nou bezwen sèlman yon douzèn. Kisa pou n' fè?" Elèv ou a gen dwa reponn, "Nou kapab divize mezi resèt la pa twa."

Pafwa ou dwe pran yon desizyon poukont ou. Kijan pou ou koupe resèt la an de, si li mande twa ze? Pale avèk elèv ou a sou diferan estrateji pou chanje yon resèt konsa.

Pase yon bon tan kwizin ansanm!

Activity 2.1

Start Thinking!

For use before Activity 2.1

You have a bag of multi-colored gumballs. What method would you use to estimate the fraction of the gumballs that are red?

Activity 2.1

Warm Up

For use before Activity 2.1

Tell what fraction is represented by the model.

1. 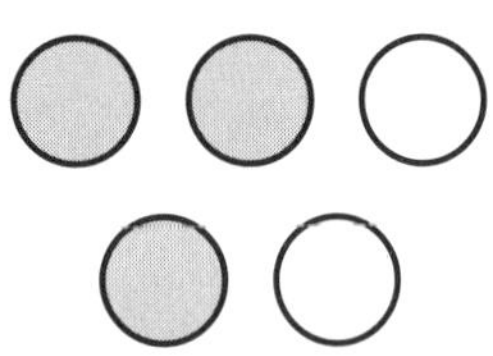

2. 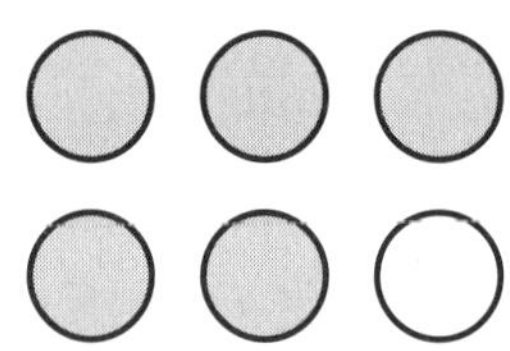

3. 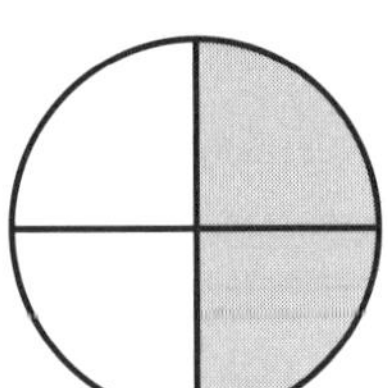

4. 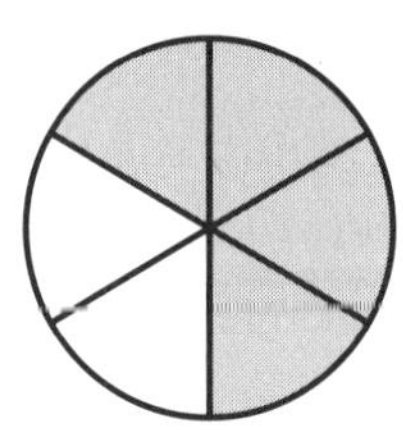

5. 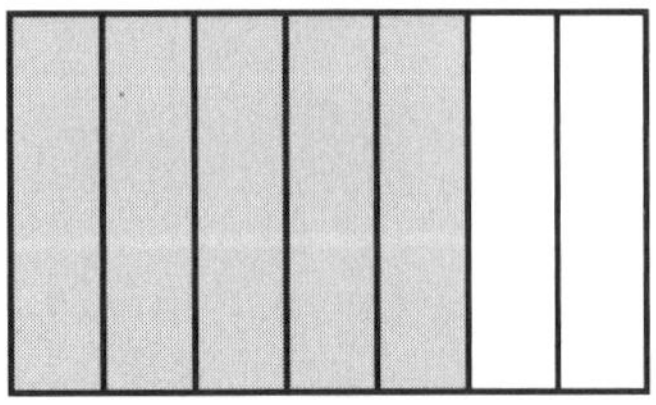

6. 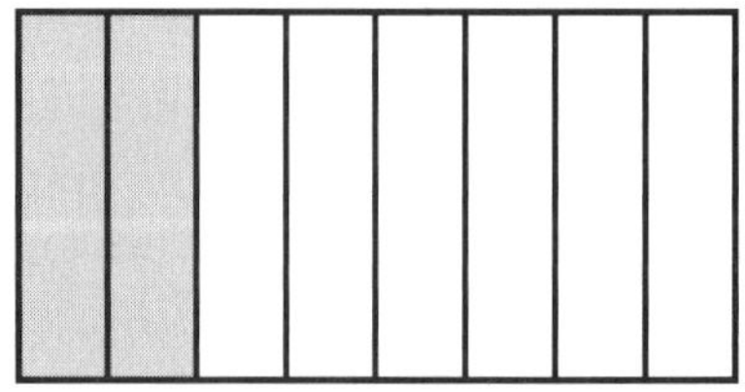

Lesson 2.1 Start Thinking!

For use before Lesson 2.1

What are some advantages of using an estimate instead of finding an exact answer?

Lesson 2.1 Warm Up

For use before Lesson 2.1

Estimate the product or quotient.

1. $\frac{7}{8} \times \frac{4}{5}$

2. $\frac{7}{16} \times \frac{9}{10}$

3. $\frac{5}{11} \times \frac{3}{5}$

4. $\frac{9}{20} \times \frac{7}{8}$

5. $\frac{14}{29} \div 2$

6. $8\frac{1}{6} \div 3\frac{3}{4}$

Name ______________________________ Date __________

2.1 Practice A

Tell whether the estimate given is *reasonable* or *not reasonable*. If the estimate is not reasonable, give a better estimate.

1. $\frac{2}{3} + \frac{1}{5} \approx 1$

2. $\frac{7}{8} - \frac{2}{3} \approx 0$

3. $1\frac{1}{7} \times \frac{9}{10} \approx \frac{1}{2}$

4. $\frac{15}{16} \div 3 \approx \frac{1}{4}$

5. $5\frac{1}{5} \times 6\frac{3}{4} \approx 30$

6. $51 \div 4\frac{7}{8} \approx 10$

Estimate the product or quotient.

7. $\frac{4}{9} \times \frac{11}{12}$

8. $\frac{4}{5} \div 2$

9. $\frac{5}{6} \times \frac{14}{15}$

10. $\frac{9}{10} \div 3$

Use compatible numbers to estimate the product or quotient.

11. $28 \times 2\frac{5}{6}$

12. $50 \div 5\frac{9}{10}$

13. $39 \times 4\frac{7}{8}$

14. $138 \div 6\frac{4}{5}$

15. $149 \times \frac{14}{15}$

16. $71 \div 8\frac{1}{6}$

17. Describe and correct the error in using compatible numbers to estimate the quotient.

> X $89 \div 2\frac{2}{3} \approx 90 \div 1 = 90$

18. A student's notebook contains 78 sheets of paper. One fourth of the notebook is math notes. About how many sheets of paper are math notes?

19. You are planning a party with about 20 guests. Should you overestimate or underestimate when buying refreshments? Why?

20. The Statue of Liberty is about 111 feet tall from heel to the top of the head. A person is five feet eight inches tall. About how many times greater is the height of the Statue of Liberty than the person?

Name ______________________________ Date __________

2.1 Practice B

Estimate the product or quotient.

1. $\frac{5}{6} \times \frac{10}{13}$

2. $\frac{15}{16} \div 2$

3. $3\frac{5}{8} \times 9\frac{1}{2}$

4. $39 \div 4\frac{1}{4}$

5. $12\frac{1}{9} \times 3\frac{7}{8}$

6. $21\frac{2}{3} \div 1\frac{9}{11}$

7. $15\frac{2}{7} \times 2\frac{5}{11}$

8. $67\frac{4}{7} \div 2$

9. $500 \div 4\frac{13}{14}$

Use compatible numbers to estimate the product or quotient.

10. $15\frac{5}{6} \times 2$

11. $54 \div 5\frac{1}{9}$

12. $57 \times 6\frac{1}{6}$

13. $175 \div 8\frac{9}{10}$

14. $105 \times \frac{7}{12}$

15. $370 \div 12\frac{1}{5}$

Estimate the value of the expression.

16. $19\frac{3}{4} \div 4\frac{2}{7} \times 15\frac{7}{8}$

17. $10\frac{3}{16} \times 8\frac{1}{5} \div 7\frac{7}{10}$

18. $117\frac{3}{10} \div \left(1\frac{4}{5} \times 9\frac{5}{6}\right)$

19. Describe and correct the error in estimating the product.

> X $\frac{4}{11} \times \frac{11}{12} \approx 0 \times 1 = 0$

20. Estimate the area of the triangle. Did you overestimate or underestimate the area? Explain.

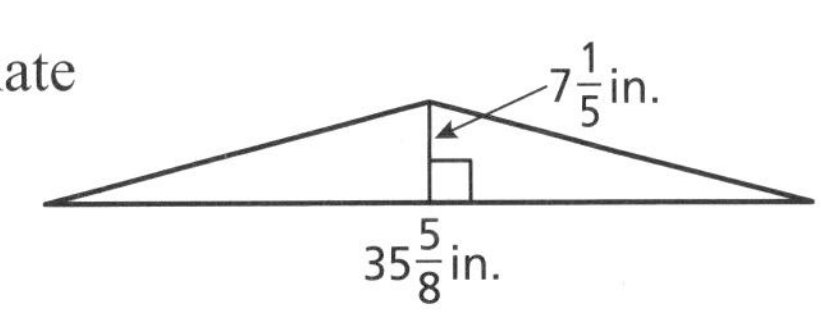

21. You cover a rectangular deck that is $27\frac{1}{2}$ feet by $9\frac{3}{4}$ feet with waterproofing sealer. One can of waterproofing sealer covers 150 square feet. Are 2 cans of waterproofing sealer enough to cover the deck? Explain.

Name ________________________________ Date __________

2.1 Enrichment and Extension

Navigating a Maze

To find an overestimate of a product, you round both numbers up. But to find an overestimate of a quotient, you round the numerator up and the denominator down.

$$15\frac{1}{3} \div 3\frac{1}{8} \approx 16 \div 3 = 5\frac{1}{3}$$

You can move left, right, up, or down in the maze.

Start	$2\frac{7}{9} \times 4\frac{1}{4}$	$48 \div 4\frac{1}{3}$	$8\frac{7}{11} \div 1\frac{1}{4}$
$\frac{5}{6} \times 8\frac{2}{9}$	$3\frac{1}{3} \times 1\frac{3}{5}$	$32\frac{1}{4} \div 3\frac{2}{9}$	$4\frac{1}{3} \times 5\frac{2}{5} \div 2\frac{1}{4}$
$3\frac{1}{3} \times 3\frac{2}{5}$	$47\frac{1}{2} \div 8\frac{6}{7}$	$3\frac{8}{9} \times 5\frac{4}{5} \div 3\frac{1}{8}$	$7\frac{1}{8} \div 2\frac{2}{3} \times 3\frac{9}{10}$
$2\frac{11}{12} \times 3$	$4\frac{1}{2} \times 4\frac{3}{4} \div 2$	$26\frac{1}{5} \div 3\frac{3}{4} \times \frac{14}{15}$	**Finish**

1. You can only move to expressions whose overestimate is less than 10. Find the path through the maze.

2. You can only move to expressions whose overestimate is greater than 10. Find the path through the maze.

Name______________________________ Date__________

2.1 Puzzle Time

How Does A Bunny Keep Cool On A Hot Day?

Write the letter of each answer in the box containing the exercise number.

Estimate the product or quotient.

1. $\frac{5}{6} \times \frac{2}{7}$ **2.** $\frac{10}{11} \times \frac{3}{4}$ **3.** $\frac{3}{4} \div 7$

4. $\frac{6}{7} \div 3$ **5.** $\frac{3}{8} \times \frac{3}{8}$ **6.** $\frac{3}{4} \times \frac{3}{8}$

7. $\frac{15}{7} \div 3$ **8.** $\frac{4}{5} \div 6$ **9.** $6\frac{1}{2} \times 4\frac{3}{4}$

10. $13\frac{7}{9} \times 5\frac{1}{8}$ **11.** $43\frac{2}{3} \div 4\frac{1}{7}$ **12.** $17\frac{3}{4} \div 5\frac{1}{2}$

13. $3\frac{5}{6} \times 8\frac{9}{10}$ **14.** $4\frac{7}{8} \times 9\frac{1}{3}$ **15.** $28\frac{2}{5} \div 3\frac{1}{2}$

Use compatible numbers to estimate the product or quotient.

16. $59 \div 4\frac{7}{8}$ **17.** $42 \div 7\frac{7}{12}$

18. $76 \times 1\frac{4}{5}$ **19.** $252 \div 5\frac{3}{11}$

20. $71 \times 4\frac{6}{7}$ **21.** $19 \times 19\frac{9}{10}$

22. $287 \div 9\frac{5}{6}$ **23.** $358 \div 3\frac{3}{4}$

Answers for Exercises 1–15

G. $\frac{1}{2}$ **R.** 35

N. 36 **A.** 1

L. 70 **I.** $\frac{1}{4}$

C. $\frac{1}{3}$ **N.** 3

T. 0 **E.** 11

R. 45 **I.** 7

N. $\frac{1}{7}$ **H.** $\frac{2}{3}$

D. $\frac{1}{6}$

Answers for Exercises 16–23

O. 50 **N.** 400

C. 5 **T.** 350

I. 12 **E.** 140

A. 30 **O.** 90

17	11	12	20	9	2	10		7	22	14	18		4	19	3	8	5	1	16	23	21	15	13	6

Activity 2.2 Start Thinking!

For use before Activity 2.2

You are helping to make pizzas for your younger brother's birthday party. How can you figure out how many pizzas to make using the product of a fraction and whole number?

Activity 2.2 Warm Up

For use before Activity 2.2

Find the sum.

1. $\frac{1}{5} + \frac{1}{5} + \frac{1}{5}$

2. $\frac{1}{6} + \frac{1}{6} + \frac{1}{6}$

3. $\frac{1}{8} + \frac{1}{8} + \frac{1}{8} + \frac{1}{8}$

4. $\frac{2}{3} + \frac{2}{3} + \frac{2}{3}$

5. $\frac{3}{4} + \frac{3}{4} + \frac{3}{4} + \frac{3}{4}$

6. $\frac{5}{6} + \frac{5}{6} + \frac{5}{6} + \frac{5}{6}$

Lesson 2.2 Start Thinking!

For use before Lesson 2.2

When is the product of a whole number and a fraction less than the whole number? greater than the whole number?

Hint: Try multiplying several different fractions by the same whole number to help you find the answer.

Lesson 2.2 Warm Up

For use before Lesson 2.2

Multiply. Write the answer in simplest form.

1. $3 \times \frac{3}{5}$

2. $\frac{2}{3} \times 9$

3. $\frac{6}{7} \times 3$

4. $2 \times \frac{3}{8}$

5. $\frac{3}{4} \times 24$

6. $\frac{2}{3} \times 13$

Name ________________________ Date ________

2.2 Practice A

Multiply. Write the answer in simplest form.

1. $7 \times \frac{1}{8}$

2. $\frac{1}{5} \times 4$

3. $\frac{2}{3} \times 8$

4. $6 \times \frac{4}{3}$

5. $\frac{5}{6} \times 9$

6. $15 \times \frac{3}{10}$

7. $\frac{5}{8} \times 32$

8. $\frac{7}{5} \times 20$

9. $14 \times \frac{8}{21}$

10. $\frac{2}{9} \times 12$

11. $25 \times \frac{2}{5}$

12. $\frac{5}{6} \times 30$

13. Describe and correct the error in finding the product.

$$\times \quad 8 \times \frac{2}{5} = \frac{8 \times 2}{8 \times 5} = \frac{16}{40} = \frac{2}{5}$$

Use compatible numbers to estimate the product.

14. $\frac{1}{4} \times 19$

15. $36 \times \frac{1}{5}$

16. $31 \times \frac{5}{8}$

17. $21 \times \frac{5}{6}$

18. You have 18 t-shirts and $\frac{2}{3}$ of them are in the laundry. How many of your t-shirts need to be washed?

19. More than two inches of rain fell in $\frac{4}{7}$ of the last 42 months. In how many months did more than two inches of rain fall?

20. You have \$20 and your friend has \$24. If you spend $\frac{2}{5}$ of your money and your friend spends $\frac{3}{8}$ of her money, who spent more? How much more?

21. A recent survey reported that four out of five people prefer Brand A over Brand B. If 500 people were surveyed, how many more people prefer Brand A?

Name______________________________ Date__________

2.2 Practice B

Multiply. Write the answer in simplest form.

1. $8 \times \frac{1}{10}$

2. $\frac{1}{6} \times 9$

3. $\frac{3}{5} \times 15$

4. $12 \times \frac{5}{3}$

5. $\frac{5}{7} \times 21$

6. $20 \times \frac{3}{8}$

7. $\frac{7}{9} \times 36$

8. $\frac{13}{12} \times 24$

9. $35 \times \frac{9}{14}$

10. Describe and correct the error in finding the product.

> X $5 \times \frac{1}{10} = \frac{5}{10} = \frac{1}{5}$

Use compatible numbers to estimate the product.

11. $\frac{1}{6} \times 29$

12. $41 \times \frac{2}{5}$

13. $28 \times \frac{7}{9}$

Evaluate the expression when $x = 12$, $y = \frac{5}{8}$, and $z = 24$.

14. $\frac{2}{3} \bullet x$

15. $10 \bullet y$

16. yz

17. $\frac{3x}{z}$

18. Your grandparents have a back yard that is 60 feet long and 48 feet wide. They are making a garden that is half the length and $\frac{2}{3}$ the width. What is the area of the garden?

19. You bought 40 packages of streamers that were red, white, and blue. One-fourth of the streamers were red and $\frac{2}{5}$ of the streamers were blue. How many packages of white streamers did you buy?

Name ______________________________ Date __________

2.2 Enrichment and Extension

What State Am I?

The table shows areas (in square kilometers) of Alabama, Florida, and Georgia. Use the table to answer the questions below.

	Alabama	Florida	Georgia
Total Area	135,000	170,000	154,000
Water Area (as a fraction of total area)	$\frac{4}{125}$	$\frac{9}{50}$	$\frac{1}{40}$
Inland Water Area (as a fraction of water area)	$\frac{23}{40}$	$\frac{2}{5}$	$\frac{17}{25}$

1. What is the water area of each state in the table?
2. What is the inland water area of each state in the table?
3. My water area is the least of the states in the table. What state am I?
4. My inland water area is the least of the states in the table. What state am I?
5. Total area consists of water area and land area. I have about 150,000 square kilometers of land area. What state am I?
6. The inland water area of the United States is about 204,000 square kilometers. My inland water area is about $\frac{3}{50}$ of the United States inland water area. What state am I?

Name ______________________________ Date __________

2.2 Puzzle Time

Why Did The Palm Tree Want To Become A Cheerleader?

Write the letter of each answer in the box containing the exercise number.

Multiply. Write the answer in simplest form.

1. $4 \times \frac{1}{9}$ **2.** $\frac{1}{7} \times 6$ **3.** $\frac{2}{5} \times 8$

4. $3 \times \frac{3}{4}$ **5.** $\frac{5}{4} \times 9$ **6.** $7 \times \frac{11}{10}$

7. $\frac{2}{3} \times 12$ **8.** $\frac{5}{6} \times 16$ **9.** $\frac{17}{18} \times 36$

10. $25 \times \frac{4}{5}$ **11.** $\frac{7}{24} \times 3$ **12.** $21 \times \frac{3}{14}$

Evaluate the expression when $x = 9$, $y = \frac{4}{15}$, and $z = 25$.

13. $\frac{6}{7} \bullet x$ **14.** $5 \bullet y$

15. $\frac{2}{5} \bullet z$ **16.** $x \bullet \frac{3}{4}$

17. $\frac{3}{5}xz$ **18.** xyz

19. $\frac{1}{5} + xy$ **20.** $\frac{1}{3} + yz$

Answers for Exercises 1–12

T. $7\frac{7}{10}$ **O.** $4\frac{1}{2}$

H. $3\frac{1}{5}$ **M.** 20

O. $\frac{4}{9}$ **E.** $11\frac{1}{4}$

R. $\frac{7}{8}$ **A.** $13\frac{1}{3}$

F. 8 **T.** $\frac{6}{7}$

O. $2\frac{1}{4}$ **R.** 34

Answers for Exercises 13–20

O. 135 **E.** $1\frac{1}{3}$

T. $2\frac{3}{5}$ **H.** 60

M. $7\frac{5}{7}$ **E.** $6\frac{3}{4}$

T. 7 **O.** 10

6	12		11	17	15	20		7	1	9		2	18	14		3	4	13	16		19	5	8	10

Activity 2.3 Start Thinking!
For use before Activity 2.3

Fold a piece of paper into thirds. Each section represents one-third of the paper.

Now fold the paper so that each third is folded in half.

Each new section represents **one-half** of **one-third** of the paper. What fraction of the paper does each new section represent?

Activity 2.3 Warm Up
For use before Activity 2.3

Write the fraction in simplest form.

1. $\frac{6}{48}$ **2.** $\frac{9}{36}$ **3.** $\frac{6}{9}$

4. $\frac{12}{27}$ **5.** $\frac{16}{38}$ **6.** $\frac{18}{21}$

Lesson 2.3 Start Thinking!

For use before Lesson 2.3

The product of what two fractions is represented by the shaded region in the diagram below? What is the product? How do you know?

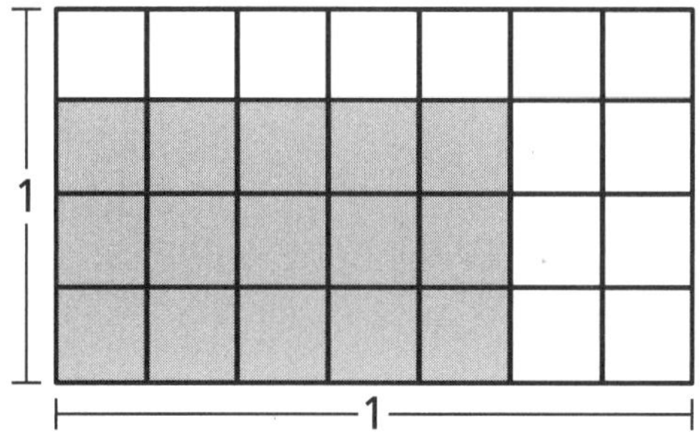

Lesson 2.3 Warm Up

For use before Lesson 2.3

Multiply. Write the answer in simplest form.

1. $\frac{2}{9} \times \frac{3}{5}$

2. $\frac{4}{5} \times \frac{3}{8}$

3. $\frac{1}{3} \times \frac{2}{3}$

4. $\frac{7}{12} \times \frac{1}{14}$

5. $\frac{4}{9} \times \frac{2}{3}$

6. $\frac{4}{11} \times \frac{1}{2}$

Name ______________________________ Date __________

2.3 Practice A

Multiply. Write the answer in simplest form.

1. $\frac{3}{5} \times \frac{1}{4}$

2. $\frac{1}{3} \times \frac{4}{7}$

3. $\frac{2}{3} \times \frac{7}{11}$

4. $\frac{6}{7} \times \frac{5}{6}$

5. $\frac{2}{5} \times \frac{5}{8}$

6. $\frac{8}{15} \times \frac{3}{10}$

7. $\frac{2}{9} \times \frac{12}{13}$

8. $\frac{7}{8} \times 12$

9. $4 \times \frac{5}{12}$

10. $\frac{4}{9} \times \frac{4}{9}$

11. $\frac{3}{14} \times \frac{7}{9}$

12. $\frac{8}{9} \times \frac{24}{11}$

13. Describe and correct the error in finding the product.

$$\text{✗} \quad \frac{3}{4} \times \frac{4}{9} = \frac{3 \times 4}{4 \times 9} = \frac{7}{13}$$

Evaluate the expression when $a = \frac{1}{3}$, $b = \frac{3}{5}$, and $c = \frac{8}{9}$.

14. $a \bullet \frac{3}{5}$

15. $\frac{10}{9} \bullet b$

16. $c \bullet \frac{3}{8}$

17. $b \bullet c$

18. Your lunch period lasts for $\frac{1}{2}$ hour. You eat your lunch for about $\frac{2}{3}$ of that time. What fraction of the hour do you spend eating?

19. In a baseball card collection, $\frac{1}{6}$ of the cards are for players who are retired. Of those, $\frac{2}{5}$ have been retired for less than five years. What fraction of the cards are for players who have been retired for less than five years?

20. Two-fifths of the students in your class are in the band. Of these, one-fourth play the saxophone.

    a. What fraction of your class plays the saxophone?

    b. There are 30 students in your class. How many students in your class play the saxophone?

Name______________________________ Date__________

2.3 Practice B

Multiply. Write the answer in simplest form.

1. $\frac{1}{6} \times \frac{5}{9}$

2. $\frac{2}{5} \times \frac{7}{8}$

3. $\frac{6}{7} \times \frac{2}{9}$

4. $\frac{11}{12} \times \frac{4}{5}$

5. $\frac{3}{4} \times \frac{16}{21}$

6. $\frac{4}{15} \times \frac{3}{8}$

7. $18 \times \frac{5}{6}$

8. $\frac{5}{24} \times \frac{8}{15}$

9. $\frac{16}{7} \times \frac{21}{8}$

10. Describe and correct the error in finding the product.

$$\times \quad \frac{5}{6} \times \frac{5}{6} = \frac{5 \times 5}{6} = \frac{25}{6} = 4\frac{1}{6}$$

Evaluate the expression when $x = \frac{2}{3}$, $y = \frac{9}{10}$, and $z = \frac{1}{8}$.

11. $\frac{3}{4} \bullet x$

12. $\frac{5}{12} \bullet y$

13. xy

14. $\frac{4}{3}y + z$

Multiply. Write the answer in simplest form.

15. $\frac{1}{4} \times \frac{5}{8} \times \frac{16}{25}$

16. $\left(\frac{2}{7}\right)^3$

17. $\left(\frac{5}{7}\right)^2 \times \left(\frac{3}{5}\right)^2$

18. One-half of your CDs are scratched. Of those, one-fifth will not play properly. You have 40 CDs. How many of them will not play properly?

19. You plant a flower garden that is 20 feet long by 18 feet wide. You plan to put roses in a section that is $\frac{3}{5}$ of the length and $\frac{2}{3}$ of the width.

a. Describe two ways to find the area of the rose section.

b. What is the area of the rose section?

Name ______________________________ Date __________

2.3 Enrichment and Extension

Finding Patterns

In Exercises 1–8, describe the pattern. Then find the next two numbers in the pattern.

1. $64, 8, 1, \frac{1}{8}, \ldots$

2. $\frac{1}{2}, \frac{1}{4}, \frac{1}{8}, \frac{1}{16}, \ldots$

3. $12, 3, \frac{3}{4}, \frac{3}{16}, \ldots$

4. $6, 4, \frac{8}{3}, \frac{16}{9}, \ldots$

5. $\frac{2}{3}, 1, \frac{3}{2}, \frac{9}{4}, \ldots$

6. $\frac{1}{5}, \frac{1}{2}, \frac{5}{4}, \frac{25}{8}, \ldots$

7. $27, 36, 48, 64, \ldots$

8. $32, 40, 50, \frac{125}{2}, \ldots$

9. After each round of a basketball tournament, half of the teams are eliminated.

a. Which pattern in Exercises 1–8 shows the fraction of the original number of teams left in the tournament after each round?

b. A tournament starts with 64 teams. How many teams are left after each round?

10. A deposit of \$1000 is made in a savings account. Each year, the value of the account is $\frac{11}{10}$ the value in the previous year.

a. Write a product that can be used to find the value of the account after 2 years.

b. What is the value of the account after 2 years?

Name____ Date__________

2.3 Puzzle Time

Why Did The Turkey Volunteer To Be The Drummer In The Popular Bird Band?

Write the letter of each answer in the box containing the exercise number.

Multiply. Write the answer in simplest form.

1. $\frac{1}{8} \times \frac{3}{5}$

2. $\frac{2}{7} \times \frac{6}{11}$

3. $\frac{1}{6} \times \frac{3}{8}$

4. $\frac{5}{9} \times \frac{1}{7}$

5. $\frac{3}{4} \times \frac{9}{13}$

6. $\frac{5}{6} \times \frac{6}{7}$

7. $\frac{5}{16} \times \frac{1}{10}$

8. $\frac{10}{11} \times \frac{1}{10}$

9. $\frac{3}{14} \times 12$

10. $8 \times \frac{9}{10}$

11. $\frac{2}{3} \times \frac{9}{16}$

12. $\frac{5}{7} \times \frac{5}{8}$

13. $\frac{2}{5} \times \frac{2}{5}$

14. $\frac{5}{8} \times \frac{4}{11}$

15. $\frac{14}{15} \times \frac{5}{7}$

Evaluate the expression when $a = \frac{5}{6}$, $b = \frac{1}{4}$, and $c = \frac{2}{3}$.

16. $a \bullet \frac{1}{15}$

17. $\frac{2}{11} \bullet b$

18. $b \bullet a$

19. $a \bullet c$

20. $\frac{3}{4}a + \frac{1}{4}$

21. $bc + \frac{1}{3}$

22. $ac + b$

Answers for Exercises 1–15

H. $2\frac{4}{7}$	**D.** $\frac{5}{22}$	**S.** $\frac{5}{7}$
R. $\frac{1}{11}$	**Y.** $\frac{3}{40}$	**M.** $\frac{5}{63}$
A. $\frac{4}{25}$	**I.** $\frac{2}{3}$	**D.** $\frac{3}{8}$
K. $\frac{12}{77}$	**L.** $\frac{1}{32}$	**E.** $\frac{27}{52}$
U. $7\frac{1}{5}$	**A.** $\frac{25}{56}$	**T.** $\frac{1}{16}$

Answers for Exercises 16–22

A. $\frac{7}{8}$	**E.** $\frac{5}{24}$	**H.** $\frac{1}{22}$
R. $\frac{29}{36}$	**S.** $\frac{1}{18}$	**C.** $\frac{5}{9}$
D. $\frac{1}{2}$		

17	5		20	7	22	18	13	11	1		9	12	21		14	8	10	4	6	3	15	19	2	16

Activity 2.4 Start Thinking!

For use before Activity 2.4

One of the games on a television game show involves rolling a ten-sided die, labeled with whole numbers 1 through 10. Each contestant rolls the die twice. The first roll gives the numerator of a fraction. The second roll gives the denominator.

$$\text{Fraction: } \frac{\bigcirc}{\bigcirc}$$

What would you hope to roll on each roll if you want to create the fraction with the greatest value? with the least value? Why?

Activity 2.4 Warm Up

For use before Activity 2.4

Write the mixed number as an improper fraction.

1. $3\frac{9}{10}$ **2.** $2\frac{3}{7}$ **3.** $7\frac{1}{2}$

4. $11\frac{7}{8}$ **5.** $8\frac{2}{5}$ **6.** $6\frac{4}{15}$

Lesson 2.4 Start Thinking!

For use before Lesson 2.4

The distance around a track is $\frac{1}{4}$ mile.

Write a word problem using the track that involves multiplying a fraction and a mixed number.

Exchange word problems with a classmate and solve your classmate's problem.

Lesson 2.4 Warm Up

For use before Lesson 2.4

Multiply. Write the answer in simplest form.

1. $3\frac{1}{2} \times \frac{3}{5}$

2. $2\frac{1}{3} \times \frac{4}{7}$

3. $1\frac{5}{6} \times \frac{3}{10}$

4. $\frac{5}{9} \times 3\frac{3}{4}$

5. $1\frac{2}{7} \times \frac{1}{4}$

6. $\frac{7}{8} \times 4\frac{6}{7}$

Name ______________________________ Date __________

2.4 Practice A

Multiply. Write the answer in simplest form.

1. $2\frac{2}{5} \times \frac{1}{4}$

2. $5\frac{3}{5} \times \frac{5}{7}$

3. $5\frac{1}{2} \times \frac{7}{11}$

4. $6\frac{2}{3} \times 12$

5. $4\frac{2}{5} \times 1\frac{7}{8}$

6. $9 \times 2\frac{2}{3}$

7. $4\frac{4}{9} \times \frac{3}{8}$

8. $3\frac{5}{8} \times 12$

9. $4\frac{2}{3} \times \frac{3}{14}$

10. $3\frac{1}{9} \times 3\frac{1}{9}$

11. $5\frac{1}{5} \times 4\frac{1}{6}$

12. $2\frac{2}{15} \times 3\frac{1}{8}$

13. Describe and correct the error in finding the product.

$$\times \quad 3\frac{7}{8} \times 6\frac{2}{5} = 18\frac{14}{40} = 18\frac{7}{20}$$

Evaluate the expression when $a = 2\frac{3}{4}$, $b = 4\frac{3}{8}$, and $c = 3\frac{2}{7}$.

14. $a \bullet 3\frac{1}{5}$

15. $6\frac{2}{7} \bullet b$

16. $1\frac{17}{18} \bullet c$

17. $b \bullet c$

18. You have 20 floor tiles that are $9\frac{5}{8}$ inches long. How long are the tiles when placed end to end?

19. Five years ago, an alligator was $2\frac{1}{6}$ feet long. Today, the alligator is three times longer. How long is the alligator now?

20. The sixth grade class consumes $5\frac{1}{3}$ cases of milk each day. How much will they consume in a five-day week?

21. A poster is $8\frac{1}{2}$ inches by 11 inches. You enlarge the poster by increasing each dimension by a factor of $2\frac{1}{2}$. What is the area of the new poster?

Name________________________ Date__________

2.4 Practice B

Multiply. Write the answer in simplest form.

1. $1\frac{3}{8} \times \frac{4}{9}$

2. $\frac{4}{15} \times 2\frac{1}{4}$

3. $\frac{5}{17} \times 3\frac{2}{5}$

4. $2\frac{11}{12} \times \frac{2}{5}$

5. $1\frac{3}{4} \times \frac{20}{21}$

6. $4\frac{9}{10} \times 1\frac{1}{7}$

7. $28 \times 8\frac{2}{7}$

8. $5\frac{5}{12} \times 2\frac{2}{5}$

9. $8\frac{1}{3} \times 4\frac{1}{2}$

10. Describe and correct the error in finding the product.

$$6\frac{1}{8} \times 8 = 6 \times \frac{1}{8} \times 8 = 6 \times \left(\frac{1}{8} \times 8\right) = 6$$

Evaluate the expression when $x = 2\frac{2}{9}$, $y = 3\frac{3}{5}$, and $z = 6\frac{2}{3}$.

11. $1\frac{4}{5} \bullet x$

12. $1\frac{8}{27} \bullet y$

13. xy

14. $yz + 8\frac{4}{9}$

Multiply. Write the answer in simplest form.

15. $\frac{3}{5} \times 6\frac{3}{4} \times 2\frac{2}{9}$

16. $1\frac{7}{8} \times 12 \times 3\frac{2}{3}$

17. $4\frac{1}{6} \times 2\frac{2}{7} \times 7\frac{1}{5}$

18. You have a ribbon that is $6\frac{7}{8}$ inches long. You use $\frac{2}{5}$ of it on a project. How much of the ribbon is left?

19. Find the area of the shaded region. Explain how you found the area.

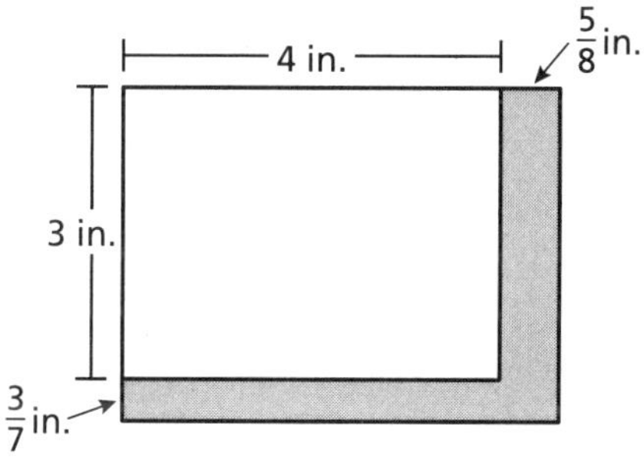

Name ______________________________ Date __________

2.4 Enrichment and Extension

Geometry

You make painting designs for your ceiling. You will paint the shaded region. Find the area of the shaded region in the design.

1.

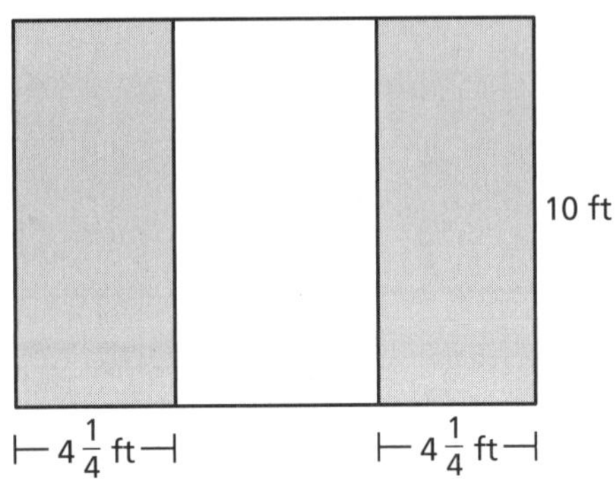

2.

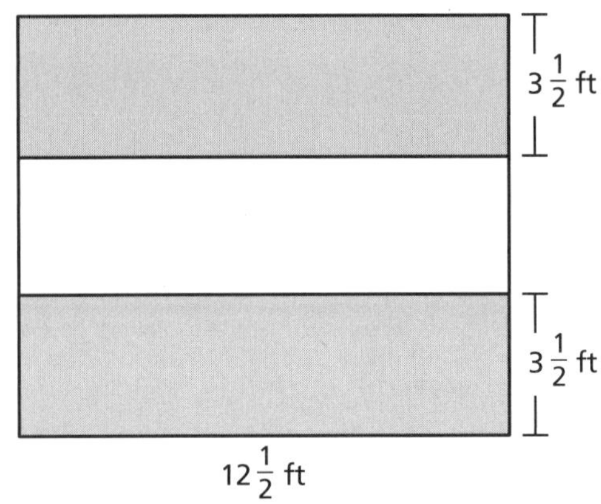

3.

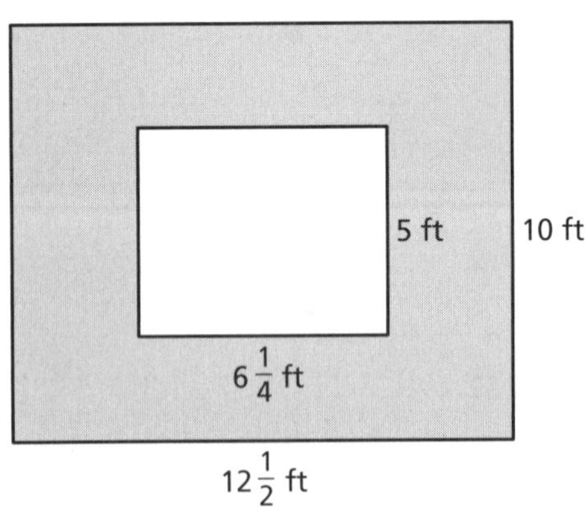

4.

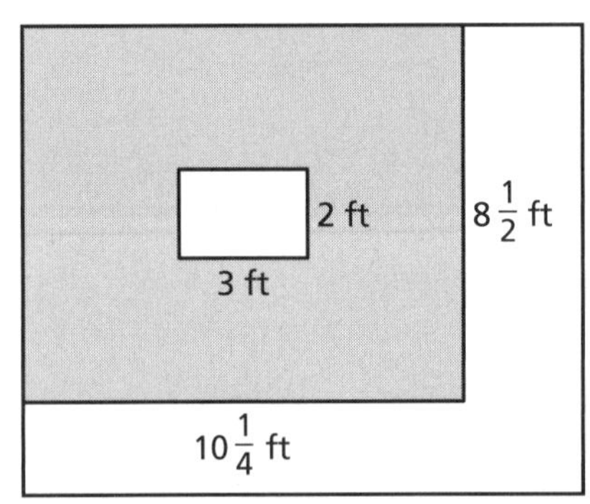

5.

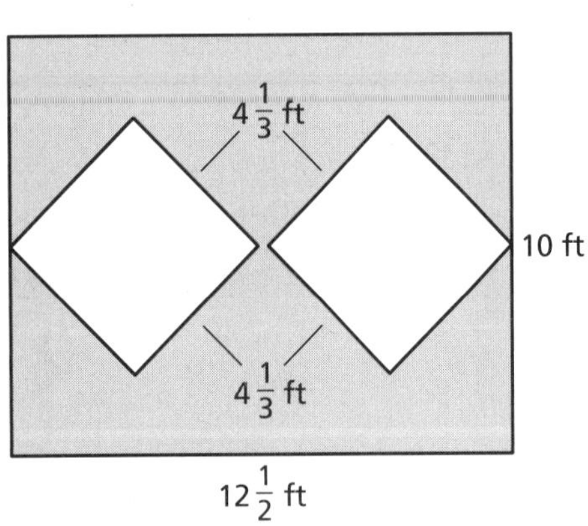

6.

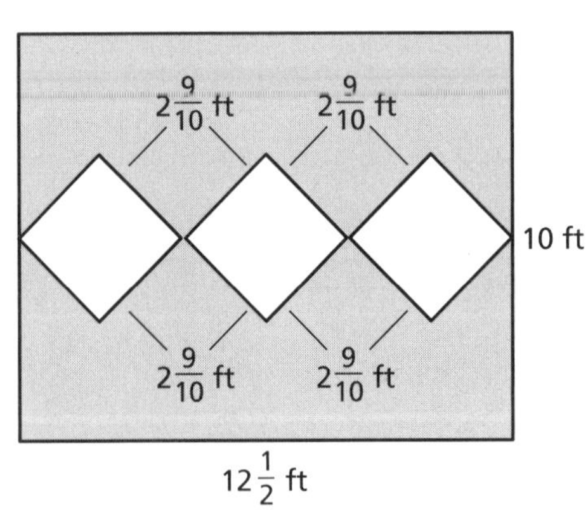

7. You choose the design that requires the least amount of paint. Which design do you choose? Explain.

Name______________________________ Date__________

2.4 Puzzle Time

Why Did The Cell Phone Ask To Go See A Psychologist?

Write the letter of each answer in the box containing the exercise number.

Multiply. Write the answer in simplest form.

1. $1\frac{1}{4} \times \frac{3}{4}$
2. $4\frac{2}{3} \times \frac{3}{4}$
3. $7\frac{1}{2} \times \frac{4}{5}$
4. $\frac{9}{10} \times 2\frac{2}{9}$
5. $4\frac{2}{3} \times \frac{6}{7}$
6. $\frac{2}{7} \times 10\frac{1}{2}$
7. $\frac{5}{8} \times 1\frac{3}{5}$
8. $6\frac{1}{4} \times \frac{2}{5}$
9. $2\frac{7}{10} \times \frac{5}{9}$
10. $\frac{2}{9} \times 3\frac{1}{6}$
11. $2\frac{5}{14} \times 21$
12. $16 \times 7\frac{5}{6}$
13. $4\frac{3}{8} \times 2\frac{2}{7}$
14. $1\frac{4}{9} \times 2\frac{7}{8}$
15. $2\frac{1}{7} \times 3\frac{9}{25}$

Evaluate the expression when $x = 3\frac{3}{4}$, $y = 2\frac{5}{7}$, and $z = 1\frac{9}{11}$.

16. $7\frac{1}{5} \bullet x$
17. $1\frac{7}{10} \bullet z$
18. $2\frac{1}{19} \bullet y$
19. xz
20. $xy + 2\frac{1}{2}$
21. $y \bullet 3\frac{7}{8} - 5\frac{1}{8}$

Answers for Exercises 1–15

T. 10	**L.** $49\frac{1}{2}$
A. 3	**I.** $2\frac{1}{2}$
M. $\frac{15}{16}$	**Y.** 2
O. $125\frac{1}{3}$	**S.** $7\frac{1}{5}$
R. $\frac{19}{27}$	**D.** $3\frac{1}{2}$
A. 1	**O.** 4
N. $4\frac{11}{72}$	**C.** $1\frac{1}{2}$
P. 6	

Answers for Exercises 16–21

T. $5\frac{4}{7}$	**L.** $5\frac{11}{28}$	**O.** $12\frac{19}{28}$
H. $3\frac{1}{11}$	**D.** 27	**A.** $6\frac{9}{11}$

8	18		17	7	2		13	5	20		1	19	14	4		16	10	12	3		9	6	21	11	15

Activity 2.5

Start Thinking!

For use before Activity 2.5

The recommended serving size listed on a box of cereal is $\frac{3}{4}$ cup.

You also know the amount of cereal left in the box. What can you find out using these two pieces of information? How?

Activity 2.5

Warm Up

For use before Activity 2.5

Find the product. Write the answer in simplest form.

1. $\frac{1}{8} \times \frac{2}{5}$

2. $\frac{4}{7} \times \frac{5}{16}$

3. $\frac{1}{3} \times \frac{1}{8}$

4. $4 \times \frac{3}{16}$

5. $15 \times \frac{2}{3}$

6. $9 \times \frac{4}{15}$

Lesson 2.5 Start Thinking!

For use before Lesson 2.5

Come up with an easy way to remember how to find the reciprocal of a fraction.

Lesson 2.5 Warm Up

For use before Lesson 2.5

Divide. Write the answer in simplest form.

1. $\frac{2}{5} \div \frac{1}{8}$

2. $\frac{4}{7} \div \frac{4}{23}$

3. $\frac{1}{3} \div \frac{1}{8}$

4. $4 \div \frac{3}{16}$

5. $\frac{2}{3} \div 8$

6. $\frac{4}{15} \div \frac{1}{9}$

Name ______________________________ Date __________

2.5 Practice A

Write the reciprocal of the number.

1. $\frac{5}{8}$ **2.** 6 **3.** $\frac{1}{3}$ **4.** $\frac{7}{4}$

Divide. Write the answer in simplest form.

5. $\frac{1}{6} \div \frac{1}{3}$ **6.** $\frac{3}{4} \div 6$ **7.** $\frac{3}{5} \div \frac{2}{15}$

8. $10 \div \frac{2}{7}$ **9.** $\frac{3}{8} \div \frac{9}{16}$ **10.** $\frac{5}{9} \div 15$

11. $\frac{10}{27} \div \frac{2}{3}$ **12.** $\frac{18}{25} \div \frac{8}{15}$ **13.** $\frac{7}{12} \div 14$

14. Describe and correct the error made in finding the quotient.

$$\times \quad \frac{3}{4} \div 6 = \frac{3}{4} \div \frac{6}{1} = \frac{18}{4} = \frac{9}{2} = 4\frac{1}{2}$$

Evaluate the expression when $c = \frac{2}{3}$, $d = \frac{4}{9}$, and $e = 6$.

15. $d \div c$ **16.** $e \div c$ **17.** $d \div e$ **18.** $c \div e$

Determine whether the numbers are reciprocals. If not, write the reciprocal of each number.

19. $\frac{5}{8}, \frac{16}{10}$ **20.** $\frac{1}{12}, 12$ **21.** $\frac{2}{3}, \frac{12}{18}$ **22.** $\frac{4}{7}, \frac{7}{8}$

23. You have $\frac{2}{3}$ of a pizza. You divide the remaining pizza into 4 equal pieces. What fraction of the pizza is each piece?

24. You have five quarts of paint. It takes $\frac{5}{8}$ quart to paint each chair you are asked to paint. How many chairs can you paint?

25. Is the reciprocal of a fraction always a whole number? Explain.

Name ____ Date __________

2.5 Practice B

Write the reciprocal of the number.

1. $\frac{7}{12}$

2. $\frac{1}{10}$

3. 15

4. $\frac{21}{5}$

Divide. Write the answer in simplest form.

5. $\frac{3}{8} \div \frac{15}{16}$

6. $9 \div \frac{6}{5}$

7. $\frac{2}{3} \div \frac{2}{9}$

8. $\frac{4}{15} \div 12$

9. $\frac{3}{7} \div \frac{9}{14}$

10. $\frac{8}{15} \div 20$

11. $\frac{3}{4} \div \frac{7}{12}$

12. $18 \div \frac{4}{7}$

13. $\frac{16}{25} \div \frac{3}{5}$

14. Describe and correct the error made in finding the quotient.

$$\times \quad \frac{3}{5} \div \frac{9}{10} = \frac{5}{3} \bullet \frac{10}{9} = \frac{50}{27} = 1\frac{23}{27}$$

Without finding the quotient, copy and complete the statement using <, >, or =. Explain your reasoning.

15. $4 \div \frac{6}{7}$ __?__ 4

16. $1 \div \frac{2}{3}$ __?__ 1

17. $\frac{4}{5} \div \frac{9}{10}$ __?__ $\frac{4}{5}$

Evaluate the expression.

18. $\frac{1}{5} \div 5 \div 3$

19. $8 \div \frac{4}{5} + \frac{7}{9}$

20. $\frac{11}{24} \bullet \left(\frac{5}{16} \div \frac{5}{12}\right)$

21. How many times larger is a 10-pound dog than a hamster weighing $\frac{5}{8}$ pound?

22. A digital camera memory card is $\frac{1}{4}$ full. The card is $\frac{2}{3}$ full when 375 more pictures have been taken.

    a. How many pictures can the memory card hold?

    b. How many pictures were originally on the memory card?

Name ______________________________ Date ________

2.5 Enrichment and Extension

Finding a Combination

9, 9	2, 7	3, 5	4, 6
1, 4	5, 7	3, 6	3, 8

You keep money in a locked box and do not want to write the combination down to open the box. So, you create the problems below and numbers above to hide the combination. Choose a set of numbers and fill in the boxes to make each equation true. Each set of numbers will be used once.

1. $\frac{\square}{2} \div \frac{3}{8} = \frac{\square}{3}$

2. $\frac{\square}{4} \div 2 = \frac{\square}{8}$

3. $\frac{\square}{2} \div \frac{1}{4} = \square$

4. $\frac{5}{14} \div \frac{2}{\square} = \frac{\square}{4}$

5. $\frac{1}{\square} \div \frac{3}{8} = \frac{1}{\square}$

6. $\frac{7}{2} \div \square = \frac{1}{\square}$

7. $\frac{5}{9} \div \frac{2}{\square} = \frac{\square}{6}$

8. $\frac{1}{\square} \div \frac{3}{2} = \frac{1}{\square}$

9. Starting at Exercise 1 and ending at Exercise 8, write down each number that was missing from the right side of the equation. Place a dash after every second digit. This series of two digit numbers will unlock the box. What is the combination to the box?

Name________________________________ Date__________

2.5 Puzzle Time

Why Was The Gentleman Who Was Selling Watches Unhappy?

Write the letter of each answer in the box containing the exercise number.

Divide. Write the answer in simplest form.

1. $\frac{1}{16} \div \frac{1}{8}$ **2.** $\frac{6}{7} \div \frac{3}{5}$ **3.** $14 \div \frac{2}{7}$

4. $\frac{5}{8} \div 10$ **5.** $\frac{14}{15} \div 7$ **6.** $\frac{5}{24} \div \frac{5}{6}$

7. $\frac{9}{20} \div \frac{3}{4}$ **8.** $\frac{1}{4} \div \frac{1}{36}$ **9.** $\frac{7}{8} \div 28$

10. $3 \div \frac{2}{3}$ **11.** $\frac{3}{14} \div \frac{9}{11}$ **12.** $18 \div \frac{9}{13}$

Evaluate the expression when $a = \frac{1}{3}$, $b = \frac{7}{8}$, and $c = 4$.

13. $c \div b$ **14.** $a \div b$ **15.** $c \div a$

16. $b \div a$ **17.** $a \div c$ **18.** $b \div c$

Evaluate the expression.

19. $\frac{1}{9} \div 9 \div 9$ **20.** $\frac{9}{16} \div 18 \div 8$

21. $3 \div \frac{9}{11} + \frac{1}{3}$ **22.** $\frac{1}{2} + \frac{7}{8} \div \frac{11}{24}$

23. $\frac{8}{21} \div \frac{2}{3} \bullet \frac{4}{9}$ **24.** $\frac{5}{12} \bullet \frac{2}{3} \div \frac{2}{9}$

Answers for Exercises 1–12

A. $\frac{11}{42}$ **N.** $4\frac{1}{2}$ **I.** 49

O. $\frac{2}{15}$ **H.** $\frac{1}{2}$ **L.** 26

S. 9 **I.** $\frac{3}{5}$ **D.** $\frac{1}{16}$

F. $1\frac{3}{7}$ **T.** $\frac{1}{32}$ **H.** $\frac{1}{4}$

Answers for Exercises 13–18

A. $\frac{1}{12}$ **M.** 12 **H.** $\frac{8}{21}$

O. $4\frac{4}{7}$ **N.** $2\frac{5}{8}$ **S.** $\frac{7}{32}$

Answers for Exercises 19–24

O. $\frac{16}{63}$ **D.** $\frac{1}{256}$ **H.** $2\frac{9}{22}$

A. $1\frac{1}{4}$ **E.** $\frac{1}{729}$ **T.** 4

6	19		14	17	4		24		12	5	9		13	2		21	7	15	19	
23	10		22	3	18		1	11	16	20	8									

Activity 2.6 Start Thinking!

For use before Activity 2.6

Amber says that $8 \div 1\frac{3}{4}$ is the same as $8 \times 1\frac{4}{3}$.

Is she correct? How do you know?

Activity 2.6 Warm Up

For use before Activity 2.6

Write the improper fraction as a mixed number.

1. $\frac{12}{5}$
2. $\frac{21}{4}$
3. $\frac{16}{9}$
4. $\frac{27}{6}$
5. $\frac{80}{3}$
6. $\frac{55}{9}$

Lesson 2.6 Start Thinking!
For use before Lesson 2.6

A beaded necklace has broken and the beads have fallen all over the floor. The diameter of each bead is $1\frac{1}{4}$ centimeters. Using the string from the broken necklace, how can you find out how many beads were on the necklace without counting them?

Lesson 2.6 Warm Up
For use before Lesson 2.6

Divide. Write the answer in simplest form.

1. $1\frac{2}{5} \div \frac{1}{2}$

2. $3\frac{1}{2} \div \frac{2}{5}$

3. $2\frac{1}{7} \div \frac{3}{7}$

4. $5\frac{1}{3} \div 2\frac{1}{4}$

5. $1\frac{5}{12} \div 2\frac{1}{10}$

6. $9\frac{1}{6} \div 2\frac{1}{5}$

Name ______________________________ Date __________

2.6 Practice A

Divide. Write the answer in simplest form.

1. $2\frac{1}{2} \div \frac{1}{4}$

2. $4\frac{1}{3} \div \frac{2}{3}$

3. $6\frac{2}{3} \div \frac{5}{6}$

4. $8\frac{1}{3} \div 2\frac{2}{9}$

5. $9\frac{4}{5} \div 7$

6. $3\frac{5}{9} \div 16$

7. $\frac{8}{27} \div 4\frac{2}{9}$

8. $21 \div 5\frac{1}{4}$

9. $9\frac{3}{5} \div 12$

10. $3\frac{3}{4} \div 5\frac{1}{4}$

11. $7\frac{1}{2} \div 1\frac{3}{4}$

12. $6\frac{2}{9} \div 6\frac{2}{3}$

13. Describe and correct the error made in finding the quotient.

$$\times \quad 8 \div 2\frac{3}{4} = \frac{1}{8} \div \frac{11}{4} = \frac{1}{8} \times \frac{4}{11} = \frac{1}{22}$$

Evaluate the expression when $a = 3\frac{3}{8}$ and $b = 4\frac{1}{2}$.

14. $a \div 4\frac{7}{8}$

15. $b \div a$

16. $3\frac{3}{5} \div b$

17. $a \div b$

18. How many $6\frac{1}{4}$-inch pieces of fabric can be cut from a 75-inch roll?

19. How many times heavier is a $16\frac{1}{2}$-pound rock than a $10\frac{1}{2}$-pound rock?

20. How many $2\frac{1}{2}$ acre plots can be made out of a 35-acre field?

21. A bag contains 36 cups of flour. How many loaves of bread can be made if each loaf takes $5\frac{1}{3}$ cups? Is there any flour left over? If so, how much?

Name______________________________ Date__________

2.6 Practice B

Divide. Write the answer in simplest form.

1. $3\frac{1}{2} \div 2\frac{1}{4}$

2. $7\frac{3}{4} \div 1\frac{5}{8}$

3. $4\frac{1}{2} \div 2\frac{2}{3}$

4. $10 \div 2\frac{2}{9}$

5. $16\frac{1}{3} \div 14$

6. $8\frac{2}{5} \div \frac{9}{10}$

7. $\frac{3}{4} \div 8\frac{1}{4}$

8. $6\frac{1}{2} \div 3\frac{3}{4}$

9. $18 \div 4\frac{1}{5}$

10. $\frac{7}{16} \div 5\frac{1}{4}$

11. $10\frac{2}{3} \div 3\frac{1}{5}$

12. $9\frac{4}{9} \div 5\frac{2}{3}$

13. Describe and correct the error made in finding the quotient.

> ✗ $8\frac{1}{3} \div 15 = \frac{25}{3} \div 15 = 15 \times \frac{3}{25} = \frac{45}{25} = \frac{9}{5} = 1\frac{4}{5}$

Evaluate the expression when $m = 4\frac{1}{5}$ and $n = 3\frac{1}{8}$.

14. $m \div 2\frac{1}{10}$

15. $10\frac{1}{8} \div n$

16. $m \div n$

17. $n \div \frac{5}{16}$

Evaluate the expression.

18. $9\frac{1}{3} \times 4\frac{1}{2} + 3\frac{2}{5}$

19. $4\frac{1}{9} \times 3\frac{2}{3} \div 1\frac{2}{9}$

20. $6\frac{3}{4} \div \left(2\frac{3}{5} \times 1\frac{2}{13}\right)$

21. How many $3\frac{3}{4}$-inch wires can be cut from a spool of wire that is 100 inches long? Will there be any wire left over? If so, how much?

22. A bag of fertilizer that weighs $18\frac{3}{4}$ pounds can cover 5000 square feet. How many pounds of fertilizer will be needed to cover 27,000 square feet? How many bags of fertilizer are needed? Explain how you found your answer.

Name ______________________________ Date __________

2.6 Enrichment and Extension

Tic-Tac-Toe

Copy the value descriptions below on pieces on paper and place them in a hat. Choose one description from the hat. Then find the expression on the board whose value matches the description. Put an "X" in the box above that expression. Choose another description and put an "O" above its matching expression. Keep alternating between X and O until the board is full. Get three Xs or three Os in a row horizontally, vertically, or diagonally to win. Did the Xs or Os win?

Do not put the descriptions back in the hat after you have chosen them.

$2\frac{11}{12} \div \frac{7}{9} \times 3\frac{1}{20}$	$10\frac{1}{2} \div 4\frac{5}{7} \div \frac{10}{11}$	$4\frac{2}{5} \times 3\frac{3}{10} \div 2\frac{14}{15}$
$15\frac{3}{4} \div 2\frac{9}{20} + 1\frac{11}{28}$	$28\frac{1}{8} \div 6\frac{3}{7} \div 4\frac{13}{20}$	$6\frac{12}{13} - 6\frac{9}{10} \div 1\frac{4}{35}$
$8\frac{1}{6} \div 1\frac{2}{5} - 1\frac{1}{6}$	$12\frac{3}{4} + 5\frac{5}{8} \div 4\frac{1}{14}$	$7\frac{3}{11} \div \frac{5}{22} \div 3\frac{5}{17}$

- The value is about $2\frac{1}{2}$.

- The value is about $9\frac{1}{2}$.
- The value is more than 14.
- The value is about $\frac{1}{2}$.

- The value is between 7 and 8.
- The value is about $11\frac{1}{2}$.
- The value is about 1.

- The value is equivalent to $4\frac{95}{100}$.
- The value is between $4\frac{1}{2}$ and $4\frac{3}{4}$.

Name______________________________ Date__________

2.6 Puzzle Time

What Does An Ant Use To Keep All Of Its Hair In Place?

Write the letter of each answer in the box containing the exercise number.

Divide. Write the answer in simplest form.

1. $1\frac{3}{5} \div \frac{4}{5}$

2. $5\frac{1}{4} \div \frac{3}{4}$

3. $12\frac{2}{5} \div \frac{1}{5}$

4. $2\frac{2}{3} \div 2\frac{2}{3}$

5. $7\frac{1}{7} \div \frac{10}{11}$

6. $3\frac{1}{6} \div \frac{5}{6}$

7. $\frac{7}{9} \div 2\frac{13}{18}$

8. $12\frac{1}{2} \div 15$

9. $14 \div 9\frac{1}{3}$

10. $5\frac{1}{8} \div 6\frac{5}{6}$

11. $3\frac{5}{8} \div 5\frac{4}{5}$

12. $16 \div 4\frac{2}{3}$

13. $4\frac{1}{4} \div \frac{1}{8}$

14. $17 \div 2\frac{4}{15} + 2\frac{5}{12}$

15. $1\frac{3}{7} \div \frac{5}{6} \div 4\frac{4}{5}$

16. $2\frac{5}{8} \div 1\frac{5}{9} \times 1\frac{1}{9}$

17. $2\frac{3}{11} + \frac{4}{9} \div 1\frac{7}{15}$

Answers

E. $1\frac{7}{8}$

A. $3\frac{3}{7}$

G. 1

D. $1\frac{1}{2}$

H. 2

P. $\frac{2}{7}$

R. 34

X. $\frac{5}{8}$

U. $7\frac{6}{7}$

Y. 7

A. $9\frac{11}{12}$

O. $2\frac{19}{33}$

T. $\frac{5}{6}$

B. $\frac{3}{4}$

L. 62

S. $3\frac{4}{5}$

R. $\frac{5}{14}$

16	11	8	13	14		1	17	3	9		10	5	4		6	7	15	12	2

Activity 2.7 Start Thinking!

For use before Activity 2.7

Decimals can be written as fractions and fractions can be written as decimals.

For example, $0.1 = \frac{1}{10}$.

Can you think of any situations in which it is preferable to use a fraction instead of the equivalent decimal?

Activity 2.7 Warm Up

For use before Activity 2.7

Write the fraction or mixed number in simplest form.

1. $\frac{36}{100}$

2. $\frac{5}{100}$

3. $\frac{165}{1000}$

4. $2\frac{42}{100}$

5. $1\frac{18}{100}$

6. $4\frac{555}{1000}$

Lesson 2.7 Start Thinking!
For use before Lesson 2.7

In gymnastics, 0.8 of a point is deducted from a gymnast's score if the gymnast falls off the apparatus during a routine. How can you write 0.8 as a fraction in simplest form?

All deductions in gymnastics scoring are given in decimal form. Why do you think that decimals are used instead of fractions when calculating gymnastics scores?

Lesson 2.7 Warm Up
For use before Lesson 2.7

Write the decimal as a fraction in simplest form.

1. 0.2 **2.** 0.3 **3.** 0.19

4. 0.25 **5.** 0.65 **6.** 0.155

Name ______________________ Date ________

2.7 Practice A

Write the decimal as a fraction or mixed number in simplest form.

1. 0.7 **2.** 0.4 **3.** 0.2 **4.** 0.5

5. 0.16 **6.** 0.44 **7.** 0.35 **8.** 0.59

9. 0.74 **10.** 0.216 **11.** 0.96 **12.** 0.624

13. 7.65 **14.** 4.08 **15.** 2.84 **16.** 8.345

17. Describe and correct the error made in writing 0.04 as a fraction.

$$\times \quad 0.04 = \frac{4}{10} = \frac{2}{5}$$

Write the decimal as an improper fraction in simplest form.

18. 2.25 **19.** 5.48 **20.** 3.625

Write the value of the money as a decimal and as a fraction of a dollar.

21. two nickels **22.** six pennies

23. eight dimes **24.** one dollar and four nickels

25. The thickness of an MP3 player is 0.32 inch. Write the thickness as a fraction in simplest form.

26. A sprinter ran 100 meters in 9.8 seconds. Write the number of seconds as a fraction in simplest form.

27. A zoom lens magnifies an image 3.25 times larger than normal. Write the magnification as a fraction in simplest form.

28. A runner is 0.45 mile from the finish line. Write the number of miles as a fraction in simplest form.

29. There were 75 students who tried out for a school play and 0.64 of the students were selected to be in the play. What fraction of the students were not selected to be in the play? How many students were not selected to be in the play?

Name______________________________ Date__________

2.7 Practice B

Write the decimal as a fraction or mixed number in simplest form.

1. 0.3 **2.** 0.8 **3.** 0.25 **4.** 0.15

5. 0.36 **6.** 0.48 **7.** 0.72 **8.** 0.63

9. 0.17 **10.** 0.225 **11.** 0.86 **12.** 0.432

13. 5.29 **14.** 6.504 **15.** 2.625 **16.** 7.056

17. Describe and correct the error made in writing 0.42 as a fraction.

$$\times \quad 0.42 = \frac{42}{1000} = \frac{21}{500}$$

Write the decimal as an improper fraction in simplest form.

18. 3.16 **19.** 4.375 **20.** 9.044

Write the value of the money as a decimal and as a fraction of a dollar.

21. three dimes and a nickel **22.** one nickel and seven pennies

23. two dollars and three dimes **24.** three dollars and eight pennies

Write the shaded part of the unit square as a decimal. Then write the decimal as a fraction.

25.
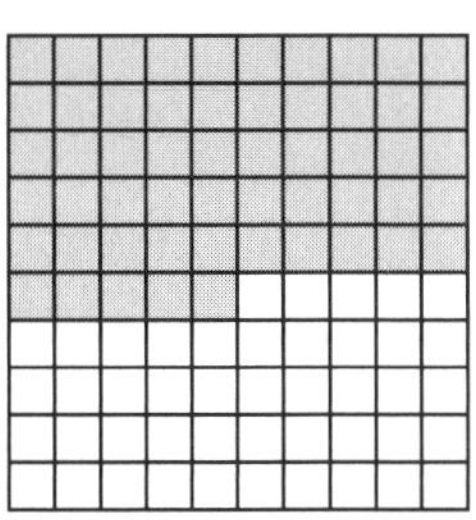

26.
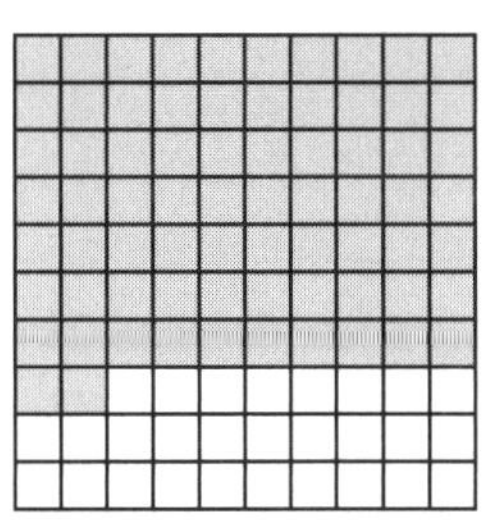

27.
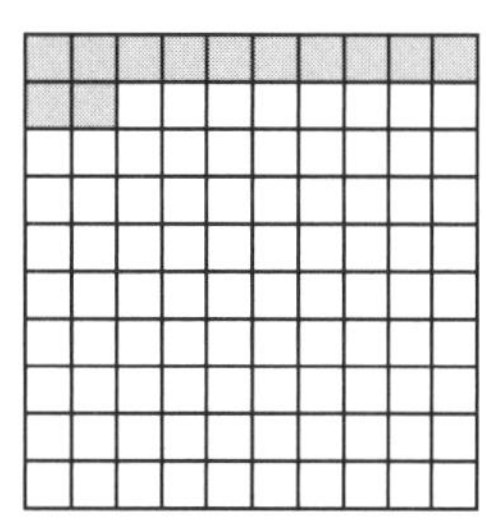

28. You have 160 trading cards. Seven-eighths of them are characters and the rest are locations. You gave 0.2 of the character cards to your friend. How many trading cards do you have left?

29. A baseball player's batting average is 0.312.

a. Write the batting average as a fraction in simplest form.

b. The batter has 500 at bats. How many hits does the player have?

Name ______________________________ Date __________

2.7 Enrichment and Extension

Writing Expressions Involving Decimals as Fractions

Write a fraction that is equivalent to the given expression without decimals.

1. $\frac{1.4}{25}$ **2.** $\frac{5.7}{19}$ **3.** $\frac{3.3}{17}$

4. $\frac{15}{2.1}$ **5.** $\frac{48}{3.72}$ **6.** $\frac{2.13}{39}$

Use your answers from Exercises 1–6 to find the product or quotient.

7. $\frac{15}{2.1} \times \frac{3.3}{17}$ **8.** $\frac{1.4}{25} \times \frac{48}{3.72}$

9. $\frac{3.3}{17} \div \frac{5.7}{19}$ **10.** $\frac{5.7}{19} \div \frac{2.13}{39}$

Name______________________________ Date__________

2.7 Puzzle Time

Why Did The Bee Go North For The Summer?

Write the letter of each answer in the box containing the exercise number.

Write the decimal as a fraction or mixed number in simplest form.

1. 0.7

2. 0.8

3. 0.2

4. 0.19

5. 0.5

6. 0.42

7. 0.14

8. 0.111

9. 0.625

10. 0.255

11. 0.91

12. 0.32

13. 0.545

14. 0.428

15. 6.25

16. 8.5

17. 7.227

18. 11.75

19. 8.246

20. 9.555

21. 1.99

Answers

T. $\frac{19}{100}$	**V.** $9\frac{111}{200}$	**N.** $11\frac{3}{4}$
A. $\frac{51}{200}$	**O.** $\frac{7}{10}$	**I.** $\frac{8}{25}$
T. $\frac{1}{2}$	**N.** $\frac{111}{1000}$	**I.** $8\frac{123}{500}$
E. $\frac{4}{5}$	**W.** $6\frac{1}{4}$	**T.** $\frac{107}{250}$
A. $\frac{21}{50}$	**R.** $\frac{91}{100}$	**O.** $\frac{1}{5}$
N. $\frac{5}{8}$	**S.** $7\frac{227}{1000}$	**I.** $\frac{109}{200}$
K. $\frac{7}{50}$	**Y.** $1\frac{99}{100}$	**N.** $8\frac{1}{2}$

5	3		20	12	17	19	4		10	16		6	18	14		13	9		8	2	15		21	1	11	7

Activity 2.8 Start Thinking!
For use before Activity 2.8

In what situations is it preferable to use a decimal instead of the equivalent fraction?

Activity 2.8 Warm Up
For use before Activity 2.8

Find the quotient.

1. $5\overline{)320}$

2. $6\overline{)162}$

3. $4\overline{)292}$

4. $7\overline{)147}$

5. $3\overline{)156}$

6. $8\overline{)184}$

Lesson 2.8 Start Thinking!

For use before Lesson 2.8

Are all three of the decimals below equivalent? Why or why not?

$0.4555 \qquad 0.4555\ldots \qquad 0.4\overline{5}$

Lesson 2.8 Warm Up

For use before Lesson 2.8

Write the fraction as a decimal.

1. $\frac{6}{25}$ **2.** $\frac{11}{20}$ **3.** $\frac{2}{5}$

4. $\frac{49}{50}$ **5.** $\frac{1}{12}$ **6.** $\frac{11}{18}$

Name ______________________________ Date __________

2.8 Practice A

Write the fraction as a decimal.

1. $\frac{3}{10}$ **2.** $\frac{2}{5}$ **3.** $\frac{16}{25}$ **4.** $\frac{5}{12}$

5. $\frac{19}{40}$ **6.** $\frac{11}{20}$ **7.** $\frac{7}{8}$ **8.** $\frac{13}{30}$

9. $\frac{7}{16}$ **10.** $\frac{7}{45}$ **11.** $\frac{27}{80}$ **12.** $\frac{54}{25}$

13. $\frac{28}{15}$ **14.** $\frac{19}{18}$ **15.** $\frac{247}{50}$ **16.** $\frac{17}{6}$

17. Describe and correct the error made in writing $\frac{2}{9}$ as a decimal.

X $2\overline{)9.0}$ = 4.5

Copy and complete the statement using <, >, or =.

18. $\frac{9}{16}$ __?__ 0.56 **19.** $\frac{5}{6}$ __?__ $0.8\overline{3}$ **20.** 1.45 __?__ $1\frac{9}{10}$

21. A photo is $5\frac{7}{8}$ inches long. Write the photo's length as decimal.

22. A basketball player made 16 shots out of 30 attempted. Write in simplest form the fraction of made shots. Write the fraction as a decimal.

23. How many pounds heavier is a $8\frac{1}{4}$-pound fish than a $6\frac{7}{8}$-pound fish? Give your answer as both a fraction and a decimal.

24. What is the perimeter of the triangle? Give your answer as both a fraction and a decimal.

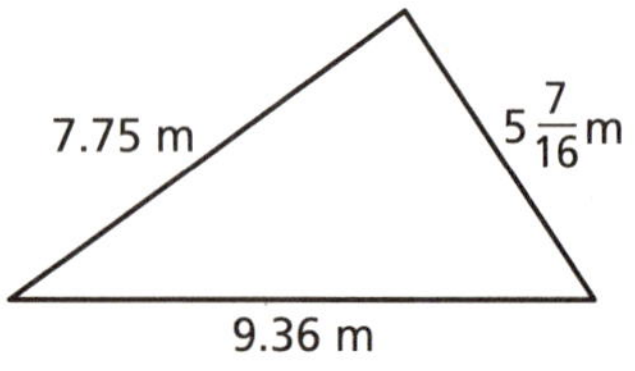

Name____ Date __________

2.8 Practice B

Write the fraction as a decimal.

1. $\frac{1}{10}$ **2.** $\frac{5}{8}$ **3.** $\frac{21}{25}$ **4.** $\frac{8}{9}$

5. $\frac{17}{20}$ **6.** $\frac{31}{80}$ **7.** $\frac{11}{12}$ **8.** $\frac{5}{18}$

9. $\frac{14}{25}$ **10.** $\frac{13}{40}$ **11.** $\frac{23}{70}$ **12.** $\frac{76}{25}$

13. $\frac{29}{30}$ **14.** $\frac{63}{40}$ **15.** $\frac{56}{25}$ **16.** $\frac{41}{15}$

17. Describe and correct the error made in writing $\frac{4}{5}$ as a decimal.

X $\begin{array}{r} 8.0 \\ 5\overline{)4.0} \end{array}$

Copy and complete the statement using <, >, or =.

18. $\frac{7}{13}$ __?__ 0.53 **19.** $\frac{4}{11}$ __?__ $0.\overline{36}$ **20.** 1.76 __?__ $1\frac{4}{5}$

Write the number as a fraction. Then write the fraction as a decimal.

21. eleven-fifteenths **22.** seventeen-fortieths **23.** seventy-one thirtieths

24. One book is $1\frac{7}{8}$ inches thick. Another book is $2\frac{7}{8}$ inches wide. Write the combined thickness of the two books as a fraction and as a decimal.

25. The table lists the portion of 600 students who chose their favorite pet.

Pet	Cat	Dog	Fish	Hamster	Snake	Turtle
Portion of students	$\frac{6}{25}$	$\frac{13}{40}$	$\frac{13}{50}$	$\frac{3}{50}$	$\frac{1}{25}$	$\frac{3}{40}$

a. Write each fraction as a decimal. Then order the decimals from least to greatest.

b. How many more students picked the most popular pet than the least popular pet?

Name ______________________________ Date __________

2.8 Enrichment and Extension

Exploring Repeating Decimals

Copy and complete the table. Explain the pattern in the repeating decimals.

1.

Fraction	Numerator	Decimal form	Repeating digit
$\frac{1}{9}$	1	$0.\overline{1}$	1
$\frac{4}{9}$			
$\frac{7}{9}$			

2.

Fraction	Numerator	Decimal form	Repeating digits
$\frac{1}{99}$	1	$0.\overline{01}$	01
$\frac{8}{99}$			
$\frac{17}{99}$			
$\frac{54}{99}$			

Use the pattern in Exercise 1 to write the repeating decimal as a fraction.

3. $0.\overline{2}$ **4.** $0.\overline{5}$ **5.** $0.\overline{8}$

Use the pattern in Exercise 2 to write the repeating decimal as a fraction.

6. $0.\overline{05}$ **7.** $0.\overline{34}$ **8.** $0.\overline{98}$

Use what you learned in the tables to write the repeating decimal as a fraction.

9. $0.\overline{001}$ **10.** $0.\overline{125}$ **11.** $0.\overline{0305}$

Name________________________________ Date__________

2.8 Puzzle Time

What Will Happen If A Banana Gets A Very Bad Sunburn?

Write the letter of each answer in the box containing the exercise number.

Write the fraction as a decimal.

1. $\frac{9}{10}$

2. $\frac{3}{4}$

3. $\frac{41}{50}$

4. $\frac{8}{9}$

5. $\frac{16}{25}$

6. $\frac{7}{20}$

7. $\frac{6}{11}$

8. $\frac{5}{6}$

9. $\frac{11}{12}$

10. $\frac{2}{3}$

11. $\frac{61}{50}$

12. $\frac{119}{25}$

13. $\frac{33}{16}$

14. $\frac{11}{18}$

15. $\frac{31}{20}$

16. $\frac{9}{5}$

17. $\frac{13}{4}$

Answers

T. 1.8	**L.** 4.76
W. $0.\overline{8}$	**I.** $0.91\overline{6}$
E. 0.9	**P.** $0.\overline{54}$
S. 2.0625	**T.** 1.22
L. 0.64	**A.** 0.75
O. 3.25	**E.** 1.55
T. $0.8\overline{3}$	**I.** $0.\overline{6}$
R. 0.82	**L.** 0.35
T. $0.6\overline{1}$	

10	16	■	4	9	12	5	■	13	14	2	3	8	■	11	17	■	7	1	15	6

Name ______________________________ Date __________

Chapter 2

School-to-Work

For use after Section 2.5

Adjusting Recipes as a Chef

A professional chef may work in a restaurant, in a bakery, or for a catering company. The chef plans meals, cooks, designs new recipes, and is involved with ordering food and ensuring that the kitchen runs smoothly. Often a chef has to adjust recipes to serve large groups of people or to avoid making too much food.

Garlic Pasta (serves 4)

1/4 cup olive oil

12 cloves of garlic

1/3 cup chopped bacon

1 1/2 cups canned tomatoes

1 pound penne pasta

1. Gather a few family recipes or some recipes of your favorite foods. Choose recipes where some of the ingredients are measured using fractions and mixed numbers.

2. How many servings does the recipe make?

3. For a family reunion you will need to feed 120 people. Rewrite each recipe so that it shows the amount of each ingredient you need to serve 120 people. Describe the process you used to find your answers.

4. For a buffet, chefs make about $\frac{1}{6}$ more of each recipe. Food will be served on a buffet at the family reunion. Adjust your recipes so that you have $\frac{1}{6}$ more food.

5. You are making dinner for you and a friend using your favorite recipes. In order to save money and not waste any food, you decide to make only two servings of your recipes. Rewrite your recipes so that they show the amount of each ingredient you need to serve two people.

Name______________________________ Date__________

Chapter 2 Study Help

You can use a **notetaking organizer** to write notes, vocabulary, and questions about a topic.

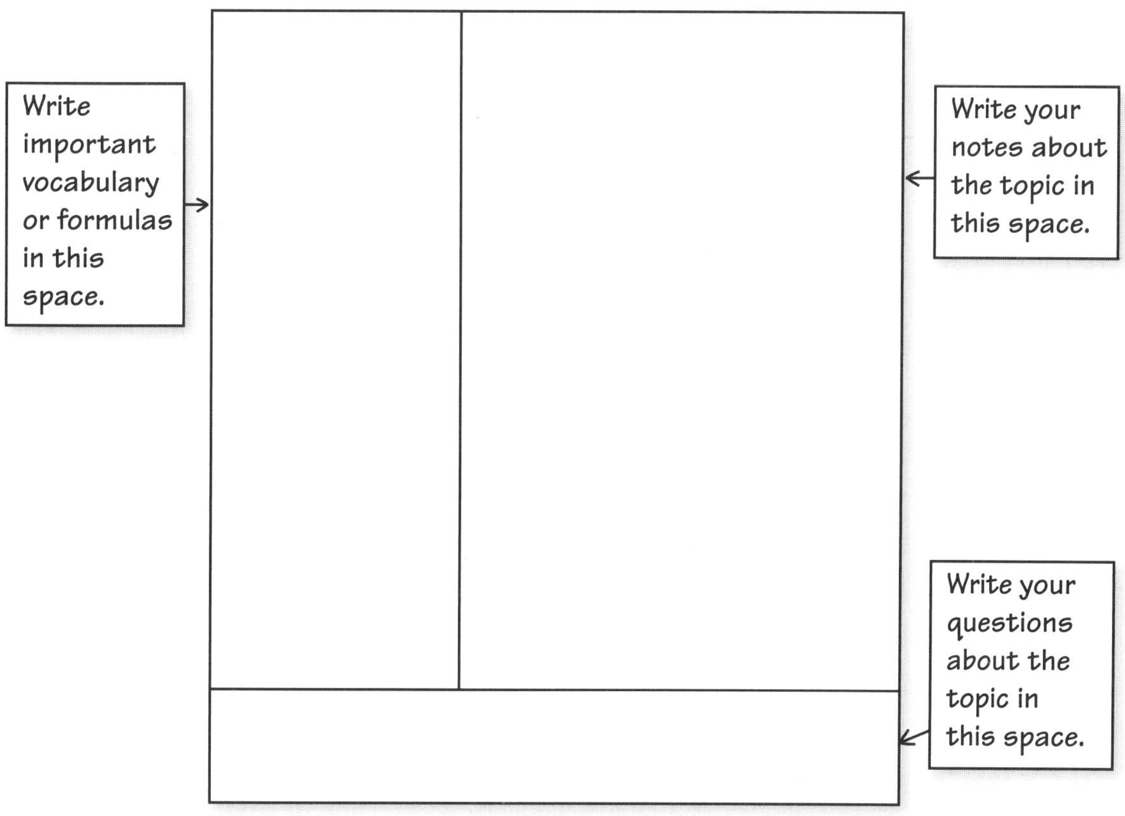

On Your Own

Make a notetaking organizer to help you study these topics.

1. estimating with mixed numbers
2. multiplying two fractions
3. multiplying mixed numbers

After you complete this chapter, make notetaking organizers for the following topics.

4. dividing by a fraction
5. dividing mixed numbers
6. writing decimals as fractions
7. writing fractions as decimals

Name ____ Date __________

Chapter 2 Financial Literacy

For use after Section 2.2

Making Change

Decimals are usually used to represent money. You use a decimal point when you have part of a dollar. Without decimals you would have no way to write down the amount of change you have. Change is made with pennies, nickels, dimes, quarters, and fifty-cent pieces.

1. How many pennies are in a dollar? What fraction of a dollar is one penny? Write this amount as a decimal.

2. How many nickels are in a dollar? What fraction of a dollar is one nickel? Write this amount as a decimal.

3. How many dimes are in a dollar? What fraction of a dollar is one dime? Write this amount as a decimal.

4. How many quarters are in a dollar? What fraction of a dollar is one quarter? Write this amount as a decimal.

5. Your bill at a store is $4.23. Some people would give the cashier $5.25. Why?

6. If you have a $10 bill, three $1 bills, two pennies, one nickel, and two dimes, how would you pay for something that costs $6.03?

7. Someone pays $21.00 for something that costs $10.98. How much change do they receive? Why do you think they paid $21.00 instead of just $20.00?

Name______________________________ Date__________

Chapter 2

Technology Connection

For use after Section 2.6

Using a Calculator with Fractions and Mixed Numbers

A calculator can be used to perform calculations with fractions and mixed numbers. Keep the order of operations in mind when you enter the expression. You can enter a mixed number in two ways:

Write the mixed number as a whole number plus a fraction.	Write the mixed number as an improper fraction.
$1\frac{2}{5} = (1 + 2 \div 5)$	$1\frac{2}{5} = \frac{5 \times 1 + 2}{5} = ((5 \times 1 + 2) \div 5)$
Calculator Keystrokes: [(] 1 [+] 2 [÷] 5 [)]	Calculator Keystrokes: [(] [(] 5 [×] 1 [+] 2 [)] [÷] 5 [)]

1. Use one of the methods above to write the calculator keystrokes for $3\frac{7}{12}$.

2. Use one of the methods above to write the calculator keystrokes for $\frac{5}{6}$.

3. Use your answers to Exercises 1 and 2 to write the calculator keystrokes for $3\frac{7}{12} \div \frac{5}{6}$. Then use a calculator to find $3\frac{7}{12} \div \frac{5}{6}$.

Evaluate the expression. Write the answer as it appears on your calculator.

4. $4\frac{1}{6} \div 1\frac{1}{3}$ **5.** $4\frac{1}{3} \div 2\frac{8}{9}$ **6.** $9\frac{4}{5} \div \frac{7}{10}$ **7.** $2\frac{1}{4} \div \frac{5}{8}$

8. Kelly and Jamie each wrote the keystrokes for $6 \div \frac{2}{3}$. Will both students' keystrokes give the same answer? What is the correct answer? Why?

Kelly	Jamie
[(] 6 [÷] [(] 2 [÷] 3 [)] [)]	[(] 6 [÷] 2 [÷] 3 [)]

Evaluate the expression. Write the answer as it appears on your calculator.

9. $2\frac{1}{2} + 1\frac{4}{5} \div \frac{3}{5} - \frac{1}{4}$ **10.** $12\frac{4}{7} \times 1\frac{8}{9} \div 2\frac{3}{7}$ **11.** $10\frac{1}{2} \div 1\frac{1}{4} - 3\frac{1}{5}$

Name ______________________________ Date __________

Chapter 2

Life Connections

For use after Section 2.1

Using Estimation to Plan Your Day

In life you often use math and don't even realize it! When you make decisions on how to use your time, you are using math. Read through the following scenario. Use estimation to find the amount of time it will take you to do each task. Use the map to determine how long it will take you to walk to each activity.

Determine: (a) the risk if you don't do one or more of the activities, (b) which of the activities you will do, (c) how much time you will spend at each with a detailed plan of your day including travel time between events, and (d) if you will be able to do them all well. You want to do them all, but can you? Don't forget to schedule time for lunch!

Saturday Events:

- You volunteer at the local animal shelter every Saturday from 9 A.M. until 1:30 P.M.
- You have a research paper due in history next week and need to use the library for at least one hour. The library is open between 8:15 A.M. and 4:45 P.M. on Saturdays and closed on Sundays.
- You have soccer practice from 2:30 P.M. to 5:00 P.M. on Saturday and Sunday.
- The 6th Grade class is working on a float on Saturday at the school. The organizers have asked for volunteers to help anytime between 9:30 A.M. and 4:30 P.M. They have to finish the float on Saturday.
- You promised your family that you would mow the grass on Saturday. It takes you 40 minutes.
- Saturday evening you and your family are going to a friend's house for a cookout and will be leaving your home at 6:00 P.M.

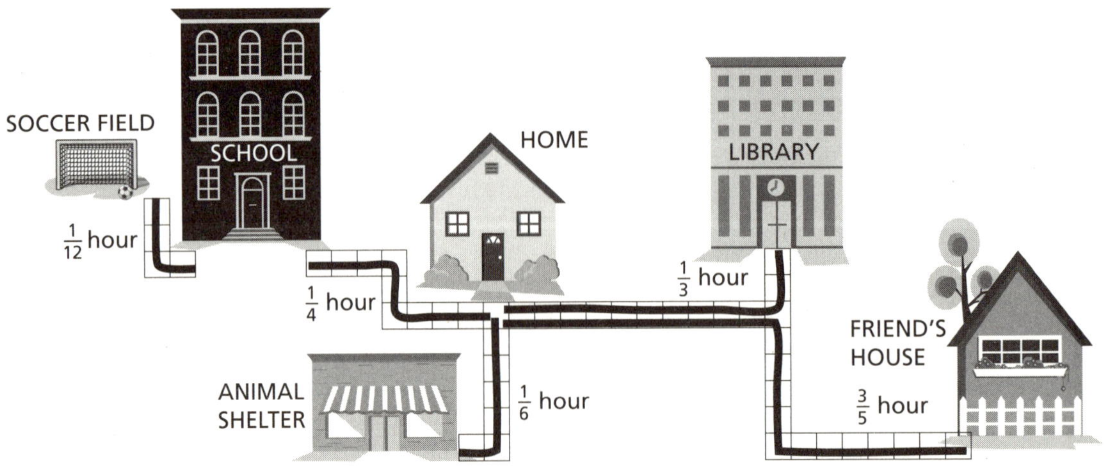

Name____ Date__________

Chapter 2

Stories in History

For use after Section 2.7

Writing Fractions

Ancient Egyptian writing used more than 2000 hieroglyphic characters to represent common objects. To represent numbers they also used hieroglyphic characters. They repeated the character for each unit of 10. We also base our number system on units of 10. Here is how 351 is represented.

Our base-10 number system

$351 = 300 + 50 + 1$

Ancient Egyptian number system

The ancient Egyptians also used hieroglyphic characters to represent fractions. With hieroglyphic fractions the numerator was always 1. They placed the symbol  above or next to the hieroglyphic number to represent the fraction.

The fraction $\frac{1}{2}$ can be represented by . To represent $\frac{3}{4}$, the Egyptians would have to write the fraction as the sum of fractions with a numerator of 1.

So, $\frac{3}{4}$ would be written as $\frac{1}{4} + \frac{1}{2}$ which would look like .

Use the hieroglyphic characters to write the fraction.

100,000 10,000 1000 100 10 1

1. $\frac{1}{100}$

2. $\frac{3}{10}$

3. $\frac{1}{1000}$

4. $\frac{2}{10{,}000}$

5. $\frac{8}{15}$

6. $\frac{13}{50}$

7. Do you think it was easier for ancient Egyptians to multiply fractions than the method we use today? Why or why not?

Name ______________________________ Date __________

Chapter 2 Cumulative Practice

Estimate.

1. $\frac{5}{11} \bullet \frac{14}{15}$

2. $30\frac{1}{2} \div 5\frac{8}{9}$

3. $8.1 - 6.9$

4. $3.8 + 7.2 - 4.1$

5. $\frac{4}{9} \bullet \frac{1}{8}$

6. $9\frac{1}{2} \div 3\frac{1}{4}$

7. $5.6 + 11.8 + 13.3$

8. $\frac{1}{16} \bullet \frac{6}{7}$

9. $\frac{9}{4} \div \frac{4}{5}$

10. A gallon of milk costs $3.89. A loaf of bread costs $2.15. Estimate the cost of buying a gallon of milk and a loaf of bread. Explain your method.

Evaluate.

11. $\frac{3}{5} \bullet \frac{5}{4}$

12. $\frac{2}{3} - \frac{1}{4}$

13. $6 \div \frac{1}{7}$

14. $\frac{9}{16} \div \frac{3}{4}$

15. $\frac{7}{8} + \frac{5}{8}$

16. $\frac{3}{7} \bullet \frac{11}{7}$

17. $1\frac{1}{4} - \frac{3}{4}$

18. $12\frac{1}{2} \div \frac{5}{8}$

19. $2\frac{7}{9} \bullet \frac{3}{10}$

20. $\frac{14}{3} \bullet 6$

21. $\frac{2}{9} \div 5$

22. $\frac{1}{3} + \frac{1}{2} + \frac{1}{6}$

23. A circular track is $\frac{1}{4}$ mile long. You run $13\frac{1}{3}$ laps around the track. How many miles did you run?

24. You have 3 books that are $\frac{1}{2}$ inch thick, 4 books that are $\frac{1}{6}$ inch thick, and 5 books that are $\frac{1}{8}$ inch thick. What is the total width of the books?

Write the numbers in order from least to greatest.

25. $2.05, \frac{9}{4}, 1, \frac{3}{8}, 1.75$

26. $\frac{22}{5}, 5, 4.45, 5.18, 4\frac{7}{10}$

27. $\frac{1}{3}, 0.68, \frac{2}{3}, 0.37, \frac{1}{2}$

28. $3, 2\frac{9}{10}, 2.89, 3.01, \frac{31}{10}$

Chapter 3

Name ______________________________ Date __________

Chapter 3 Multiplying and Dividing Decimals

Dear Family,

Budgeting for a family vehicle can be challenging because of fluctuating prices and maintenance costs. Fuel prices are usually at the top of our minds because we deal with them nearly every day.

Because the price of fuel might change every day, you have to make estimates when you budget. How much will a gallon of fuel cost this month: \$2.29, \$3.57, \$4.24? You have to make your best guess when budgeting. Usually you choose an "easy" number to work with, such as \$3.50 a gallon.

Estimating how much fuel will cost is only part of the problem. We also have to guess how much we will use. You probably have a good idea of how often you buy fuel—perhaps once or twice a week. If you put in "about 8 gallons" each time, you might budget 1.5 times 8 gallons each week—about 12 gallons a week.

You and your student could work on a fuel budget for a family vehicle. Ask your student to find out what you need to know for your budget.

- What will you budget for the price of fuel this month?
- How much fuel do you think you will use this month?
- What if you had a car that got more miles to the gallon—how would that affect the budget?

Each time you buy fuel, have your student estimate the cost of the fuel and use the actual cost to check their estimate. Is the cost of fuel close to what you planned in your budget? What factors may have contributed to any differences?

There are several alternatives that families use to cut their transportation costs. If public transit is available, it is sometimes cheaper than using a family vehicle. When doing your budget, discuss alternative forms of transportation with your student. Are there other alternatives, such as bike riding or walking? Would these save your family money?

Getting your student involved with family budgeting is a great learning experience for them—and it just might save you time and money as well.

Enjoy your planning!

Nombre ______________________________ Fecha __________

Capítulo 3 Multiplicando y Dividiendo Decimales

Estimada Familia:

Sacar el presupuesto para un vehículo familiar puede representar un desafío debido a los precios fluctuantes y los costos de mantenimiento. Los precios de la gasolina generalmente son los que más nos preocupan porque tratamos con ellos todos los días.

Porque el precio de la gasolina puede cambiar cada día, se deben hacer cálculos para obtener el presupuesto. ¿Cuánto costará un galón de gasolina este mes: $2.29, $3.57, $4.24? Para obtener el presupuesto, hay que adivinar lo mejor que podamos. Generalmente se escoge un número "fácil" con el cual se puede trabajar, como por ejemplo $3.50 por galón.

Estimar cuánto costará la gasolina es solo una parte del problema. También debemos adivinar cuánto usaremos. Probablemente tiene una buena idea de la frecuencia con que compra gasolina—quizás una o dos veces por semana. Si anota "unos 8 galones" cada vez, puede obtener 1.5 veces 8 galones cada semana—unos 12 galones por semana.

Usted y su estudiante podrían trabajar sacando un presupuesto de gasolina para un vehículo familiar. Pida a su estudiante que busque lo que necesita saber para su presupuesto.

- ¿Cuál será su presupuesto para el precio de la gasolina este mes?
- ¿Cuánta gasolina cree que usará este mes?
- ¿Qué pasa si tuviera un vehículo que obtiene más millas por galón—cómo se vería afectado su presupuesto?

Cada vez que compre gasolina, haga que su estudiante calcule el costo de la gasolina y use el costo real para verificar su cálculo. ¿El costo de la gasolina se aproxima a lo que planeó en su presupuesto? ¿Qué factores pueden haber contribuido a hacer alguna diferencia?

Existen muchas alternativas que las familias usan para recortar sus gastos de transporte. Si hay transporte público disponible, a veces es más barato que usar el vehículo de la familia. Cuando saque su presupuesto, converse acerca de formas alternativas de transporte con su estudiante. ¿Existen otras alternativas, como por ejemplo montar bicicleta o caminar? ¿Éstas ahorrarían dinero para su familia?

Hacer que su estudiante se involucre en el presupuesto familiar es una gran experiencia de aprendizaje para él—y también puede hacerle ahorrar tiempo y dinero.

¡Disfrute su planificación!

Non ______________________________ Dat __________

Chapít 3 Ann Miltipliye ak Divize Desimal

Chè Fanmi:

Fè bidjè pou yon machin familyal kapab difisil poutèt pri k'ap varye ak kou mentnans. Pri gaz abityèlman nan tèt panse nou, poutèt nou fè fas ak yo prèske chak jou.

Kòm pri gaz gen dwa chanje chak jou, ou oblije fè estimasyon lè w'ap fè bidjè. Konbyen yon galon gaz pwal koute mwa sa a $2,29, $3,57, $4,24? Ou dwe devine chif ki pi pwòch la lè w'ap fè bidjè. Abityèlman ou chwazi yon chif "fasil" pou travay avèk li, tankou $3,50 pa galon.

Estime konbyen gaz la pwal koute se sèlman yon pati nan pwoblèm nan. Nou dwe devine tou ki kantite nou pwal itilize. Ou siman gen yon bon ide sou konbyen fwa pa semèn ou achte gaz—petèt youn oswa de fwa. Si ou mete "apeprè 8 galon" chak fwa, ou gen dwa bezwen fè bidjè pou 1,5 fwa 8 galon chak semèn—apeprè 12 galon pa semèn.

Ou menm ak elèv ou a ta kapab travay sou yon bidjè gaz pou yon machin familyal. Mande elèv ou a jwenn sa ou bezwen konnen pou bidjè ou an.

- Kisa bidjè ou ap ye pou pri gaz la mwa sila a?
- Ki kantite gaz ou panse w'ap itilize mwa sila a?
- E si ou te gen yon machin ki fè plis mil pa galon—kijan sa t'ap afekte bidjè a?

Chak fwa ou achte gaz, fè elèv ou a estime kou gaz la epi sèvi ak vre pri gaz la pou tcheke estimasyon li an. Èske pri gaz la pre sa ou te planifye nan bidjè ou an? Ki faktè ki te gen dwa kontribye nan diferans yo si genyen?

Gen plizyè altènativ ke fanmi itilize pou bese kou transpò yo. Si transpò piblik disponib, li pafwa pi bon mache pase sèvi ak yon machin familyal. Lè w'ap fè bidjè ou an, diskite sou fòm altènatif transpò avèk elèv ou a. Èske gen lòt altènativ, tankou monte bisiklèt oswa mache? Èske yo t'ap fè fanmi ou ekonomize lajan?

Fè elèv ou a enplike nan fè bidjè familyal la se yon bon esperyans aprantisaj pou li—epi sa gen dwa tou fè ou ekonomize tan ak lajan.

Fè bon planifikasyon!

Activity 3.1

Start Thinking!

For use before Activity 3.1

How can you use estimation when you are shopping online?

Activity 3.1

Warm Up

For use before Activity 3.1

Round the decimal to the nearest whole number.

1. 16.3 **2.** 98.9 **3.** 7.878

4. 1.02 **5.** 6.592 **6.** 0.24

Lesson 3.1 Start Thinking!

For use before Lesson 3.1

Find a number you can use in place of 29 to estimate the quotient in each division problem below. Tell what number you used.

$29 \div 3$ $29 \div 4$ $29 \div 8$ $29 \div 9$

Then write a division problem in which the number used in place for 29 is 33.

Lesson 3.1 Warm Up

For use before Lesson 3.1

Estimate by rounding each factor to the nearest whole number.

1. 2.21×6

2. 5.06×8

3. 11.89×5

4. 3.74×9

5. 5.14×3

6. 7.67×11

Name________________________________ Date__________

3.1 Practice A

Estimate by rounding each factor to the nearest whole number.

1. 2.23×9 **2.** 4.79×8 **3.** 7.41×4

4. 6.48×5 **5.** 7.4×10 **6.** 4.51×3

7. 11.71×4.96 **8.** $6.4 \div 1.9$ **9.** $35.84 \div 5.92$

10. 3.3×7.9 **11.** 5.1×17.8 **12.** $53.61 \div 8.77$

Use compatible numbers to estimate the product or quotient.

13. 19.4×7.87 **14.** 5.66×16.75 **15.** 23.73×4.31

16. $20.89 \div 6.67$ **17.** $47.59 \div 5.78$ **18.** $67.44 \div 6.19$

Estimate the area of the figure by rounding to the nearest whole number.

19.

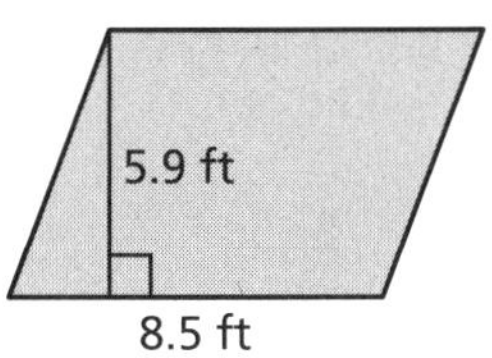

20.

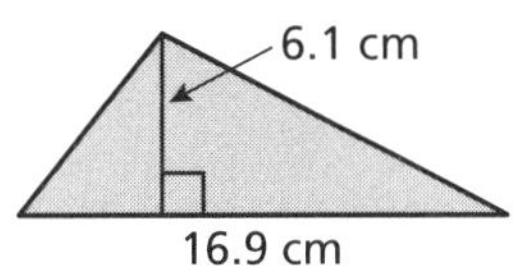

21.

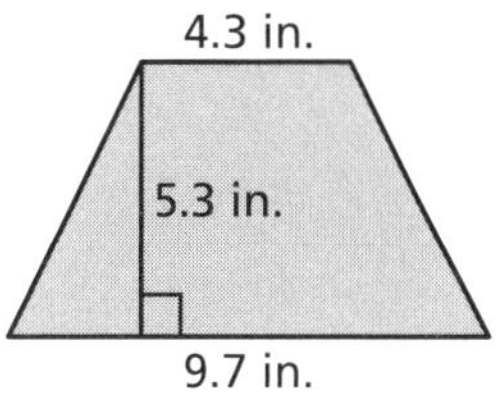

22. Show another way to estimate the quotient. Explain your steps.

$$57.9 \div 8.1 \approx 56 \div 8 = 7$$

23. A part-time job pays \$28.31 per week. Estimate the number of weeks it will take to earn \$150.

24. It takes 32 jars to fill a water jug. Each jar contains 0.94 liter of water. Estimate how many liters of water there are in a full jug.

25. Your fish tank has a rectangular base that is 50.8 centimeters long and 25.4 centimeters wide. You want to estimate the amount of pebbles needed for your fish tank.

a. Estimate the area *A* of the fish tank floor. Is your estimate too little or too much?

b. Use the expression 3*A* to estimate how much pebbles (in cubic centimeters) are needed for a 3-centimeter layer.

Name ______________________ Date ________

3.1 Practice B

Estimate by rounding each factor to the nearest whole number.

1. 3.62×7 **2.** 5.46×9 **3.** 6.72×7

4. 2.98×5 **5.** 9.68×13 **6.** 3.47×8

7. 5.7×8.3 **8.** $9.39 \div 2.7$ **9.** $51.71 \div 12.67$

10. 14.82×4.75 **11.** 6.2×16.8 **12.** $27.81 \div 6.62$

Use compatible numbers to estimate the product or quotient.

13. 62.3×5.85 **14.** 49.7×7.44 **15.** 87.91×4.94

16. $23.3 \div 3.89$ **17.** $35.77 \div 3.61$ **18.** $71.18 \div 8.19$

Estimate the area of the figure by rounding to the nearest whole number.

19.
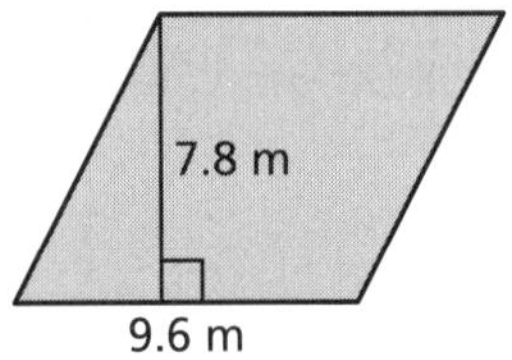

20.
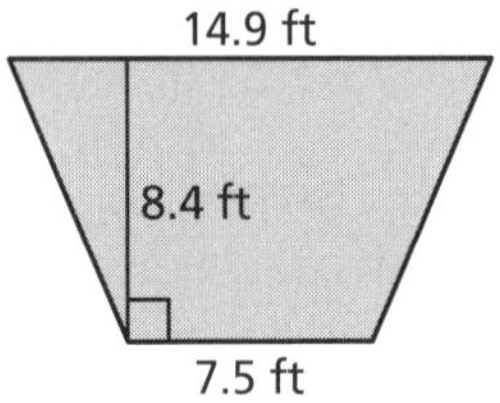

21.
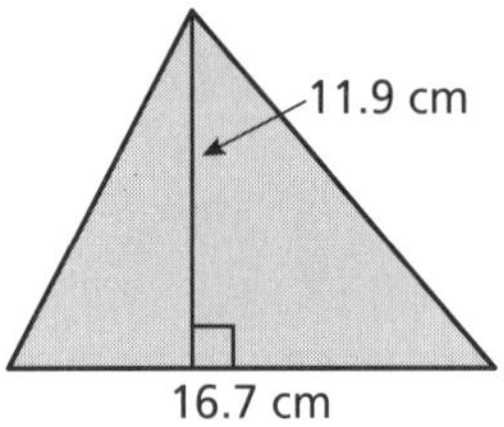

Estimate the value of the expression.

22. $36.37 \div 3.79 \times 8.7$ **23.** $66.5 \div (4.23 \times 2.86)$

24. Show two ways to estimate the quotient. Explain your steps.

$39.7 \div 4.55$

25. Four students each save an average of $5.25 a week. About how much will they save in 25 weeks? About how much will they save in a year (52 weeks)?

26. Drinking cups hold about 7.5 fluid ounces of water.

a. Estimate the fluid ounces of water you will need for 25 drinking cups.

b. A quart contains 32 fluid ounces. Estimate the quarts of water you will need for 25 drinking cups.

Name ______________________________ Date __________

3.1 Enrichment and Extension

Math Riddle

To find an underestimate of a product, you round both numbers down. But to find an underestimate of a quotient, you round the numerator down and the denominator up.

$$10.6 \div 4.5 \approx 10 \div 5 = 2$$

Find the underestimate of the expression.

1. 100.9×15.7 R

2. 2.62×15.51 L

3. $15.4 \div 4.9$ A

4. $22.1 \div 7.2$ H

5. $55.62 \div 4.75$ D

6. $\dfrac{40.05}{5.79}$ F

7. $5.1 \times 36.3 \div 7.7$ O

8. $10.125 \times 10.503 \div 19.88$ L

9. $240.3 \div 14.6 \times 2.5$ A

10. $80.82 \div 19.05 \times 7.5$ L

11. Order the underestimates in Exercises 1–10 from least to greatest. Arrange the letters corresponding to each quotient in the same order to answer the riddle.

What coin is worth twice as much when half is removed?

Name ______________________________ Date __________

3.1 Puzzle Time

Where Is The Best Place To Find A Terrified Hummingbird?

Write the letter of each answer in the box containing the exercise number.

Estimate by rounding each factor to the nearest whole number.

1. 4.21×9
2. 6.06×3
3. 3.8×8
4. 3.57×4
5. 8.9×13
6. 4.42×5
7. 6.6×6.1
8. 1.5×7.4
9. $10.1 \div 2.1$
10. 13.83×7.91
11. $45.11 \div 4.85$
12. $50.31 \div 24.55$

Use compatible numbers to estimate the product or quotient.

13. 48.3×7.53
14. $36.54 \div 3.93$
15. $283.67 \div 57.9$
16. 6.85×28.26
17. 67.23×1.69
18. $68.23 \div 9.54$

Answers for Exercises 1–12

I. 20	**F.** 9
R. 112	**N.** 32
D. 36	**O.** 117
E. 2	**I.** 14
T. 42	**P.** 18
A. 16	**E.** 5

Answers for Exercises 13–18

I. 140

F. 210

E. 7

T. 10

S. 400

R. 5

17	3		4		2	12	14	10	8	16	6	9	1		11	5	15	18	13	7

Activity 3.2 Start Thinking!

For use before Activity 3.2

Describe a real-life situation that involves multiplying a decimal by a whole number.

Activity 3.2 Warm Up

For use before Activity 3.2

Find the sum.

1. $0.4 + 0.4 + 0.4$

2. $0.12 + 0.12 + 0.12$

3. $2.7 + 2.7 + 2.7$

4. $0.41 + 0.41 + 0.41$

5. $0.6 + 0.6 + 0.6 + 0.6$

6. $0.15 + 0.15 + 0.15 + 0.15$

Lesson 3.2 Start Thinking!

For use before Lesson 3.2

The table shows the maximum speeds of various slow animals.

Animal	Maximum Speed (ft/sec)
Spider	1.716 ft/sec
Three-toed sloth	0.249 ft/sec
Garden snail	0.044 ft/sec

How far can each animal travel in 10 seconds? in 100 seconds? in 1000 seconds? in 10,000 seconds?

Lesson 3.2 Warm Up

For use before Lesson 3.2

Multiply. Use estimation to check your answer.

1. 1.3×6 **2.** 2.6×7 **3.** 3.61×4

4. 5.9×15 **5.** 1.23×8 **6.** 5.16×7

Name ______________________ Date __________

3.2 Practice A

Find the product. Use an estimate to place the decimal point.

1. $\begin{array}{r} 0.4 \\ \times\ 8 \\ \hline \end{array}$

2. $\begin{array}{r} 0.6 \\ \times\ 9 \\ \hline \end{array}$

3. $\begin{array}{r} 3.4 \\ \times\ 6 \\ \hline \end{array}$

Multiply. Use estimation to check your answer.

4. 3.2×6

5. 5.1×8

6. 0.34×14

7. 8.4×3

8. 6.2×7

9. 1.3×3

10. 5.4×2.2

11. 6.13×15

12. 0.92×24

Multiply.

13. 0.036×3

14. 0.015×6

15. 0.022×8

16. 9×0.031

17. 16×0.004

18. 7×0.0092

19. Describe and correct the error in finding the product.

X $\begin{array}{r} 0.0004 \\ \times\quad\ 8 \\ \hline 0.032 \end{array}$

Use mental math to find the product.

20. 1.054×10

21. 0.0047×100

22. 56.204×1000

23. An international long-distance phone call costs \$0.79 per minute. How much will a 22-minute call cost?

24. The weight of an object on Jupiter is about 2.3 times its weight on Earth. How much does a 98-pound person weigh on Jupiter?

25. An ad in an old newspaper advertised jeans on sale for \$7.50. Now they cost 6.66 times as much. How much do the jeans cost today?

26. New basketballs must bounce back to between 0.68 and 0.75 of the starting height. A ball is dropped from 6 feet. Within what range should the ball bounce?

Name ______________________________ Date __________

3.2 Practice B

Find the product. Use an estimate to place the decimal point.

1. $\begin{array}{r} 0.5 \\ \times\ 9 \\ \hline \end{array}$

2. $\begin{array}{r} 1.7 \\ \times\ 8 \\ \hline \end{array}$

3. $\begin{array}{r} 5.8 \\ \times\ 7 \\ \hline \end{array}$

Multiply. Use estimation to check your answer.

4. 4.5×8

5. 6.1×6

6. 0.48×15

7. 9.2×3

8. 7.4×5

9. 9.25×16

10. 2.84×21

11. 5.59×3

12. 7.175×8

Multiply.

13. 0.042×4

14. 0.0019×5

15. 0.035×7

16. 9×0.065

17. 18×0.012

18. 4×0.0137

19. Describe and correct the error in finding the product.

X $\begin{array}{r} 0.0021 \\ \times \quad\ 8 \\ \hline 0.068 \end{array}$

Use mental math to find the product.

20. 4.512×1000

21. 0.0249×100

22. $27.895 \times 10{,}000$

Evaluate the expression.

23. $2.6 \bullet (3.7 + 7.3)$

24. $5 + 8.4 \times 6.6$

25. $9.93 \times 6 \times 5$

26. Use the Distributive Property to rewrite $6(a + 0.25)$.

27. Which is greater, $7.8 \bullet (6.9 + 3.5)$ or $7.8 + (6.9 \bullet 3.5)$? Explain your answer.

28. The table shows the cost (in cents) of producing and distributing each coin for two years.

Coin	2000	2007
Quarter	4.38	9.78
Dime	1.92	4.09
Nickel	3.16	9.53
Penny	0.82	1.67

a. Which coins cost less to make than they are worth? Which coins cost more to make than they are worth?

b. How much more does it cost to produce 50 of each coin in 2007 than in 2000?

Name____ Date__________

3.2 Enrichment and Extension

Money, Money, Money

Coin	Thickness (mm)	Weight (oz)	Coins per Roll
Penny	1.55	0.088	50
Nickel	1.95	0.176	40
Dime	1.35	0.08	50
Quarter	1.75	0.2	40

In Exercises 1–4, use the table.

1. How much taller is a roll of nickels than a roll of dimes?

2. How much taller is a roll of pennies than a roll of quarters?

3. You have 6 rolls of pennies and 4 rolls of quarters.

a. How many pennies do you have? How many quarters do you have?

b. How much money do you have?

c. How much do the coins weigh?

4. You have 1 roll of nickels, 4 rolls of dimes, and 2 rolls of quarters. Your friend has 3 rolls of pennies, 2 rolls of nickels, 1 roll of dimes, and 3 rolls of quarters.

a. Who has more money? By how much?

b. Whose coins weigh more? By how much?

Name ______________________________ Date __________

3.2 Puzzle Time

How Did The Goblin Football Player Score The Winning Touchdown?

Write the letter of each answer in the box containing the exercise number.

Multiply.

1. 3.8×8

2. 5.1×8

3. 0.26×15

4. 8.43×4

5. 2.7×13

6. 8.5×6

7. 5.08×7

8. 2.24×3

9. 3.8×6

10. 7.3×9

11. 6.19×14

12. 0.78×12

13. 2.59×10

14. 4.98×4

15. 2.563×3

16. 7.177×11

17. 0.013×2

18. 0.024×8

19. 0.072×3

20. 8×0.0008

21. 0.0029×6

Answers

R. 33.72	**E.** 0.026
A. 0.216	**N.** 65.7
O. 30.4	**H.** 9.36
E. 35.1	**V.** 6.72
T. 0.192	**G.** 40.8
H. 7.689	**O.** 19.92
E. 86.66	**N.** 3.9
R. 22.8	**L.** 78.947
I. 25.9	**E.** 35.56
L. 0.0064	**H.** 51
U. 0.0174	

6	17		9	19	3		1	8	11	4		18	12	5	2	15	14	21	16		20	13	10	7

Activity 3.3

Start Thinking!

For use before Activity 3.3

You can use a model to find the product of decimals.

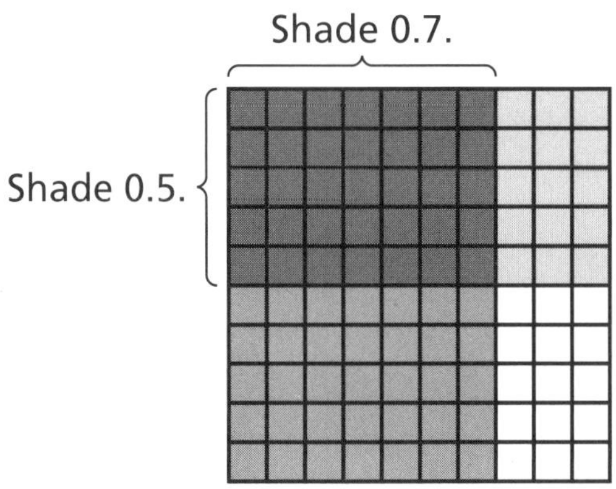

Explain why the model above shows that $0.5 \times 0.7 = 0.35$.

Activity 3.3

Warm Up

For use before Activity 3.3

Estimate by rounding to the nearest whole number.

1. 1.8×12.2

2. 6.78×3.12

3. 6.11×4.5

4. 9.09×8.91

5. 0.058×1.922

6. 15.215×2.964

Lesson 3.3 Start Thinking!
For use before Lesson 3.3

How can you apply multiplying decimals at a gas station?

Lesson 3.3 Warm Up
For use before Lesson 3.3

Multiply. Use estimation to check your product.

1. $\begin{array}{r} 0.6 \\ \times\ 0.2 \\ \hline \end{array}$

2. $\begin{array}{r} 0.04 \\ \times\ \ 0.5 \\ \hline \end{array}$

3. $\begin{array}{r} 0.008 \\ \times\ \ 0.03 \\ \hline \end{array}$

4. $\begin{array}{r} 0.0007 \\ \times\ \ \ 0.04 \\ \hline \end{array}$

5. $\begin{array}{r} 0.005 \\ \times\ \ \ 0.9 \\ \hline \end{array}$

6. $\begin{array}{r} 0.0002 \\ \times\ \ 0.003 \\ \hline \end{array}$

Name____ Date__________

3.3 Practice A

Copy the problem and place the decimal point in the product.

1. $\begin{array}{r} 2.6 \\ \times\ 7.9 \\ \hline 2\,0\,5\,4 \end{array}$

2. $\begin{array}{r} 8.8\,3 \\ \times\ 0.4\,2 \\ \hline 3\,7\,0\,8\,6 \end{array}$

3. $\begin{array}{r} 9.4\,4 \\ \times\ 1.9\,5 \\ \hline 1\,8\,4\,0\,8\,0 \end{array}$

Multiply. Use estimation to check your product.

4. $\begin{array}{r} 0.3 \\ \times 0.4 \\ \hline \end{array}$

5. $\begin{array}{r} 0.05 \\ \times\ 0.9 \\ \hline \end{array}$

6. $\begin{array}{r} 0.34 \\ \times\ 0.7 \\ \hline \end{array}$

7. 1.46×0.5

8. 19.4×4.8

9. 5.12×0.23

10. 2.024×1.35

11. 0.0032×16.4

12. 2.965×3.208

13. Describe and correct the error in finding the product.

X $\begin{array}{r} 0.008 \\ \times\ 0.004 \\ \hline 0.00032 \end{array}$

Evaluate the expression when $a = 2.5$, $b = 3.16$, and $c = 0.049$.

14. $9a$

15. $11b$

16. $14c$

17. ac

18. ab

19. $8b - 5.28$

20. A baseball card originally cost $1.25. Twenty years later, it is worth 15.6 times as much. How much is the baseball card worth twenty years later?

21. A baby weighs 8.5 pounds at birth. At the age of five months, the baby weighs 1.9 times as much. How much does the baby weigh?

22. The area of Pennsylvania is 46,055 square miles. The area of Florida is about 1.428 times larger than Pennsylvania. What is the area of Florida to the nearest square mile?

23. You buy three baskets of fruit at $5.65 per basket. You hand the cashier a $20 bill. How much change will you get back?

24. One car travels 35.4 miles on one gallon of gas. Another car travels 28.5 miles on one gallon of gas. How much farther can the first car travel on 12 gallons of gas than the second car?

Name ______________________________ Date __________

3.3 Practice B

Multiply. Use estimation to check your product.

1. $\begin{array}{r} 0.5 \\ \times\, 0.3 \\ \hline \end{array}$

2. $\begin{array}{r} 0.49 \\ \times\ \ 0.8 \\ \hline \end{array}$

3. $\begin{array}{r} 0.0015 \\ \times\ \ 0.004 \\ \hline \end{array}$

4. 2.37×0.8

5. 21.6×5.7

6. 6.25×0.25

7. 3.128×2.45

8. 0.0059×19.8

9. 3.659×3.126

10. Describe and correct the error in finding the product.

$$\begin{array}{r} 0.0004 \\ \times\, 0.0004 \\ \hline 0.0016 \end{array}$$

Evaluate the expression when $d = 3.6$, $e = 6.5$, and $f = 0.068$.

11. $6d$

12. $13e$

13. $17f$

14. df

15. $fe - 0.042$

16. $2.3e + 12.6f$

Describe the pattern. Find the next three numbers.

17. 0.7, 2.1, 6.3, 18.9, …

18. 0.13, 0.65, 3.25, 16.25, …

19. 2, 0.04, 0.0008, 0.000016, …

20. 6, 1.8, 0.54, 0.162, …

Convert the fractions or mixed numbers to decimals. Then multiply.

21. $\frac{7}{16} \times \frac{9}{50}$

22. $\frac{19}{25} \times \frac{2}{5}$

23. $4\frac{3}{8} \times 2\frac{17}{20}$

24. Ham costs \$4.29 per pound and cheese costs \$5.89 per pound. You buy 3.25 pounds of ham and 0.75 pound of cheese. You hand the cashier a \$20 bill. How much change will you get back?

25. A store is selling sculptures. The first week, it prices 6 sculptures at \$80 each. After each week, if a sculpture is not sold, it will be priced to sell for 0.85 times the previous week's price. The store needs to sell the 6 sculptures for a total of at least \$270 to make a profit. What is the last week in which all 6 sculptures could be sold so that the store makes a profit?

Name________________________________ Date__________

3.3 Enrichment and Extension

Which Factors Did I Multiply?

You write an example in your notebook during class. When you go to do your homework, you realize that you forgot to write the decimal points in the factors.

1. You ask your brother for help. He says that decimal points should be placed between the 3 and 9 in the first factor and between the 7 and 1 in the second factor. Is his answer correct? Check his answer.

2. Is there more than one way to place the decimal points? If so, show every possibility. Do not add zeroes to any of the factors.

Fill in the blanks and place a decimal point in the first factor to complete the statement.

3.

```
  □6
× 9.□
 33□
5□4
5□.76
```

4.

```
  78□
× □.4
 312□
23□6
26.□88
```

5.

```
   2□15
 × □.3□
   846□
  □345
1□920
17.6□910
```

Name ____________________ Date ________

3.3 Puzzle Time

Why Did The Suspected Robber Decide To Go Into The Kitchen?

Write the letter of each answer in the box containing the exercise number.

Multiply.

1. 0.8×0.3

2. 0.07×0.2

3. 0.006×0.04

4. 0.0009×0.08

5. 0.003×0.9

6. 0.05×0.6

7. 0.0007×0.005

8. 0.0002×0.07

9. 11.5×0.3

10. 19.7×4.8

11. 2.25×4.46

12. 7.589×29.32

13. 2.042×6.408

14. 0.0031×16.4

Evaluate the expression when $x = 2.6$, $y = 7.27$, and $z = 0.064$.

15. $4x$

16. $9z$

17. $3.31y$

18. xy

19. $xz + 4.3$

20. $5.7y - 6.21$

Answers for Exercises 1–14

B. 13.085136	**E.** 0.03
A. 0.000014	**O.** 0.000072
R. 0.05084	**T.** 0.24
L. 10.035	**P.** 94.56
E. 0.014	**G.** 0.0000035
I. 0.0027	**U.** 222.50948
V. 3.45	**A.** 0.00024

Answers for Exercises 15–20

S. 35.229	**E.** 0.576
R. 24.0637	**A.** 4.4664
T. 18.902	**I.** 10.4

18	4		20	6	14	9	16		12	10		8		7	17	2	19	1		3	11	5	13	15

Activity 3.4 Start Thinking!
For use before Activity 3.4

How are multiplication and division related? Be sure to include examples in your explanation.

Activity 3.4 Warm Up
For use before Activity 3.4

For each decimal, draw a model using base ten blocks. Recall the value of each block:

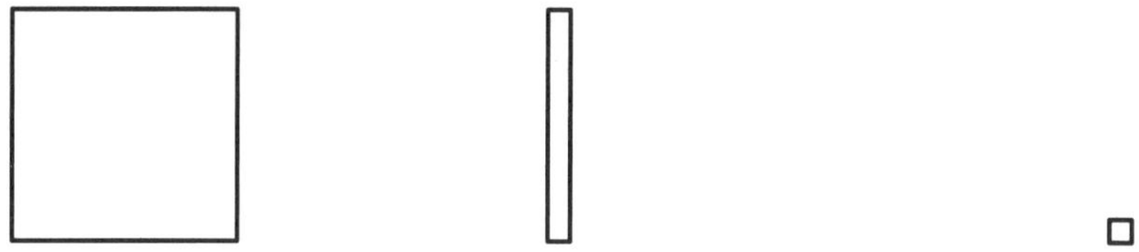

one (1) tenth (0.1) hundredth (0.01)

1. 2.2 **2.** 0.8 **3.** 1.3

4. 2.32 **5.** 1.23 **6.** 0.16

Lesson 3.4 Start Thinking!
For use before Lesson 3.4

How could dividing a decimal by a whole numbcr apply to going to a rock conccrt?

Lesson 3.4 Warm Up
For use before Lesson 3.4

Divide. Check your answer.

1. $5\overline{)30.5}$ **2.** $9\overline{)23.4}$ **3.** $4\overline{)15.2}$

4. $69.3 \div 7$ **5.** $17.1 \div 3$ **6.** $4.2 \div 6$

Name______________________________ Date__________

3.4 Practice A

Divide. Check your answer.

1. $7\overline{)1.4}$ **2.** $5\overline{)2.65}$ **3.** $6\overline{)24.6}$

4. $1.28 \div 4$ **5.** $21.6 \div 9$ **6.** $34.3 \div 7$

7. $53.6 \div 8$ **8.** $6.025 \div 5$ **9.** $0.256 \div 4$

10. $29.7 \div 11$ **11.** $4.2 \div 12$ **12.** $34.8 \div 20$

13. $177.8 \div 7$ **14.** $28.7 \div 14$ **15.** $248.4 \div 8$

16. Describe and correct the error in finding the quotient.

X

$$\begin{array}{r} 2.7 \\ 5\overline{)1.3\ 5} \\ \underline{-1\ 0} \\ 3\ 5 \\ \underline{-3\ 5} \\ 0 \end{array}$$

Evaluate the expression.

17. $20.6 + 19.4 \div 2$ **18.** $23.43 \div 3 - 2.5$ **19.** $18.6 \div 3 \times 6$

20. $9.3 \div 3 + 7.6$ **21.** $48.02 \div 7 \div 2$ **22.** $17.5 \times (2.4 \div 3)$

23. A story in a building is about 10 feet tall. The Sears Tower in Chicago, Illinois is 1451 feet tall. To the nearest whole number, about how many stories tall is the Sears Tower?

24. You buy 65 oranges for $22.75. How much does each orange cost?

25. You and three of your friends go parasailing for $128.40. You split the cost evenly. How much does each person pay?

26. The yearly precipitation of a city is 38.94 inches. About how much precipitation falls each month?

27. A 12-fluid ounce bottle of shampoo costs $8.40. A 16-fluid ounce bottle of shampoo costs $10.24. Which is the better buy? Explain.

Name ______________________________ Date __________

3.4 Practice B

Divide. Check your answer.

1. $9\overline{)7.2}$ **2.** $6\overline{)38.4}$ **3.** $5\overline{)32.5}$

4. $1.08 \div 4$ **5.** $77.6 \div 8$ **6.** $60.2 \div 7$

7. $62.4 \div 3$ **8.** $260.7 \div 11$ **9.** $0.252 \div 9$

10. $158.4 \div 12$ **11.** $105.3 \div 13$ **12.** $42.66 \div 18$

13. $16.616 \div 8$ **14.** $280.8 \div 12$ **15.** $72.96 \div 16$

16. Describe and correct the error in finding the quotient.

X
$$\begin{array}{r} 3.60 \\ 4\overline{)12.24} \\ \underline{-\,12} \\ 24 \\ \underline{-\,24} \\ 0 \end{array}$$

Evaluate the expression.

17. $23.4 + 34.4 \div 4$ **18.** $28.65 \div 5 - 4.85$ **19.** $21.5 \div 5 \times 8$

20. $7.3 \times 3.6 \div 9$ **21.** $79.8 \div 7 \div 3$ **22.** $18.1 \times (9.6 \div 6)$

23. You raise $550.50 for a charity in 25 weeks. About how much did you collect each week?

24. The table shows the ounces and cost of three boxes of the same cereal. Which box of cereal is the best buy? Explain.

Ounces	Cost
12	$2.88
18	$3.96
28	$6.44

25. A store has a sale that when you buy 4 DVDs that cost $15.95 each, you will get a fifth DVD that costs $9.95 for free. You buy 4 DVDs and get a fifth one free. What is your cost per DVD?

Name____ Date__________

3.4 Enrichment and Extension

What Factors Did I Divide?

Find the value of the variable in the expression.

1. $r9\overline{)396.9}$ with quotient 8.1

2. $21a\overline{)759.5}$ with quotient 3.5

3. $h7\overline{)710.816}$ with quotient 7.328

4. $1y\overline{)115.56}$ with quotient 6.42

5. $5t2\overline{)224.664}$ with quotient 0.407

6. $7b6\overline{)1459.26}$ with quotient 2.01

7. $10e1\overline{)9856.36}$ with quotient 9.56

8. $2n\overline{)779.94}$ with quotient 37.14

9. $3p75\overline{)176.4}$ with quotient 0.048

Use the values of the variables from Exercises 1–9 to find the answer to the riddle.

10. *I have a head and a tail, but no body. What am I?*

____   ____ ____ ____ ____ ____
7   6 3 1 1 8

11. *I am light as a feather, but no one can hold me for long. What am I?*

____   ____ ____ ____ ____ ____ ____
7   2 4 3 7 5 9

Name ______________________________ Date __________

3.4 Puzzle Time

Why Did The Pilot Decide To Go See A Psychiatrist?

Write the letter of each answer in the box containing the exercise number.

Divide.

1. $5\overline{)39.5}$

2. $8\overline{)33.6}$

3. $9\overline{)8.1}$

4. $7\overline{)11.2}$

5. $34.8 \div 6$

6. $29.2 \div 4$

7. $55.25 \div 5$

8. $0.183 \div 3$

9. $3.3 \div 22$

10. $74.88 \div 16$

11. $19.618 \div 17$

12. $52.8 \div 12$

13. $45.87 \div 6$

14. $51.288 \div 4$

15. $15.75 \div 18$

16. $29.6 \div 8$

Answers

Z. 7.3	**A.** 0.15
S. 7.645	**E.** 4.2
P. 4.68	**H.** 5.8
A. 1.6	**Y.** 12.822
E. 7.9	**R.** 4.4
C. 3.7	**A.** 1.154
N. 0.875	**L.** 0.9
W. 11.05	**S.** 0.061

8	5	2		7	4	13		10	3	9	15	1		16	12	11	6	14

Activity 3.5 Start Thinking!

For use before Activity 3.5

How would you plan a trip overseas? How could decimal division be used in your planning?

Activity 3.5 Warm Up

For use before Activity 3.5

Use base ten blocks to complete the statement.

1. 3 tenths = __?__ hundredths

2. 5 ones = __?__ tenths

3. 2 ones = __?__ hundredths

4. 20 hundredths = __?__ tenths

5. 37 hundredths = __?__ tenths and __?__ hundredths

6. 2 ones and 4 tenths = __?__ tenths

Lesson 3.5 Start Thinking!
For use before Lesson 3.5

List all of the ways that you can check your answer for a division problem.

Lesson 3.5 Warm Up
For use before Lesson 3.5

Divide.

1. $7.2 \div 0.9$

2. $8.5 \div 0.5$

3. $0.81 \div 0.03$

4. $0.54 \div 0.06$

5. $20.7 \div 0.9$

6. $1.44 \div 0.12$

Name______________________ Date__________

3.5 Practice A

Rewrite the problem so that the divisor is a whole number.

1. $23.1 \div 4.2$ **2.** $12.87 \div 0.15$ **3.** $8.76 \div 0.535$

Divide.

4. $4.2 \div 0.2$ **5.** $6.4 \div 0.4$ **6.** $0.54 \div 0.09$

Divide. Check your answer.

7. $0.5\overline{)13.5}$ **8.** $2.5\overline{)45.5}$ **9.** $5.6\overline{)16.8}$

10. $1.764 \div 0.07$ **11.** $8.82 \div 0.6$ **12.** $1.46 \div 0.04$

13. $4.08 \div 1.2$ **14.** $75 \div 2.5$ **15.** $90.2 \div 0.22$

16. $357 \div 0.007$ **17.** $2.94 \div 0.21$ **18.** $86.4 \div 3.2$

19. Describe and correct the error in rewriting the problem.

> X $0.05\overline{)25} \rightarrow 5\overline{)250}$

Divide. Round to the nearest hundredth if necessary.

20. $25.6 \div 0.7$ **21.** $19.43 \div 0.003$ **22.** $75.6 \div 0.34$

23. $23.45 \div 2.3$ **24.** $56.22 \div 7.2$ **25.** $938.5 \div 0.35$

26. A sold out movie theater earns \$4875. Each ticket costs \$6.50. How many tickets were sold?

27. A piece of paper is 0.0075 inch thick. How many sheets of paper will be in a stack that is 2.25 inches high?

28. A 9.2-ounce can of tomato juice costs \$3.48. What is the price per ounce of the tomato juice to the nearest cent?

29. A runner completed the 100-meter dash in 9.7 seconds. How far does the runner travel in one second? Round your answer to the nearest tenth of a meter.

30. An MP3 player can hold 350 hours of music. Each song on an MP3 player is about 2.5 minutes in length. About how many songs can the MP3 player store?

Name ______________________________ Date __________

3.5 Practice B

Divide.

1. $2.8 \div 0.7$ **2.** $8.6 \div 0.02$ **3.** $1.89 \div 0.03$

Divide. Check your answer.

4. $1.4\overline{)28.7}$ **5.** $0.25\overline{)3.15}$ **6.** $6.3\overline{)107.1}$

7. $10.16 \div 0.08$ **8.** $39.42 \div 0.06$ **9.** $75.25 \div 0.05$

10. $0.272 \div 1.6$ **11.** $78 \div 0.26$ **12.** $1.836 \div 0.27$

13. Describe and correct the error in rewriting the problem.

X $0.24\overline{)960} \rightarrow 24\overline{)960.00}$

Divide. Round to the nearest hundredth if necessary.

14. $30.48 \div 2.4$ **15.** $198.2 \div 4.4$ **16.** $4.657 \div 0.35$

17. $60.258 \div 7.3$ **18.** $458.12 \div 6.8$ **19.** $618.27 \div 15.2$

Without finding the quotient, copy and complete the statement using $<$, $>$, or $=$.

20. $4.58 \div 0.57$ __?__ $45.8 \div 0.57$ **21.** $24.5 \div 0.4$ __?__ $24.5 \div 4$

22. $156.26 \div 28.7$ __?__ $15.626 \div 28.7$ **23.** $407.3 \div 19.4$ __?__ $40.73 \div 1.94$

24. A 14.5-ounce can of peaches costs $2.19. A 20.5-ounce can of peaches costs $2.99. Which can is the better buy? How many cents cheaper per ounce is it? Give your answer to the nearest tenth of a cent.

25. You are saving your money to buy a guitar that costs $275.75. You have $40 and plan to save $7.50 each week. Your uncle decides to give you an additional $8 each week.

a. How many weeks will you have to save until you have enough money to buy the guitar?

b. How many more weeks would you have to save to buy a guitar that costs $339.75? Explain how you found your answer.

Name______________________________ Date__________

3.5 Enrichment and Extension

What is the Missing Length?

In Exercises 1–4, the figure has the same area as the area of the rectangle below. Use this information to find the length of *x*. Explain how you found your answer.

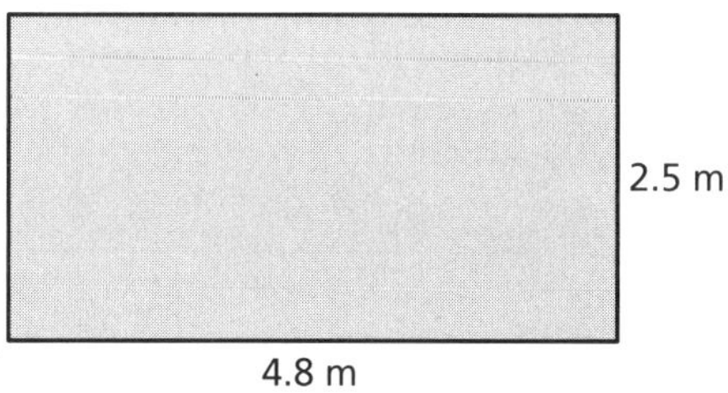

1.

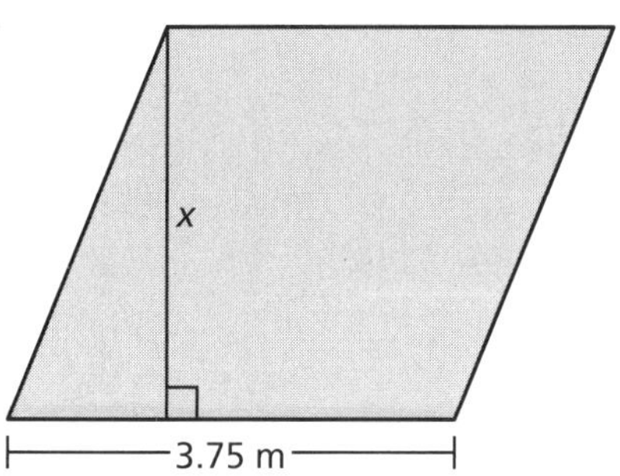

2.

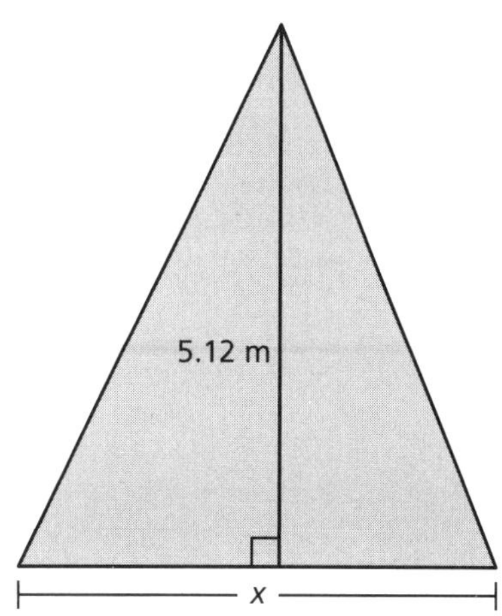

3.

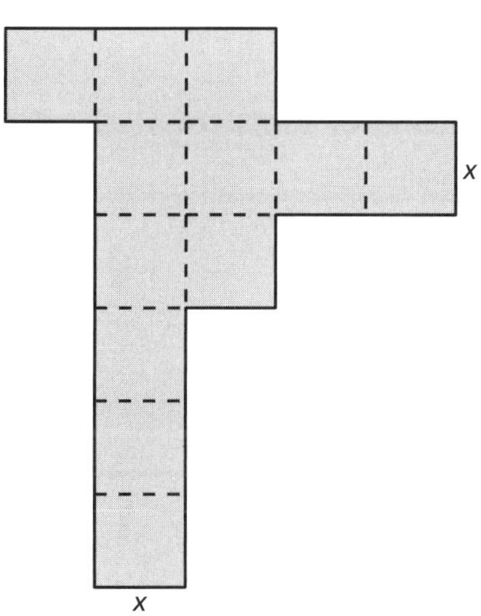

4.

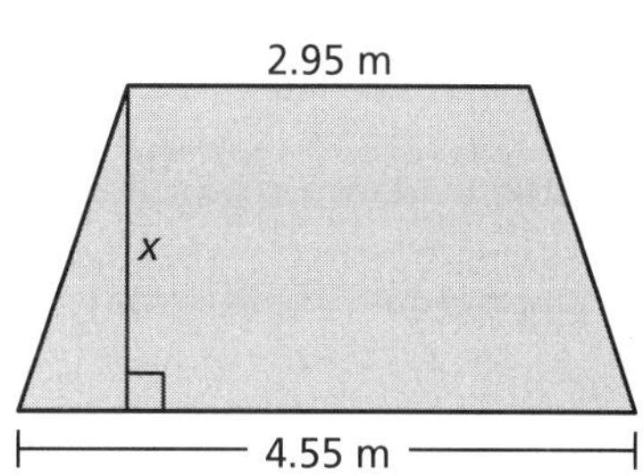

Name ______________________________ Date __________

3.5 Puzzle Time

Why Did The Young Lady Go Buzz Buzz In The Hallway?

Write the letter of each answer in the box containing the exercise number.

Divide.

1. $3.2 \div 0.4$

2. $2.7 \div 0.9$

3. $0.84 \div 0.07$

4. $0.33 \div 0.03$

5. $3.2\overline{)41.6}$

6. $4.9\overline{)68.6}$

7. $0.5\overline{)17.7}$

8. $9.75\overline{)39.39}$

9. $61.38 \div 3.1$

10. $13.68 \div 2.28$

11. $50.56 \div 0.8$

12. $0.912 \div 11.4$

13. $0.84 \div 0.0014$

14. $76.23 \div 12.1$

15. $22.4 \div 0.04$

16. $33.6 \div 0.3$

17. $0.861 \div 0.7$

Answers

S. 1.23	**E.** 63.2
H. 11	**A.** 19.8
E. 8	**T.** 35.4
B. 600	**U.** 0.08
N. 13	**S.** 3
E. 6.3	**D.** 112
W. 4.04	**E.** 6
A. 12	**T.** 14
S. 560	

15	4	11		8	3	17		9		13	10	1		2	7	12	16	14	5	6

Name______________________________ Date __________

Chapter 3

School-to-Work

For use after Section 3.5

Calculating Dosages as a Nurse

Nurses examine and treat patients who are ill or injured. They work in hospitals, schools, doctor's offices, nursing homes, laboratories, and private homes. One of their jobs is to give patients their medicines.

Liquid medicines come in different *concentrations*—a teaspoon of infant medication often has a different strength of medicine than a teaspoon of adult medication. Tablets come in different sizes—a dose might be two tablets, one tablet, or half a tablet. The correct dosage of a medicine will also depend on the nature of the illness and on the patient's weight and age.

A nurse needs to use math to calculate the correct amount of medicine to give a patient in one dose.

1. A medication in tablet form contains 2.5 milligrams of a medicine. A patient needs 10 milligrams of medicine in each dose.

   a. Write a formula to calculate how many tablets to give the patient.

   b. Use your formula to find the correct dose.

In Exercises 2–7, find the correct number of tablets or teaspoons (tsp) to give the patient. All units are in milligrams (mg).

2. Medicine in tablet: 5 mg
Desired dose: 7.5 mg

3. Medicine in tablet: 5 mg
Desired dose: 12.5 mg

4. Medicine in tablet: 20 mg
Desired dose: 10 mg

5. Medicine in liquid tsp: 50 mg
Desired dose: 25 mg

6. Medicine in liquid tsp: 20 mg
Desired dose: 25 mg

7. Medicine in liquid tsp: 40 mg
Desired dose: 70 mg

8. A small container holds 1 cubic centimeter of liquid with 40 milligrams of medicine dissolved in it. You shake the container to evenly distribute the medicine. How much of the liquid do you need for a dose of 30 milligrams?

Name ______________________________ Date __________

Chapter 3 Study Help

You can use a **process diagram** to show the steps involved in a procedure.

Example

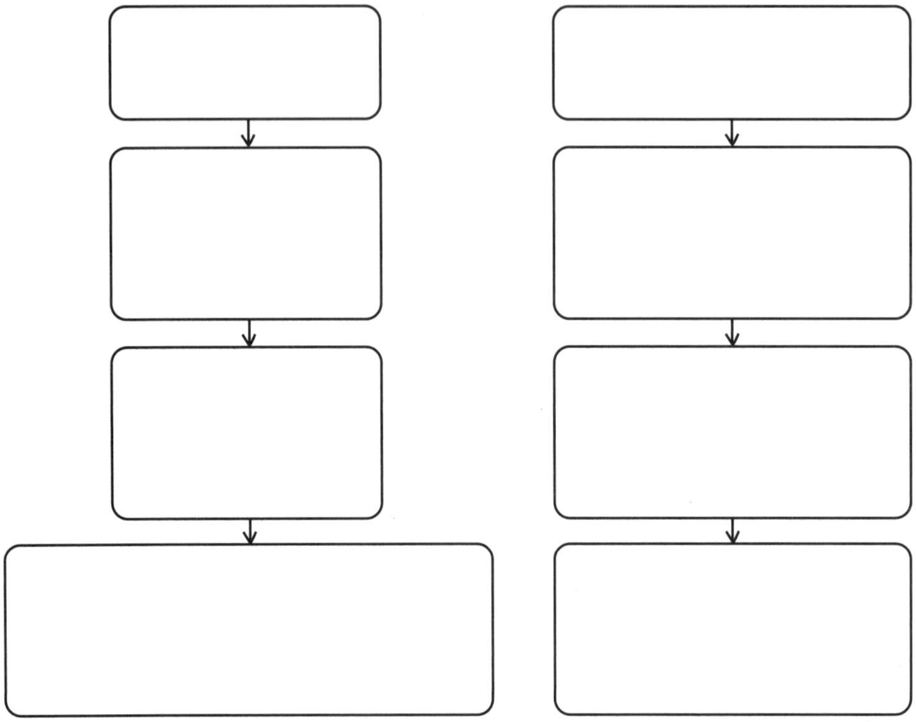

On Your Own

Make a process diagram with an example to help you study these topics.

1. multiplying decimals and whole numbers
2. estimating with decimals

After you complete this chapter, make process diagrams for the following topics.

3. dividing decimals by whole numbers
4. dividing decimals
5. Describe the steps in the process that are similar for multiplying decimals and dividing decimals.

Name ______________________________ Date __________

Chapter 3

Financial Literacy

For use after Section 3.1

Planning a Celebration

Families gather for holidays or a special birthday. New homeowners host a party to celebrate a new house. Baseball teams have an end-of-season barbecue. The class that raises the most money in a fundraiser has a pizza party. All of these events are gatherings or celebrations where a large meal is served. For this worksheet, you will plan a large gathering and estimate the cost.

1. What event would you like to plan? You can use one of the examples or choose a new one. Use an actual gathering you've attended or one that sounds like fun. There should be more than a dozen people attending.

2. Who would you invite? How many people would you need to serve? (Rounding is fine—for example, 52 people are about 50.)

3. What food and beverages will you serve? There should be a small number of dishes prepared in large amounts, not many small dishes. List the amount of each item you will need to buy.

    - Because you cannot purchase fractions of items, round amounts up to whole numbers. For example, if you will need $1\frac{2}{3}$ cartons of eggs, you need to round up to 2 cartons.
    - You may have some items—like 4 teaspoons of cinnamon or salt—so you don't need to estimate costs for these ingredients.

4. Do you need anything else, such as disposable plates and cups? If so, add them to the list.

5. Use your lists from Questions 3 and 4. Using the grocery store or an online resource, find out how much one of each item costs and list the price next to the item. (Example: 40 ears of corn, $0.79 per ear)

6. Estimate the total cost for each item and write this amount to the right of the item. Then estimate the cost for the entire meal.

7. Suppose your budget is the amount you estimated, rounded up to the nearest $50. How much extra money is this? What would you spend it on?

Name ______________________________ Date ________

Chapter 3 Cumulative Practice

Estimate.

1. $2.1 - 1.9$

2. 6.2×3.1

3. $13.8 \div 6.75$

4. $98.9 - 75.4$

5. $28.79 + 45.14$

6. $42.325 \div 5.95$

7. 93.05×1.11

8. $54.2 + 6.7$

9. 4.82×8.13

10. $7.8 - 0.012$

11. $100.01 + 99.99$

12. $25.3 \div 5.4$

13. $10.99 + 26.76 + 11.15$

14. $6.8 \times 12.1 \div 4.12$

15. One pound of almonds costs $5.19. Estimate the cost of 13.1 pounds of almonds.

16. You buy 2.2 yards of fabric for $25.80. Estimate the cost per yard.

Evaluate the expression.

17. $\begin{array}{r} 9.4 \\ \times\ 7 \\ \hline \end{array}$

18. $1.35 + 19.78$

19. $3.1\overline{)46.5}$

20. $9.75 \div 1.3$

21. $5.08 - 0.82$

22. $\begin{array}{r} 10.3 \\ \times\ 1.2 \\ \hline \end{array}$

23. $62.79 + 2.038$

24. 28.76×1.06

25. $0.23\overline{)1.4375}$

26. $18 \div 7.2$

27. $19 - 3.01$

28. $\begin{array}{r} 4.003 \\ \times\ \ \ 9 \\ \hline \end{array}$

29. $12.04 \times 8.03 \times 1.1$

30. $1.684 + 5.8 \times 21.05$

31. $(9.8 - 2.3) \div 1.5$

32. $46.5 \div (4.5 \div 1.2)$

33. $(10.24 + 2.56) \div 128$

34. $98 - 4.56 \times 1.5$

35. A $14 shirt is on sale.

- **a.** Multiply by 0.75 to find the new price.
- **b.** Multiply the price in part (a) by 1.06 to find the price including the tax.
- **c.** You give the cashier $15. How much change do you receive?

36. A 16-ounce box of cereal costs $3.29. A 20-ounce box of cereal costs $4.32.

- **a.** Find the cost per ounce for each size box to the nearest cent.
- **b.** Which box is the better buy? Explain your answer.

Name______________________________ Date __________

Unit 1 Project: Sports and Mathematics

For use after Unit 1

Objective **Explain and use some of the mathematical formulas used in sports.**

Materials A source of sports data such as a Sports Almanac or the websites of sports organizations or teams; a calculator.

Investigation When you listen to a sports broadcast, numbers are everywhere: the score, the time remaining, the statistics about players and teams. In this project, you will research various formulas used in sports.

Write a report that contains the following:

- an introduction to how math is used in sports
- a section on calculating scores. Use basketball, football, and two other sports which you choose. For each sport:
  - **a.** Write a brief description of how points are earned.
  - **b.** Write the formula to find the score for each team.
  - **c.** Use your formula to work out an example computing the score for each team in a game, or for 2 players in a competition. Use real data, if possible; if not, choose reasonable numbers.
- a section on calculating sports statistics. Choose 8 statistics from at least 4 sports. Examples might include the earned run average in baseball, the average yards gained on an offensive play in football, or assists per game in basketball or hockey. You may use the same sports as in the scoring section, all new sports, or a mix. For each sports statistic:
  - **a.** Write a brief description of what the statistic measures and how it is calculated.
  - **b.** Write the formula to compute the statistic.
  - **c.** Give a real-life example of each statistic. For example, the average yards gained on an offensive play by a football team in a recent game. You do not need to do the computation.
- Finally, calculate a sports statistic based on your own performance.

Think About It

What sports will you use? How well do you understand the statistics you hear in a broadcast? How can you look up the meaning and method of calculating a statistic you don't understand?

Completing the Project

Check that you have included all the pieces asked for, and that your examples are clear and easy to understand.

Name ____________________ Date ________

Unit 1 Student Grading Rubric

For use after Unit 1

		Student Score	Teacher Score
Cover Page	10 points		
	a. Name (2 points)	______	______
	b. Class (2 points)	______	______
	c. Project Name (4 points)	______	______
	d. Due Date (2 points)	______	______
Introduction	20 points		
	a. Touches on math to be covered in the report. (8 points)	______	______
	b. Clear, interesting, and well-written. (12 points)	______	______
Sports Scores	60 points (15 apiece for 4 sports)		
	a. Identify and explain each sport. (3 points each, 12 total)	______	______
	b. Explanation of scoring is clear and easy to understand. (4 points each, 16 total)	______	______
	c. Formula for scoring is correct. (5 points each, 20 total)	______	______
	d. Example is present, clearly labeled, and math is correct. (3 points each, 12 total)	______	______
Sports Statistics	96 points (12 apiece for 8 statistics)		
	a. Identify and explain each statistic. (2 points each, 16 total)	______	______
	b. Explanation of what the statistic measures and how to calculate it is clear and easy to understand. (4 points each, 32 total)	______	______
	c. Formula for computation is correct. (4 points each, 32 total)	______	______
	d. Example is present and clearly labeled. (2 points each, 16 total)	______	______
Personal Stat	14 points		
	a. Personal sports statistic is present. (4 points)	______	______
	b. Explanation is thorough. (10 points)	______	______
FINAL GRADE		______	______

Unit 1 Teacher's Project Notes

For use after Unit 1

Resources Sports almanacs should be available at the library. The website for each professional sports team is a good source of statistics: www.nba.com, www.nfl.com, www.mlb.com, www.mlsnet.com, www.nhl.com, www.wnba.com. You can also check collegiate or local sports sites.

Alternatives Allow students to work in pairs or groups. Grouping students who may be more familiar with sports statistics with those who may be less familiar will help each group.

As an extension, students could create their own game. There should be many ways to score points (similar to football's 3-, 6-, and 2-point scores, rather than baseball's 1-point-only scores), and they should be creative in deciding what statistics would make sense to track.

Rather than national competition, focus on your own school's sports teams. Use records from last year for statistics, and have the students compute the statistics themselves if they don't exist.

Common Errors Watch for errors in explaining averages. Usually the number of points/assists/yards/hits, etc., per game is added to the total for the season and then divided by the number of games.

Timed sports, such as track and swimming, don't lend themselves to the algebraic model rubric, because the scoring is by ordered list rather than points earned. They could be used for the sports statistics section. The same goes for distance sports, such as the shot put or pole vault.

For the statistics section, it is acceptable to give the statistic example without doing the math, as raw data may not be available or may be too lengthy to list. For the scoring section, they should do the math to find the score.

Suggestions Most statistics will either be addition (yards in previous plays + yards gained in this play) or division (yards gained in each offensive play ÷ offensive plays). Assure students that they do not need to find more complicated statistics—totals and averages are to be expected.

The two sections do not need to use the same 4 sports. Students may want to include a more unusual sport in the scoring section (especially if it has "interesting" scoring, rather than 1 point per goal), but use an easier-to-research popular sport for the statistics section.

For the statistics section, students could use 2 statistics for each sport, or they could use 4 or 5 statistics from a sport they are interested in and know well, and 3 or 4 from other sports.

Encourage students to include a sport they play: soccer, cross country, gymnastics, bowling, golf, volleyball, lacrosse.

Unit 1

Grading Rubric

For use after Unit 1

Scoring Rubric	
A	179-200
B	159-178
C	139-158
D	119-138
F	118 or below

Cover Page 10 points

- **a.** Namc (2 points)
- **b.** Class (2 points)
- **c.** Project Name (4 points)
- **d.** Due Date (2 points)

Introduction 20 points

- **a.** Touches on math to be covered in the report. (8 points)
- **b.** Clear, interesting, and well-written. (12 points)

Sports Scores 60 points, 15 points per sport

- **a.** Identify and explain each sport. (3 points each, 12 total)
- **b.** Explanation of scoring is clear and easy to understand. (4 points each, 16 total)
- **c.** Formula for scoring is correct. (5 points each, 20 total)
- **d.** Example is present, clearly labeled, and math is correct. (3 points each, 12 total)

Sports Statistics 96 points, 12 points per statistic

- **a.** Identify and explain each statistic. (2 points each, 16 total)
- **b.** Explanation of what the statistic measures and how to calculate it is clear and easy to understand. (4 points each, 32 total)
- **c.** Formula for computation is correct. (4 points each, 32 total)
- **d.** Example is present and clearly labeled. (2 points each, 16 total)

Personal Stat 14 points

- **a.** Personal sports statistic is present. (4 points)
- **b.** Explanation is thorough. (10 points each)

Chapter 4

Name ______________________________ Date __________

Chapter 4 Fractions, Decimals, and Percents

Dear Family,

Our days are filled with tasks: getting ready for school or work, making meals, running errands, visiting family and friends, just to name a few. Many of us keep a calendar of our tasks to help us budget our time.

Adults might work for about half of their waking hours on a given day. Preparing and eating meals may take a sixth of the day. Your student probably spends about a third of each weekday at school and perhaps another 10% of the week on homework.

You and your student may want to work out a family schedule to help make the best use of your time. You and your student should consider these questions:

- What percent of your week is spent at school? What percent is spent on homework?
- What percent of the week is spent in other activities, such as sports or music lessons? How much time do you spend practicing?
- How much of your week is spent traveling in the car?
- What other tasks does each person have? What percent of the week does each task take up?

You can use graph paper to help plan your schedule for each day. Use different colors for each of the activities. You and your student may want to make a family schedule.

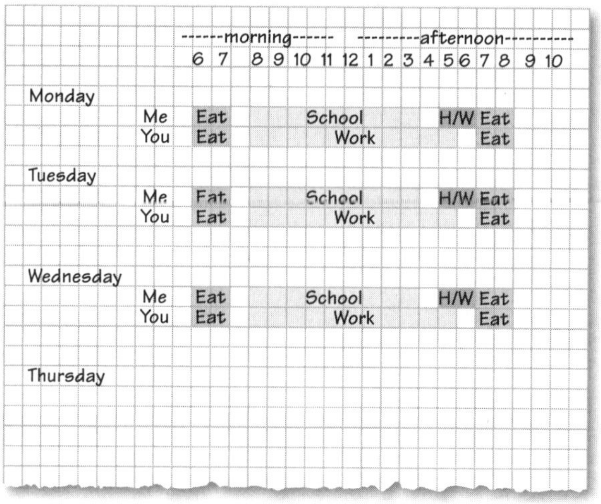

Are there tasks that can be assigned to other people to help free up more family time? Not all weeks are the same—sometimes a holiday or vacation changes the schedule. You and your student might work on strategies for making your family schedule more flexible.

Making a family schedule with your student will be "time" well spent!

Nombre ______________________________ Fecha __________

Capítulo 4 Fracciones, Decimales y Porcentajes

Estimada Familia:

Nuestros días están llenos de tareas: alistarse para ir a la escuela o al trabajo, preparar comidas, hacer mandados, visitar a familiares y amigos, por nombrar unos cuantos. Muchos de nosotros tenemos un calendario de nuestras tareas para ayudarnos a estimar nuestro tiempo.

Los adultos pueden trabajar aproximadamente la mitad de sus horas despiertos en un día dado. Preparar y comer las comidas puede tomar un sexto del día. Su estudiante probablemente se la pase aproximadamente un tercio de cada día de la semana en la escuela y tal vez otro 10% de la semana haciendo la tarea escolar.

Usted y su estudiante pueden trabajar en un horario familiar para hacer un mejor uso de su tiempo. Usted y su estudiante deben considerar estas preguntas:

- ¿Qué porcentaje de su semana se la pasa en la escuela? ¿Qué porcentaje se la pasa haciendo la tarea escolar?
- ¿Qué porcentaje de la semana se la pasa en otras actividades, como por ejemplo en deportes o lecciones de música? ¿Cuánto tiempo se la pasa practicando?
- ¿Cuánto tiempo de su semana se la pasa viajando en el auto?
- ¿Qué otras tareas tiene cada persona? ¿Qué porcentaje de la semana dura cada tarea?

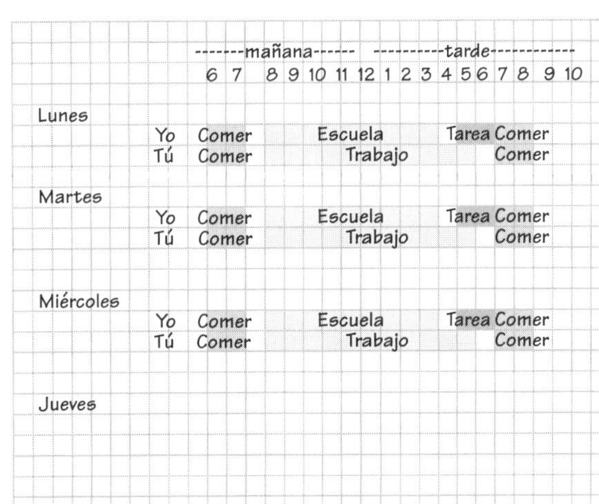

Puede usar papel para gráficos, para ayudarse a planear su horario de cada día. Use colores distintos para cada una de las actividades. Usted y su estudiante querrán hacer un horario familiar.

¿Hay tareas que pueden asignarse a otras personas para ayudar a generar más tiempo con la familia? No todas las semanas son iguales—a veces un feriado o unas vacaciones cambian nuestros horarios. Usted y su estudiante pueden trabajar con estrategias para flexibilizar más su horario familiar.

¡Hacer un horario familiar con su estudiante será un "tiempo" bien usado!

Non ____________________ Dat __________

Chapít 4 Fraksyon, Desimal, ak Pousantaj

Chè Fanmi:

Jounen nou ranpli ak tach: prepare pou al lekòl oubyen travay, prepare manje, fè komisyon, al wè fanmi ak zanmi, pou site detwa sèlman. Anpil nan nou kenbe yon kalandriye tach nou pou ede nou fè bidjè tan nou.

Granmoun gen dwa travay pou apeprè mwatye tan yo leve a pandan yon jounen. Prepare ak manje repa gen dwa pran yon sizyèm jounen an. Elèv ou a siman pase apeprè yon tyè chak jou lasemèn lekòl epi petèt yon lòt 10% semèn nan ap fè devwa.

Ou menm ak elèv ou a gen dwa vle tabli yon orè familyal pou ede itilize tan nou pi byen. Ou menm ak elèv ou a ta dwe konsidere keksyon sila yo:

- Ki pousantaj nan semèn nan ou pase lekòl? Ki pousantaj pase nan fè devwa?
- Ki pousantaj nan semèn nan pase nan lòt aktivite, tankou espò oswa leson mizik? Ki kantite tan ou pase ap antrene?
- Ki kantite nan semèn nan ou pase nan vwayaje nan machin nan?
- Ki lòt tach chak moun genyen? Ki pousantaj nan semèn nan chak tach pran

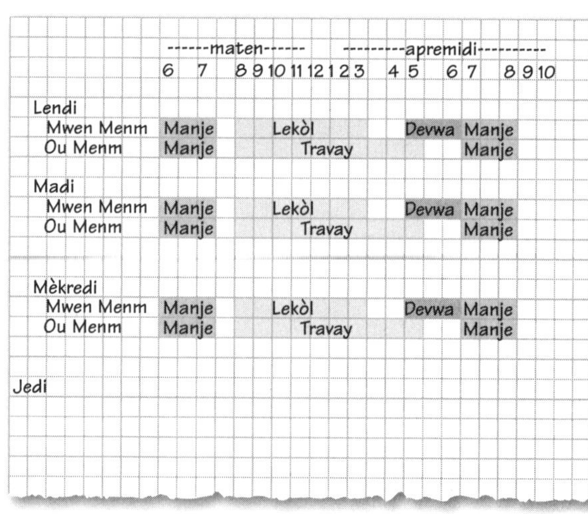

Ou kapab sèvi ak papye kadriye pou ede nan planifikasyon orè ou pou chak jou. Sèvi ak diferan koulè pou chak aktivite. Ou menm ak elèv ou a gen dwa vle fè yo orè familyal.

Èske gen tach nou kapab bay lòt moun pou ede libere plis tan familyal? Tout semèn pa menm—pafwa yon konje oswa vakans chanje orè a. Ou menm ak elèv ou a gen dwa travay sou estrateji pou rann orè familyal nou an pi fleksib.

Fè yon orè familyal ak elèv ou a sa ap "tan" ki byen itilize!

Activity 4.1 Start Thinking!
For use before Activity 4.1

Percents are used commonly in day-to-day life. What are some instances in which you have come across percents in your day-to-day life?

Activity 4.1 Warm Up
For use before Activity 4.1

Rewrite the fraction with a denominator of 100.

1. $\frac{12}{20}$

2. $\frac{1}{4}$

3. $\frac{15}{25}$

4. $\frac{33}{50}$

5. $\frac{27}{20}$

6. $\frac{54}{10}$

Lesson 4.1 Start Thinking!

For use before Lesson 4.1

Which models have an equivalent portion shaded? How do you know?

A. 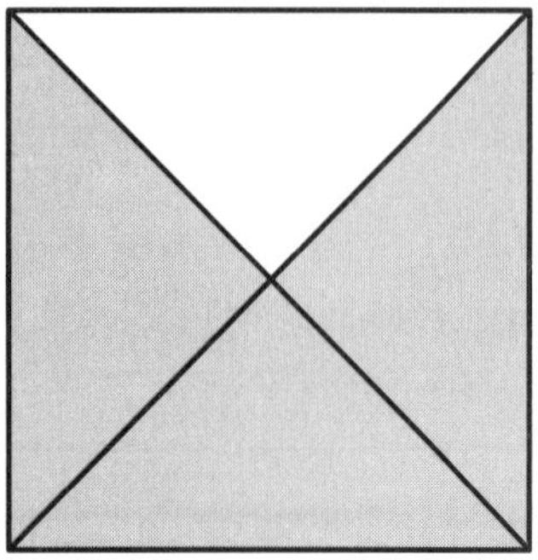

B. 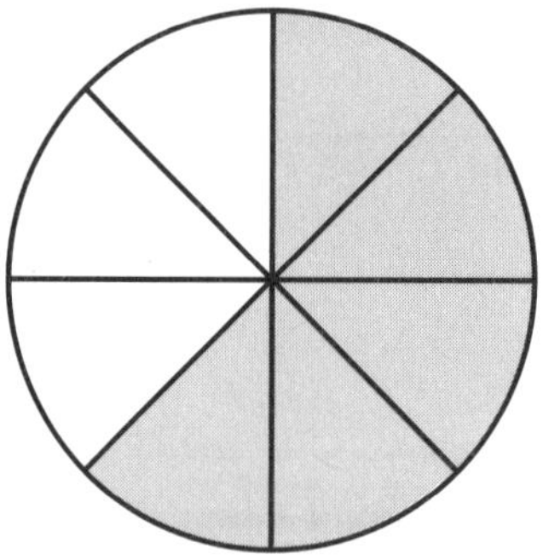

C. 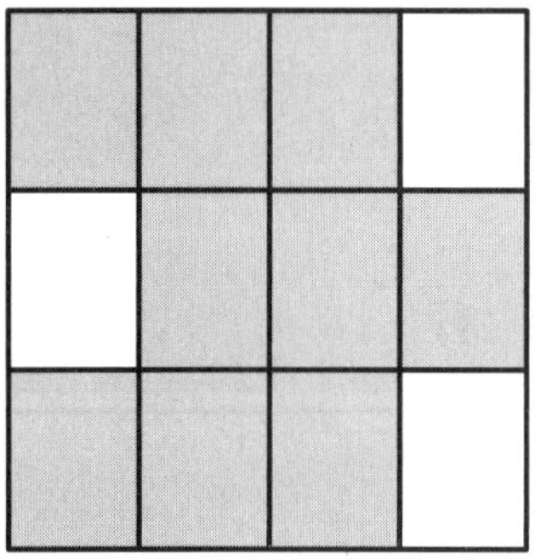

D. 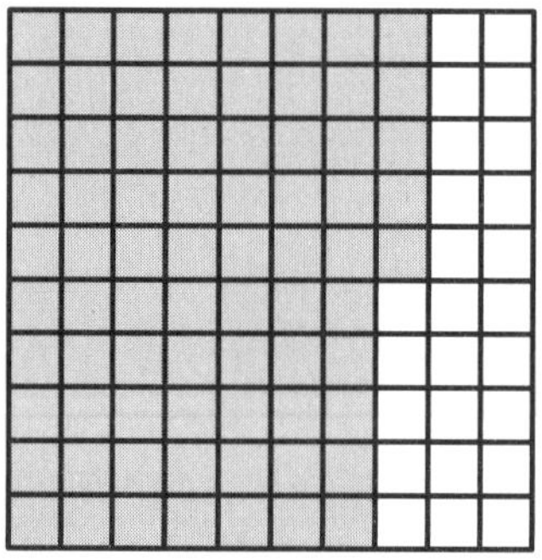

Lesson 4.1 Warm Up

For use before Lesson 4.1

Write the percent as a fraction in simplest form.

1. 12%

2. 4%

3. 78%

Write the fraction as a percent.

4. $\frac{1}{2}$

5. $\frac{9}{25}$

6. $\frac{19}{20}$

Name____ Date__________

4.1 Practice A

Write the percent as a fraction or mixed number in simplest form.

1. 20% **2.** 25% **3.** 65% **4.** 44%

5. 32% **6.** 9% **7.** 250% **8.** 120%

9. 44.4% **10.** 204.4% **11.** 0.5% **12.** 0.75%

13. Describe and correct the error in writing 8% as a fraction.

X $8\% = \frac{8}{10} = \frac{4}{5}$

Write the fraction as a percent. Which method did you use?

14. $\frac{6}{25}$ **15.** $\frac{3}{20}$ **16.** $\frac{7}{10}$ **17.** $\frac{3}{4}$

18. $\frac{9}{50}$ **19.** $\frac{12}{25}$ **20.** $\frac{5}{8}$ **21.** $\frac{5}{16}$

22. $\frac{2}{5}$ **23.** $\frac{9}{40}$ **24.** $\frac{23}{80}$ **25.** $\frac{13}{20}$

26. Describe and correct the error in writing $\frac{39}{40}$ as a percent.

X $\frac{39}{40} = 0.975 = \frac{9.75}{100} = 9.75\%$

27. Of the students in your class, 25% walk to school. Write this percent as a fraction in simplest form.

28. At the age of 4 months, a puppy's weight is 275% of its weight at the age of 2 months. How many times heavier is the puppy at 4 months than at 2 months? Write your answer as a mixed number in simplest form.

29. Four out of every five visitors at an amusement park buy day passes. What percent of the visitors buy day passes?

30. Only 6 of the 75 trees in a park are at least 30 feet tall. What percent of the trees are under 30 feet tall?

Write a fraction and a percent to represent the shaded portion of the model.

31.

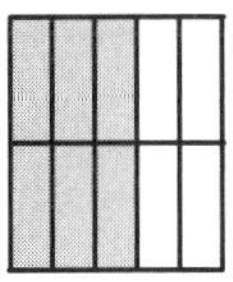

32.

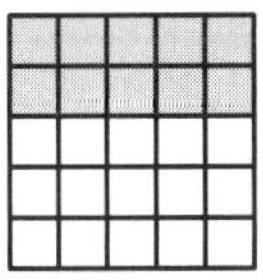

33.

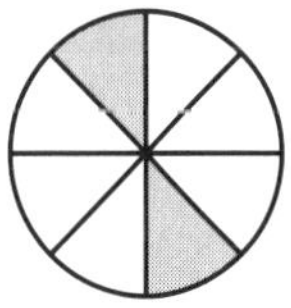

Name ______________________________ Date __________

4.1 Practice B

Write the percent as a fraction or mixed number in simplest form.

1. 35% **2.** 81.4% **3.** 210% **4.** 0.8%

Write the fraction as a percent. Which method did you use?

5. $\frac{37}{50}$ **6.** $\frac{21}{25}$ **7.** $\frac{3}{16}$ **8.** $\frac{19}{40}$

Write the mixed number as a percent.

9. $3\frac{1}{4}$ **10.** $2\frac{1}{2}$ **11.** $1\frac{4}{5}$ **12.** $5\frac{7}{50}$

13. $4\frac{3}{25}$ **14.** $6\frac{11}{20}$ **15.** $2\frac{1}{8}$ **16.** $3\frac{5}{16}$

17. On a school bus, 22 of the 40 students are in window seats. What percent of the students are in window seats?

18. You and a friend are selling lemonade. You sell three times as many cups as your friend. What percent of the cups sold were sold by your friend? Explain.

19. The United States flag is actually 105% as tall as the state flag of Florida. Write this percent as a mixed number and explain why the perspective in the figure may be misleading.

20. At a zoo, an anaconda is 118% as long as a Burmese python and $1\frac{3}{20}$ times as long as a reticulated python. Which is longer, the Burmese python or the reticulated python? Explain.

21. Copy the square. Then draw another square with sides that are $\frac{1}{2}$ as long. What percent of the area of the original square is the area of the smaller square?

22. The weight of a tree is doubling every 3 years. In how many years will the weight of the tree be 1600% of its weight now?

Name______________________________ Date__________

4.1 Enrichment and Extension

Geometry

Write a fraction and a percent to represent the shaded portion of the model.

1. 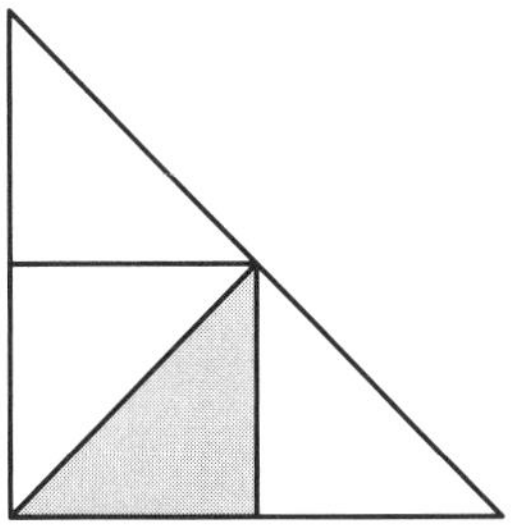

2. 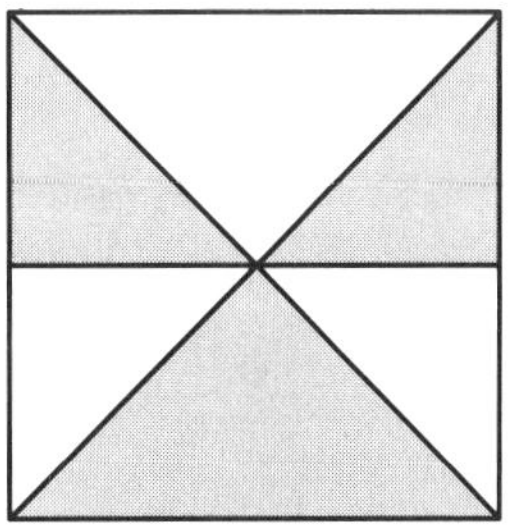

3. 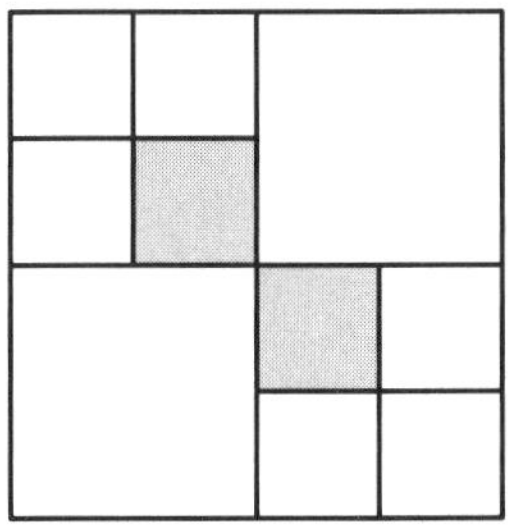

4. 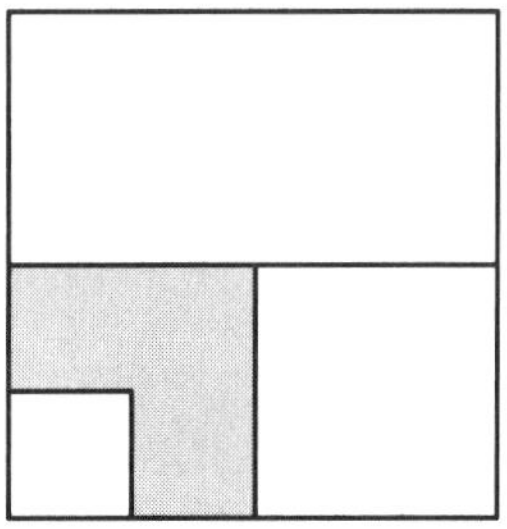

5. 

6. 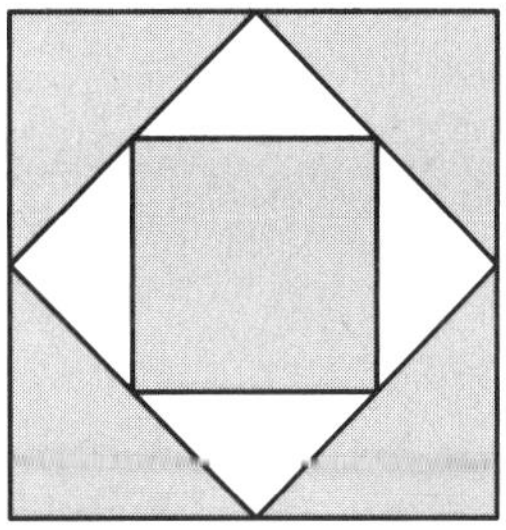

7. 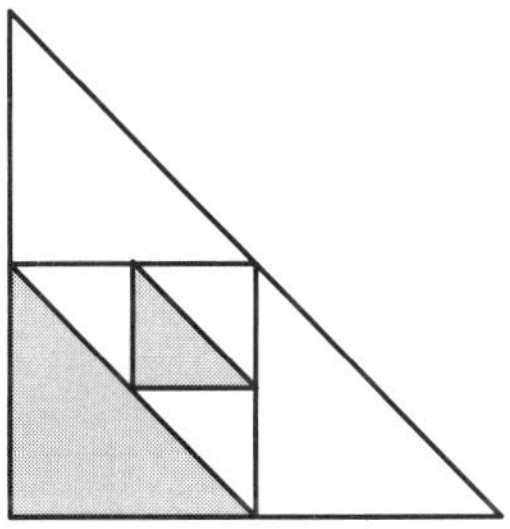

8. 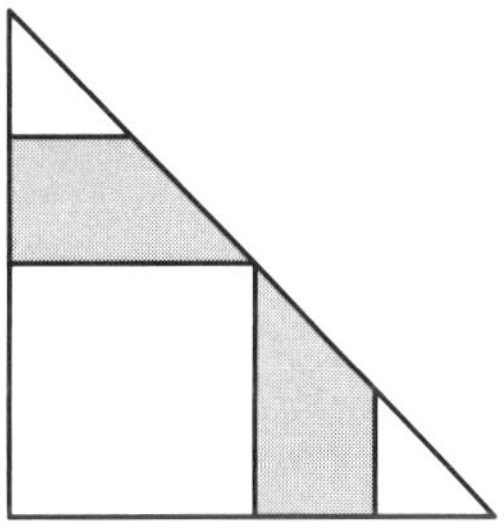

Name ______________________________ Date __________

4.1 Puzzle Time

What Did The Alien From Outer Space Say To The Green Book?

Write the letter of each answer in the box containing the exercise number.

Write the percent as a fraction or mixed number in simplest form.

1. 35%

2. 30%

3. 55%

4. 84%

5. 59%

6. 43.9%

7. 144%

8. 0.75%

9. 334%

10. 132.6%

11. 0.8%

Write the fraction as a percent.

12. $\frac{2}{10}$

13. $\frac{2}{5}$

14. $\frac{11}{25}$

15. $\frac{3}{25}$

16. $\frac{29}{50}$

17. $\frac{16}{25}$

18. $\frac{5}{8}$

19. $\frac{7}{16}$

Answers for Exercises 1–11

E. $1\frac{163}{500}$

R. $1\frac{11}{25}$

D. $\frac{439}{1000}$

A. $\frac{11}{20}$

O. $\frac{7}{20}$

L. $\frac{3}{400}$

T. $\frac{59}{100}$

E. $\frac{1}{125}$

R. $3\frac{17}{50}$

U. $\frac{3}{10}$

S. $\frac{21}{25}$

Answers for Exercises 12–19

R. 58%

O. 40%

E. 62.5%

A. 64%

M. 20%

Y. 43.75%

D. 12%

E. 44%

8	14	3	6		12	18		5	1		19	13	2	9		16	10	17	15	11	7	4

Activity 4.2 Start Thinking!
For use before Activity 4.2

Study the decimal and percent equivalents below. What do you notice?

$0.35 = 35\%$	$0.27 = 27\%$	$0.03 = 3\%$
$0.2 = 20\%$	$0.5 = 50\%$	$1.38 = 138\%$
$2.1 = 210\%$	$6 = 600\%$	$0.886 = 88.6\%$

Andrew says that $0.98 = 9.8\%$. Is he correct? Why or why not?

Activity 4.2 Warm Up
For use before Activity 4.2

Write the fraction represented by the model.

1. 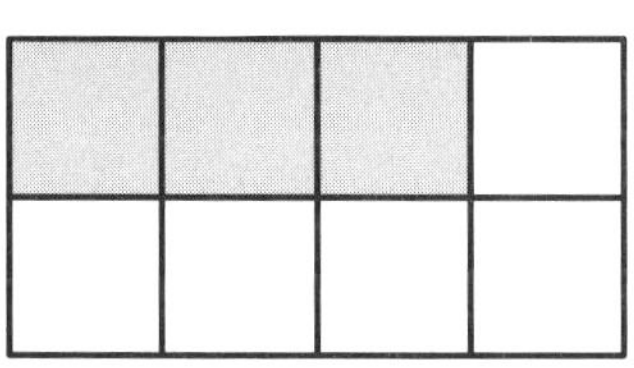

2. 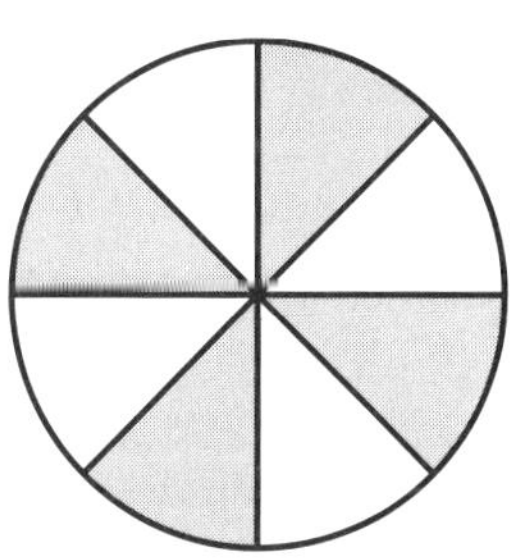

3. 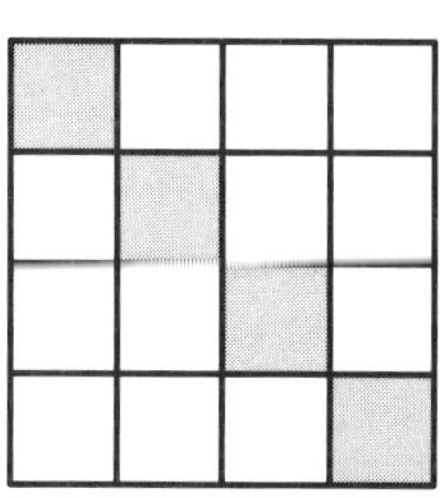

4. 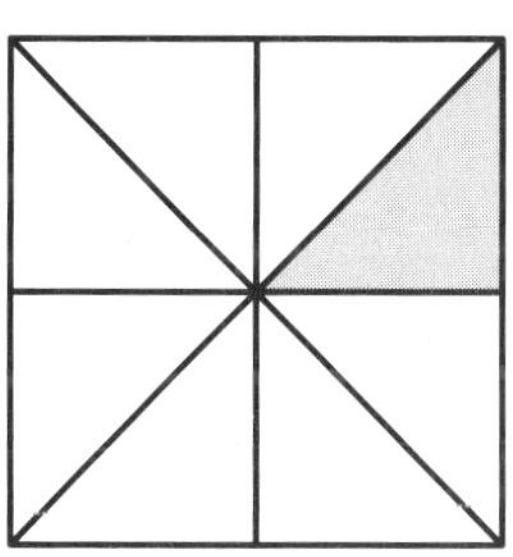

5. 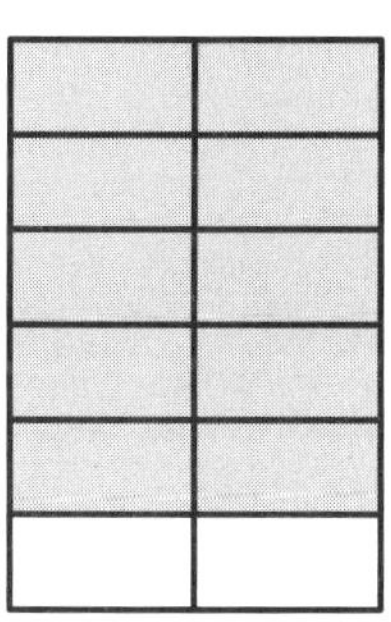

6. 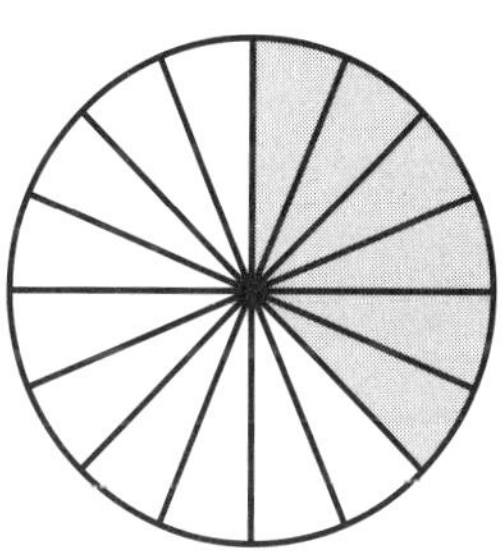

Lesson 4.2 Start Thinking!

For use before Lesson 4.2

Describe a situation in sports when decimals and percents arc used interchangeably.

Lesson 4.2 Warm Up

For use before Lesson 4.2

Write the percent as a decimal.

1. 19% **2.** 2% **3.** 7.5%

Write the decimal as a percent.

4. 0.89 **5.** 0.54 **6.** 0.1

Name ______________________________ Date __________

4.2 Practice A

Write the percent as a decimal.

1. 81% **2.** 78% **3.** 5% **4.** 8%

5. 40% **6.** 60% **7.** 23.7% **8.** 16.75%

9. 150% **10.** 210% **11.** 186% **12.** 416%

13. 100.8% **14.** 5.17% **15.** 0.4% **16.** 0.04%

17. Describe and correct the error in writing 1.475% as a decimal.

✗ $1.475\% = 1.475\% = 147.5$

Write the decimal as a percent.

18. 0.66 **19.** 0.32 **20.** 0.51 **21.** 0.97

22. 0.01 **23.** 0.04 **24.** 0.312 **25.** 0.468

26. 0.5 **27.** 1.2 **28.** 1.08 **29.** 1.16

30. 0.003 **31.** 0.025 **32.** 0.0245 **33.** 2.025

34. Describe and correct the error in writing 1.8 as a percent.

✗ $1.8 = 1.8 = 18\%$

35. Fifty-four percent of the students in your class have moved at least one time. Write this percent as a decimal.

36. Only 0.15 of the total number of vehicles in your school parking lot are buses. What percent of the vehicles are buses?

37. You spent 0.88 of your allowance this week. What percent of your allowance did you spend?

38. On a history test, you get 86 out of a possible 100 points. Write a decimal and a percent that represent a score of 86 out of 100.

39. Of the fluids that you drink on a typical day, $\frac{1}{10}$ is milk and 50% is water. How many times more water do you drink than milk?

Write each percent as a fraction in simplest form and as a decimal.

40. 21% **41.** 75% **42.** 64% **43.** 85%

Name ______________________________ Date __________

4.2 Practice B

Write the decimal as a percent.

1. 0.54 **2.** 0.37 **3.** 0.222 **4.** 0.929

5. 1.4 **6.** 2.5 **7.** 20 **8.** 0.005

Write each percent as a fraction in simplest form and as a decimal.

9. 68% **10.** 9% **11.** 55% **12.** 26%

13. 42.4% **14.** 73.6% **15.** 31.25% **16.** 44.65%

17. About 36% of the students at a middle school are sixth graders. What percent are *not* in sixth grade?

18. The percents of three types of tickets collected at the gate for a high school football game are shown.

Ticket type	Student	Adult	Senior (65 and older)
Percent	48%	28%	14%

a. Write the percents as decimals and as fractions.

b. There is one other type of ticket that is not shown. It is a ticket for a child under 5. What percent of the tickets were of this type?

c. Make a bar graph to represent the percents for all four ticket types.

19. Students in an after-school enrichment program chose one of five subjects areas.

Subject Chosen

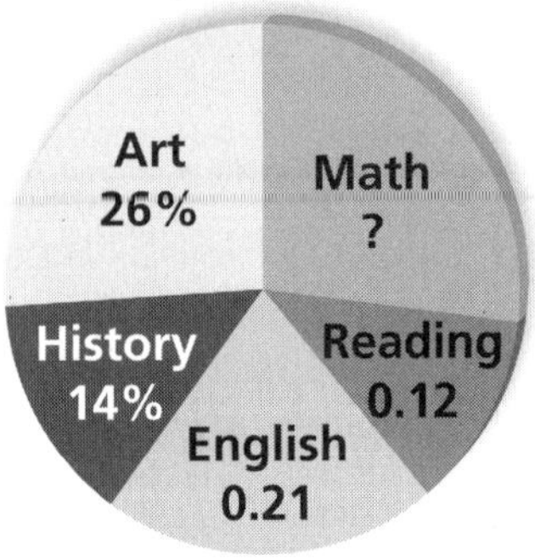

a. What percent chose English or reading?

b. What percent chose English or history?

c. How many times more students chose English than reading?

d. What percent chose math? Write the percent as a decimal.

20. At one school, half of the students live within 1 mile, 78% live within 2 miles, and 0.1 of the students live between 2 and 3 miles from the school.

Make a table to show the percent of students who live at each distance from the school.

a. within 1 mile **b.** between 1 and 2 miles

c. between 2 and 3 miles **d.** more than 3 miles

Name______________________________ Date __________

4.2 Enrichment and Extension

Playing Checkers

You set up a game of checkers as shown. Write a decimal for each percent you find.

Your friend's side

Your side

1. What percent of the checker board is shaded?

2. What percent of the checker board is *not* shaded?

3. Your friend arrives and places pieces appropriately on the board.

   a. What percent of the checker board squares have pieces?

   b. What percent of the checker board squares do *not* have pieces?

   c. What percent of the shaded checker board squares have pieces?

   d. What percent of the shaded checker board squares do *not* have pieces?

4. At one point during the game, 4 of your pieces have been removed and 5 of your friend's pieces have been removed.

   a. What percent of the original number of pieces have been removed?

   b. What percent of the original number of pieces are remaining?

5. The checker piece near the middle of the board can move diagonally in any direction between shaded squares.

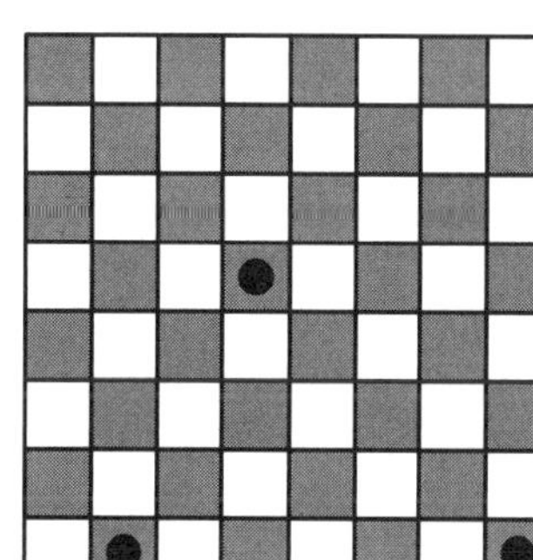

   a. How many squares can the piece move to in one turn?

   b. On how many different squares can the piece be located at the end of two turns? What percent of the shaded checker board is this? Write this percent as a decimal.

Name ______________________________ Date __________

4.2 Puzzle Time

Why Are Math Assignments Like The Water That Is Found On The Ground In The Early Morning?

Write the letter of each answer in the box containing the exercise number.

Write the percent as a decimal.

1. 67% **2.** 44%

3. 29.6% **4.** 46.3%

5. 22% **6.** 8%

7. 58.74% **8.** 80.14%

9. 277% **10.** 106%

11. 0.05% **12.** 0.045%

Write the decimal as a percent.

13. 0.85 **14.** 0.41

15. 0.98 **16.** 0.657

17. 0.77 **18.** 0.51

19. 0.376 **20.** 0.239

21. 2.57 **22.** 0.0482

Answers for Exercises 1–12

T. 2.77	**E.** 0.00045
S. 0.296	**O.** 0.22
B. 0.67	**H.** 1.06
R. 0.8014	**P.** 0.5874
O. 0.463	**D.** 0.44
E. 0.0005	**U.** 0.08

Answers for Exercises 13–22

B. 37.6%	**P.** 65.7%
S. 4.82%	**A.** 23.9%
T. 85%	**E.** 98%
U. 257%	**D.** 51%
O. 77%	**E.** 41%

19	17	9	10		20	8	12		3	6	7	16	5	22	15	2		13	4		1	11		18	21	14

Activity 4.3 Start Thinking!

For use before Activity 4.3

What are two methods that you can use to compare two fractions?

Use one of the methods you described to tell whether $\frac{21}{40}$ or $\frac{7}{12}$ is greater. Which method did you use? Why?

Activity 4.3 Warm Up

For use before Activity 4.3

Order the decimals from least to greatest.

1. 1.4, 1.44, 4.1, 4.3, 3.3

2. 0.5, 1.75, 0.7, 1.57, 1.5

3. 0.44, 0.02, 0.2, 0.04, 0.24

4. 6.6, 0.6, 6.63, 6.06, 0.06

5. 0.2, 0.17, 0.02, 0.72, 2.27

6. 8.1, 0.81, 0.18, 1.8, 1.88

Lesson 4.3 Start Thinking!

For use before Lesson 4.3

Design a survey with three yes or no questions to ask the students in your class. Take turns in class asking the survey questions. Record your results in a table like the one below. You can use fractions, percents, or decimals. Which of your questions had the greatest percent of students answering yes?

Question	Yes	No
Do you play a sport?	19/28	9/28
Do you play a musical instrument?	20%	80%
Do you have any siblings?	0.85	0.15

Lesson 4.3 Warm Up

For use before Lesson 4.3

Tell which number is greater.

1. $\frac{5}{13}$, 45%

2. 30%, 0.4

3. $\frac{9}{17}$, 50%

4. 0.6, $\frac{4}{7}$

5. $\frac{2}{5}$, 0.225

6. 240%, 0.24

Name____ Date__________

4.3 Practice A

Tell which number is greater.

1. $\frac{3}{4}$, 70%

2. $\frac{1}{2}$, 0.54

3. 0.21, 21%

4. $\frac{2}{3}$, 66%

5. 0.482, 49%

6. 16%, 0.108

7. $\frac{12}{25}$, 48%

8. $\frac{1}{10}$, 12%

9. 1.2, 11%

10. 58%, $\frac{31}{50}$

11. 5020%, $50\frac{1}{4}$

12. 12.25%, $\frac{1}{8}$

13. Describe and correct the error in comparing 0.7% and $\frac{17}{25}$.

> ✗ $\frac{17}{25} \overset{\times 4}{=} \frac{68}{100} = 0.68\%$
>
> 0.7% is greater than 0.68%, so 0.7% is the greater number.

Use a number line to order the numbers from least to greatest.

14. 0.64, $\frac{13}{20}$, 63%

15. 45%, 0.46, $\frac{11}{25}$

16. 0.12, $\frac{1}{8}$, 0.135, 13%

17. $\frac{15}{16}$, 90%, 0.925, $\frac{7}{8}$, 0.93

18. $3\frac{2}{3}$, 362%, 3.66, $3\frac{3}{5}$, 36

19. 0.3, 27.3%, $\frac{11}{40}$, 28%, 0.27

20. You use 8 fluid ounces of fruit juice in a recipe to make 64 fluid ounces of fruit punch. A fruit punch you can buy at the store has 10% real fruit juice. Which has a higher percent of fruit juice?

21. While shooting baskets at a basketball hoop, you make 36 out of 80 shots. Your friend makes 43.75% of the shots. Who made a higher percent?

22. To earn a bonus in a video game, you must find at least 60% of the hidden gems. You find 25 out of 40 gems. Do you get the bonus? Explain.

23. The table shows the portion of students at a middle school that are in each grade. Order the grades from the least to the greatest number of students.

Grade	6	7	8
Portion of students	$33\frac{1}{3}\%$	0.3125	$\frac{17}{48}$

Name ______________________________ Date __________

4.3 Practice B

Tell which number is greater.

1. $\frac{1}{4}$, 22%　　**2.** $\frac{5}{9}$, 55%　　**3.** 3.2, 32%　　**4.** 99.9%, 1

Order the numbers from least to greatest.

5. $\frac{1}{3}$, 0.3, 33%, $\frac{8}{25}$, 33.6%　　**6.** 210%, 2.2, $2.\overline{2}$, $\frac{43}{20}$

Tell which letter shows the graph of the number.

7. 0.884　　**8.** $\frac{8}{9}$　　**9.** $\frac{22}{25}$　　**10.** 0.89

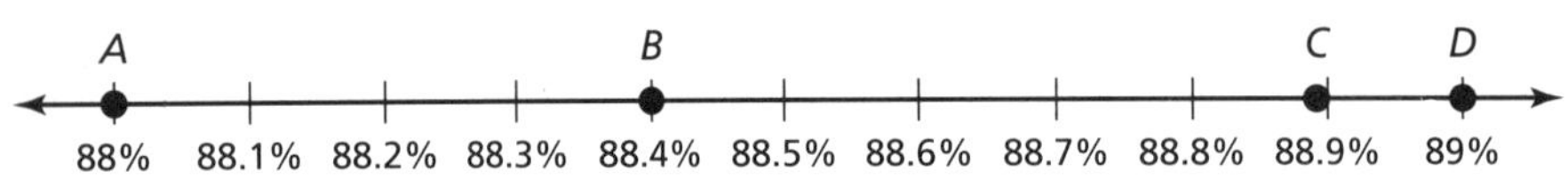

11. Describe a process that you can use to find a decimal whose value is between 31% and 32%.

12. Is 6 centimeters greater than 5% of a meter? Explain.

13. Does 6% of a pound weigh more than an ounce? Explain.

14. Order the periods of time from least to greatest:

1% of an hour　　$\frac{2}{3}$ of a minute　　0.0004 of a day

15. The table shows the portions of the U.S. population that were in Florida in certain years.

Year	1860	1910	1960	2010
Portion of U.S. Population in Florida	0.45%	0.0082	$\frac{1}{36}$	$\frac{1}{16}$

a. Order the portions from least to greatest.

b. Since 1860, how has the population of Florida increased compared to the population of the United States? Why do you think this happened?

c. Do you think this will always happen? Explain your reasoning.

16. Arsenic is toxic to humans. The greatest amount of arsenic that is allowed in drinking water is 10 parts per billion. A test shows that a source of drinking water contains 0.000002% arsenic. Is this an allowable amount? Explain.

4.3 Enrichment and Extension

Treasure Map

You want to find the treasure shown on the map. To get to the treasure, start at 1% and move to 100%. The only way you can move on the map is if a number is greater than the number you are currently on. List the numbers in order that make up your path to the treasure.

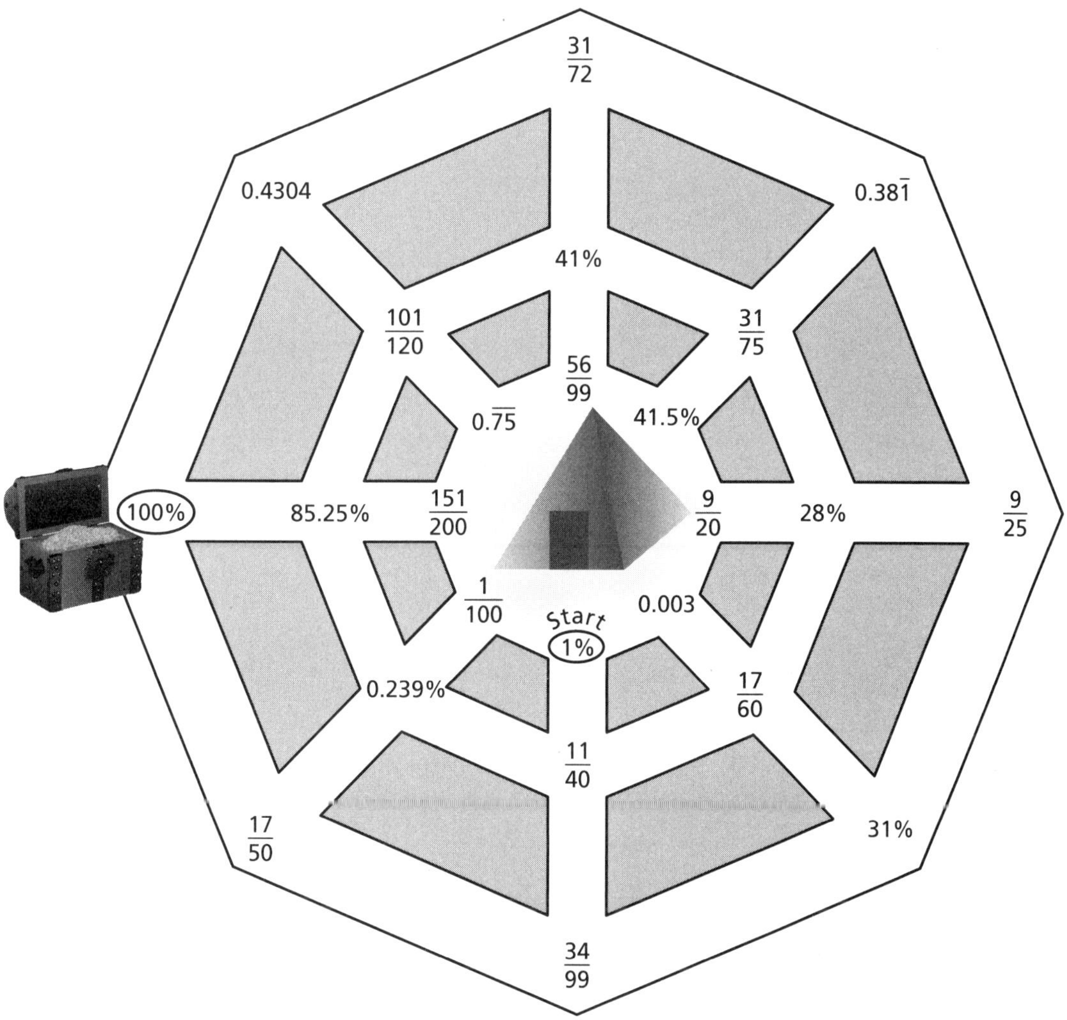

Name ______________________________ Date __________

4.3 Puzzle Time

Why Did The Peach Need To Hire An Attorney?

Write the letter of each answer in the box containing the exercise number.

Tell which number is greater.

1. $\frac{17}{20}$, 95%

2. 30%, 0.03

3. $\frac{27}{50}$, 27%

4. $\frac{14}{25}$, 60%

5. 0.097, 97%

6. 65%, 0.56

7. 80%, $\frac{7}{8}$

8. 0.13, 13.5%

9. 11%, $\frac{3}{25}$

10. 150%, 0.15

11. $\frac{1}{4}$, 20%

12. 66%, $\frac{2}{3}$

Which number is the greatest?

13. 28%, $\frac{3}{5}$, 0.31

14. 79%, 0.52, $\frac{17}{20}$

15. $\frac{17}{25}$, 0.81, $\frac{5}{8}$, 64%

16. 0.14%, $\frac{7}{20}$, 0.014, 1.4

Answers for Exercises 1–12

W. $\frac{3}{25}$	**D.** $\frac{14}{25}$	**G.** 80%
E. 66%	**R.** $\frac{1}{4}$	**L.** $\frac{27}{50}$
Y. 97%	**I.** 20%	**E.** 11%
A. 95%	**T.** 150%	**C.** 0.15
J. $\frac{2}{3}$	**N.** $\frac{7}{8}$	**O.** 27%
I. 60%	**S.** 30%	**K.** 0.56
U. 13.5%	**M.** 0.03	**A.** 65%
P. 0.097	**B.** $\frac{17}{20}$	**F.** 0.13

Answers for Exercises 13–16

E. $\frac{5}{8}$	**D** 0.014	**T.** 0.81
W. 0.52	**A.** $\frac{3}{5}$	**S.** 0.31
M. $\frac{17}{20}$	**Y.** 79%	**I.** 1.4
M. 28%	**R.** 0.14%	**B.** 64%

4	15		9	13	2		10	11	8	3	5		16	7		6		12	1	14

Activity 4.4 Start Thinking!
For use before Activity 4.4

1% of 500 is 5. Use this fact to answer the following questions. Be sure to explain your reasoning.

What is 2% of 500?

What is 3% of 500?

What is 4% of 500?

What is 5% of 500?

What is n% of 500?

Activity 4.4 Warm Up
For use before Activity 4.4

Multiply.

1. $\frac{1}{10} \times 120$

2. $\frac{2}{5} \times 150$

3. $\frac{1}{2} \times 26$

4. $\frac{3}{10} \times 100$

5. $\frac{1}{4} \times 20$

6. $\frac{1}{100} \times 920$

Lesson 4.4 Start Thinking!
For use before Lesson 4.4

Research the sales tax rates in various U.S. states.

Write a word problem that compares the cost of buying an item in one state with buying the same priced item in a state with a different sales tax rate.

Exchange word problems with a classmate and solve your classmate's problem.

Lesson 4.4 Warm Up
For use before Lesson 4.4

Find the percent of the number.

1. 25% of 40

2. 70% of 50

3. 150% of 16

4. 10% of 46

5. 15% of 80

6. 2% of 500

Name____ Date__________

4.4 Practice A

Use a fraction to find the percent of the number.

1. 25% of 80 **2.** 60% of 20 **3.** 10% of 210 **4.** 20% of 50

5. 50% of 64 **6.** 75% of 160 **7.** 30% of 70 **8.** 40% of 35

Use a decimal to find the percent of the number.

9. 26% of 200 **10.** 35% of 80 **11.** 82% of 150 **12.** 8% of 75

13. 62% of 50 **14.** 36% of 25 **15.** 3% of 90 **16.** 71% of 400

17. Describe and correct the error in finding 4% of 65.

✗ $4\% \text{ of } 65 = 0.4 \times 65 = 26$

Find the percent of the number.

18. 40% of 90 **19.** 45% of 80 **20.** 24% of 65 **21.** 40% of 85

22. 66% of 250 **23.** 52% of 950 **24.** 52% of 44 **25.** 72% of 65

26. 91% of 60 **27.** 8% of 275 **28.** 13% of 58 **29.** 27% of 92

30. Yesterday, 5% of the 120 sixth graders at a school were late. How many sixth graders were late?

31. You spent 20% of the 70 minutes it took you to do your homework on math. How much time did you spend on your math homework?

32. The sale price of a pair of pants is 65% of the regular price of $25. How much do you save by buying the pants on sale?

33. Your family drives 126 miles from Key West to the Florida mainland. About 15% of this distance is spent on bridges. How many miles of bridges do you travel over?

34. About 3.5% of the weight of 20 pounds of sea water is salt. How many pounds of salt are in the water?

35. Before a century plant dies, a tall flower stem grows from its center. The flower stem grows for only two months to reach 40% of a height of 25 feet. How tall is the flower stem after two months?

Find the percent of the number.

36. 110% of 40 **37.** 250% of 18 **38.** 0.4% of 80 **39.** 1.2% of 180

Name ______________________ Date __________

4.4 Practice B

Find the percent of the number.

1. 6% of 15 **2.** 65% of 60 **3.** 78% of 81 **4.** 14% of 106

5. 160% of 75 **6.** 230% of 45 **7.** 514% of 205 **8.** 115% of 130

9. 7.6% of 110 **10.** 3.9% of 240 **11.** 0.6% of 500 **12.** 11.6% of 150

Copy and complete the statement using <, >, or =.

13. 55% of 60 __?__ 60% of 65 **14.** 20% of 80 __?__ 80% of 20

15. 36% of 150 __?__ 27% of 200 **16.** 110% of 3 __?__ 0.9% of 300

17. How many hours is 75% of 3 days? **18.** How many feet is 20% of 4 miles?

19. A restaurant serves you a 16-fluid ounce glass of juice that is 30% ice. How many fluid ounces of juice do you actually get?

20. The table shows the grading scale for one of your classes.

Letter grade	A	B	C	D
Percent range	90–100%	80–89%	70–79%	60–69%

Tell the letter grade that you earn for each score.

a. You earn 14 out of a possible 20 points on a quiz.

b. You earn 66 out of a possible 80 points on a test.

c. You earn 216 out of a possible 250 points for a report.

21. A 15% discount saves you \$5 off the price of an electronic game. How much do you save off the regular price if the discount is raised to 45%? Explain your reasoning.

22. Draw two different rectangles with perimeters that are each 80% of the perimeter of the rectangle shown. Show the length and width of each rectangle.

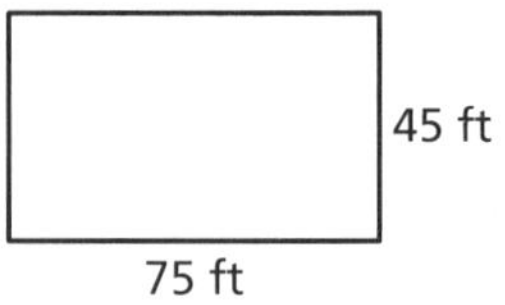

23. A monitor that regularly costs \$100 is on sale for 15% off. The salesperson offers you 20% off the sale price. What percent of the original price is the salesperson's price?

24. A store pays a manufacturer \$20 for a hat. To make a profit, the store prices the hat 60% higher than the amount they paid. Later, a sale reduces the price of the hat by 10%. Finally, a clearance sale reduces the sale price by 40%. What is the clearance price? Will this price give the store a profit?

Name________________________________ Date __________

4.4 Enrichment and Extension

Better Buy?

In Exercises 1–3, which store has the better buy? Explain. Assume no sales tax.

1. Identical sweaters are sold in two different stores.

- The sale price in Store A is 25% off the regular price of \$34.96.
- The sale price in Store B is 20% off the regular price of \$29.75.

2. An identical pair of sneakers is sold in two different stores.

- The sale price in Store A is 30% off the regular price of \$69.99.
- The sale price in Store B is 15% off the regular price of \$63.80 plus an additional 10% off of the sale price.

3. Identical video game systems are sold in two different stores.

- The sale price in Store A is 10% off the regular price of \$250 plus an additional 15% off of the sale price.
- The sale price in Store B is 15% off the regular price of \$250 plus an additional 10% off of the sale price.

4. Your aunt is buying a new couch. The regular price of the couch is \$799. The store sale price is 20% off the regular price. Your aunt uses a newspaper coupon for an additional 5% off of any sale price item. For having an account with the store, she also gets an additional 10% off her final price.

a. How much does your aunt pay for the couch?

b. Does it make a difference in what order you use the percents when calculating the price your aunt pays for the couch?

Name ______________________________ Date ________

4.4 Puzzle Time

Why Was The Math Textbook Feeling Sad?

Write the letter of each answer in the box containing the exercise number.

Find the percent of the number.

1. 10% of 50

2. 20% of 30

3. 16% of 80

4. 42% of 20

5. 7% of 80

6. 15% of 10

7. 25% of 40

8. 4% of 50

9. 40% of 60

10. 50% of 38

11. 56% of 56

12. 82% of 18

13. 60% of 70

14. 28% of 33

15. 90% of 20

16. 130% of 50

17. 110% of 44

18. 125% of 45

19. 2.5% of 75

20. 0.6% of 90

21. 3.9% of 250

22. 5.1% of 125

Answers

P. 31.36	**B.** 0.54
M. 5.6	**F.** 9.75
L. 5	**T.** 48.4
E. 24	**L.** 9.24
O. 1.5	**P.** 6
I. 14.76	**E.** 56.25
A. 65	**H.** 19
D. 12.8	**S.** 10
E. 1.875	**G.** 6.375
L. 42	**E.** 8.4
R. 18	**S.** 2

2	15	6	20	13	18	5	8		21	12	1	14	9	3		17	10	19		11	16	22	4	7

Activity 4.5 Start Thinking!
For use before Activity 4.5

Estimate the percent of the square that is shaded. Describe the method you used to estimate.

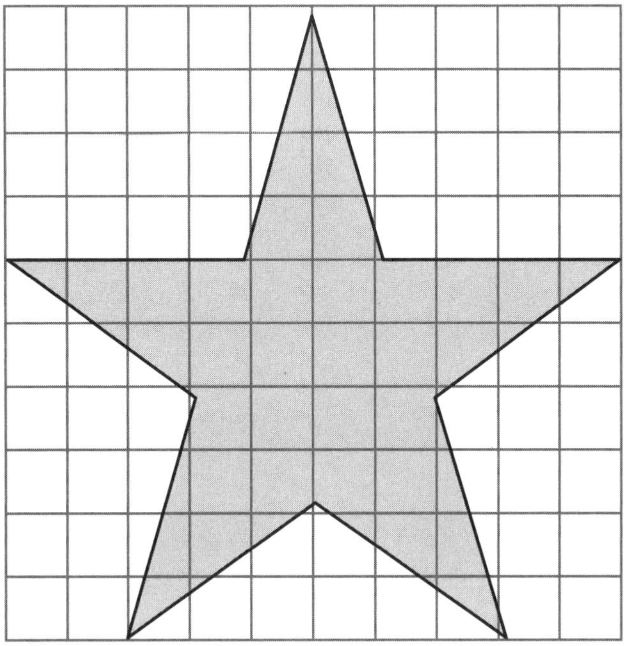

Activity 4.5 Warm Up
For use before Activity 4.5

Find the percent of the number.

1. 10% of 50

2. 12% of 60

3. 2% of 45

4. 30% of 90

5. 15% of 40

6. 5% of 200

Lesson 4.5 Start Thinking!

For use before Lesson 4.5

When solving real-life problems involving percents, sometimes an exact answer is needed and other times an estimate is sufficient.

In which of the following situations is an estimate sufficient?

- Calculating the tip at a restaurant
- Comparing basketball players' field goal percentages
- Comparing student survey results

Lesson 4.5 Warm Up

For use before Lesson 4.5

Estimate the percent of the number.

1. 61% of 105

2. 73% of 390

3. 19% of 22

4. 33% of 550

5. 98% of 45

6. 82% of 69

Name____ Date __________

4.5 Practice A

Estimate the percent of the number.

1. 31% of 200 **2.** 58% of 400 **3.** 52% of 59 **4.** 9% of 80

5. 40% of 197 **6.** 20% of 148 **7.** 26% of 78 **8.** 74% of 42

9. 19% of 53 **10.** 87% of 304 **11.** 68% of 212 **12.** 29% of 41

13. 73% of 81 **14.** 2% of 190 **15.** 5% of 220 **16.** 49% of 363

17. About 62% of the 279 songs on your MP3 player are less than 5 minutes long. About how many songs is this?

18. In a study, a fisherman catches, tags, and releases 1513 greater amberjack fish. About 11% of the fish are recaptured. Estimate the number of fish that are recaptured.

19. A twelve-year-old girl weighs 93 pounds. About 53% of her body weight is water. About how many pounds of water are there in her body?

20. A mall has 142 stores. The table shows the percent of stores that sell products of certain types.

Product type	Percent
Sporting goods	9
Toys	8
Clothing/shoes	53
Food/restaurants	24
Jewelry	19

a. About how many stores sell clothing or shoes?

b. About how many stores sell jewelry?

c. About how many stores sell toys?

21. A middle school has 914 students. The table shows the percent of families of students who own various numbers of vehicles.

Number of vehicles	Percent
0	9
1	22
2	41
3	17
4 or more	11

a. About how many students do not have any vehicles in their family?

b. About how many students have at least three vehicles in their family?

c. About how many more students are there in families with two vehicles than in families with one vehicle?

Estimate the percent of the number.

22. 117% of 20 **23.** 208% of 70 **24.** 194% of 87 **25.** 251% of 72

Name ______________________________ Date __________

4.5 Practice B

Estimate the percent of the number.

1. 37% of 600 **2.** 76% of 24 **3.** 7% of 81 **4.** 98% of 57

Use compatible numbers to estimate the percent of the number.

5. 23% of 108 **6.** 35% of 58 **7.** 18% of 392 **8.** 45% of 1023

9. 64% of 248 **10.** 26.3% of 152 **11.** 86% of 514 **12.** 32% of 878

Estimate the percent of the number.

13. 120% of 42 **14.** 150% of 59 **15.** 142% of 68 **16.** 162% of 77

17. 252% of 59 **18.** 273% of 98 **19.** 346% of 51 **20.** 388% of 212

Determine whether the statement is *sometimes*, *always*, or *never* true. Explain your reasoning.

21. If the percent is rounded down and the number is rounded up, then the estimate will be greater than the actual answer.

22. If both the percent and the number are rounded up, then the estimate will be less than the actual answer.

23. The circle graph shows the results of a survey that asked 444 students their favorite type of ethnic restaurant.

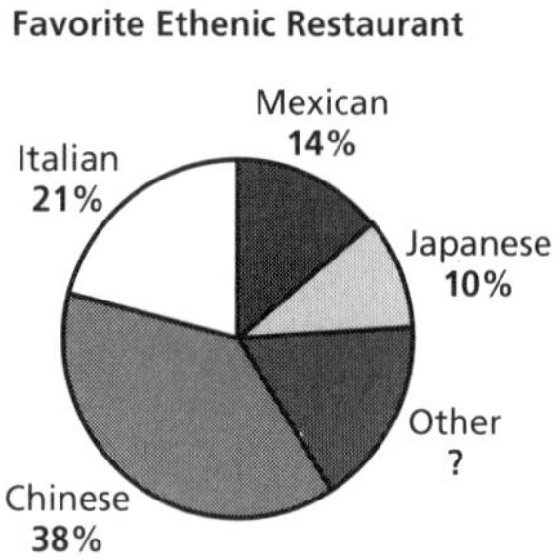

a. Estimate the number of the students whose favorite is either Italian or Japanese.

b. Estimate the number of students who chose *other*. Explain your reasoning.

24. A survey asked 187 people in a town what their favorite FM station is. The top four choices are shown.

FM Station	Percent
101.3	28
95.9	21
90.5	11
106.3	8

a. About how many people favored FM 101.3?

b. Estimate how many more people chose FM 101.3 than FM 106.3.

c. Add the percents of the top four choices. How can you use the result to show that the town has at least eight FM radio stations?

Name ______________________________ Date __________

4.5 Enrichment and Extension

Estimating Percents That Are Less Than 1

Example: Estimate $\frac{1}{5}\%$ of 394.

$\frac{1}{5}\%$ is one fifth of 1%. For 394, use the compatible number 400.

$$1\% \text{ of } 400 = \frac{1}{100} \times 400 \qquad \text{Write 1\% as a fraction.}$$

$$= \frac{1 \times \cancel{400}^{4}}{_{1}\cancel{100}} \qquad \text{Divide out the common factor 100.}$$

$$= 4 \qquad \text{Multiply.}$$

One fifth of 4 is $\frac{1}{5} \times 4 = \frac{4}{5}$.

So, $\frac{1}{5}\%$ of 394 is about $\frac{4}{5}$.

Estimate the percent of the number.

1. $\frac{1}{2}\%$ of 108

2. $\frac{1}{4}\%$ of 789

3. $\frac{3}{5}\%$ of 495

4. $\frac{3}{4}\%$ of 62

5. $\frac{2}{5}\%$ of 610

6. $\frac{1}{8}\%$ of 1590

7. 0.7% of 247

8. 0.9% of 193

9. 0.15% of 1997

10. Explain how you would estimate $0.3\overline{4}\%$ of 2380.

Name ______________________________ Date __________

4.5 Puzzle Time

Why Do Fish Have Such Huge Phone Bills?

A	B	C	D	E	F
G	H	I	J	K	L
M	N				

Complete each exercise. Find the answer in the answer column. Write the word under the answer in the box containing the exercise letter.

21 LINE
80 WHEN
35 SEEM
300 ON
20 GET
3 NOT
30 THE

Estimate the percent of the number.

A. 18% of 42

B. 81% of 106

C. 29% of 27

D. 42% of 48

E. 97% of 288

F. 64% of 51

G. 33% of 70

H. 66% of 400

I. 53% of 198

J. 8% of 30

Use compatible numbers to estimate the percent of the number.

K. 71% of 52

L. 49% of 186

M. 24% of 97

N. 21% of 343

280 THEY
25 GET
100 CAN
9 FISH
95 TO
70 OFF
8 BECAUSE

Name________________________________ Date__________

Chapter 4

School-to-Work

For use after Section 4.4

Data Analyst

Data Analysts are used in finance, research, advertising, manufacturing, retail, and government to compile and track all sorts of data. A data analyst also needs to be able to review data to identify what is unusual, and determine why.

You work for a small tool manufacturing company. The company ships 1000 orders in one month. Of these, 968 orders arrive on time, 20 arrived one day late, and the rest of the orders arrived more than one day late.

1. What percent of the orders arrived on time?

2. What percent of the orders arrived late?

3. The advertising department makes a flier that claims more than 98% of orders arrive on time. They ask you if this is accurate. What do you say?

4. How many orders out of 1000 would need to move from "late" to "on time" for advertising to claim 99% "on time" delivery?

The company has 400 employees. The Human Resources department asks you to analyze some data.

5. This year 80 employees used more than a week of sick leave. What percent is this?

6. The company gets a discount on dental insurance if more than 78% of the people in the company sign up for the insurance. There are 320 employees that get dental insurance through the company. Will the company get a discount? Why or why not?

7. What other percentages might you need to find when keeping track of employee benefits?

Name ______________________________ Date __________

Chapter 4 Study Help

You can use a **summary triangle** to explain a concept.

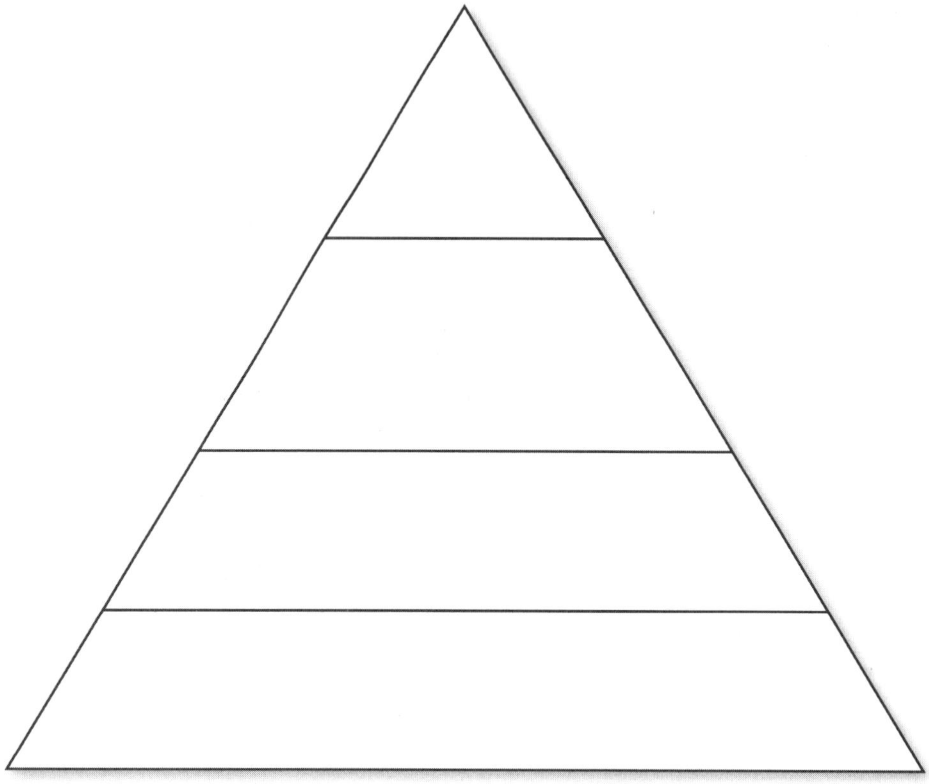

On Your Own

Make a summary triangle to help you study these topics.

1. writing a fraction as a percent
2. writing a percent as a decimal
3. writing a decimal as a percent
4. comparing and ordering fractions, decimals, and percents

After you complete this chapter, make summary triangles for the following topics.

5. finding the percent of a number
6. estimating the percent of a number

Name_______________________________ Date__________

Chapter 4

Financial Literacy

For use after Section 4.5

Tipping

A **tip** is a payment to a service worker. The tip is in addition to the advertised cost of the service (such as a haircut) or item (such as the meal served by a waiter or waitress). A tip is usually calculated as a percentage of the cost of the item or service.

Tipping varies between industries and between countries. People tip a higher percentage than the norm for exceptionally good service, and a lower percentage for poor service.

Tips are normally calculated using mental math. To avoid accidentally over- or under-tipping, you should learn to estimate a reasonable tip.

1. Tip amounts are based on 10%. Explain how to find 10% of the bill using mental math.

2. Apply your method to a restaurant bill of $36.57.

3. Explain how you would use your 10% value to estimate a 20% tip.

4. Explain how you would use your 10% value to estimate a 15% tip.

5. Using mental math, determine the range of values that would be a reasonable 15% to 20% tip on a $36.57 bill.

6. Research tipping customs in at least two other countries besides the United States. How do they compare?

7. Describe a situation in which an adult, such as a relative, would need to calculate the tip for an expensive service or item. Explain how you would help them calculate the tip.

8. Think of a sit-down restaurant in which you enjoy eating. Estimate how many tables the waitperson serves in an hour, and how much the bill would be for a typical table. How much could the server earn in tips?

9. At some restaurants the servers give a portion of their tips to the bussing staff, often 10% of their tips. Using the amount from Exercise 8, how much would the server pay the bussing staff?

10. Some restaurants require the server to tip part of their assumed tips, rather than actual tips. Suppose a server must give 2% of total sales to be split between the 2 busboys. Explain how to estimate this amount.

Name ______________________________ Date __________

Chapter 4 Cumulative Practice

Write the number as a decimal, percent, and fraction or mixed number in simplest form.

1. $\frac{3}{8}$

2. 2.25

3. 95%

4. 10%

5. 0.75

6. $3\frac{1}{2}$

7. 0.06

8. $\frac{4}{5}$

9. 115%

10. A shirt is on sale for 75% of the original price. What fraction of the original cost is the sale price?

Use a number line to order the numbers from least to greatest.

11. $\frac{1}{5}$, 0.21, $\frac{3}{10}$, 0.02, 2.1%

12. 56%, $\frac{11}{20}$, 0.567, $\frac{3}{5}$

13. $2\frac{7}{8}$, 280%, 2.87, 2.78

14. 98%, $\frac{99}{100}$, 0.97, $\frac{110}{99}$, 1

Evaluate the expression when $a = \frac{2}{3}$, $b = 2\frac{4}{5}$, and $c = \frac{9}{4}$.

15. $3a$

16. $a + b$

17. ac

18. $c + 10$

19. $b - c$

20. $b \div 7$

Find the percent of the number.

21. 20% of 300

22. 7% of 25

23. 180% of 360

24. 250% of 4

25. 12.5% of 16

26. 84% of 55

27. A tank can hold 10 gallons of water. The tank is 35% full. How much water is in the tank?

Estimate the percent of the number.

28. 408% of 976

29. 24% of 59

30. 9% of 99

31. 201% of 396

32. 41% of 789

33. 61% of 24

34. Texas is 268,601 square miles. Florida is about 24.5% the size of Texas. Estimate the number of square miles in Florida.

Chapter 5

Name ______________________________ Date __________

Chapter 5 Ratios, Rates, and Data Analysis

Dear Family,

Sports and games provide an opportunity to relax and have fun with our families and friends. The nature of competition gives us an opportunity to explore mathematics at the same time.

When we are competing, we are often thinking about how we are doing. Are we hitting the ball as well as we did last year? Are we running faster now than at the beginning of the season? Are we currently winning, or is our opponent? Even if we are only watching a game, many of us tend to obsess over our favorite player's and team's performance.

Spend some time with your student talking about your family's favorite sport or game. What kinds of "stats" are kept about the players and events? How does that help you understand the game? For example, you and your student might talk about the following.

- How are batting averages figured out in baseball and softball? What does this tell you about the next time your favorite player is at bat?
- What does the ratio of red pieces to black pieces tell you about how a game of checkers is going? Who's winning?
- How fast can you run a 100-meter sprint? Do you think you could run the same speed in the 200-meter or the 400-meter?

Next time your team is playing their big rival, ask your student how he or she could predict who will win. Do you think it matters more what each team's average score is, or what the win-loss ratio is for the two teams? What kind of information could help you decide which team is better?

Enjoy the game!

Nombre ______________________________ Fecha__________

Capítulo 5 Relaciones, Índices y Análisis de Datos

Estimada Familia:

Los deportes y juegos ofrecen una oportunidad para relajarse y divertirse con nuestras familias y amigos. La naturaleza de la competencia nos da una oportunidad para explorar las matemáticas a la vez.

Cuando estamos compitiendo, a menudo estamos pensando acerca de cómo vamos. ¿Estamos golpeando la pelota tan bien como lo hicimos el año anterior? ¿Estamos corriendo más veloz ahora que al inicio de la temporada? ¿Estamos ganando actualmente o lo está haciendo nuestro contrincante? Incluso si sólo estamos observando un juego, muchos de nosotros solemos obsesionarnos con el rendimiento de nuestro equipo y jugador favoritos.

Pase algo de tiempo con su estudiante conversando acerca del deporte o juego favorito de su familia. ¿Qué tipo de "estadísticas" se tienen sobre los jugadores y los eventos? ¿Cómo te puede ayudar eso a entender el juego? Por ejemplo, usted y su estudiante pueden hablar acerca de lo siguiente:

- ¿Cómo se calculan los promedios de bateos en béisbol y softbol? ¿Qué les dice esto acerca de la próxima vez que su jugador favorito tenga que batear?
- ¿Qué les dice la relación de pieza rojas con respecto a las negras acerca de cómo está yendo un juego de damas? ¿Quién está ganando?
- ¿Qué tan rápido pueden correr una carrera de 100 metros de velocidad? ¿Creen que pueden correr a la misma velocidad en carreras de 200 ó 400 metros?

La próxima vez que su equipo esté jugado con su gran rival, pregunte a su estudiante cómo puede predecir quién va a ganar. ¿Crees que importa más el puntaje promedio de cada equipo o la relación ganar-perder de los dos equipos? ¿Qué tipo de información podría ayudarte a decidir cuál es el mejor equipo?

¡Disfruten el juego!

Non ______________________________ Dat __________

Chapit 5 Rapò, To, ak Analiz Done

Chè Fanmi:

Espò ak match ba nou opòtinite rilaks ak amize nou avèk fanmi ak zanmi nou. Nati konpetisyon an ba nou yon opòtinite esplore matematik an menm tan.

Lè n'ap fè konpetisyon, nou souvan ap reflechi sou èske nou fè byen. Èske nou frape balon an menm jan nou te fè li ane pase? Èske nou kouri pi vit kounyeya pase nan kòmansman sezon an? Èske n'ap genyen aktyèlman, oubyen se advèsè nou an? Menm si nou sèlman ap gade yon match, anpil nan nou gen tandans obsede ak pèfòmans jwè ak ekip prefere nou.

Pase yon ti tan ak elèv ou a ap pale sou espò oswa match prefere fanmi nou. Ki kalite "estatistik" nou kenbe sou jwè yo ak evènman yo? Kijan sa ede nou konprann match la? Paregzanp, ou menm ak elèv ou a gen dwa pale sou bagay sila yo.

- Kijan yo jwenn mwayèn eskò baton nan bezbòl ak sòfbòl? Kisa sa di ou sou pwochen fwa jwè prefere ou a gen baton an?
- Kisa rapò pyon wouj sou pyon nwa a di ou sou kijan yon pati dam pwale? Kiyès k'ap genyen?
- Ak ki vitès ou kapab kouri yon kous 100 mèt? Èske ou panse ou ta kapab kouri ak menm vitès la nan 200 mèt la oswa 400 mèt la?

Pwochen fwa ekip ou ap jwe kont gran advèsè li a, mande elèv ou a kijan li ta kapab predi kiyès ki pwal genyen. Èske ou panse sa ki pi enpòtan se eskò mwayen chak ekip, oswa sa rapò viktwa-pèt la ye pou de ekip yo? Ki kalite enfòmasyon ta kapab ede ou deside ki ekip ki pi bon?

Pase bon match!

Activity 5.1 Start Thinking!

For use before Activity 5.1

A **ratio** is a comparison of two quantities using division. For example, $\frac{2}{3}$ is a ratio of 2 to 3.

How do ratios relate to mixing paint colors?

Hint: Think about trying to mix a color that is the exact same color as your friend has mixed.

Activity 5.1 Warm Up

For use before Activity 5.1

Write the fraction in simplest form.

1. $\frac{3}{9}$

2. $\frac{15}{60}$

3. $\frac{36}{72}$

4. $\frac{20}{28}$

5. $\frac{18}{30}$

6. $\frac{34}{51}$

7. $\frac{6}{16}$

8. $\frac{48}{80}$

Lesson 5.1 Start Thinking!
For use before Lesson 5.1

A soccer ball is made up of 12 pentagons and 20 hexagons. What is the ratio of pentagons to hexagons?

Write another ratio using things that you can find in your classroom.

Lesson 5.1 Warm Up
For use before Lesson 5.1

Write two equivalent ratios for the given ratio.

1. $\frac{4}{7}$

2. $\frac{2}{15}$

3. $\frac{2}{4}$

4. $\frac{20}{25}$

5. $\frac{12}{16}$

6. $\frac{18}{27}$

Name______________________________ Date__________

5.1 Practice A

Write the ratio in three ways. Explain what the ratio means.

1. flies to lizards

2. cars : trucks

3. $\dfrac{\text{notebooks}}{\text{pencils}}$

4. hamburgers to hot dogs

Write the ratio in simplest form.

5. $\dfrac{10}{15}$ **6.** $\dfrac{16}{8}$ **7.** 33 : 6 **8.** 12 to 40

Write two equivalent ratios for the given ratio.

9. $\dfrac{2}{10}$ **10.** $\dfrac{4}{16}$ **11.** $\dfrac{27}{9}$ **12.** $\dfrac{18}{6}$

13. 1 : 6 **14.** 3 to 4 **15.** 66 to 22 **16.** 25 : 60

17. There are 3 buses to carry 96 students on a field trip. Write the ratio of buses to students in simplest form.

18. There are 10 students at the chalkboard and only 6 pieces of chalk. Write the ratio of students to chalk in simplest form.

19. Four friends have 20 fluid ounces of milk to share. Write the ratio of persons to fluid ounces of milk in simplest form.

20. You buy an electric scooter. The first five times you charge the battery you are able to ride a total distance of 30 miles on the scooter. How many miles can you ride on one charge?

21. Each student taking a standardized test is given 3 pencils. Tell how many pencils are needed for each number of students.

a. 15 students **b.** 18 students **c.** 31 students **d.** 61 students

22. You walk six miles every two days. How many miles do you walk in one week? Explain how you got your answer.

Name ______________________________ Date __________

5.1 Practice B

Write the ratio in simplest form.

1. 42 to 18 **2.** 14 : 35 **3.** $\frac{55}{65}$ **4.** $\frac{90}{70}$

Write two equivalent ratios for circles to triangles.

5. 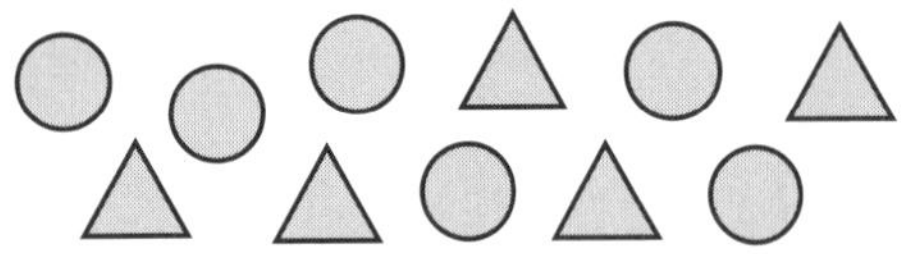

6. 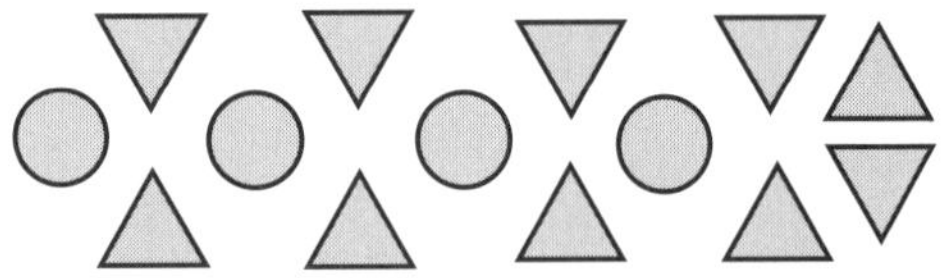

The ratios of wins to losses are given for two teams. Tell which ratio is greater.

7. 5 to 10, 2 to 5 **8.** 8 : 2, 6 : 3 **9.** 15 to 5, 24 to 10 **10.** 12 : 20, 5 : 8

11. Your boat needs a fuel ratio of 16 fluid ounces of oil to 6 gallons of gasoline. You plan to use 18 gallons of gasoline in the boat this weekend. How many pints of oil do you need? Explain how you got your answer.

12. A middle school band has 45 sixth and seventh graders. The ratio of sixth graders to seventh graders is 2 to 3. How many sixth graders are in the band? Explain how you got your answer.

13. Copy the rectangle shown. Using a reasonable number a, draw a second rectangle with a width and length that are each a times as long.

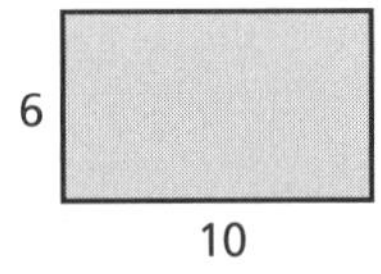

a. Find each ratio in each rectangle: width to length, width to perimeter, and length to perimeter.

b. Which ratios in part (a) are equal? Explain why this is true.

c. Find each ratio from the first rectangle to the second rectangle: width to width, length to length, perimeter to perimeter, area to area.

d. Which ratio in part (c) is different? Explain why this ratio is different.

14. The material in your favorite T-shirt is a blend of cotton and polyester. The ratio of cotton to polyester is 7 to 13.

a. You want to find the percent of the material that is cotton. Explain why finding a ratio equivalent to 7 to 13 doesn't help.

b. Write the ratio of cotton to the total material blend in the shirt.

c. Explain how you can use the ratio in part (b) to find the percent of the material that is cotton. Then find the percent that is cotton.

Name____ Date __________

5.1 Enrichment and Extension

The Golden Ratio

1. Which of the rectangles below do you think is most visually pleasing?

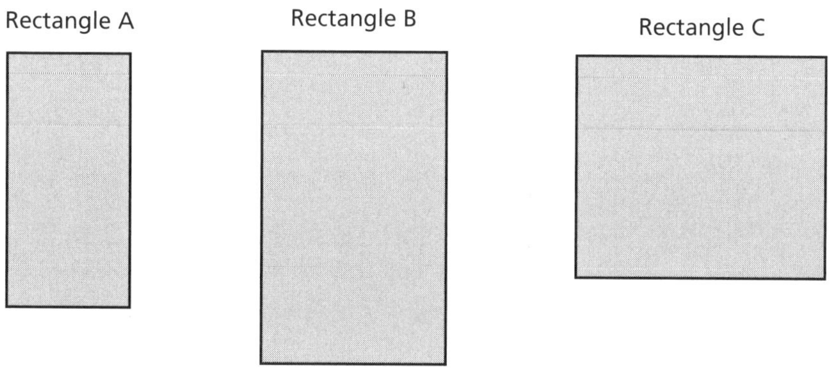

2. In Exercise 1, many people will choose Rectangle B. This rectangle is an example of a *golden rectangle*. The ratio of the length to the width in a golden rectangle is called the *golden ratio*. Use a ruler to approximate the golden ratio.

3. Consider the following pattern, called the *Fibonacci sequence*.

$$0, 1, 1, 2, 3, 5, 8, 13, 21, 34, 55, 89, \ldots$$

a. Describe the pattern.

b. Find the next three numbers in the pattern.

c. Starting at the third number in the pattern, find the ratio of each number to the previous number.

d. Write each ratio in part (c) as $a : 1$, where a is rounded to the nearest thousandth, if necessary.

e. The ratios are approaching the golden ratio. Make a new approximation of the golden ratio.

4. List several common rectangular objects that are close to the shape of a golden rectangle.

Name ______________________________ Date __________

5.1 Puzzle Time

Why Did The Bear Decide To Tiptoe Around The Campfire?

A	B	C	D	E	F
G	H	I	J	K	L
M	N				

Complete each exercise. Find the answer in the answer column. Write the word under the answer in the box containing the exercise letter.

$\frac{7}{8}$	WAKE
$\frac{1}{8}$	SLEEPING
3 to 7, $\frac{3}{7}$, 3 : 7	WANT
$1\frac{6}{25}$	OF
$\frac{2}{5}$	THE
1 to 2, $\frac{1}{2}$, 1 : 2	TO
3	ANY

Write the ratio in three ways.

A. circles to squares

□○□

□○○□

B. triangles to rectangles

△□□△□

□△△□□

C. $\dfrac{\text{times signs}}{\text{equal signs}}$

× × = = =

= = = = ×

D. $\dfrac{\text{arrows}}{\text{dollar signs}}$

\$ \$ \$ ↗

\$ \$ \$ ↗ ↗

Write the ratio in simplest form.

E. $\frac{35}{40}$

F. $\frac{21}{28}$

G. $\frac{27}{9}$

H. $\frac{31}{25}$

Write an equivalent ratio for the given ratio.

I. $\frac{3}{9}$

J. $\frac{2}{16}$

K. $\frac{6}{24}$

L. $\frac{14}{20}$

M. $\frac{10}{25}$

N. $\frac{15}{9}$

$\frac{1}{3}$	THE
$\frac{5}{3}$	FIRE
3 to 4, $\frac{3}{4}$, 3 : 4	SHE
$\frac{3}{4}$	UP
$\frac{1}{4}$	BAGS
2 to 3, $\frac{2}{3}$, 2 : 3	DIDN'T
$\frac{7}{10}$	AROUND

Activity 5.2 Start Thinking!

For use before Activity 5.2

What is the speed limit on your street?

A speed limit is an example of a ratio using two different units.

What are the two units in the speed limit?

What is another example of a ratio using two different units that you use in your day-to-day life?

Activity 5.2 Warm Up

For use before Activity 5.2

Complete the statement.

1. 1 hour = __?__ minutes

2. 1 week = __?__ days

3. 1 foot = __?__ yard

4. 1 foot = __?__ inches

5. 1 year = __?__ days

6. 1 year = __?__ weeks

Lesson 5.2 Start Thinking!

For use before Lesson 5.2

A grocery store has three options for your favorite hot breakfast cereal: single serving containers, a box of several single serving packets, and a large container. How can you decide which is the best buy for your money?

Lesson 5.2 Warm Up

For use before Lesson 5.2

Write a unit rate for the situation.

1. $60 earned in 4 hours

2. 110 miles in 2 hours

3. 14 cups in 7 servings

4. $16.20 for 4 pounds

5. 15 feet in 3 seconds

6. 24 degrees in 6 hours

Name____ Date __________

5.2 Practice A

Write a rate that represents the situation.

1. 50 candles in 4 boxes

2. 6 miles in 7 minutes

3. \$25 in 10 minutes

4. 52 students in 3 classrooms

5. 5 pounds in 3 months

6. 4 days for \$350

7. 10 kilograms in 9 cans

8. 87 strokes in 18 holes

Write a unit rate for the situation.

9. \$44 in 4 days

10. 12 haircuts in 4 hours

11. 256 heartbeats in 4 minutes

12. 15 liters in 3 minutes

13. 12 cans for 6 people

14. 27 outs in 9 innings

15. 85 drops in 5 minutes

16. 36 children from 12 families

Decide whether the rates are equivalent.

17. $\frac{17 \text{ heartbeats}}{15 \text{ seconds}}, \frac{68 \text{ heartbeats}}{60 \text{ seconds}}$

18. $\frac{96 \text{ miles}}{4 \text{ gallons}}, \frac{380 \text{ miles}}{15 \text{ gallons}}$

19. Describe and correct the error in writing the unit rate of gallons per hour.

✗

$$\frac{70 \text{ gallons}}{5 \text{ minutes}} \overset{\div 5}{=} \frac{14 \text{ gallons}}{1 \text{ hour}} \quad (\div 5 \text{ top and bottom})$$

20. You receive 9 text messages in 12 minutes. What is the rate of text messages per hour?

21. You studied for 14 hours last week. What was your unit rate of hours of study per day?

22. In a walk for charity, you walk at a rate of 100 meters per minute. How long does it take you to walk 2 kilometers?

23. A deli sells a sandwich spread for \$6.40 per pound. How much do you have to pay for 24 ounces of the spread?

24. You can buy 3 sandwiches for \$4.95 or 4 sandwiches for \$6.72. Which is the better buy?

25. You can buy 20 ounces of cereal for \$4.40 or 16 ounces of the same brand for \$3.68. Which is the better buy?

Name ______________________________ Date __________

5.2 Practice B

Write a rate that represents the situation. Then write a unit rate for the situation.

1. 6 kittens in 3 boxes
2. $96 for 16 hours of work
3. 72 biscuits from 9 batches
4. 1800 revolutions in 50 seconds

Decide whether the rates are equivalent.

5. $\frac{35 \text{ kilometers}}{25 \text{ minutes}}, \frac{14 \text{ kilometers}}{10 \text{ minutes}}$
6. $\frac{25 \text{ minutes}}{\$3}, \frac{1 \text{ hour}}{\$6}$

Decide which is the better buy.

7. $\frac{\$3.20}{16 \text{ ounces}}, \frac{\$5.04}{28 \text{ ounces}}$
8. $\frac{\$0.24}{6 \text{ fluid ounces}}, \frac{\$0.46}{10 \text{ fluid ounces}}$
9. $\frac{\$0.68}{340 \text{ milliliters}}, \frac{\$0.85}{500 \text{ milliliters}}$
10. $\frac{\$9.60}{1 \text{ dozen}}, \frac{\$14.76}{18}$
11. An aquarium is leaking water at a rate of three quarts per day. How many fluid ounces of water is this each hour?
12. A glacier flows at a rate of 20 meters per day. What is the flow rate in kilometers per year?
13. A teacher keeps track of how many books are read by students in each class. Which grade has read a higher rate of books per student? How many more books does the other grade need to read to have the same rate?

	Grade 6		Grade 7	
	Class A	**Class B**	**Class C**	**Class D**
Students	25	31	21	23
Books Read	181	155	116	126

14. Charles Lindbergh made the first solo airplane flight from New York to Paris. His flight covered about 3610 miles in 33.5 hours.

    a. Find the unit rate in $\frac{\text{miles}}{\text{hour}}$.

    b. Find the unit rate in $\frac{\text{hours}}{\text{mile}}$.

    c. Which is a better description of Lindbergh's rate, *about two miles per minute* or *about two minutes per mile*? Explain your reasoning.

Name______________________________ Date__________

5.2 Enrichment and Extension

How much does it cost to travel one mile?

Your aunt and uncle each fill their cars' gas tanks at the gas station whose sign is shown. Your aunt gets 18 gallons of regular unleaded. Her car averages 32 miles per gallon. Your uncle gets 12 gallons of premium unleaded. His car averages 30 miles per gallon.

Regular Unleaded	**3.09**
Mid-grade Unleaded	**3.19**
Premium Unleaded	**3.29**

1. Find your aunt's unit cost for each mile she drives.

2. Find your uncle's unit cost for each mile he drives.

3. Who is getting the better value for their vehicle? Explain.

4. Your uncle switches to mid-grade unleaded and his car still averages about 30 miles per gallon. How does his unit cost change? Is he getting the better value for his vehicle compared to your aunt? Explain.

5. Suppose over time, gas prices decrease to about 80% of their current prices. Does this change your answer to Exercise 3? Explain.

6. The United States uses about 146,000,000,000 gallons of gasoline each year.

   a. How many gallons does the United States use every day?

   b. Use compatible numbers to estimate how many gallons the United States uses every minute.

   c. Use compatible numbers to estimate how many gallons the United States uses every second.

Name ______________________________ Date __________

5.2 Puzzle Time

Where Does An Umpire Like To Sit When He Is Eating Dinner?

Write the letter of each answer in the box containing the exercise number.

Write a rate that represents the situation.

1. 45 soccer balls for 6 players
2. 3 skateboards for 4 friends
3. $2.80 for 5 apples
4. 12 first downs in 3 quarters
5. 35 text messages in 20 minutes
6. 10 laps in 60 minutes

Write a unit rate for the situation.

7. $45.00 for 9 pizzas
8. 24 friendship bracelets in 3 hours
9. 390 miles in 6 hours
10. 42 songs in 21 minutes
11. 180 oranges in 10 bags
12. 864 rubber bands in 8 bags

Decide whether the rates are equivalent.

13. $\frac{9 \text{ miles}}{3 \text{ hours}}, \frac{27 \text{ miles}}{6 \text{ hours}}$

14. $\frac{152 \text{ points}}{8 \text{ games}}, \frac{171 \text{ points}}{9 \text{ games}}$

Answers

T. $\frac{108 \text{ rubberbands}}{1 \text{ bag}}$

P. $\frac{35 \text{ text messages}}{20 \text{ minutes}}$

E. $\frac{45 \text{ soccer balls}}{6 \text{ players}}$

B. $\frac{12 \text{ first downs}}{3 \text{ quarters}}$

E. $\frac{8 \text{ friendship bracelets}}{1 \text{ hour}}$

T. $\frac{3 \text{ skateboards}}{4 \text{ friends}}$

I. $\frac{\$2.80}{5 \text{ apples}}$ **A.** $\frac{18 \text{ oranges}}{1 \text{ bag}}$

L. $\frac{\$5}{1 \text{ pizza}}$ **D.** $\frac{10 \text{ laps}}{60 \text{ minutes}}$

H. $\frac{65 \text{ miles}}{1 \text{ hour}}$ **N.** $\frac{2 \text{ songs}}{1 \text{ minute}}$

E. Yes **H.** No

4	1	9	3	10	6		12	13	8		5	7	11	2	14

Activity 5.3 Start Thinking!

For use before Activity 5.3

Write a word problem that involves a unit rate. Exchange problems with a classmate and solve your classmate's problem.

Activity 5.3 Warm Up

For use before Activity 5.3

Complete the statement.

1. 5 weeks = __?__ days

2. 2.5 feet = __?__ inches

3. 18 yards = __?__ feet

4. 48 hours = __?__ days

5. 11 years = __?__ months

6. 4 years = __?__ days

Lesson 5.3 Start Thinking!

For use before Lesson 5.3

How fast can you walk? Estimate your walking rate in feet per second.

Measure the length of a hallway in your school.

Have a classmate record the number of seconds it takes you to walk the entire length of the hallway.

Find your actual rate in feet per second. How close was your estimate?

Lesson 5.3 Warm Up

For use before Lesson 5.3

Find how far the object travels in the given amount of time.

1. A toy train travels 5 feet every 2 seconds. How far does it travel in 10 seconds?

2. A car travels 60 miles in 2 hours. How far does it travel in 3 hours?

3. A roller coaster travels 500 feet every 10 seconds. How far does it travel during a 140-second ride?

4. A cheetah can travel up to 300 feet in 3 seconds. How far can it travel in 10 seconds?

Name____ Date __________

5.3 Practice A

Find the distance.

1. $d = \underline{\ ?\ }, r = 10 \text{ ft/sec}, t = 6 \text{ sec}$

2. $d = \underline{\ ?\ }, r = 25 \text{ mi/h}, t = 3 \text{ h}$

3. $d = \underline{\ ?\ }, r = 14 \text{ m/day}, t = 12 \text{ days}$

4. $d = \underline{\ ?\ }, r = 72 \text{ ft/min}, t = 14 \text{ min}$

5. $d = \underline{\ ?\ }, r = 102 \text{ km/h}, t = 5 \text{ h}$

6. $d = \underline{\ ?\ }, r = 26 \text{ in./sec}, t = 44 \text{ sec}$

Find the speed.

7. 396 miles in 6 hours

8. 902 kilometers in 11 hours

9. 984 inches in 24 seconds

10. 4844 feet in 4 minutes

11. 1225 feet in 49 minutes

12. 320 yards in 16 seconds

Find how far the object travels in the given amount of time.

13. 15 days

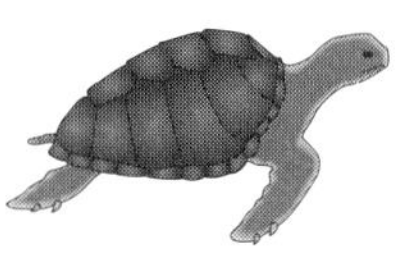

Moves 129 miles every 3 days

14. 64 seconds

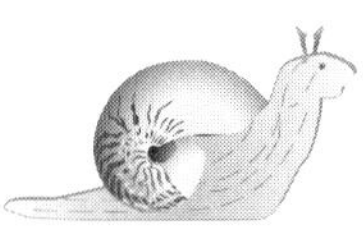

Moves 9 millimeters every 8 seconds

15. 40 seconds

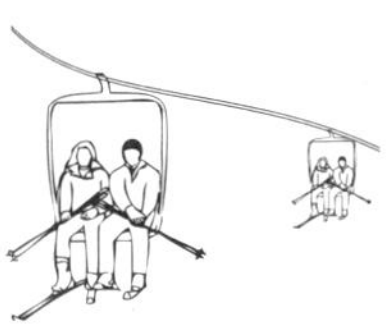

Moves 132 feet every 15 seconds

16. 8 hours

Moves 240 miles every 3 hours

17. Your friend has an electric wheelchair that can travel at a speed of 352 feet per minute. How far can your friend travel in 15 minutes?

18. Some people want to bring high speed "bullet trains" to Florida. A bullet train could take you 340 miles from Jacksonville to Miami in about 2.5 hours. At what speed would it travel?

19. At a grocery store, you buy four cans of soup for \$5. How much would you pay for six cans of soup?

20. At a fruit stand, you pay \$3.87 for three oranges. How much would you pay for five oranges?

21. It took you 20 minutes to jog 2 miles to your friend's house at a steady rate. How far could you jog in 30 minutes? Explain your method.

Name ______________________________ Date __________

5.3 Practice B

Find the distance or speed.

1. $d = \underline{\ ?\ }, r = 24 \text{ ft/sec}, t = 6 \text{ sec}$

2. $d = 275 \text{ mi}, r = \underline{\ ?\ }, t = 5 \text{ h}$

3. $d = 189 \text{ m}, r = \underline{\ ?\ }, t = 27 \text{ h}$

4. $d = \underline{\ ?\ }, r = 214 \text{ km/h}, t = 6 \text{ h}$

5. A bus leaves St. Augustine at 12:30 P.M. How fast should the bus drive to arrive at Ft. Lauderdale at 5:00 P.M.?

6. The total weight of 4 baseballs is about 20 ounces. How many baseballs weigh 10 pounds?

7. Each team in a 12-mile bicycle relay race can have 3 riders. Each rider is allowed to complete any part of the 12 miles. The table shows the greatest distance each rider on your team can cover, and the amount of time it takes.

	Distance (miles)	Time (minutes)
You	4	18
Your friend	6	36
Your cousin	5	24

What is the least amount of time your team will take to finish the race?

8. You and a friend are 2 miles apart. You start walking toward each other at the same moment. You travel 22 feet every 5 seconds. Your friend travels 32 feet every 8 seconds.

a. How close are you to each other after 10 minutes?

b. Will you and your friend meet in 20 minutes? Explain your reasoning.

9. The basic calling rate for your pre-paid cell phone is \$0.20 per minute. A second option is to pay \$6.99 per month to have a rate of \$0.10 per minute. A third option is to buy a chunk of minutes (good for only one month):

\$20 for 200 minutes	\$30 for 400 minutes	\$50 for 1000 minutes

Suppose that the least, the greatest, and the *typical* number of minutes that you use your cell phone in a month are 60 minutes, 275 minutes, and 100 minutes, respectively.

a. Find the best option and least cost for each number of minutes.

b. Choose an option to use each month. Explain your reasoning.

Name_______________________________________ Date__________

5.3 Enrichment and Extension

More Difficult Rate Problems

1. It takes a bus driver 2.5 hours to travel from St. Petersburg to Daytona Beach at a speed of 64 miles per hour. The bus driver travels 60 miles per hour on the return trip.

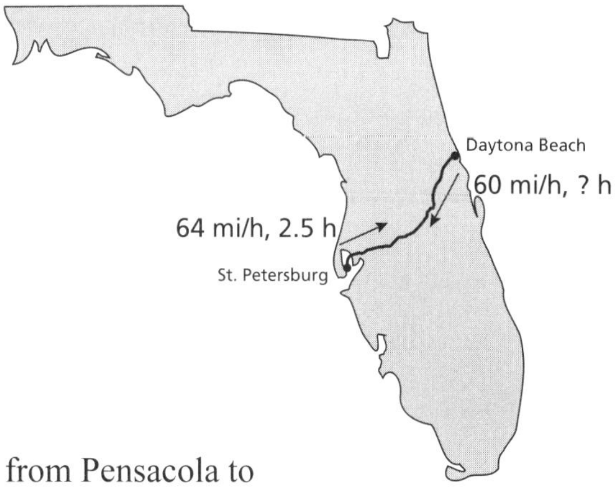

   a. What is the distance from St. Petersburg to Daytona Beach?

   b. How long did the return trip take?

2. A driver travels 57 miles per hour for 190 miles from Pensacola to Tallahassee. The driver travels 60 miles per hour for 160 miles from Tallahassee to Jacksonville.

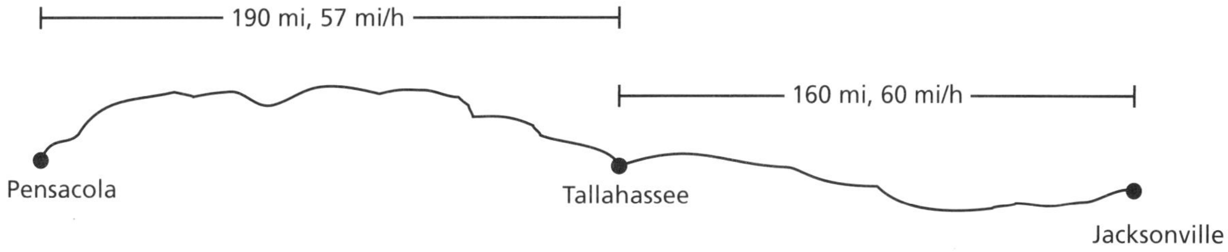

   a. What is the distance from Pensacola to Jacksonville?

   b. Which part of the trip took less time? By how many minutes?

   c. What is the speed for the entire trip?

3. Two buses travel 330 miles from Atlanta, GA to a championship game in Gainesville, FL. The speed of Bus A is 55 miles per hour. Bus B leaves 30 minutes after Bus A. The buses arrive in Gainesville at the same time.

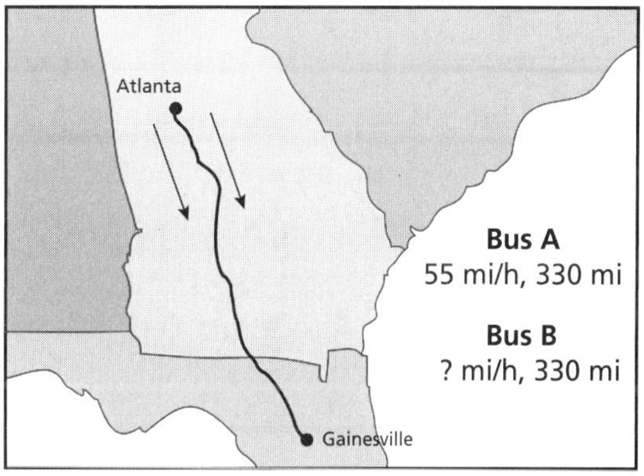

   a. How much time does it take Bus A to get to Gainesville?

   b. How much time does it take Bus B to get to Gainesville?

   c. What is the speed of Bus B?

Name ______________________________ Date __________

5.3 Puzzle Time

If All The Letters Of The Alphabet Were Invited To A Tea Party, What Letters Would Arrive Late?

A	B	C	D	E	F
G	H	I	J	K	L
M	N				

Complete each exercise. Find the answer in the answer column. Write the word under the answer in the box containing the exercise letter.

780 cm X
15 m COME
286 in. V
82 cm/sec BECAUSE
150 km LETTERS
65 mi/h AND
80 cm T

Find the distance.

A. $r = 4 \text{ mi/h}, t = 5 \text{ h}$

B. $r = 25 \text{ km/h}, t = 6 \text{ h}$

C. $r = 27 \text{ ft/sec}, t = 14 \text{ sec}$

D. $r = 11 \text{ in./min}, t = 26 \text{ min}$

E. $r = 45 \text{ m/h}, t = 3.5 \text{ h}$

F. $r = 60 \text{ cm/sec}, t = 13 \text{ sec}$

Find the speed.

G. 525 in. in 35 sec

H. 780 mi in 12 h

I. 2250 ft in 45 mi

J. 1804 cm in 22 sec

Find how far the object travels in the given amount of time.

K. 20 sec; A fish swims 2 feet every 4 seconds.

L. 25 sec; A fly moves 3 meters every 5 seconds.

M. 45 min; A plane travels 1250 kilometers every 9 minutes.

N. 12 sec; A balloon rises 20 centimeters every 3 seconds.

378 ft U
50 ft/min Z
20 mi THE
15 in./sec Y
6250 km AFTER
157.5 m W
10 ft THEY

Activity 5.4 Start Thinking!

For use before Activity 5.4

What comes to mind when you think of the word *average*? How does it relate to math?

Activity 5.4 Warm Up

For use before Activity 5.4

The bar graph shows the number of home runs hit by 10 players on the Tampa Bay Rays during the 2008 regular season.

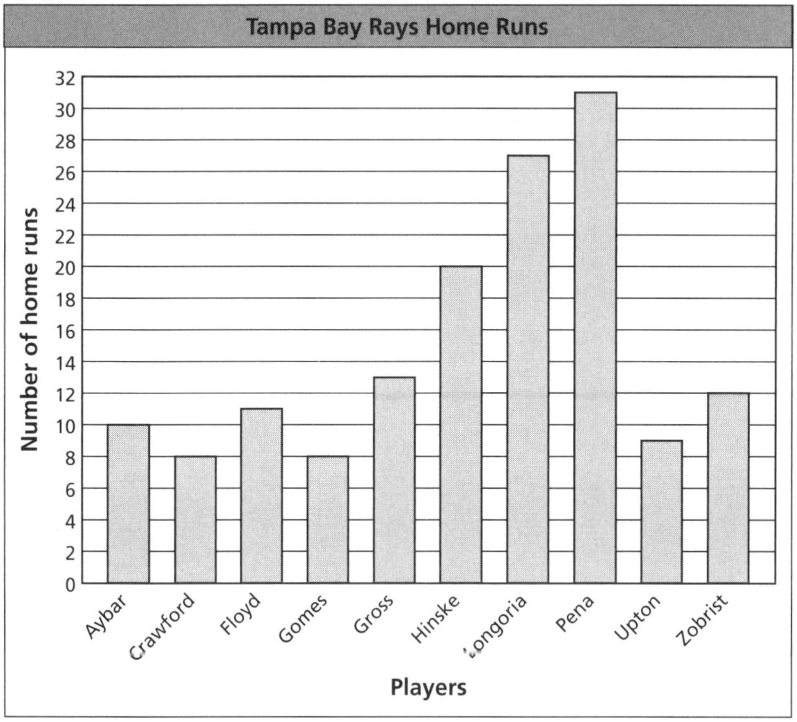

1. Which player hit the most home runs? How many did he hit?

2. Of the ten players shown, what percent hit more than 20 home runs?

3. Of the ten players shown, what percent hit fewer than 12 home runs?

Lesson 5.4 Start Thinking!
For use before Lesson 5.4

What do you think is the average height of all of the people in your class?

Measure everyone in your class (including your teacher). Round to the nearest inch and record everyone's height in inches.

Find the average by adding all of the heights and dividing by the number of people.

Was your guess close to the average?

Lesson 5.4 Warm Up
For use before Lesson 5.4

Describe the "average" of the data.

1. TV shows watched each day this week: 2, 2, 3, 4, 1, 0, 2

2. Points scored by each starter on a basketball team: 22, 18, 15, 6, 4

3. New messages in your inbox: 2, 5, 7, 2, 3, 2

4. Minutes spent getting ready for school: 45, 53, 65, 37, 29

Name____ Date__________

5.4 Practice A

Find the mean of the data.

1. 7, 5, 9, 6, 3

2. 8, 10, 15, 7, 18, 14

3. 66, 93, 76, 101, 88, 81, 72, 95

4. 31, 73, 82, 58, 44, 60

5.

<table>
<tr><th colspan="2">Movies Watched in Class</th></tr>
<tr><td>Language Arts</td><td>||||</td></tr>
<tr><td>Social Studies</td><td>𝍸 |</td></tr>
<tr><td>Math</td><td>|</td></tr>
<tr><td>Science</td><td>|||</td></tr>
<tr><td>Art</td><td>|</td></tr>
</table>

6.

<table>
<tr><th colspan="2">Students Absent</th></tr>
<tr><td>Monday</td><td>||||</td></tr>
<tr><td>Tuesday</td><td>|</td></tr>
<tr><td>Wednesday</td><td>|</td></tr>
<tr><td>Thursday</td><td>||</td></tr>
<tr><td>Friday</td><td>|||</td></tr>
</table>

7.

Marching Band Members	
Wilson M.S.	44
East M.S.	62
Central M.S.	75
Seminole M.S.	39
Gator M.S.	55

8.

Problems Completed	
Kara	18
Josh	12
Dana	13
Robert	8
Katie	19

9.

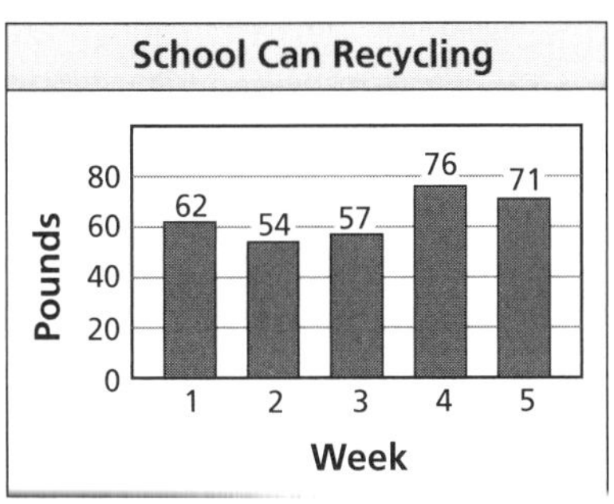

10.

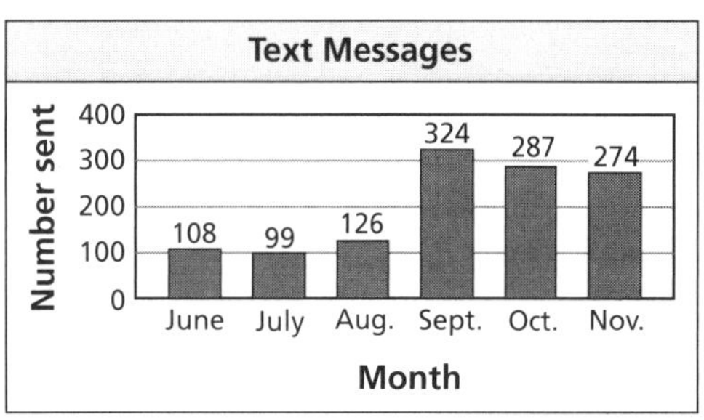

11. The throwing speeds (in miles per hour) of five 12-year old baseball pitchers are 55, 51, 48, 53, and 58. What is the mean throwing speed?

12. For ten school days, the numbers of bikes parked at a school bike rack are 10, 12, 8, 11, 13, 9, 2, 1, 9, and 12.

a. What is the mean number of bikes per day?

b. Describe two outliers for the data.

c. Describe how the outliers affect the mean.

13. You spend 100 minutes on homework each night from Monday through Friday. You spend 190 minutes on homework on Saturday. What is your mean daily number of minutes on homework?

Name ______________________________ Date __________

5.4 Practice B

Find the mean of the data.

1. 84, 89, 99, 78, 67, 98, 100, 71, 88

2. 234, 201, 311, 189, 404, 255, 276, 333

3. 0.34, 0.27, 0.11, 0.45, 0.32, 0.29, 0.32

4. 62.0, 62.8, 62.1, 62.8, 62.2, 62.3, 62.4, 62.8

5. 34, 101, 255, 87, 66, 75, 222

6. 11.2, 67.1, 8.5, 30.0, 44.4, 56.3, 38.8, 3.7

7. Compare your mean test scores for your science and math classes.

Science: 88, 96, 93, 99   Math: 91, 92, 96

8. The body temperatures (in degrees Fahrenheit) of some students are shown.

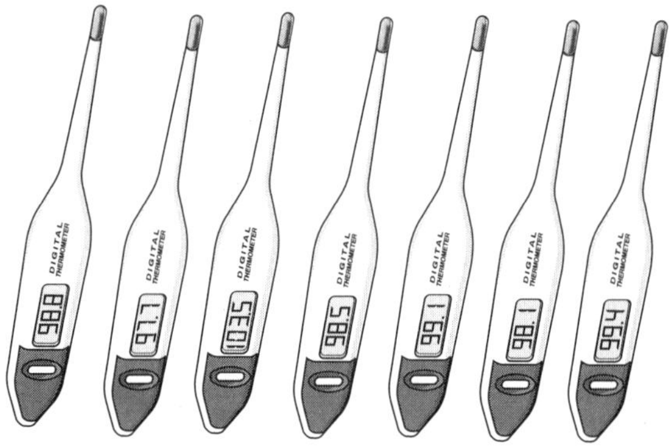

a. Which data value is an outlier? Explain.

b. Find the mean with and without the outlier. Then describe how the outlier affects the mean.

c. Describe a situation that could have caused the outlier in the problem.

9. The table shows the radii of the four planets closest to the Sun. Estimate the mean radius. Explain your method. Then find the actual mean.

Planet	Radius (km)
Mercury	2439
Venus	6052
Earth	6378
Mars	3398

10. The table shows the radii of the five best-known objects in the outer solar system.

Object	Radius (km)
Jupiter	71,492
Saturn	60,268
Uranus	25,559
Neptune	24,764
Pluto	1151

a. Which data value is an outlier? Explain.

b. Find the mean with and without the outlier. Describe how the outlier affects the mean.

c. Look up information about the outer solar system. Based on how the five objects are classified, do you think that an average of the data has more meaning *with* the outlier or *without* it? Explain your reasoning.

11. There are six boys and six girls in a class. The mean height of the girls is 59 inches and the mean height of the boys is 60 inches. One of the boys is 72 inches tall. Is the mean height of the other five boys greater than the mean height of the six girls? Explain.

Name________________________________ Date__________

5.4 Enrichment and Extension

Which Class Wins a Prize?

Students in three classes are collecting food to donate to charity. The number of pounds of food each student collected is shown in a table.

Mrs. Rodriguez's Class									
4.2	0	1.7	3.0	14.0	2.8	0	27.7	4.3	9.3
2.5	2.1	0	7.0	52.8	0	15.5	29.5	0	1.3
7.2	6.6	44.2	0	25.8	9.0	0	9.5		

Mr. Woo's Class									
2.6	0.7	0	3.8	10.5	50.0	14.2	8.5	0	6.3
7.5	1.8	0	14.3	0	0	43.1	57.5	12.8	78.1
0	22.8	5.5	0						

Mrs. Smith's Class									
2.6	11.5	1.6	14.2	30.3	7.2	3.6	0	10.2	15.2
9.8	0	20.2	3.2	11.4	23.5	4.5	18.4	8.2	48.6
18.6	10.0	6.3	10.2	0	1.5	13.5	4.0	22.9	10.8

1. Find the number of pounds of food collected for each class. The class with the greatest number of pounds of food collected wins one week of free lunches. Which class wins?

2. Find the mean number of pounds of food collected per student for each class. The class with the greatest mean number of pounds of food collected per student wins free yearbooks. Which class wins?

3. Each class also collected the amount of money shown in the table. The money buys 5-pound bags of potatoes for \$2.50 each.

Mrs. Rodriguez's Class	\$120
Mr. Woo's Class	\$80
Mrs. Smith's Class	\$65

   a. Add the weights of the potatoes to the food collection totals for each class.

   b. Do the winners in Exercises 1 and 2 change if the potatoes are included in the calculations? Explain.

Name ____ Date __________

5.4 Puzzle Time

What Is Really Easy To Get Into, But Really Hard To Get Out Of?

Write the letter of each answer in the box containing the exercise number.

Find the mean of the data.

1.

Number of Text Messages Per Day	
Jill	22
Dylan	15
Bill	18
Bella	20
Drew	10

2.

Number of Cats Owned	
Louise	|||
Ted	||
Mark	||
Alexis	|

3.

Number of TVs in Home	
Spencer	□□□□□
Megan	□□□□
Tyler	□□□
Ann	□□
Beth	□□□□□
Ashley	□□□□
Mike	□□□□□

4.

Number of Visits to the Stadium	Matt	Brady	Olivia	Ellie	Riley	Noah	Sam
	○○○○	○○	○○○○○	○○○	○○	○	○○○○

5. 12, 15, 18, 22, 25, 28

6. 2.6, 2.9, 3.2, 4.2, 5.6

Find the outlier of the data.

7. 60, 55, 65, 8, 57, 62

8. 2, 2, 3, 3, 4, 32

9. 11, 13, 13, 15, 15, 76

10. 18, 17, 1, 15, 19, 23

Answers

G. 4

O. 3.7

E. 76

I. 2

R. 20

B. 32

U. 17

L. 3

T. 8

B. 1

8	2	3		7	5	6	1	10	4	9

Activity 5.5 Start Thinking!

For use before Activity 5.5

Record the age of everyone in your class (including your teacher).

Then find the mean.

Is the mean a good representation of the average age of the people in your class?

Activity 5.5 Warm Up

For use before Activity 5.5

Find the mean of the numbers.

1. 5, 6, 2, 8, 3, 6

2. 12, 16, 16, 28, 50

3. 63, 68, 45, 13, 68, 31

4. 38, 2, 8, 45, 78, 23, 23

5. 80, 81, 82, 83, 84, 85, 86, 87

6. 4, 7, 12, 46, 12, 26, 18, 3

Lesson 5.5

Start Thinking!

For use before Lesson 5.5

The **mode** of a data set is the value that occurs most often.

Do you think it's possible for a set of data to have no mode? If so, give an example. If not, explain why not.

Do you think it's possible for a set of data to have more than one mode? If so, give an example. If not, explain why not.

Lesson 5.5

Warm Up

For use before Lesson 5.5

Find the mean of the data. Is the mean a good "average"? If not, what would be a better "average"? Explain.

1. 48, 49, 37, 41, 43, 44, 46

2. 18, 35, 46, 234, 65, 34

3. 14, 11, 10, 14, 15, 13, 14, 17

4. 6, 6, 6, 4, 5, 7, 6, 7

Name____________________ Date__________

5.5 Practice A

Find the median, mode(s), and range of the data.

1. 1, 3, 5, 6, 7, 9, 11

2. 2, 2, 6, 8, 10, 14

3. 12, 15, 10, 12, 21, 9, 12

4. 15, 22, 21, 22, 12, 16

5. 52, 61, 56, 55, 72, 64

6. 71, 51, 37, 44, 50, 49, 55

7. 91, 96, 89, 97, 89, 98, 94, 93

8. 55, 60, 60, 40, 60, 75, 40, 25

9. 110, 210, 160, 80, 70, 210, 160, 190

10. 1, 9, 21, 4, 9, 5, 32, 5, 7

11. Describe and correct the error in finding the median and mode of the data.

> ✗ **data:** 25, 28, 24, 22, 27, 24, 29, 26
>
> **ordered data:** 22, 24, 25, 26, 27, 28, 29
>
> ↑
>
> The median is 26.
> There is no mode.

Find the mode(s) of the data.

12.

Breeds of Dogs in a Pet Hotel		
lab	poodle	beagle
schnauzer	lab	pug
pit bull	boxer	pit bull
beagle	akita	poodle
boxer	pit bull	lab

13.

Students' Initials				
TC	AT	RC	CC	LD
DT	LB	TR	SS	TL
JG	RL	JC	MB	GR
FM	MS	DR	BB	JT
RG	PS	MJ	WS	FC

14. Can you find the range of the data in Exercises 12 and 13? Explain.

15. The data are the numbers of pancakes eaten by the contestants in a pancake eating contest. Find the mean, median, mode(s), and range of the data.

8, 16, 6, 12, 9, 15, 13, 9, 12, 9

16. The data are the numbers of hours the pancake-eating contestants waited before eating again. Find the mean, median, mode(s), and range of the data.

3, 5, 3, 4, 4, 8, 12, 6, 18, 3

Name ______________________________ Date __________

5.5 Practice B

Find the mean, median, mode(s), and range of the data.

1. 22, 32, 27, 21, 31, 29

2. 19, 30, 12, 27, 33, 13, 10, 23, 18, 27

3. 0.39, 0.52, 0.30, 0.33, 0.44, 0.42

4. 2.4, 1.6, 1.2, 2.4, 2.7, 1.9, 2.5, 2.1

5. $\frac{1}{3}, \frac{1}{2}, \frac{1}{6}, \frac{2}{3}, \frac{5}{6}, \frac{1}{2}$

6. $4\frac{1}{8}, 2\frac{1}{4}, 4\frac{3}{4}, 1\frac{1}{4}, 3\frac{1}{2}, 3\frac{5}{8}, 4\frac{1}{8}, 2\frac{3}{8}$

Find the mode(s) of the data.

7. A, B, D, C, D, B, A, B, C, B, A

8. X, Y, X, X, Y, Y, Y, X, X, Y, X, Y, X

9. Find the mean, median, mode, and range of the numbers of minutes the hall pass was used.

Rm. 112 Hall Pass		
Name	**Time out**	**Time in**
S. Brady	8:57	9:02
Jordyn	9:24	9:27
L. Loftus	10:09	10:20
A. Torchio	10:52	11:04
Alec F.	11:08	11:12
M. Quinn	1:31	1:55

10. In each of six basketball games, you made 50% of the free throws you attempted. Complete the table.

Game	Free throw attempts	Free throws made
1	8	4
2	4	?
3	6	?
4	4	?
5	2	?
6	6	?

a. Find the mean, median, mode(s), and range of your free throws attempted per game.

b. Find the mean, median, mode(s), and range of your free throws made per game.

c. How are your answers in part (a) related to your answers in part (b)? Explain why.

11. There are 15 players on your baseball team. Which set of the players' statistics is more likely to have a mode, their numbers of *hits* or *home runs*? Explain your reasoning.

12. Would the weights (in pounds) of the students in your school be likely to have a greater range than the weights (in pounds) of the cars in a convenience store parking lot? Explain your reasoning.

Name______________________________ Date__________

5.5 Enrichment and Extension

Using Measures of Central Tendency

1. The mean age of a family of five is 22 years. The median is 13, the mode is 42, and the range is 39 years.

   a. Find the ages of the five family members.

   b. Will the range of the ages always be the same throughout any given year? Explain.

2. The mean age of a family of seven is 23 years. The median is 16, the modes are 12 and 45, and the range is 35 years.

   a. Find the ages of the seven family members.

   b. Give a scenario that would result in the two given modes.

3. A Fish and Wildlife Conservation release three bald eagles that have been rehabilitated. Before release, they measure the wingspan (in feet) and weight (in pounds) of each eagle.

   - The mean of the wingspans is $6\frac{1}{4}$ feet, the median is 6, and the range is $1\frac{3}{4}$ feet.
   - The mean of the weights is 12.75 pounds, the median is 12.75, and the range is 5 pounds.

   a. What are the wingspans of the three bald eagles?

   b. What are the weights of the three bald eagles?

   c. How do you think you would complete the table that shows the wingspan and weight for each bald eagle? Explain.

	Wingspan (feet)	Weight (pounds)
Bald Eagle 1		
Bald Eagle 2		
Bald Eagle 3		

Name ______________________________ Date ________

5.5 Puzzle Time

What Should Always Go Up When The Rain Is Pouring Down?

Write the letter of each answer in the box containing the exercise number.

Find the median, mode(s), and range of the data.

1. 4, 6, 8, 10, 12, 4, 9
2. 13, 18, 15, 12, 15, 13
3. 94, 82, 95, 72, 90, 92, 95, 100
4. 43, 12, 35, 51, 18, 26, 31
5. 14, 35, 20, 30, 31, 14, 19, 6, 4
6. 50, 39, 35, 50, 44, 39, 53, 62

Find the mode of the data.

7.

Favorite Sport	
Soccer	Gymnastics
Baseball	Baseball
Swimming	Soccer
Football	Gymnastics
Gymnastics	Basketball
Golf	Lacrosse
Basketball	Golf
Hockey	Football

8.

Favorite Food	
Pizza	Pizza
Cheeseburger	Spaghetti
Apples	Cheeseburger
Spaghetti	Pizza
Hot dog	Spaghetti
Pizza	Apples
Cheeseburger	Hot dog

Answers

U. 19, 14, 31

F. Soccer

T. Apples

R. 14, 13 and 15, 6

P. Spaghetti

L. Gymnastics

A. 31, none, 39

B. 8, 4, 8

W. Football

L. 47, 39 and 50, 27

M. 93, 95, 28

E. Pizza

5	3	1	2	8	7	6	4

Activity 5.6 Start Thinking!

For use before Activity 5.6

Write a survey question in which the answer is a number.

Record all of your classmates' answers to the question.

Make a chart, table, or graph to display your data.

Activity 5.6 Warm Up

For use before Activity 5.6

Make a bar graph of the data.

1.

Siblings	Tally
0	||
1	||||
2	|||
3	
4	|

2.

Favorite Sport	Tally
Basketball	𝍸
Baseball	𝍸 ||
Hockey	|||
Football	𝍸 |||

3.

Languages Spoken	Tally
1	𝍸 𝍸
2	𝍸 ||
3	||

4.

Favorite Juice	Tally
Apple	𝍸
Orange	𝍸
Grape	𝍸 |||
Other	𝍸 |

Lesson 5.6

Start Thinking!

For use before Lesson 5.6

Use the results from your survey question in the Start Thinking! for use before Activity 5.6. Calculate the mean, median, and mode of the data. Which measure of central tendency best describes the data? Explain why.

Lesson 5.6

Warm Up

For use before Lesson 5.6

Make a tally chart and a bar graph of the data.

1. 4, 7, 6, 5, 5, 7, 6

2. 3, 1, 0, 5, 4, 3, 0

3. 5, 15, 15, 10, 5, 10

4. 3, 1, 8, 1, 5, 2, 13

Name____ Date__________

5.6 Practice A

Find the mean, median, and mode(s) of the data. Choose the measure that best represents the data. Explain your reasoning.

1. 8, 6, 1, 10, 7, 10, 11, 9

2. 66, 57, 66, 16, 2, 35, 66, 114

3. 6, 35, 8, 20, 6, 16, 7

4. 100, 31, 100, 100, 82, 92, 100, 11

Find the mean, median, and mode(s) of the data with and without the outlier. Describe the effect of the outlier on the measures of central tendency.

5. 99, 82, 95, 99, 8, 92, 85

6. 12, 10, 16, 10, 12, 55, 18

7. 74, 62, 74, 66, 75, 63, 20

8. 4, 29, 81, 3, 1, 22, 24, 2, 23

9. You sent the same email message to 10 of your friends. The numbers of hours it took them to reply were 1, 1, 1, 2, 2, 3, 4, 5, 5, and 25.

a. Find the mean, median, and mode of the data.

b. Which measure best represents the data? Explain your reasoning.

c. Which data value is an outlier? Explain your answer.

d. Predict how the mean, median, and mode would change if you omit the outlier in the data list.

10. The mean and median of the weight of a book in the stack are shown.

a. Which measure is more useful in finding the weight of a typical book?

b. Which measure is more useful in finding the total weight of the books?

c. Suppose you know how many books are in the stack. How could you find the total weight of the books?

Median: 3 lb

Mean: 5 lb

11. Look up the weather forecast for your area over the next seven to ten days. Record the predicted high and low temperatures for each day.

a. Find the mean, median, and mode(s) of the predicted daily high temperatures. Do the same for the predicted daily low temperatures.

b. Which measure best represents each set of data? Explain your reasoning.

Name ______________________ Date __________

5.6 Practice B

Find the mean, median, and mode(s) of the data. Choose the measure that best represents the data. Explain your reasoning.

1. 83, 68, 56, 100, 78, 94, 74

2. 55, 65, 49, 22, 56, 65, 61, 57

3. 104, 128, 72, 73, 75, 73, 119

4. 20.1, 13.4, 9.8, 21.3, 20.8, 19.1, 68.1, 22.6

Find the mean, median, and mode(s) of the data with and without the outlier. Describe the effect of the outlier on the measures of central tendency.

5. 72, 75, 26, 65, 72, 67, 71

6. 70, 85, 150, 70, 85, 65, 65, 85

7. Each of five children picks out a birthday present for their mother. They share the total cost of the gifts equally.

Gift prices
\$9.95
\$10.25
\$9.80
\$9.95
\$24.95

a. Find the mean, median, and mode.

b. Which measure best represents the typical price of a gift? Explain.

c. Which measure best represents each child's share of the cost? Explain.

8. A punter's first four punts in a football game are shown. After a fifth punt, the punter's mean was 30 yards. How long was the fifth punt? Explain how you found your answer.

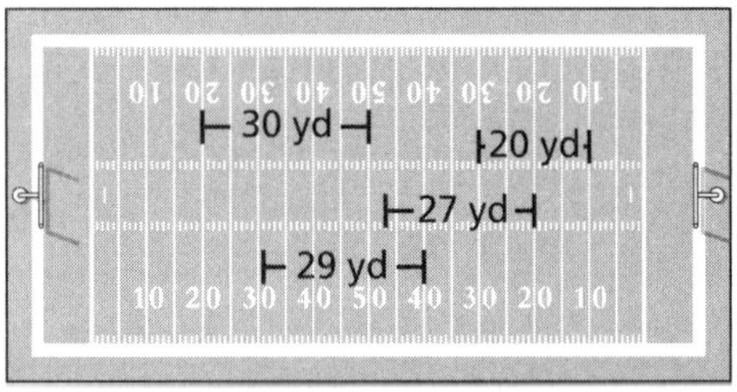

9. Each dot (•) shows the spot where a shot was taken by a team in a basketball game.

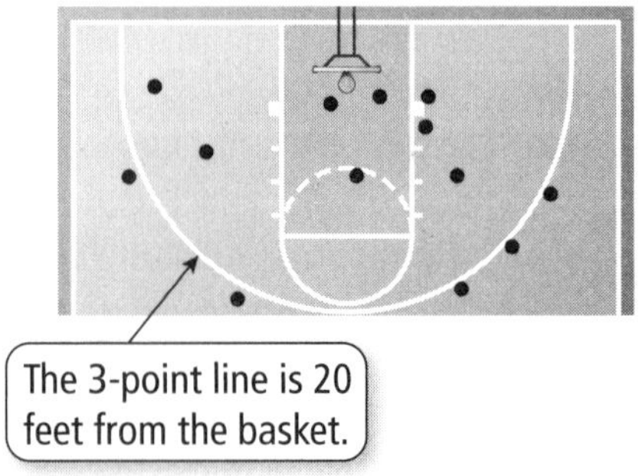

a. What is the mode of the distance from the basket for the shots?

b. Explain how the basketball rules caused this mode to occur.

c. How do the mean and median compare to the mode for the situation? Explain your reasoning.

Name____________________ Date__________

5.6 Enrichment and Extension

Analyzing a Data Set

Use the graphs that show the heights of players on three basketball teams.

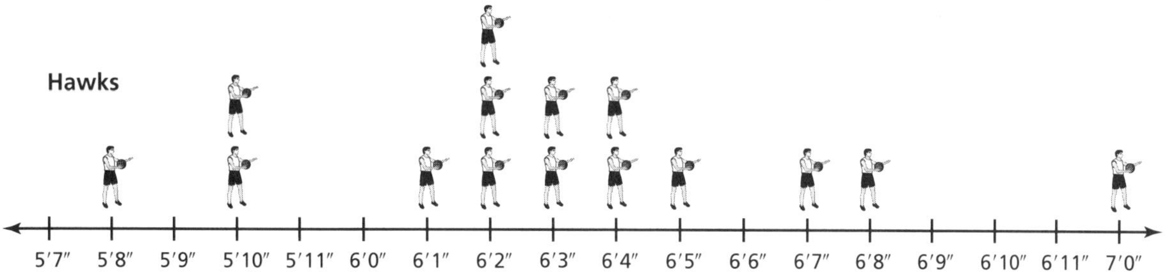

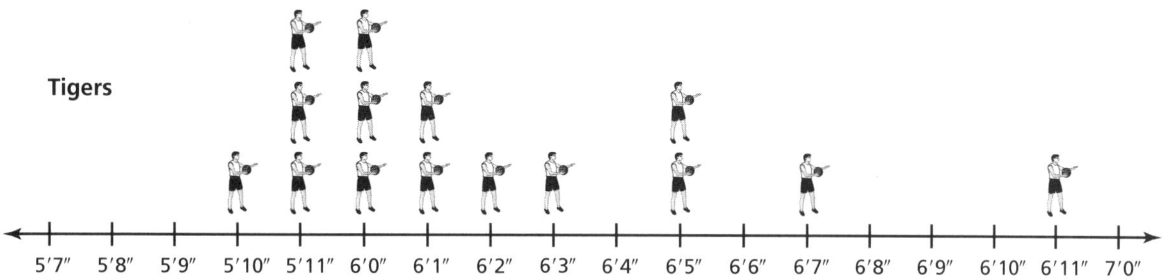

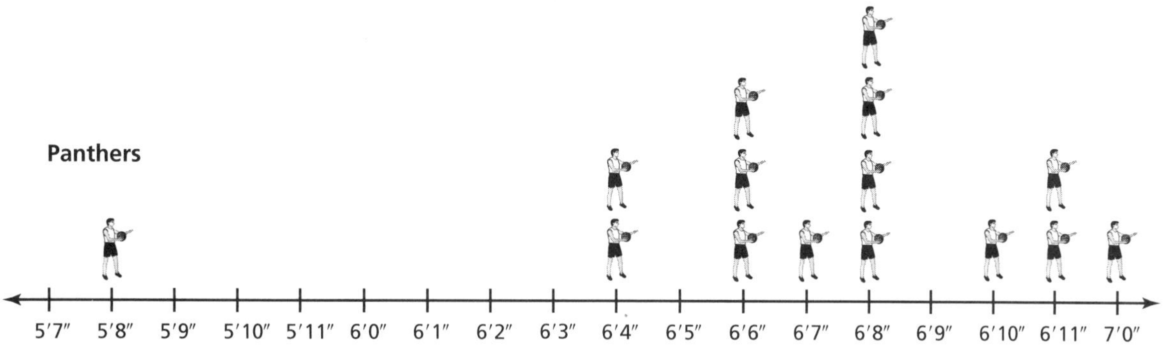

1. Without doing any calculations, estimate the mean height of each team. Explain how you found each estimate.

2. Calculate the actual mean height of each team. How close were your estimates in Exercise 1?

3. Find the mode(s) of each data set. How are these related to your answers in Exercise 1?

4. Find the range of each data set. Which two teams have the same range? Your friend concludes that because the ranges are equal, their players must be similar in height. Is your friend's conclusion accurate? Explain.

5. Describe the variability of each data set.

6. Are there any gaps in the data sets? If so, where do they occur? What could cause gaps in these data sets?

Name ______________________________ Date __________

5.6 Puzzle Time

What Fish Leaves Footprints On The Sea-bed?

Write the letter of each answer in the box containing the exercise number.

Find the mean, median, and mode(s) of the data.

1.

Penalty Shots		
3	1	6
2	7	1
5	3	8

2.

Stem	Leaf
4	4 4 4 7 8
5	2 4 6
6	1

Key: 5 | 2 = 52

3.

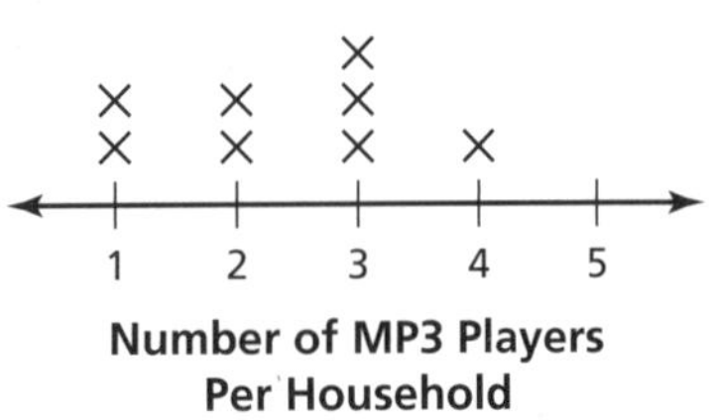

4.

Chapter 4 Test Scores		
95	82	98
91	94	40
	95	

5. Identify the outlier in the data set in Exercise 4. Find the mean, median, and mode(s) without the outlier.

Answers

L. mean = 50, median = 48, mode = 44

O. mean = 2.375, median = 2.5, mode = 3

S. mean = 4, median = 3, mode = 1, 3

E. mean = 92.5, median = 94.5, mode = 95

A. mean = 85, median = 94, mode = 95

4		1	3	2	5

Name______________________________ Date__________

Chapter 5

School-to-Work

For use after Section 5.6

Small Business Manager

When you run a business it's important to understand your customers. What do they want? How can you reach them? What determines what they buy? How can you keep customers?

Divide into groups of about 5 people. Pretend you are going to turn your classroom into a store during a large holiday fair held at the school. What would you like to sell? (Be creative. If you have an exciting idea for a store, use it, even if you wouldn't usually find that product in a fair.)

1. Make a list of questions about the classroom—the store space. You can use a survey to answer them, or do experiments. You may want to work as a class on this section. Some examples:

- When people enter the classroom, do they turn left, turn right, or walk straight ahead? Do an experiment and find the mode.
- How many people can comfortably fit in the classroom? Try some crowd levels (1 person per square yard, 2 people per square yard). Take a vote and find the mean and median.

2. Create a list of questions about your possible store. Give the survey to the other groups in the class, and answer their surveys. Then use the data they provide to help determine what you would sell, how much you would charge, and so on. Some sample questions:

- Would you buy this product?
- For this particular product (a sweatshirt, a CD, a brownie), what do you think is a reasonable price? What price is too expensive?
- How long would you wait in line to buy this product? For example, if you sell snacks, would people give up after 5 minutes? 15 minutes?

Name ____ Date __________

Chapter 5 Study Help

You can use a **definition and example chart** to organize information about a concept.

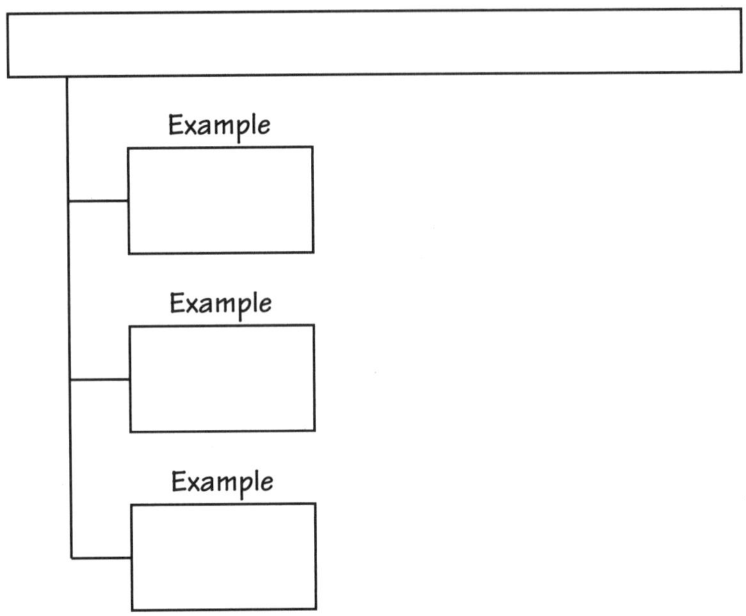

On Your Own

Make a definition and example chart to help you study these topics.

1. equivalent ratios

2. rate

3. unit rate

4. unit cost

After you complete this chapter, make definition and example charts for the following topics.

5. mean

6. outlier

7. measure of central tendency

8. median

9. mode

10. range

Name____ Date__________

Chapter 5

Financial Literacy

For use after Section 5.2

Unit Price

The unit price is the price for a single unit of a product. You can use the unit price to compare the prices of items sold in different sizes. For example, a store charges \$1.39 for a 12-ounce box of tri-color pasta and \$1.49 for a 1-pound box of plain pasta. The box of plain pasta costs more. However, you compare the unit price of the pasta to determine which one is the better buy.

Tri-color pasta

$$\frac{\$1.39}{0.75 \text{ lb}} \approx \frac{\$1.85}{1 \text{ pound}}$$

Plain pasta

$$\frac{\$1.49}{1 \text{ pound}}$$

The plain pasta costs less per pound, and is therefore the better buy.

Go to the grocery store and find 10 examples of an item that comes in different types and sizes. (For example dry pasta, shampoo, bottled salad dressing, crackers, drinks.) Don't use items that are already sold by a unit price, such as corn priced per ear or oranges priced per pound. Be sure to include different sizes.

1. Record the item, cost, and size in a table.
2. Decide on a unit, and calculate the unit rate for each item. Round to the nearest cent, if needed. (*Note*: 16 ounces = 1 pound.)
3. Which item has the highest unit rate? Which has the lowest? Would you have guessed that based on the price for each item as a whole?
4. The store may mark the unit price on the shelf for each item.
   a. Why do you think stores do this?
   b. Why would you still need to be able to calculate the unit price?
5. Suppose two grocery stores run the following promotions for a brand of shampoo that you like. Which one is the better buy? Show your work.

   Store A: A 13-ounce bottle of shampoo is normally \$3.99. This weekend, buy two and get the third bottle free!

   Store B: A 9-ounce bottle of shampoo is normally \$2.99; take 50 cents off!

Name ____________________ Date ________

Chapter 5

Technology Connection

For use after Section 5.1

Understanding Internet Sources

The Internet is a free and open global forum that allows people to access information quickly and easily; but the openness of the Internet means anyone can write their opinions as fact, misrepresent information, or present outright wrong information. To use the Internet correctly, you need to validate and evaluate the information you find.

The information in a website address contains *meta-web* information. *Meta* means "about" so you can tell a lot about the source of the information from the Internet address. Let's look at an Internet address:

The **edu** tells us this site is hosted by a school of higher learning (Union College). Schools in the United States include k12 in the address.

A **tilde (~)** tells us that this is a personal posting, not a posting by Union College. The **dpvc** tells us the name of the person who posted this information.

This shows the path to this page. Working backwards, homework policies is in a folder of all policies for the Fall 2008 semester for Math Course 053 during 2008-09. The main level is called "Courses."

http://**www.math.union.edu/** ~dpvc/**courses/2008-09/MTH053-FA08/policies/hw.**html

In Exercises 1–6, use the URLs.

http://www.nctm.org/

http://academic.reed.edu/math/faculty.html

http://www.k12.hi.us/~konawahs/homework_policy.htm

http://www.bigideaslearning.com/

http://www.ed.gov/students/homework/resources/edpicks.htm

http://people.reed.edu/~davidp/111/

1. Which site(s) are hosted by an institution of higher learning? What college hosts these sites? Which of these pages was created by the college? How do you know?

2. Which site(s) are hosted by a K-12 school? What state is the school in? What is posted at this site? Who posted this information?

3. Which site(s) are hosted by a government organization? What information would you find at this site?

4. Which site(s) are hosted by an organization? a company? How do you know?

5. Which sites represent information created by one person? If you wanted to learn more about the author of one of these pages, what could you do?

6. Look at some of the websites you use as sources of information. Do some provide more reliable information than others? Explain. Give some examples.

Name________________________________ Date__________

Chapter 5 Cumulative Practice

Write two equivalent ratios for the given ratio.

1. $\frac{7}{21}$ **2.** $\frac{3}{30}$ **3.** $\frac{1}{5}$ **4.** $\frac{4}{16}$

5. $\frac{18}{6}$ **6.** $\frac{3}{8}$ **7.** $\frac{13}{26}$ **8.** $\frac{7}{6}$

9. $\frac{9}{25}$ **10.** $\frac{16}{24}$ **11.** $\frac{35}{20}$ **12.** $\frac{21}{45}$

13. $\frac{18}{42}$ **14.** $\frac{12}{9}$ **15.** $\frac{16}{20}$ **16.** $\frac{39}{52}$

17. You use 30 pieces of fruit to make smoothies. The ratio of bananas to strawberries is 2 : 8. How many strawberries are used?

Write a unit rate for the situation.

18. 32 miles in 2 minutes **19.** $13.75 for 11 pounds **20.** 87 words in 3 minutes

21. 28 fish in 4 aquariums **22.** $1.50 for 10 grams **23.** 99 chairs in 3 rooms

24. You can buy 6 bottles of water for $3.60 or 30 bottles of water for $13.20. Which is the better buy?

Find the distance or rate.

25. Distance: ?
Rate: 45 m/sec
Time: 13 sec

26. Distance: 91 in.
Rate: ?
Time: 7 min

27. Distance: ?
Rate: 22 mm/h
Time: 15 h

28. You buy 8 boxes of pens for $25.60. What do 10 boxes cost?

Evaluate the expression.

29. $\frac{14 + 16}{2}$ **30.** $\frac{11 + 19 + 36}{3}$ **31.** $\frac{1 + 2 + 3}{3}$

32. $\frac{96 + 22}{2}$ **33.** $\frac{11 + 53}{2}$ **34.** $\frac{86 + 42 + 31}{3}$

Find the mean, median, and mode(s) of the data. Choose the measure that best represents the data. Explain your reasoning.

35. 2, 5, 8, 7, 13, 10, 15, 4 **36.** 70, 69, 70, 71, 30

37. 9, 2, 11, 10, 11, 11 **38.** 30.1, 30.2, 30, 30.4, 39.3

Name ______________________________ Date __________

Unit 2 Project: Race Course Rates

For use after Unit 2

Objective **Measure rates and analyze data**

Materials Masking tape, yardstick or ruler, stopwatch or clock with a second hand, calculator, (optional: camera to take pictures of people completing the course.)

Investigation Work with a partner. Use tape to mark a starting point on the floor or sidewalk. From the starting point, measure a straight line 10 feet long. Mark this distance with the tape. With the tape, label each 1-foot interval.

Think of 10 different methods you can use to "walk" the course. Some possibilities include spinning, walking backwards, crawling, hopping on one foot, etc. List the methods you choose in a table like the one below.

One partner should complete the course while the other partner measures the time it takes to complete the course to the nearest second. For each method, record the time it takes to complete the course in the table. Then calculate the unit rate for each method. Switch roles and repeat the process.

	Partner 1		Partner 2	
Method	**Time (sec)**	**Unit rate (ft/sec)**	**Time (sec)**	**Unit rate (ft/sec)**
walk toe-to-heel				
spin				
hop on one foot				
....				

Data Analysis

- For each method, what are some things that affect the time?
- Order the rates for each person from slowest to fastest. Which method took the most time? Which method took the least amount of time? Were the slowest and fastest methods the same for both people?
- Find the mean, median, mode, and range of the data for each person and for the combined data. Are there any significant differences? Explain.
- What measure of central tendency best represents each person's data? the combined data? Why?
- Identify any outliers in the data sets. Does an outlier affect the mean? the median? Explain.
- Use your own data and the mean of your data. What percent of your rates were slower than your mean? faster than your mean?

Summary Make a poster to summarize your results. Include sketches, the answers to the questions above, and any other patterns or observations you made. If you took photos of someone completing the course, include them.

Name____ Date__________

Unit 2 Student Grading Rubric

For use after Unit 2

		Student Score	Teacher Score
Poster Information 10 points			
	a. Name (4 points)	_____	_____
	b. Class (2 points)	_____	_____
	c. Project Name (2 points)	_____	_____
	d. Due Date (2 points)	_____	_____
Introduction 20 points			
	a. Explain how you performed the experiment and collected the data, so that someone not in the class would understand what you did. (20 points)	_____	_____
Data Table 30 points			
	a. Data table is present, completely filled out, and is neat and easy to read. (10 points)	_____	_____
	b. Rates are calculated correctly. (20 points)	_____	_____
Analysis 100 points			
	a. Rates are ordered correctly and the fastest and slowest rates are correctly identified. (10 points)	_____	_____
	b. Means are correctly calculated. (10 points)	_____	_____
	c. Medians are correctly calculated. (10 points)	_____	_____
	d. Modes are correctly calculated. (10 points)	_____	_____
	e. Ranges are correctly calculated. (10 points)	_____	_____
	f. Analysis of parts (b) (e). (20 points)	_____	_____
	g. Outliers identified and discussed. (10 points)	_____	_____
	h. Percents calculated and discussed. (20 points)	_____	_____
Poster 50 points			
	a. The poster includes a brief explanation of how you gathered the data, your table, an analysis of your data, visuals, answers to the questions above, and other observations and comparisons. (35 points)	_____	_____
	b. Present the information in a neat and organized way. (15 points)	_____	_____
FINAL GRADE		_____	_____

Unit 2

Teacher's Project Notes

For use after Unit 2

Materials Blue painting tape may work better than masking tape. You will need a stopwatch, watch, or wall clock that measures to the nearest second, enough for each pair of students. You may want to give students a calculator to make the calculations easier.

Alternatives Work in larger groups, or as an entire class. Create a larger data set by timing many more methods of completing the course, perhaps a different one for each student.

Or choose two or three methods and measure the time for every student to complete the course using each method. Perform data analysis for each method: for example, find the mean, median, mode, and range of the rates for rolling along the course, for toe-to-heel walking, and for hopping. Then compare the analyses for different motions—e.g. which one has the greatest range? Which has the least?

Working with a large group, you could create an obstacle course and measure each student's time to complete the course. It would also be instructive to measure each student's time on 2 trials, and compare the two means—does practice help?

If you have the space, you can create a longer course. If you have both slow methods (heel-to-toe walk) and fast (sprint), the range of the rates will be greater. A shorter course encourages clustering of times and rates.

Common Errors Watch for students to set up the fraction properly to find the rates: $\frac{10 \text{ ft}}{\text{time}}$.

The longest time will have the smallest rate, and the shortest time will have the greatest rate.

Many data sets will not have a mode.

Suggestions Marking each foot along the track will make it easier for students to estimate the rate, e.g. 1 foot in 2 seconds while hopping.

Before students calculate the percents, you may want to discuss what percent of data students think will be above the median (50%) versus above the mean (could be 50%, or more, or less). Then compare students' predictions with their actual results.

Unit 2 Grading Rubric
For use after Unit 2

Scoring Rubric	
A	179-200
B	159-178
C	139-158
D	119-138
F	118 or below

Poster 10 points

a. Name (4 points)

b. Class (2 points)

c. Project Name (2 points)

d. Due Date (2 points)

Introduction 20 points

a. Explain how you performed the experiment and collected the data, so that someone not in the class would understand what you did. (20 points)

Data Table 30 points

a. Data table is present, completely filled out, and is neat and easy to read. (10 points)

b. Rates are calculated correctly. (20 points)

Analysis 100 points

a. Rates are ordered correctly and the fastest and slowest rates are correctly identified. (10 points)

b. Means are correctly calculated. (10 points)

c. Medians are correctly calculated. (10 points)

d. Modes are correctly calculated. (10 points)

e. Ranges are correctly calculated. (10 points)

f. Analysis of parts (b)–(e). (20 points)

g. Outliers identified and discussed. (10 points)

h. Percents calculated and discussed. (10 points)

Poster 50 points

a. The poster includes a brief explanation of how the data was gathered, a completed table, an analysis of the data, visuals, and answers the questions above. (35 points)

b. The information is presented in a neat and organized way. (15 points)

Chapter 6

Name____________________________________ Date__________

Chapter 6 Circles and Area

Dear Family,

Many families enjoy making crafts. Sewing, scrap-booking, landscaping, gardening, and doing woodwork are just a few activities that use mathematics to form pleasing shapes.

When you lay out a circular garden in a landscape, you cut the edging to match the circumference of the circle. Then you curve the edging around the circular plot to define the area. The area of the circle is used to figure out how much mulch, plants, and flowers you will need.

Similarly, basting the edge of a placemat involves figuring out the length of the edge. For a placemat with semicircles on each end, you use the circumference to find the lengths around the curve. You then add those lengths to the sides of the rectangular portion in the middle.

Find a craft to enjoy with your student. Ask your student to help with the following.

- Draw a plan of your project. You can use combinations of geometric shapes (like the rectangle and circle in the placement above) to help find the measurements.
- Find the parts of the project that use measurements around the perimeter and then make a list of the lengths of each edge. Use these lengths to plan for outlining and edging material.
- Find the parts of the project that fill in an area and make a list of the areas for each part. You will use these lengths to plan for interior material, such as the amount of fabric or wood.

As you work with your student, ask him or her whether to cut material exactly to the measurements or leave a little extra—different crafts have different needs. Work out strategies for changing your plans to match the materials you have readily available.

You and your student will have a lasting record of your work together—have fun showing it off to your friends and family!

Nombre ______________________ Fecha ________

Capítulo 6 Círculos y Área

Estimada Familia:

Muchas familias disfrutan haciendo manualidades. Costura, álbumes de recortes, diseño de jardines, jardinería y trabajos en madera son solo algunas de las actividades que usan las matemáticas para hacer formas agradables.

Cuando se dispone un jardín circular en un diseño, se cortan los bordes para hacer coincidir la circunferencia del círculo. Luego se curvan los bordes alrededor del plano circular para definir el área. Se usa el área del círculo para averiguar cuánta composta, plantas y flores se necesitarán.

De forma similar, al hacer el hilván de un mantel individual se averigua el largo del borde. Para un individual con semicírculos en cada extremo, se puede usar la circunferencia para hallar los largos alrededor de la curva. Luego se pueden sumar esos largos a los lados de la porción rectangular en el medio.

Busque una manualidad que disfrute hacer con su estudiante. Pida a su estudiante ayuda con lo siguiente:

- Dibujen un plano de su proyecto. Pueden usar combinaciones de formas geométricas (como el rectángulo y círculo en la disposición anterior) para ayudar a encontrar las medidas.
- Encuentren las partes del proyecto que usan las medidas alrededor del perímetro y luego hagan una lista de los largos de cada borde. Usen estos largos para planear el delineado y el material del reborde.
- Encuentren las partes del proyecto que encajan en un área y hagan una lista de las áreas para cada parte. Usarán estos largos para planear el material interior, como por ejemplo la cantidad de tela o madera.

A medida que va trabajando con su estudiante, pregúntele si debe cortar el material exacto según las medidas o si debe dejar un poquito extra—las diversas manualidades tienen necesidades diferentes. Diseñen estrategias para cambiar sus planos y hacer que sus materiales se adapten a los materiales que ya tiene disponibles.

Usted y su estudiante tendrán un registro duradero de su trabajo juntos—diviértanse mostrándolo a sus amigos y familiares!

Non ____________________ Dat __________

Chapít 6 Sèk ak Sifas

Chè Fanmi:

Anpil fanmi renmen fè travay manyèl. Kouti, ranje albòm, amenajman peyizaj, jadinay, ak travay bwa se sèlman detwa aktivite ki sèvi ak matematik pou bay bèl fòm.

Lè w'ap desine yon jaden won nan yo peyizaj, ou koupe bòdi a pou fè li koresponn ak sikonferans sèk la. Apre sa ou koube bòdi a toutotou trase koub la pou defini sifas la. Sifas sèk la sèvi pou jwenn ki kantite pay, plant, ak flè w'ap bezwen.

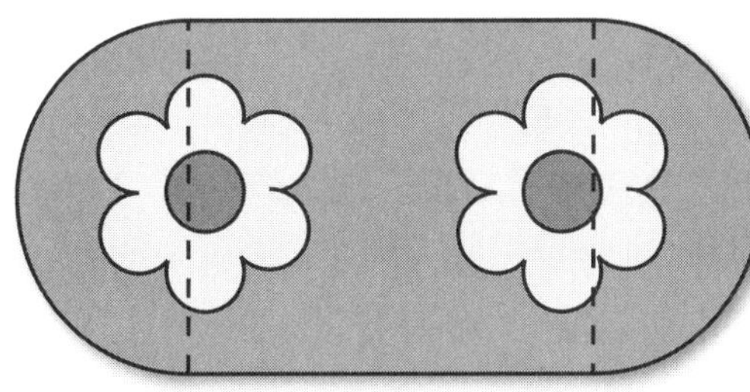

Menm jan an, fofile bòdi yon soupla enplike jwenn longè bòdi a. Pou yon soupla ak demisèk nan chak bò, ou sèvi ak sikonferans la pou jwenn longè yo toutotou koub yo. Apre sa ou ajoute longè sa yo sou kote pòsyon rektang ki nan mitan an.

Jwenn yon travay manyèl pou pran plezi ak elèv ou a. Mande elèv ou a ede ak bagay sila yo.

- Desine yon bouyon plan pou pwojè nou an. Nou kapab sèvi ak konbinezon fòm jeyometrik (tankou rektang ak sèk ki nan soupla ki pi wo a) pou ede jwenn mezi yo.
- Jwenn pati nan pwojè a ki sèvi ak mezi toutotou perimèt la epi fè yon lis longè chak bò. Sèvi ak longè sa yo pou planifye materyèl pou kontou ak bòdi.
- Jwenn pati nan pwojè a ki ranpli yon sifas epi fè yon lis sifas pou chak pati. Nou pwal sèvi ak sifas sa yo pou planifye materyèl enteryè, tankou kantite twal oswa bwa.

Pandan w'ap travay ak elèv ou a, mande li si pou nou koupe materyèl la ak mezi egzak yo oubyen si pou nou kite yon ti diplis—diferan travay manyèl gen diferan bezwen. Tabli estrateji pou chanje plan nou pou yo ka ale ak materyèl nou gen ki disponib deja yo.

Ou menm ak elèv ou a ap gen souvni dirab travay nou fè ansanm nan—amize nou fè zanmi ak fanmi nou admire li!

Activity 6.1 Start Thinking!
For use before Activity 6.1

Pi, written π, is the ratio of a circle's circumference to its diameter. The digits of π do not repeat and continue on without end.

"Piems" are poems that can help you remember π to a certain number of digits. The length of each word in a "piem" represents a digit of π. For example:

Piem: **Can I find a trick recalling pi easily?**

$\pi \approx$ **3 . 1 4 1 5 9 2 6**

Write you own "piem" to represent at least the first five digits of π (3.1415).

Activity 6.1 Warm Up
For use before Activity 6.1

Find the perimeter of the regular polygon.

1.

2.

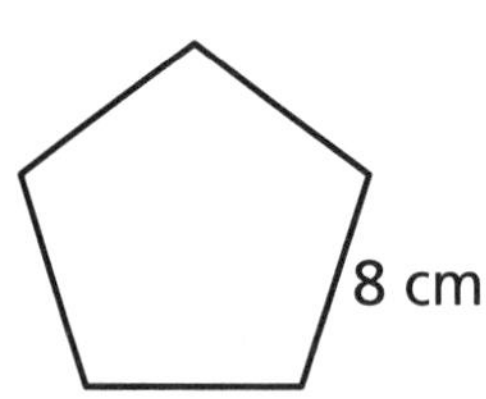

3.

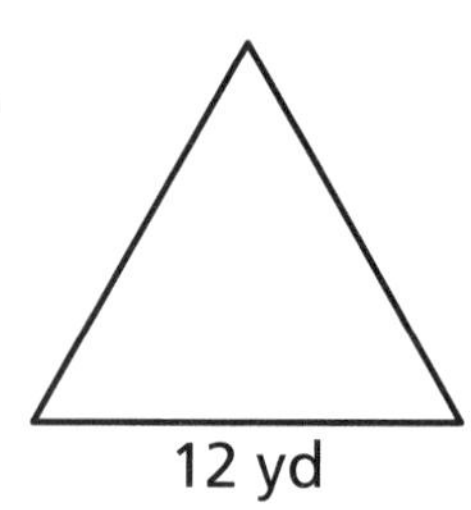

4.

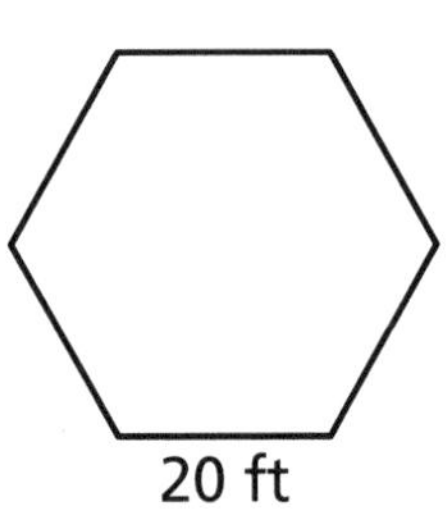

5.

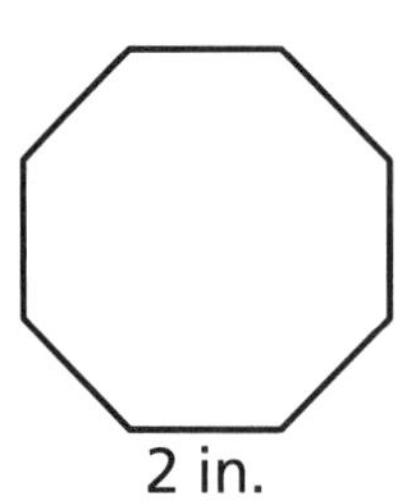

6.

Lesson 6.1 Start Thinking!
For use before Lesson 6.1

Is it easier to measure the diameter or circumference of a tree trunk? Why?

Is it easier to measure the diameter or circumference of a quarter? Why?

In general, when is it easier to measure the diameter of a circular object? When is it easier to measure the circumference?

Lesson 6.1 Warm Up
For use before Lesson 6.1

Find the circumference of the circle. Use 3.14 for π.

1.

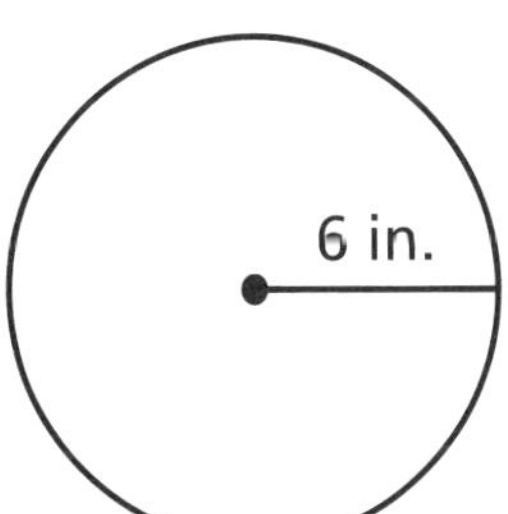

2.

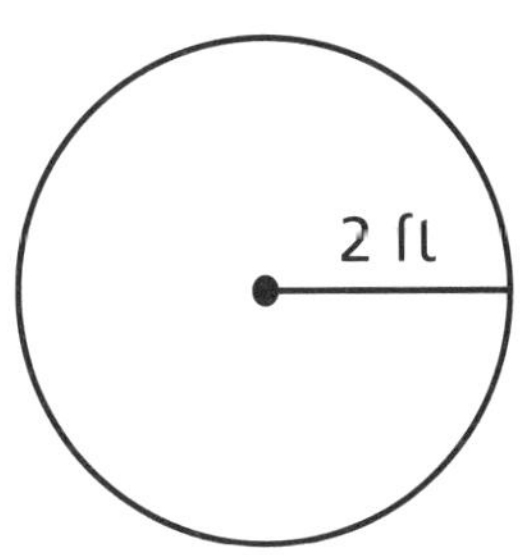

3.

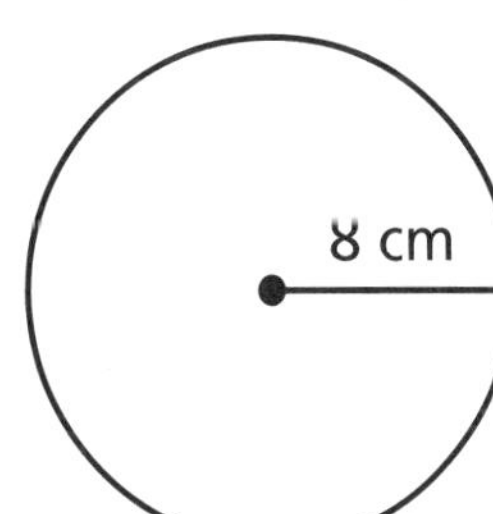

4.

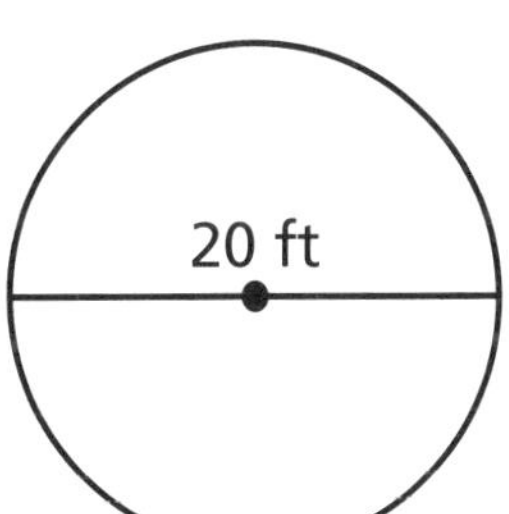

5.

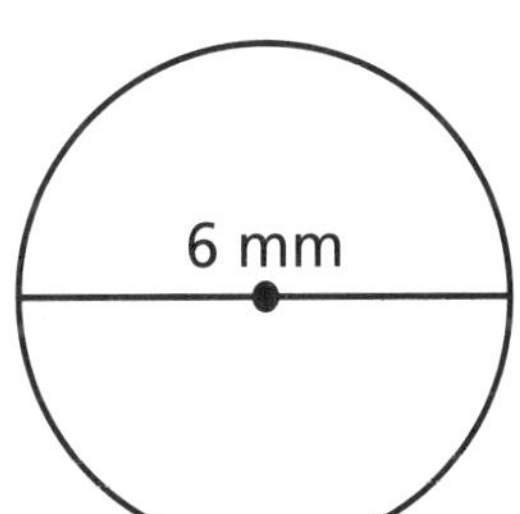

6.

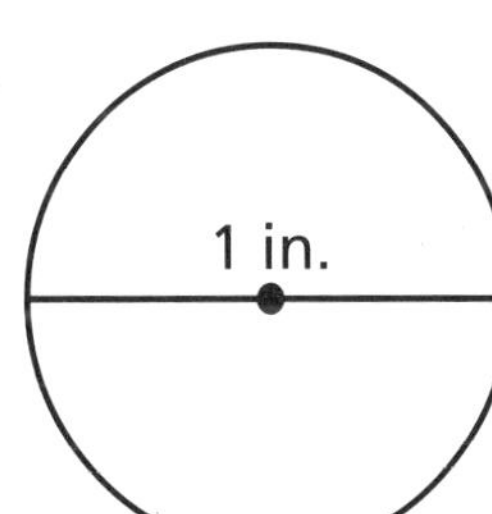

Name ______________________________ Date __________

6.1 Practice A

Find the diameter of the circle.

1.
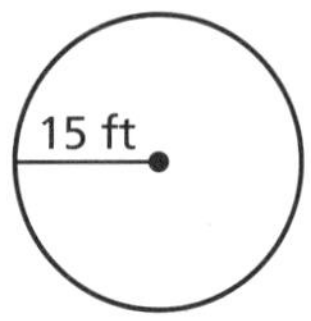

2.
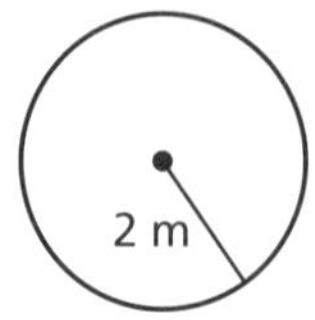

3.

Find the radius of the circle.

4.
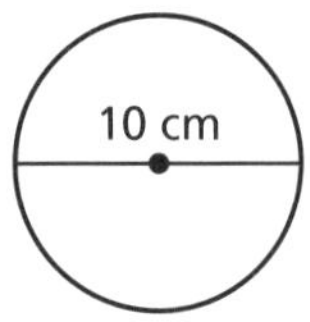

5.

6.
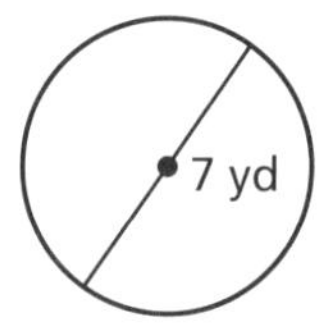

Find the circumference of the circle. Use 3.14 or $\frac{22}{7}$ for π.

7.
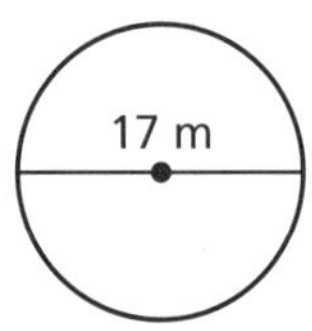

8.
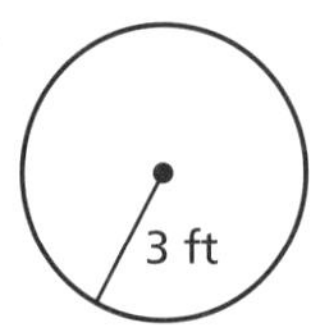

9.
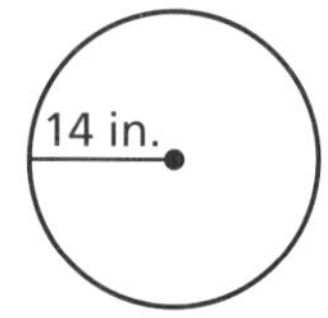

Find the perimeter of the semicircular region.

10.
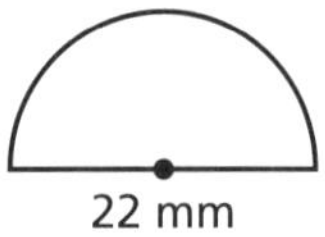

11.
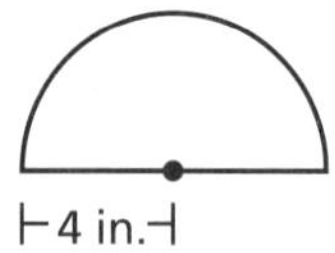

12.
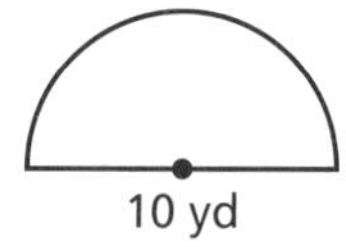

13. Describe and correct the error in finding the circumference of the circle.

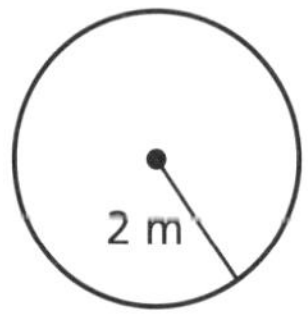

✗
$$C = \pi d$$
$$= 3.14 \bullet 2$$
$$\approx 6.28 \text{ m}$$

14. A United States quarter has a diameter of 24.26 millimeters. Find the circumference to the nearest millimeter.

15. You are enclosing a circular flower garden with a fence that costs \$2.99 per foot. The radius of the garden is 7 feet. How much will it cost to buy the fence? $\left(\text{Use } \frac{22}{7} \text{ for } \pi.\right)$

16. Find the perimeter of the semicircular tabletop shown at the right.

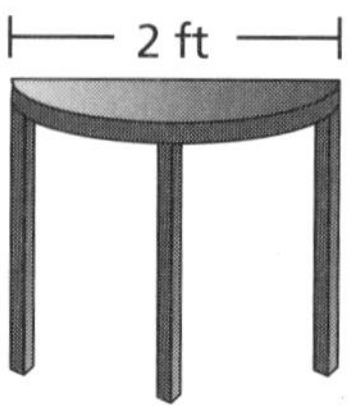

Name______________________________ Date__________

6.1 Practice B

Find the circumference of the circle. Use 3.14 or $\frac{22}{7}$ for π.

1. 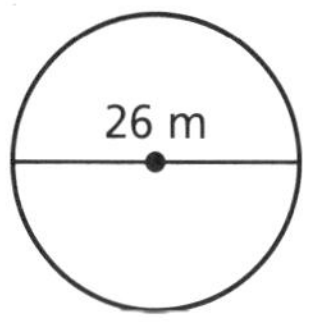

2. 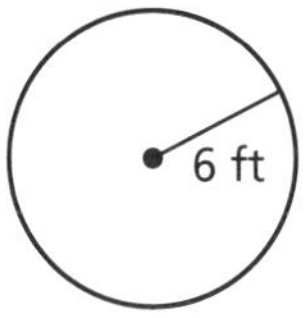

3. 

Find the perimeter of the semicircular region.

4. 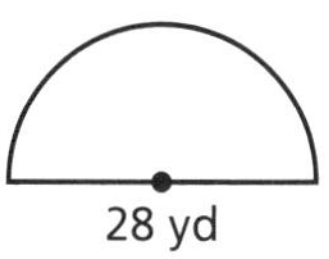

5. 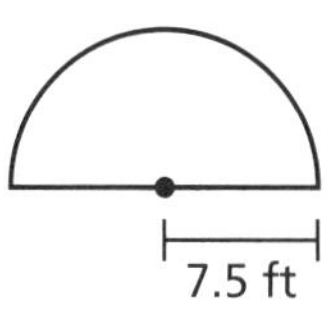

6. 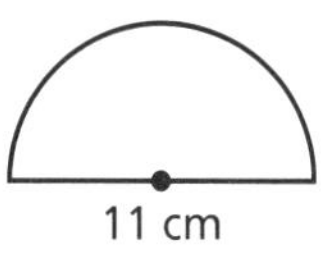

7. Copy and complete the table for circles *A*, *B*, *C*, and *D*.

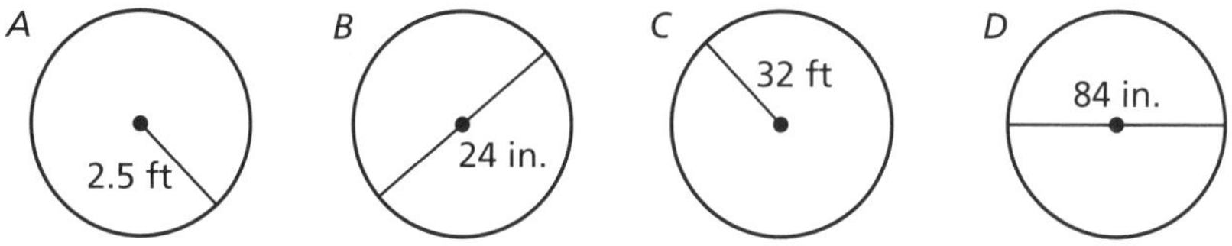

Circle	*A*	*B*	*C*	*D*
Radius	2.5 ft	? ft	32 ft	? ft
Diameter	? in.	24 in.	? in.	84 in.

8. A coaster has a diameter of 4 inches. Suppose the same amount of coaster is visible around the bottom of a glass as shown. What is the circumference of the glass?

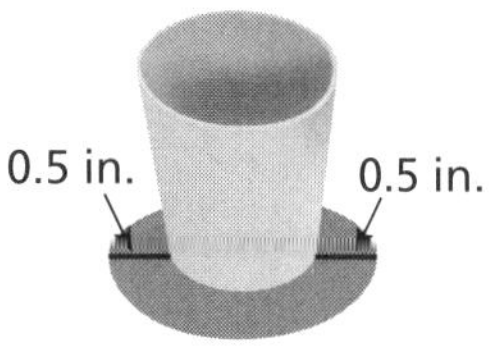

9. Is the side length of the squares in diagram A and diagram B equivalent? Explain your reasoning?

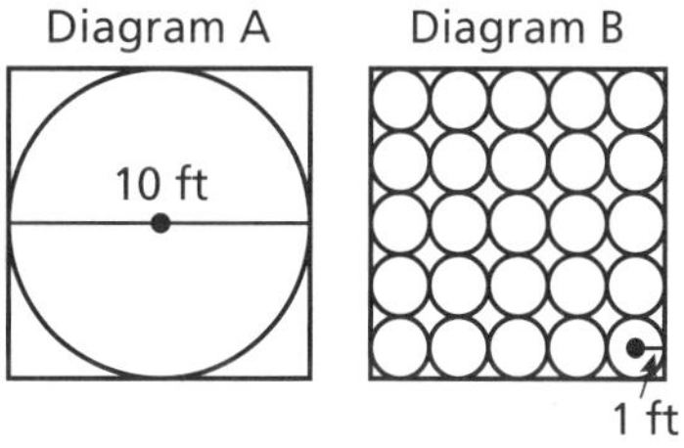

10. You release a ball with a radius of 1 inch into a pipe as shown. How many times will the ball rotate before it falls out of the other end of the pipe?

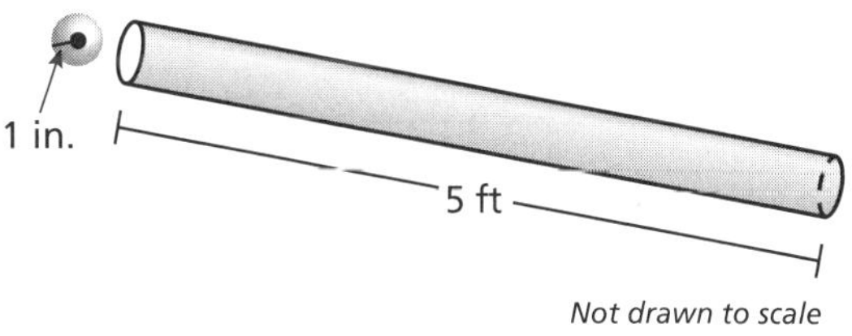

Name ______________________ Date __________

6.1 Enrichment and Extension

Changing Dimensions

Find the circumference of the circle. Then find the circumference if the radius is multiplied by 2. Use 3.14 for π.

1.

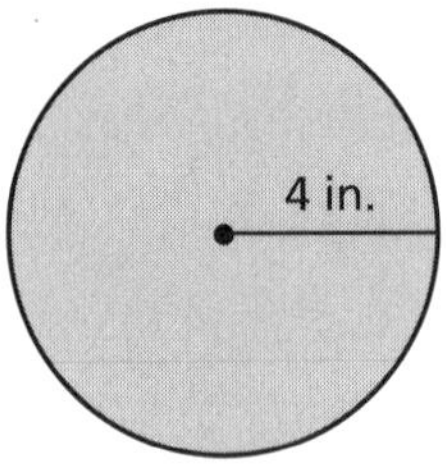

2.

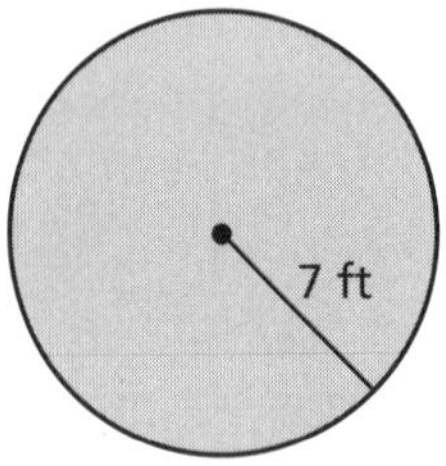

3.

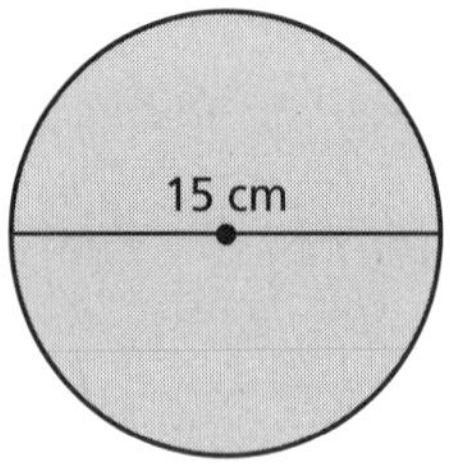

4. What happens to the circumference of a circle when its radius is multiplied by 2?

5. What happens to the circumference of a circle when its radius is multiplied by a positive number n?

Find the perimeter of the semicircle. Then find the perimeter if the radius is multiplied by $\frac{1}{2}$. Use 3.14 for π.

6.

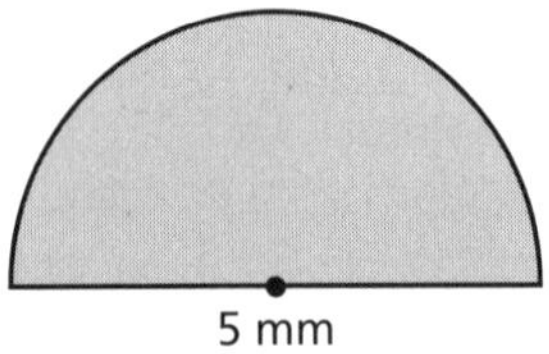

7.

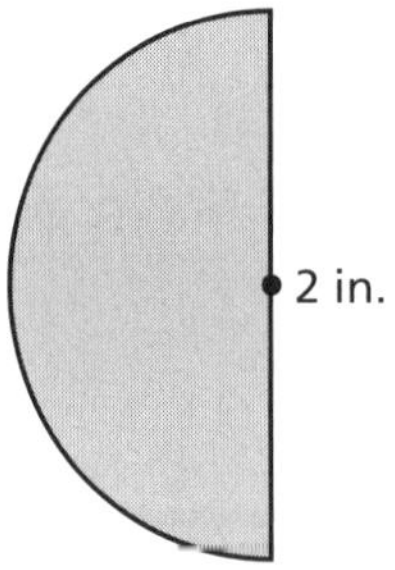

8.

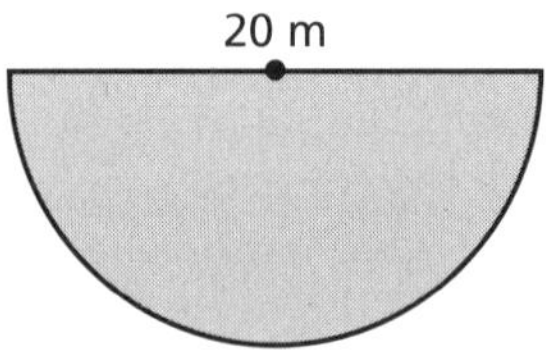

9. What happens to the perimeter of a semicircle when its diameter is multiplied by $\frac{1}{2}$?

10. What happens to the perimeter of a semicircle when its radius is multiplied by a positive number n?

Name____ Date__________

6.1 Puzzle Time

Why Did The Golfer Wear Two Pairs Of Pants?

A	B	C	D	E	F
G	H				

Complete each exercise. Find the answer in the answer column. Write the word under the answer in the box containing the exercise letter.

62.8 mm GET
15 m BECAUSE
44 cm A
37.68 in. MIGHT

A. The diameter of a circle is 30 meters. Find the radius.

B. The radius of a circle is 11 millimeters. Find the diameter.

Find the circumference of the circle. Use 3.14 or $\frac{22}{7}$ for π.

C.

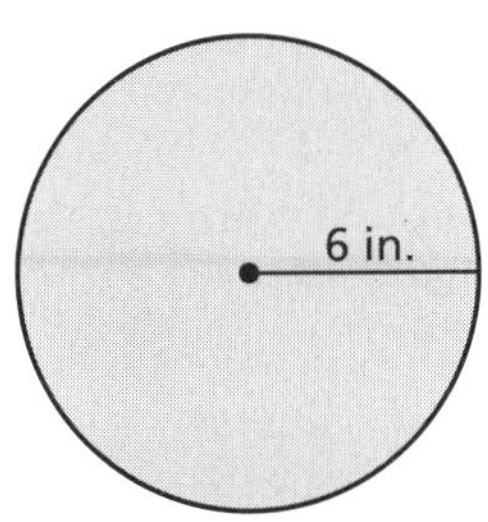

D.

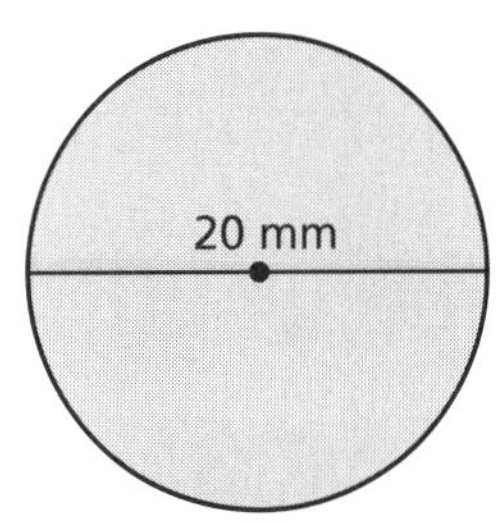

E.

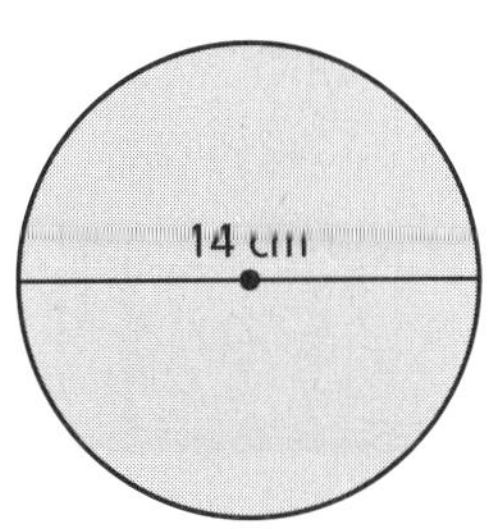

F.

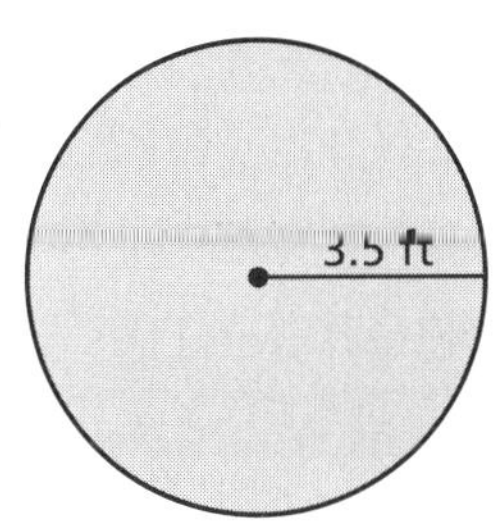

Find the perimeter of the semicircle. Use 3.14 for π.

G.

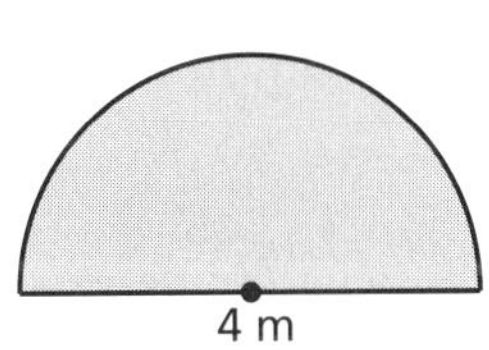

H.

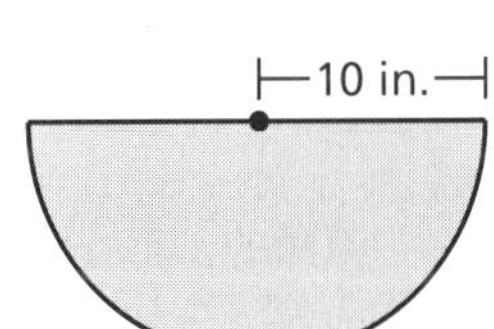

51.4 in. ONE
22 ft HOLE
10.28 m IN
22 mm HE

Activity 6.2 Start Thinking!
For use before Activity 6.2

Have you ever used a map to find the distance you need to travel to get from one place to another?

How is using a map similar to finding the perimeter of an irregular shape?

Activity 6.2 Warm Up
For use before Activity 6.2

Find the perimeter or circumference of the figure described.

1. square with side length 4 cm

2. rectangle with length 5 ft and width 3.5 ft

3. rectangle with length 19 in. and width 7 in.

4. triangle with side lengths 7 m, 8 m, and 10 m

5. circle with radius 9 in.

6. circle with diameter 20 ft

Lesson 6.2 Start Thinking!

For use before Lesson 6.2

Is a race track an example of a composite figure? Why or why not?

What are some other objects that are composite figures?

Lesson 6.2 Warm Up

For use before Lesson 6.2

Each square on the grid paper is 1 square inch. Estimate the perimeter of the figure.

1. 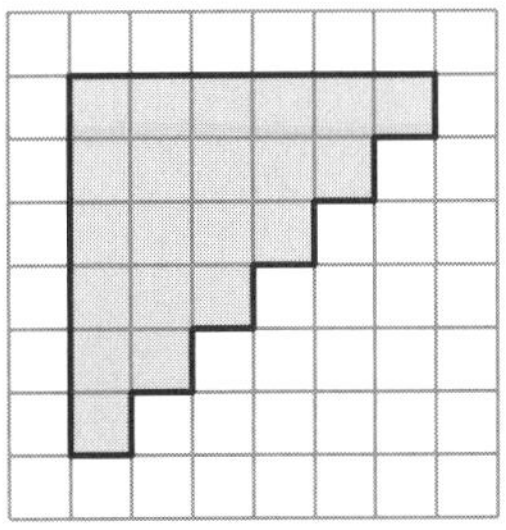

2. 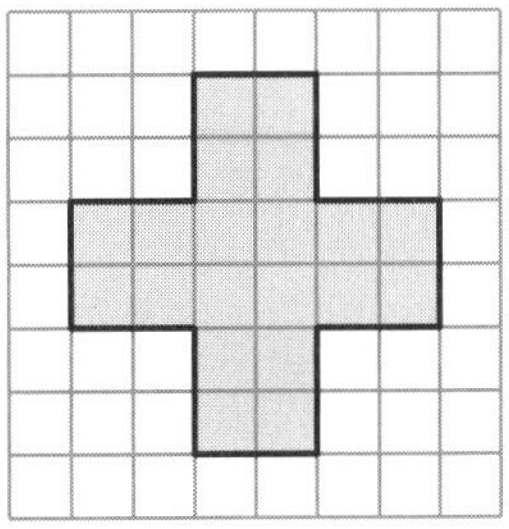

3. 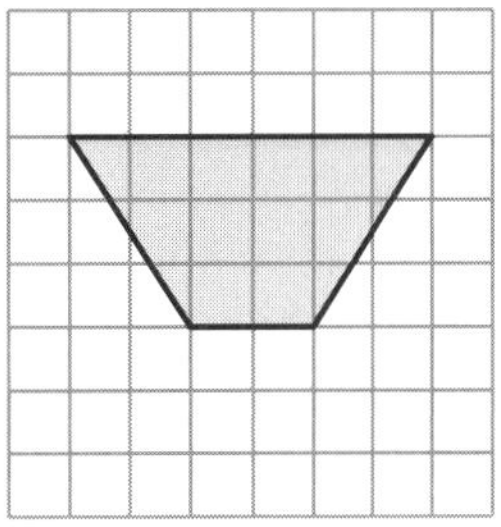

4. 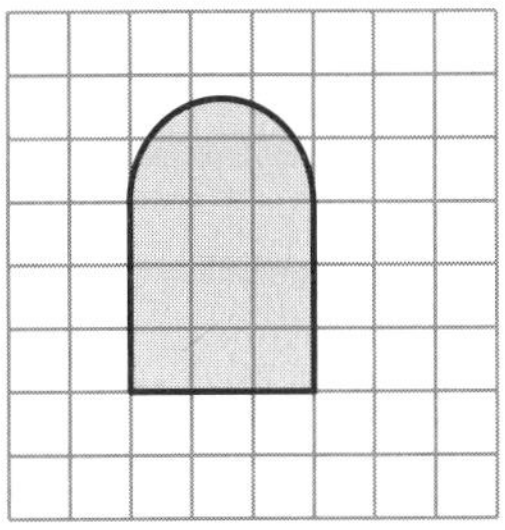

5. 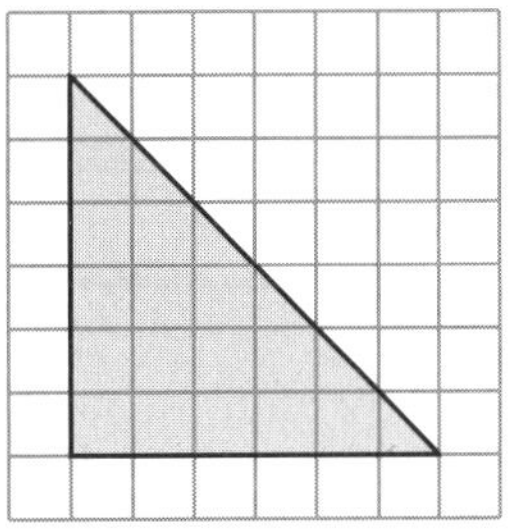

6. 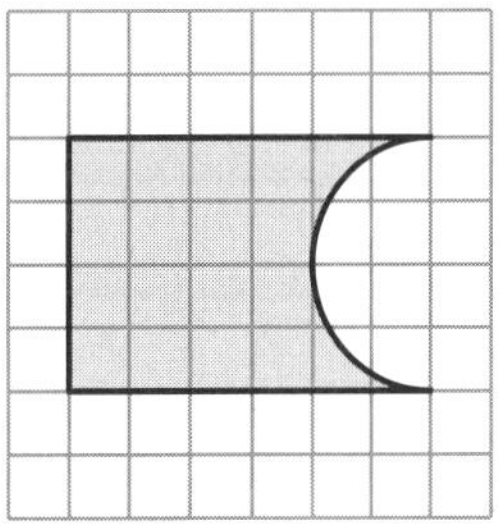

Name ______________________ Date ________

6.2 Practice A

Each square on the grid paper is 1 square inch. Estimate the perimeter of the figure.

1. 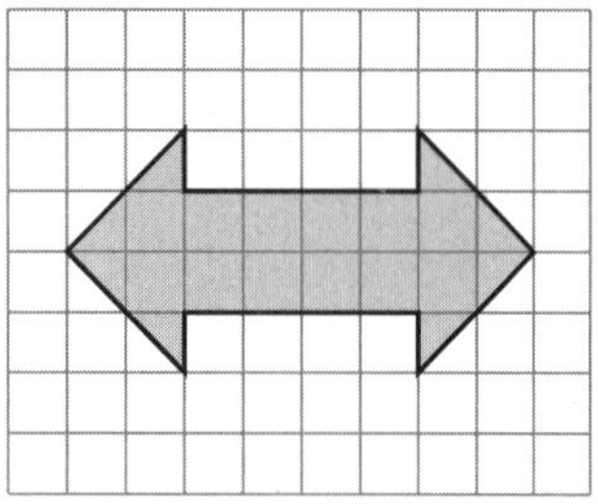

2. 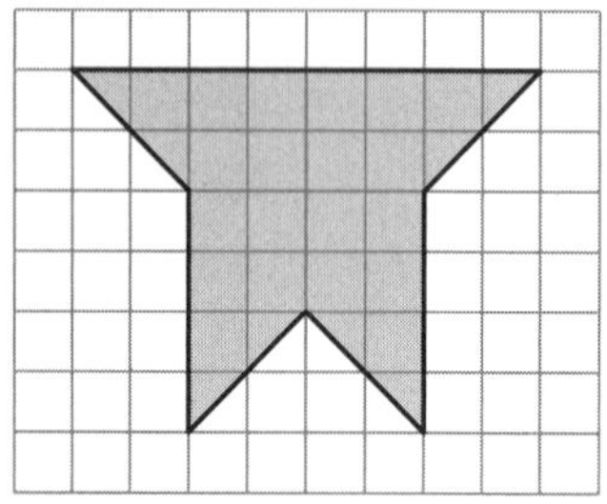

3. 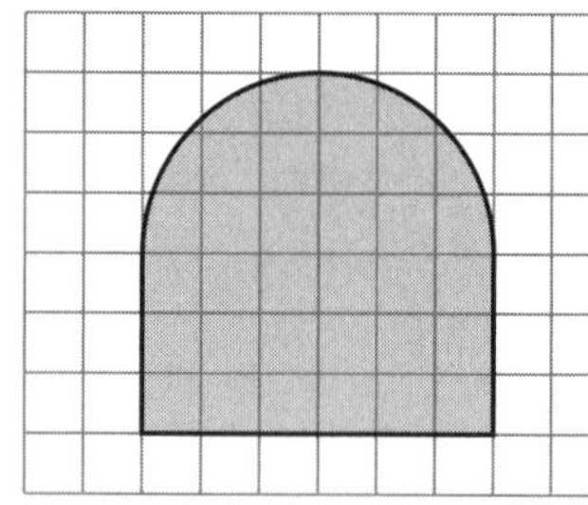

Find the perimeter of the figure.

4. 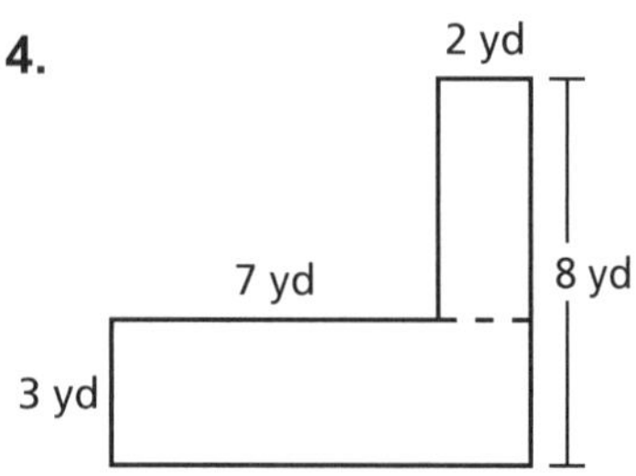

5. 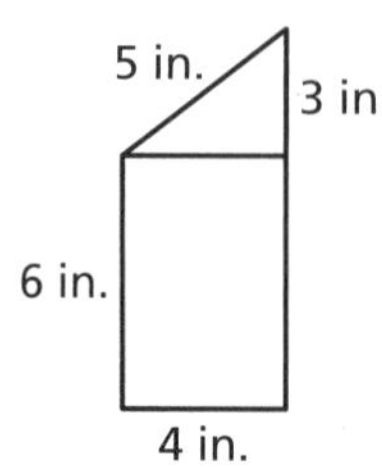

6. 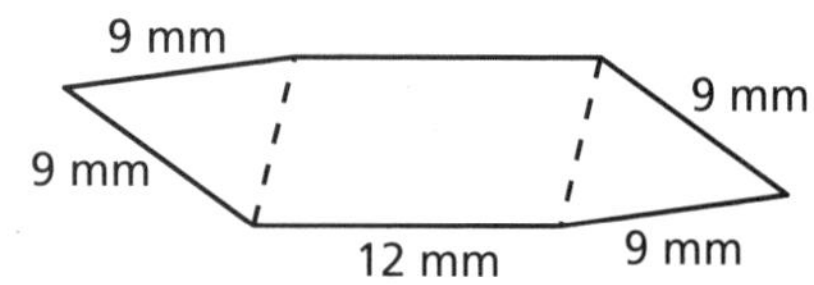

7. 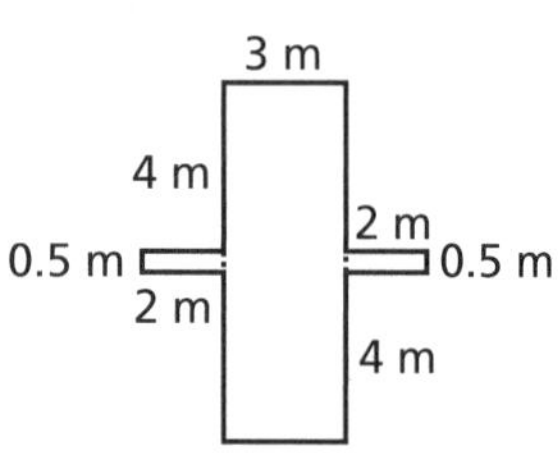

8. 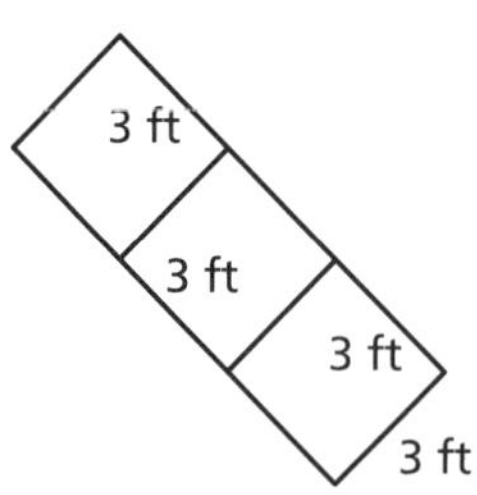

9. 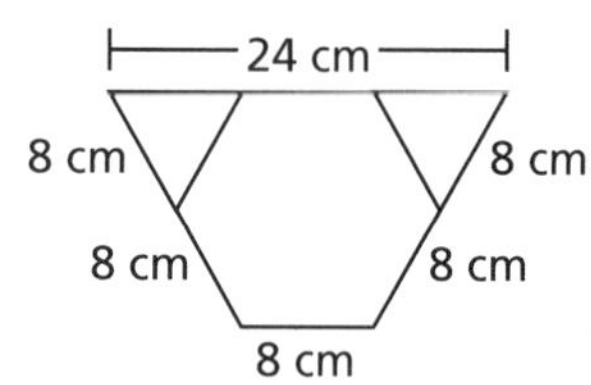

10. A stained glass window has the dimensions shown. What is the perimeter of the hole that should be cut in the wall in order for the window to be installed?

11. The dimensions of a new city park basketball court are shown at the right. A fence is to be built around the court and bleachers. The fence costs $8.99 per foot. How much will it cost to install the fence?

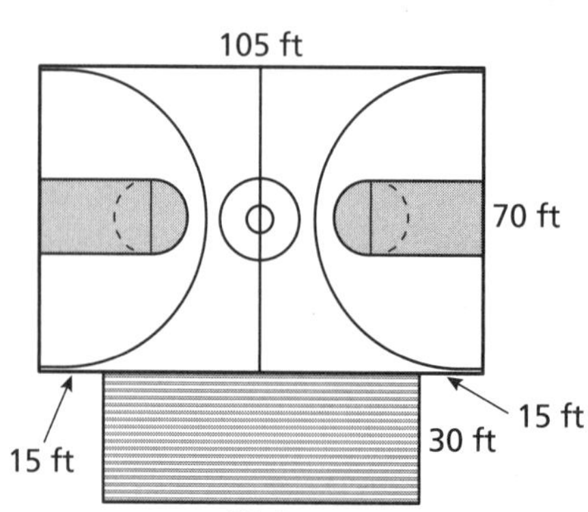

Name______________________________ Date __________

6.2 Practice B

Each square on the grid paper is 1 square inch. Estimate the perimeter of the figure.

1.

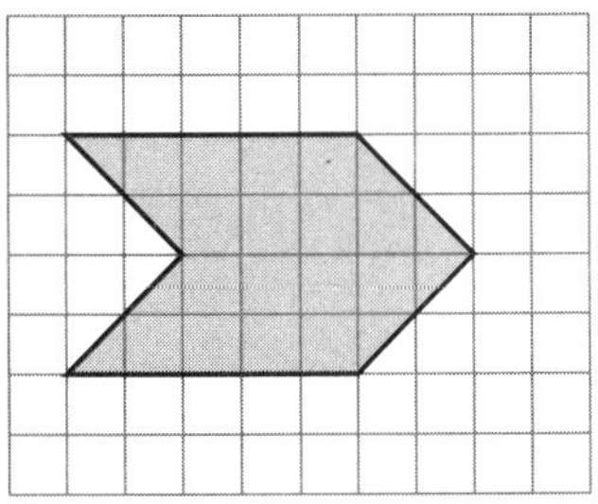

2.

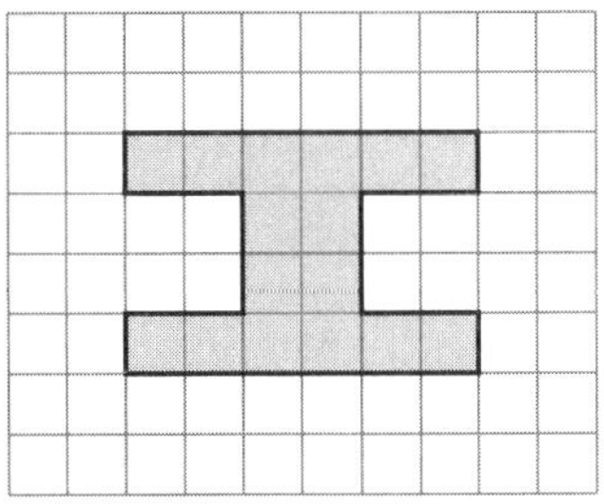

3.

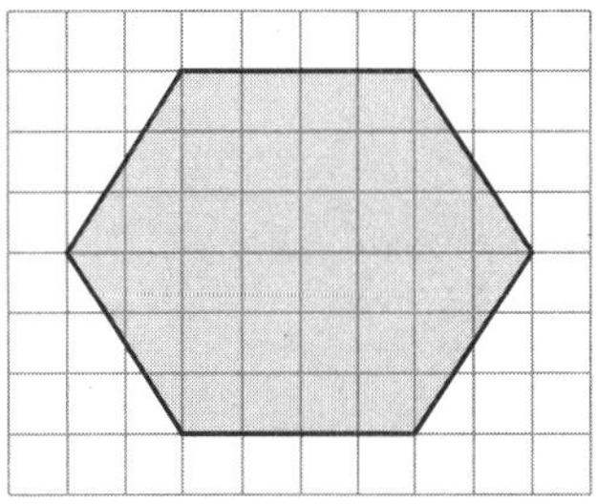

Find the perimeter of the figure.

4.

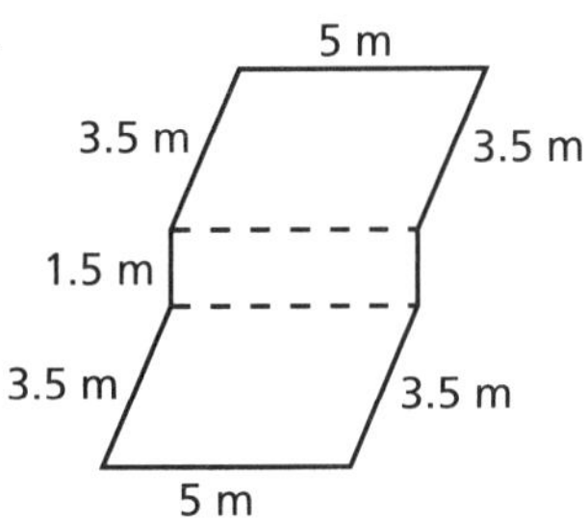

5.

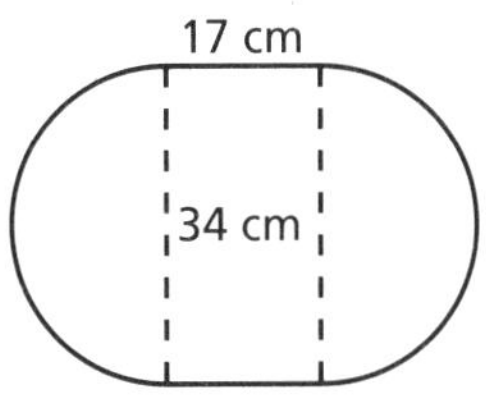

6.

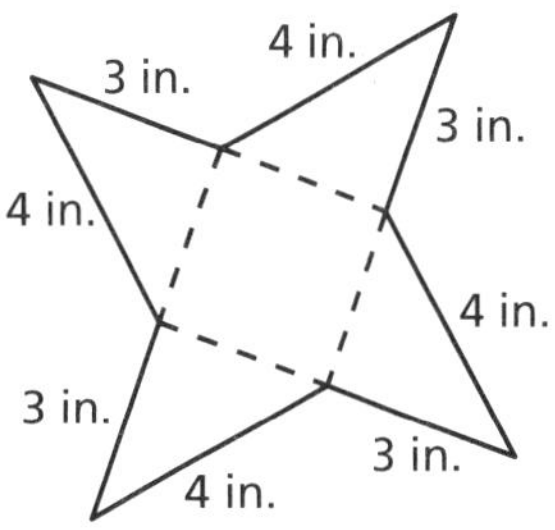

7. Describe and correct the error in finding the perimeter of the figure.

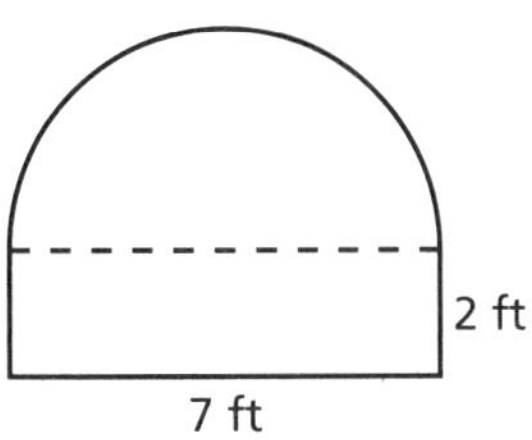

✗	Perimeter $\approx 2 + 7 + 2 + 21.98$
	≈ 32.98 ft

8. A school has a garden in the shape of a pencil. A fence is to be built around the garden. The fence costs $2.75 per foot. How much will it cost to install the fence?

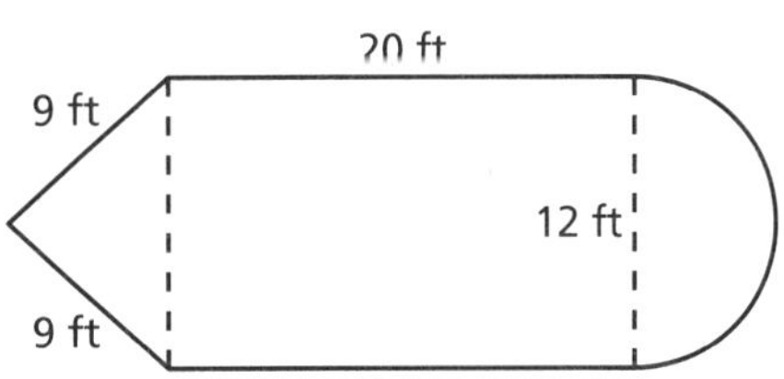

9. A shrub has been cut and trimmed into the shape of an "F." The owner has hired a landscaper to decrease the perimeter of the shrub by 10 feet. Draw a diagram of how the landscaper might do this. Is there more than one way? Explain.

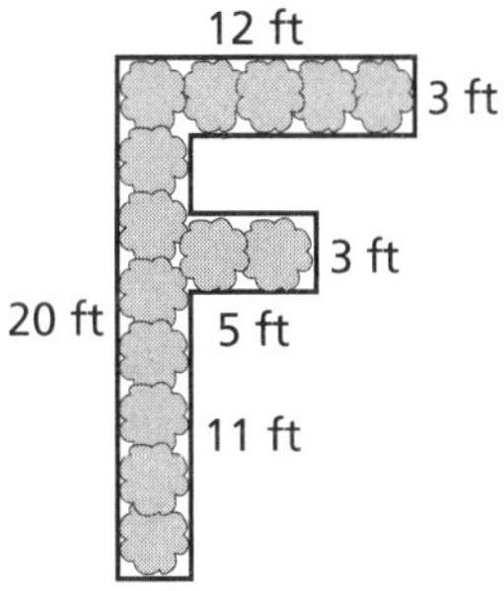

Name ______________________________ Date ________

6.2 Enrichment and Extension

Geometry

Find the perimeter of the figure. Use 3.14 for π.

1.

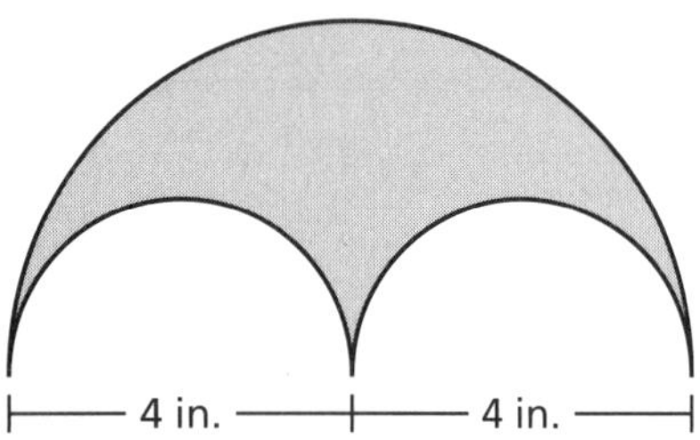

2.

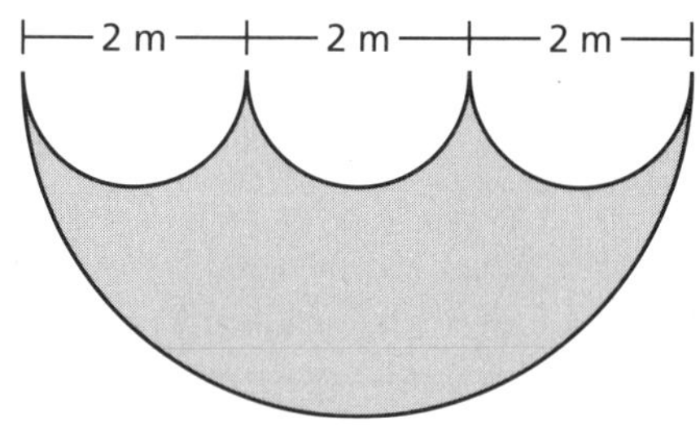

3.

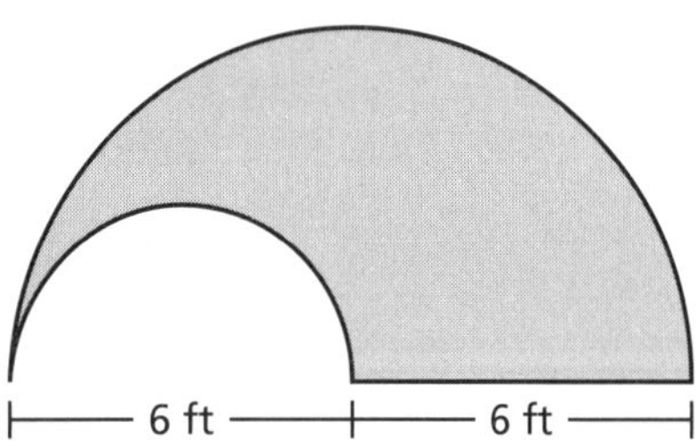

4.

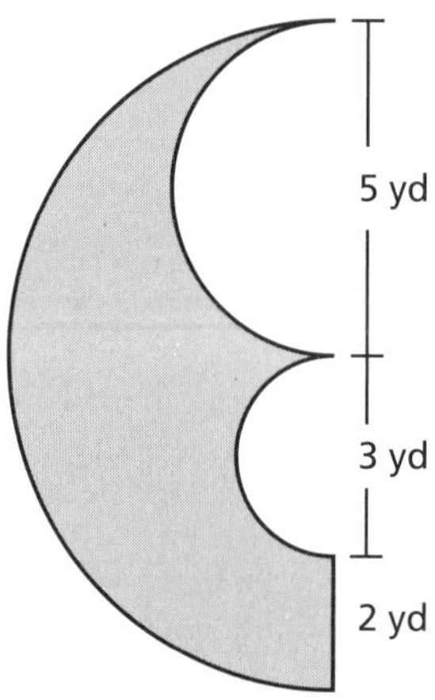

5.

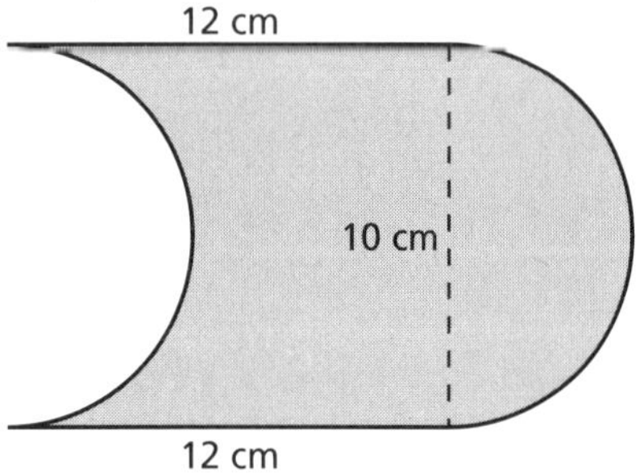

6.

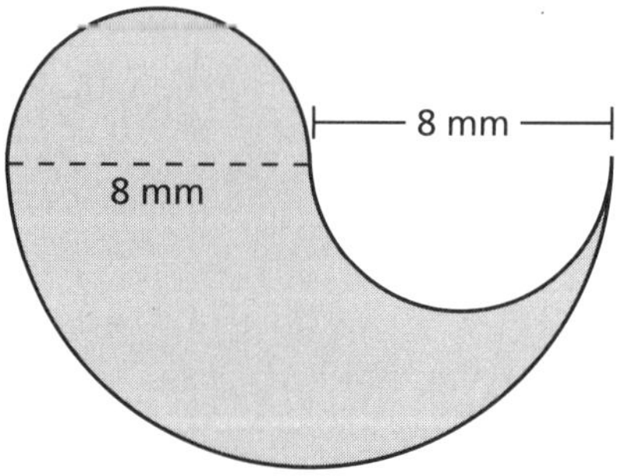

Name________________________________ Date__________

6.2 Puzzle Time

What Is The Building In Your City That Has The Most Stories?

Write the letter of each answer in the box containing the exercise number.

Find the perimeter of the figure.

1.

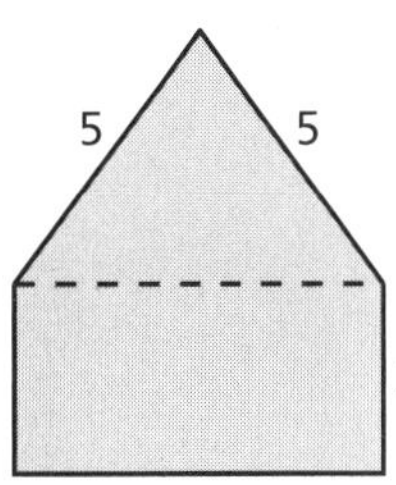

2.

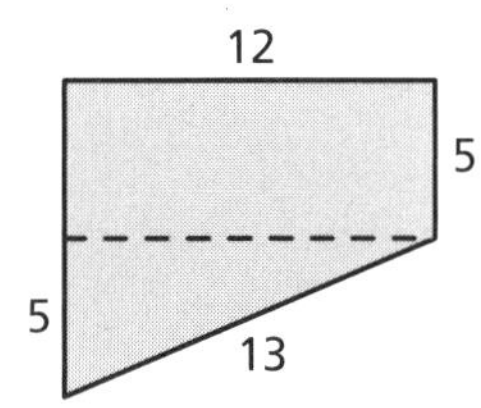

3.

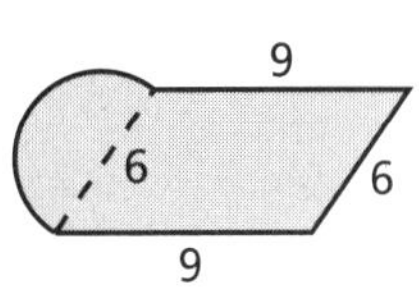

4.

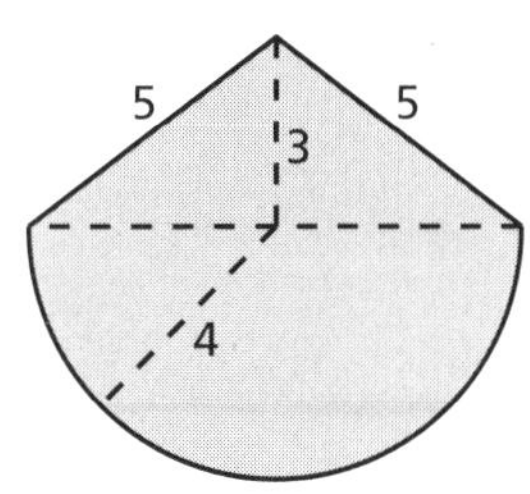

5.

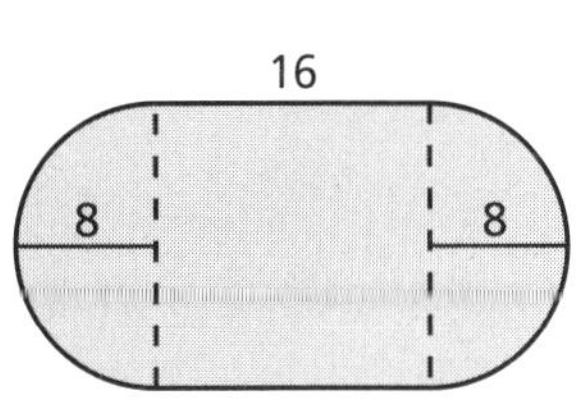

6.

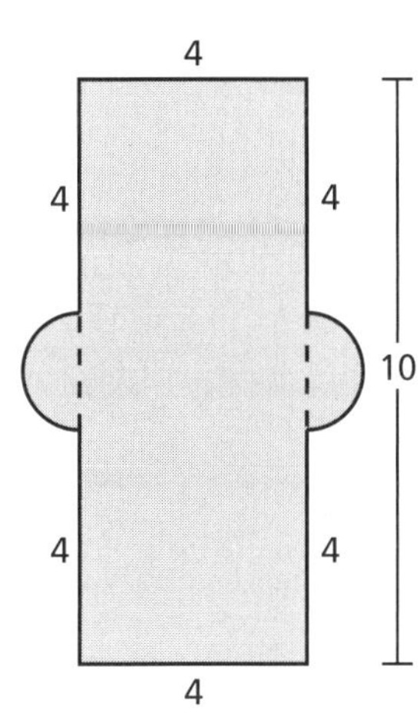

7.

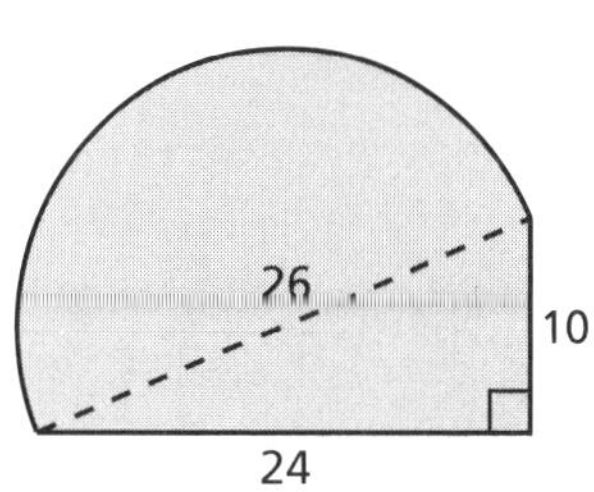

Answers

R. 22.56

B. 40

I. 30.28

Y. 33.42

R. 82.24

L. 22

A. 74.82

1	6	2	5	7	4	3

Activity 6.3 Start Thinking!
For use before Activity 6.3

You know how to find the area of squares, rectangles, triangles, trapezoids, and parallelograms.

Describe three different methods you could use to estimate the area of a circle.

Activity 6.3 Warm Up
For use before Activity 6.3

Find the area of the triangle.

1.

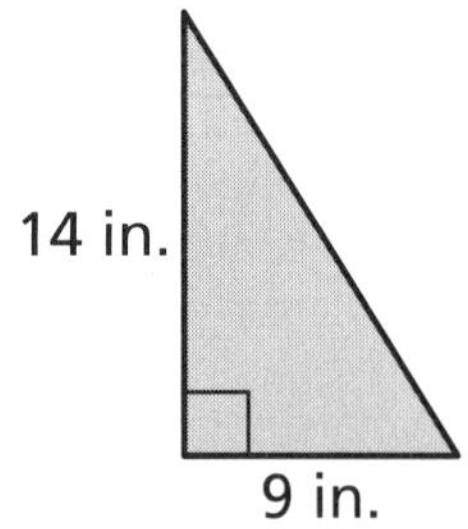

2.

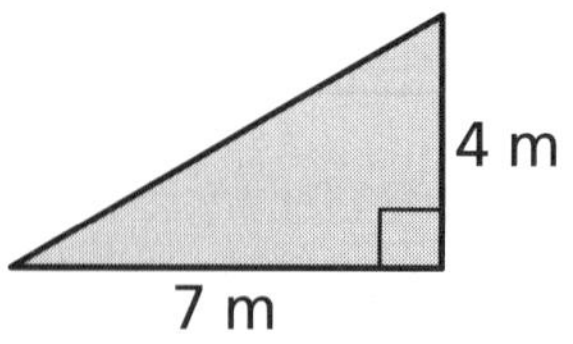

3.

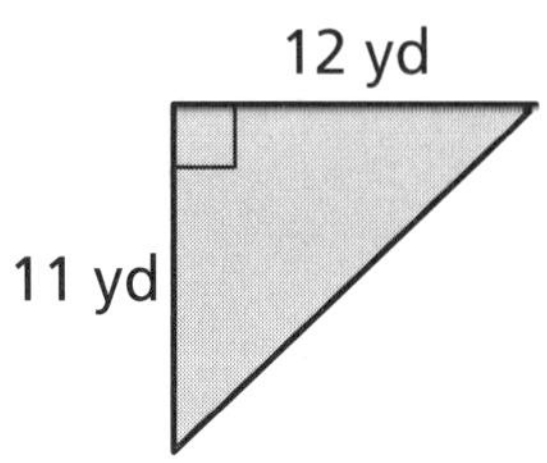

4.

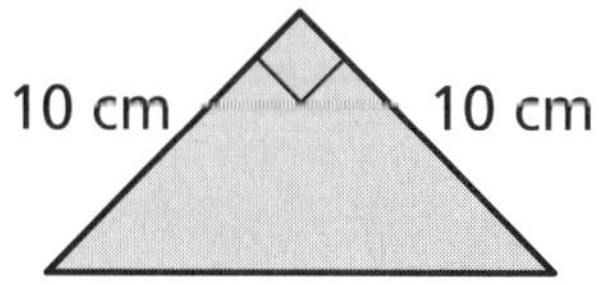

5.

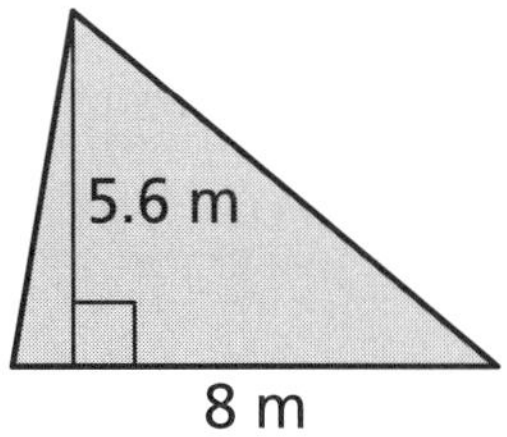

6.

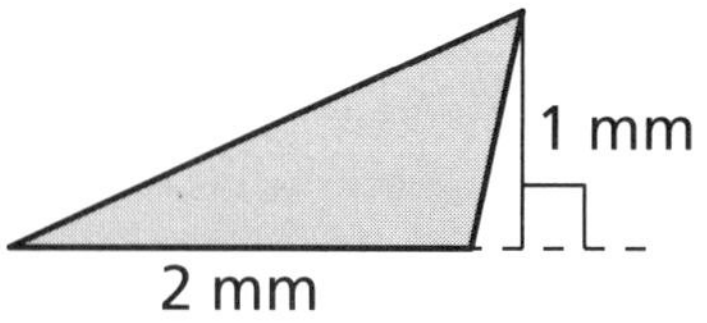

Lesson 6.3 Start Thinking!

For use before Lesson 6.3

Two approximations for π are $\frac{22}{7}$ and 3.14. In finding the area of a circle, when is it easier to use $\frac{22}{7}$? When is it easier to use 3.14?

Write a word problem involving the area of a circular object. Exchange problems with a classmate and solve your classmate's problem. Is it easier to use $\frac{22}{7}$ or 3.14 for π to solve your classmate's problem?

Lesson 6.3 Warm Up

For use before Lesson 6.3

Find the area of the circle. Use 3.14 or $\frac{22}{7}$ for π.

1.

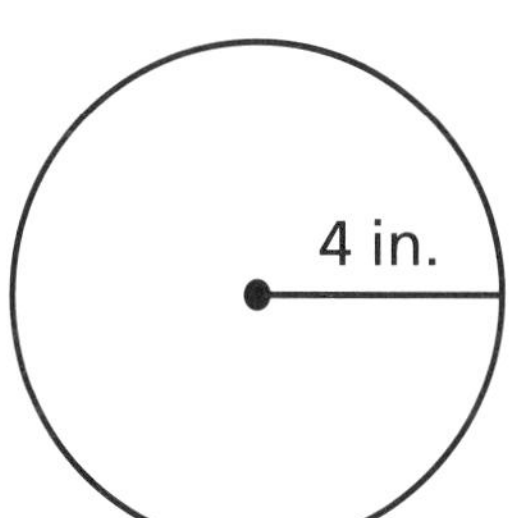

2.

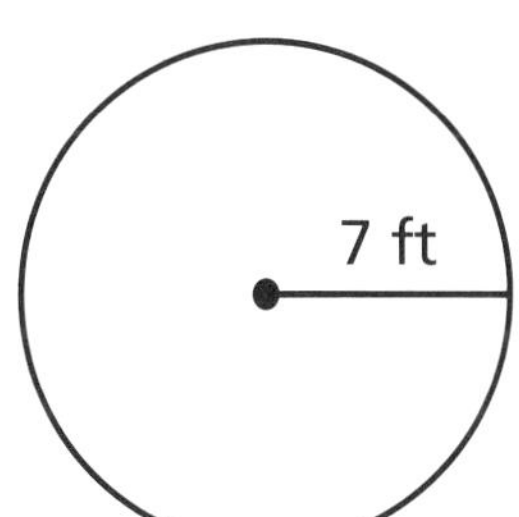

3.

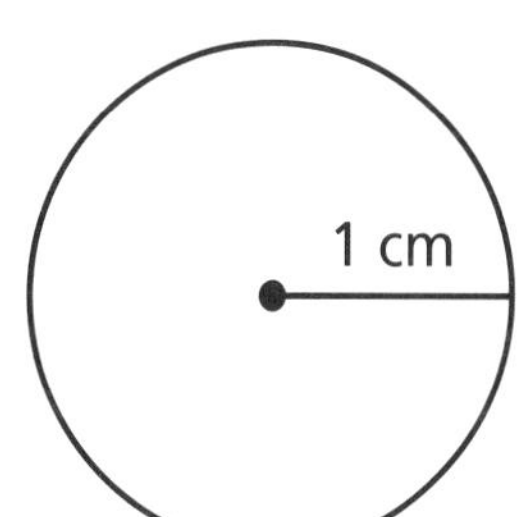

Name ______________________________ Date __________

6.3 Practice A

Find the area of the circle. Use 3.14 or $\frac{22}{7}$ for π.

1. 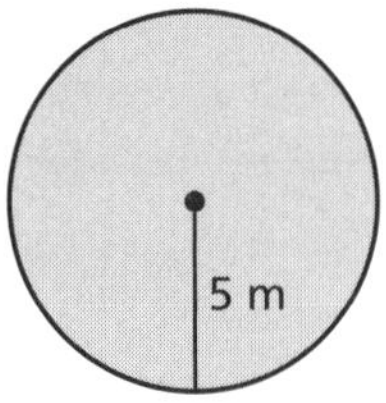

2. 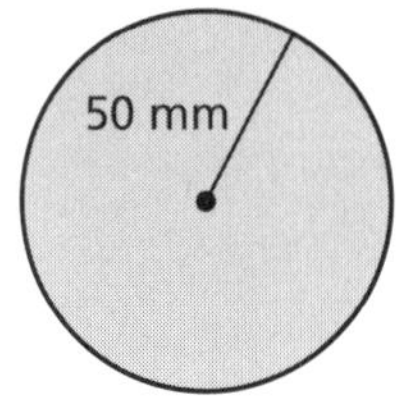

3. 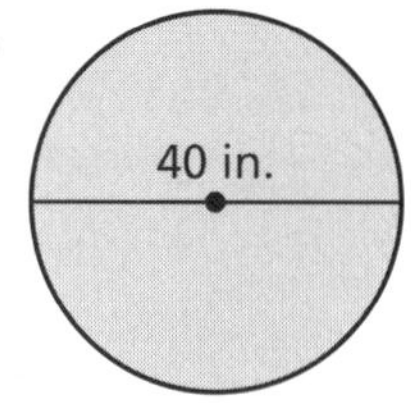

4. 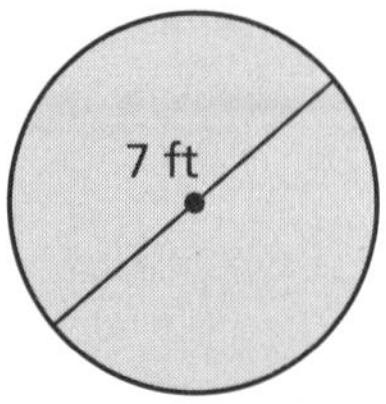

5. 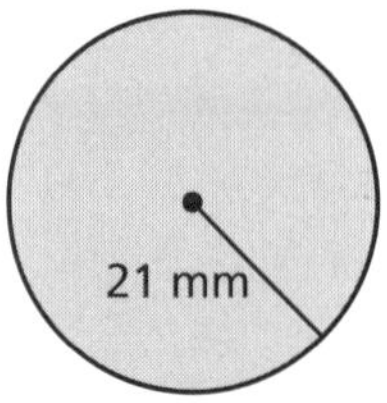

6. 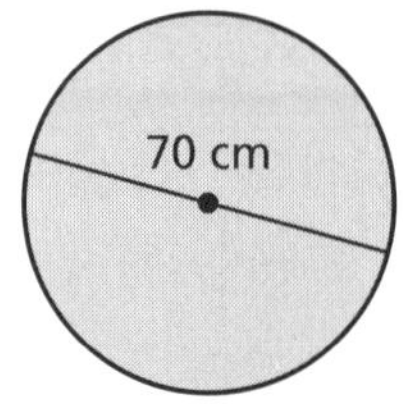

Find the area of the object.

7. 

8. 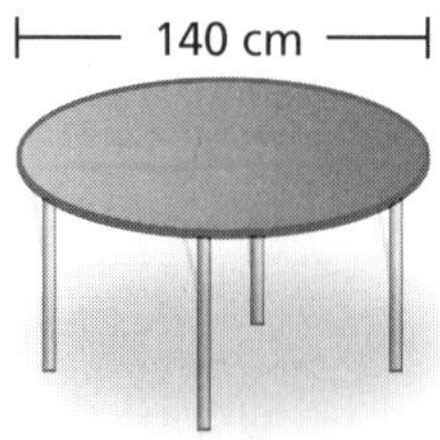

9. 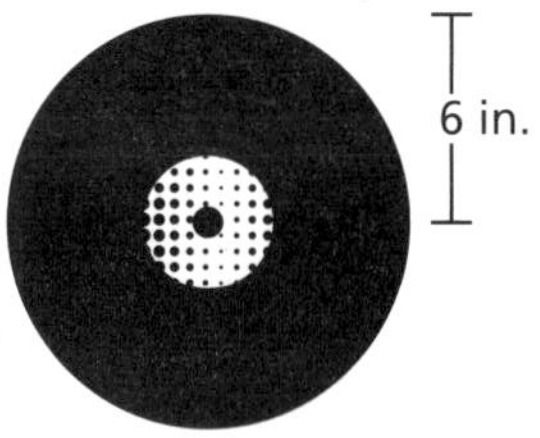

10. You use the compass to draw a circle. What will be the area of the circle?

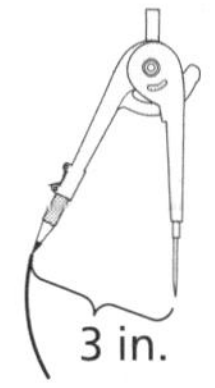

11. What fraction of a square inch is the area of one side of a penny? Use $\frac{22}{7}$ for π. Write your answer in simplest form.

12. To make a pizza, you spread pizza sauce over all but a 1-inch area around the outside edge as shown. What area of the crust is covered with sauce?

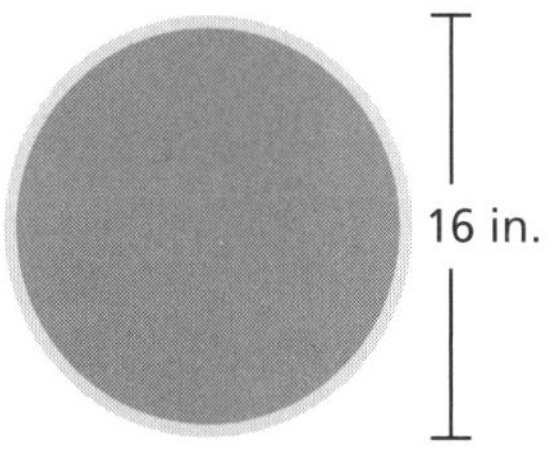

Name____ Date__________

6.3 Practice B

Find the area of the circle. Use 3.14 or $\frac{22}{7}$ for π.

1.

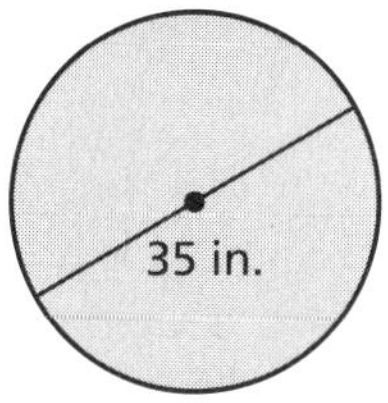

2.

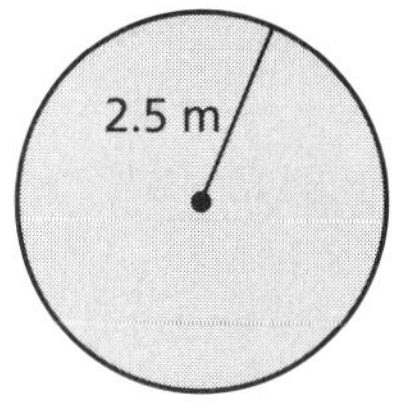

3.

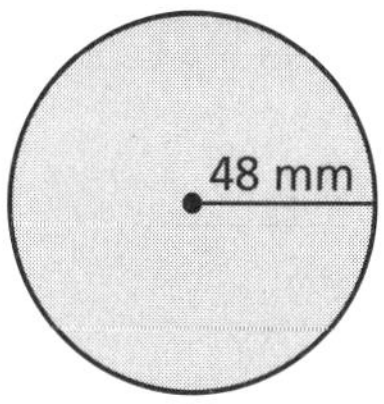

Find the area of the semicircle.

4.

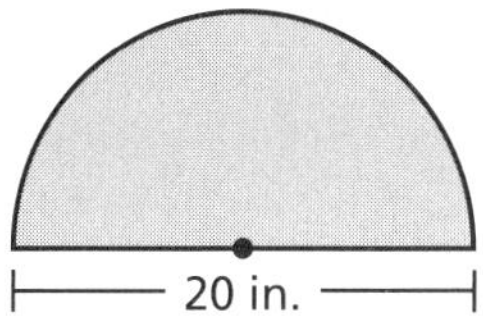

5.

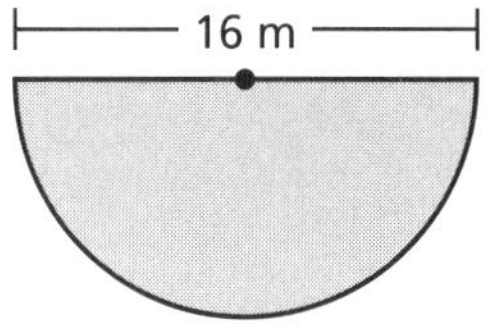

6.

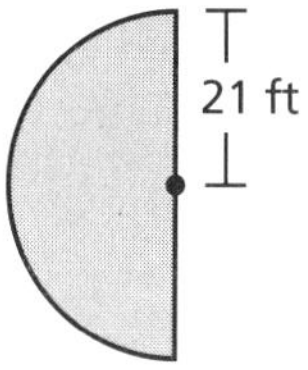

7. The shadow of an object is roughly the same size as the object. What is the area of the circular shadow of the hot air balloon?

8. How many *square feet* of the ground are sprayed by the beach shower?

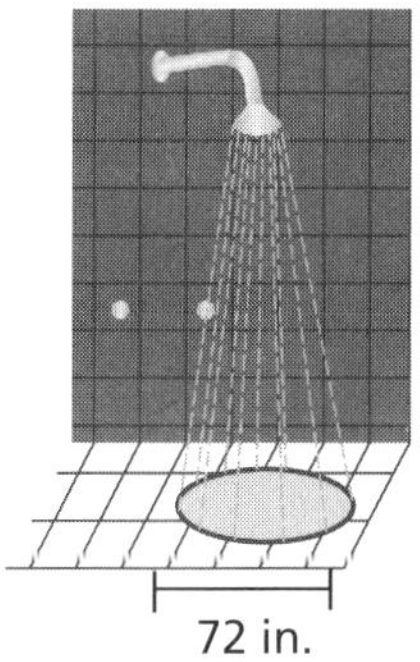

9. The radius of the small circle is half the radius of the large circle.

a. Use the radius r to write a formula for the area of the large circle.

b. Use the radius $\frac{r}{2}$ to write a formula for the area of the small circle.

c. How does the area of a circle compare to the area of another circle whose radius is twice as large? Explain your reasoning.

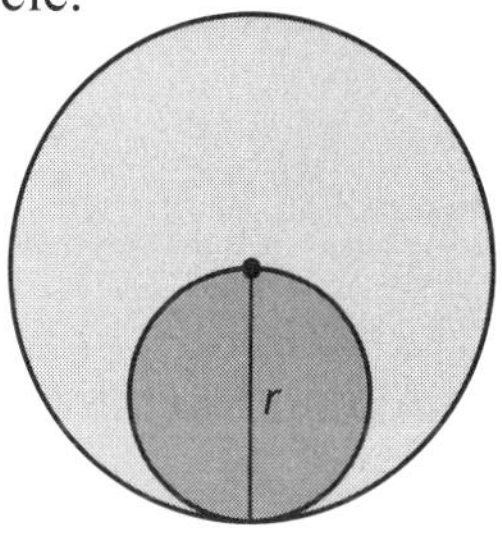

10. The number of square inches of a circle's area is equal to the number of inches of its circumference. What is the radius of the circle? Explain how you found your answer.

Name ____________________ Date __________

6.3 Enrichment and Extension

Where Are You Located?

The circular broadcast areas of two television stations and two radio stations are shown in the diagram. Use 3.14 for π.

	Broadcast radius (mi)
TV Channel 19	150
TV Channel 36	120
Radio 105.5 FM	60
Radio 94.7 FM	40

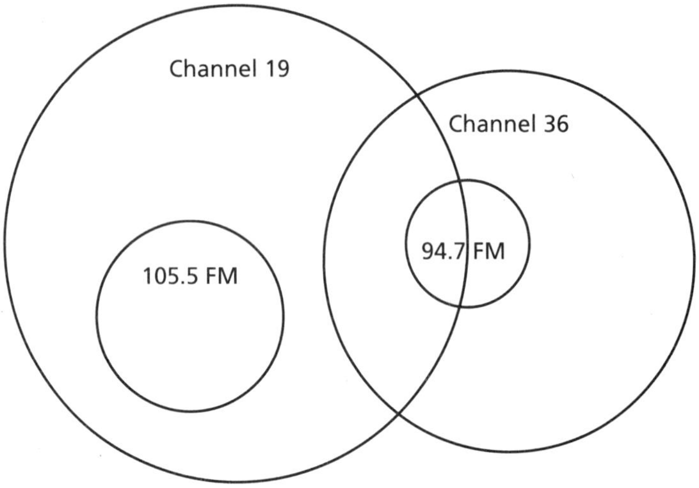

1. How many times larger is the broadcasting area of Channel 19 than the broadcasting area of Channel 36?

2. How many times larger is the broadcasting area of 105.5 FM than the broadcasting area of 94.7 FM?

3. What percent of the broadcast area of Channel 19 is also in the broadcast area of 105.5 FM?

4. What fraction of the broadcast area of Channel 36 is also in the broadcast area of 94.7 FM?

5. Half of the broadcast area of 94.7 FM can watch both television channels.

   a. What percent of the broadcast area of Channel 19 can listen to 94.7 FM?

   b. You can watch one of the television channels and can listen to one of the radio stations. Draw a diagram and shade the area where you may be located. What is the total area of the shaded region?

   c. Your friend can watch Channel 19 but cannot listen to either of the radio stations. You do not know if your friend can watch Channel 36. Draw a diagram and shade the area where your friend may be located. What is the total area of the shaded region?

Name______________________________ Date__________

6.3 Puzzle Time

What Do Little Piggies Do As Soon As They Get Home From School?

Write the letter of each answer in the box containing the exercise number.

Find the area of the circle. Use 3.14 or $\frac{22}{7}$ for π.

1.

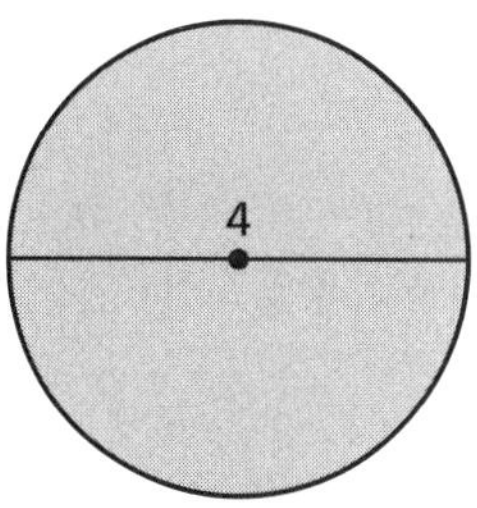

2.

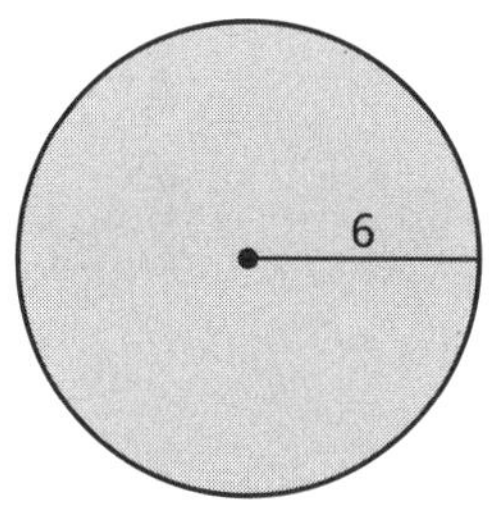

3.

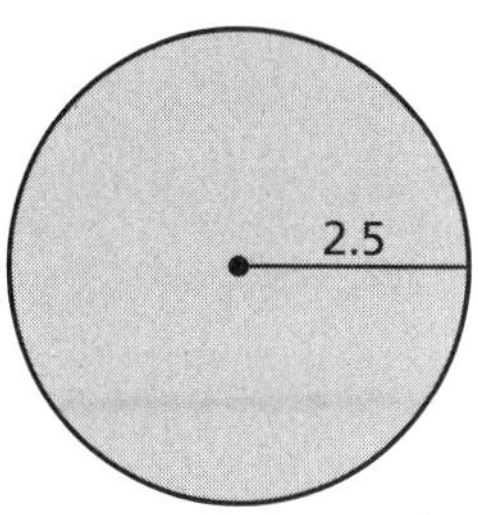

4.

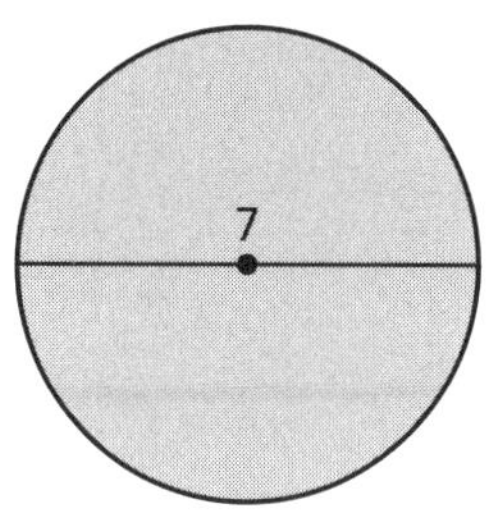

5.

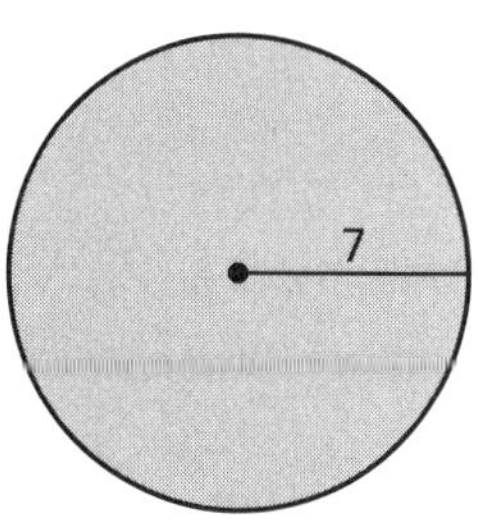

Answers

K. 38.465 square units

H. 12.56 square units

M. 19.625 square units

R. 127.17 square units

W. 154 square units

O. 113.04 square units

A. 157 square units

Find the area of the semicircle. Use 3.14 for π.

6.

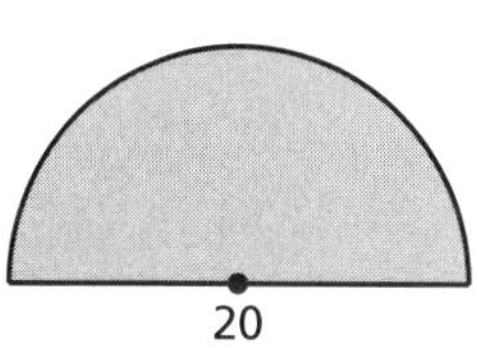

7.

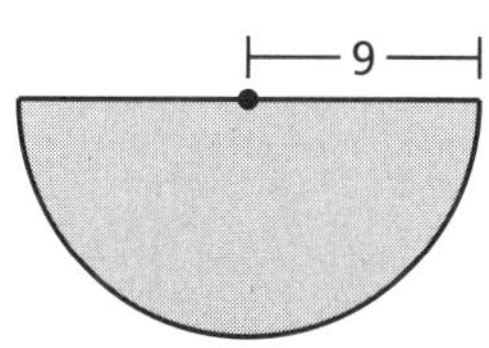

1	6	3	5	2	7	4

Activity 6.4 Start Thinking!
For use before Activity 6.4

When might it be useful to know how to find the area of a real-life composite figure?

Activity 6.4 Warm Up
For use before Activity 6.4

Find the area of the figure described.

1. square with side length 10 ft
2. square with side length 16 in.
3. rectangle with length 15 m and width 10 m
4. triangle with base 5 cm and height 12 cm
5. circle with radius 10 ft
6. circle with diameter 100 yd

Lesson 6.4 Start Thinking!
For use before Lesson 6.4

Draw a picture of a house with a roof and a chimney.

How can you use a ruler and some calculations to find the area covered by your drawing?

Use a ruler and the method you described to estimate the area of the house.

Lesson 6.4 Warm Up
For use before Lesson 6.4

Each square on the grid paper is 1 square inch. Find the area of the figure.

1. 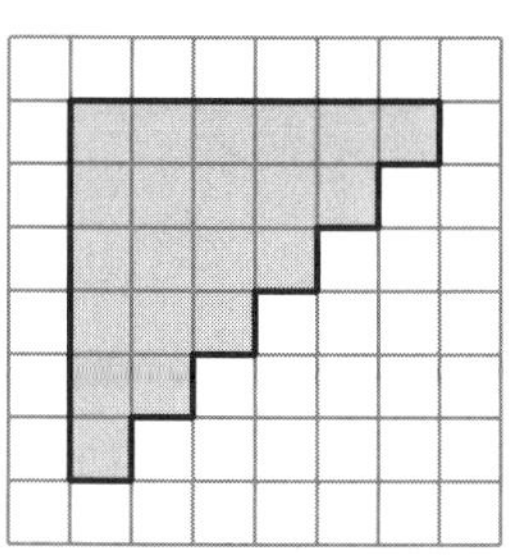

2. 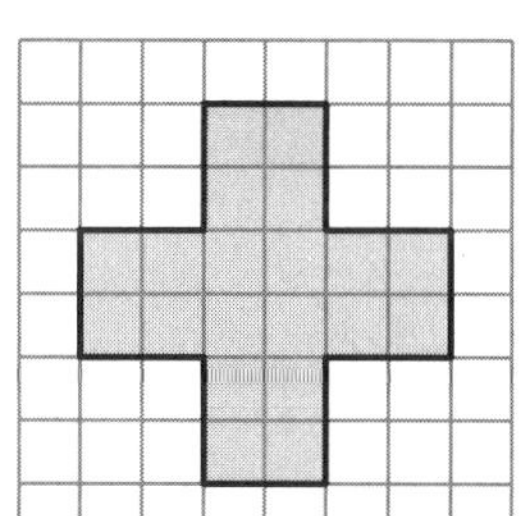

3. 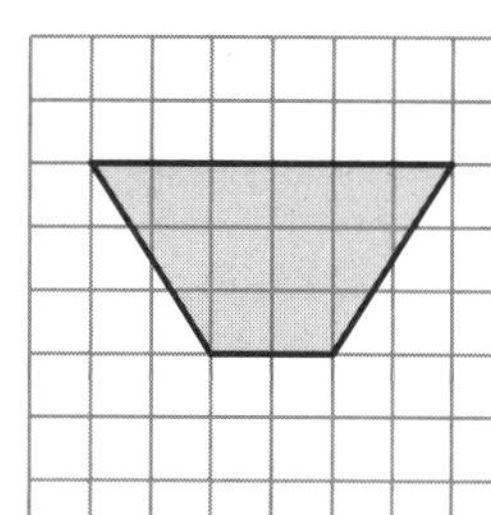

4. 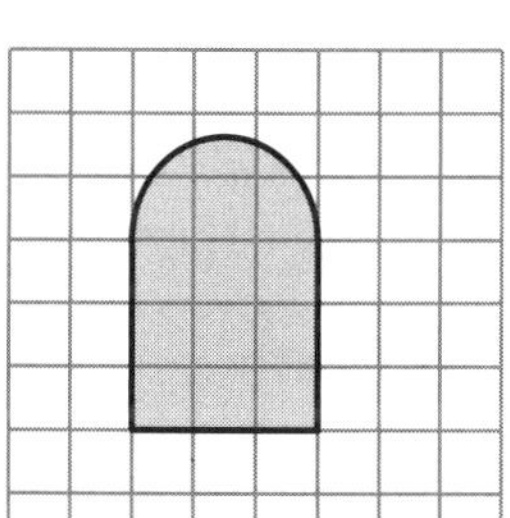

5. 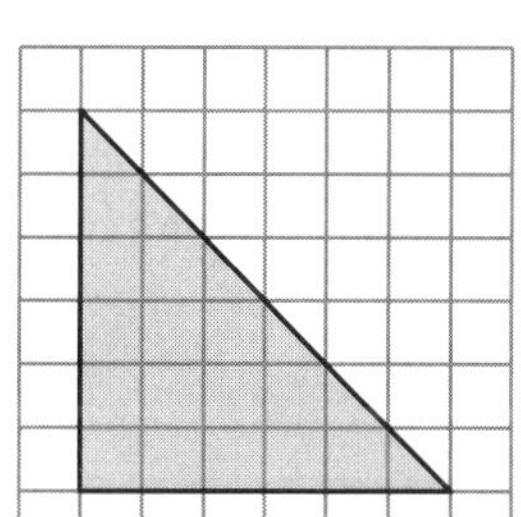

6. 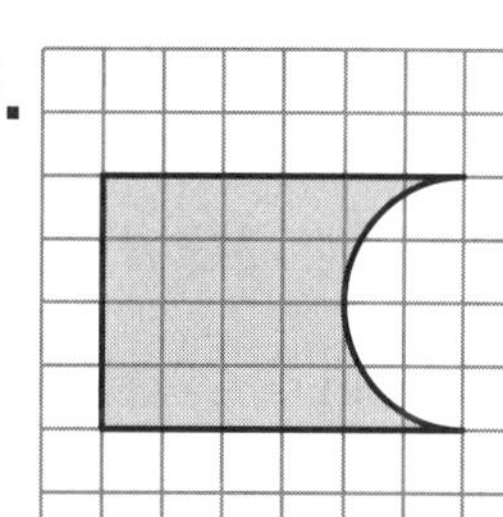

Name ______________________________ Date __________

6.4 Practice A

Each square on the grid paper is 1 square meter. Find the area of the figure.

1. 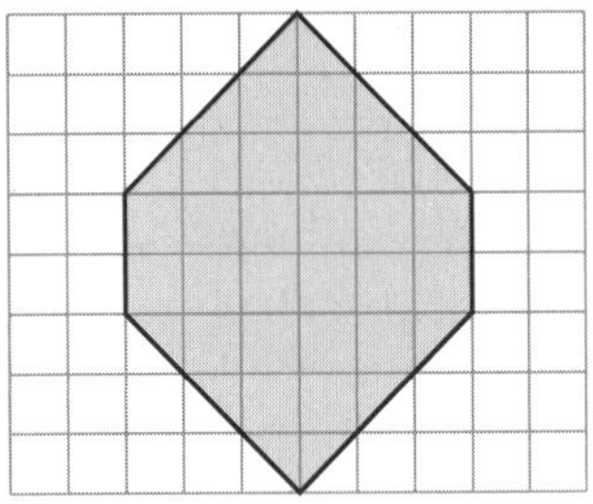

2. 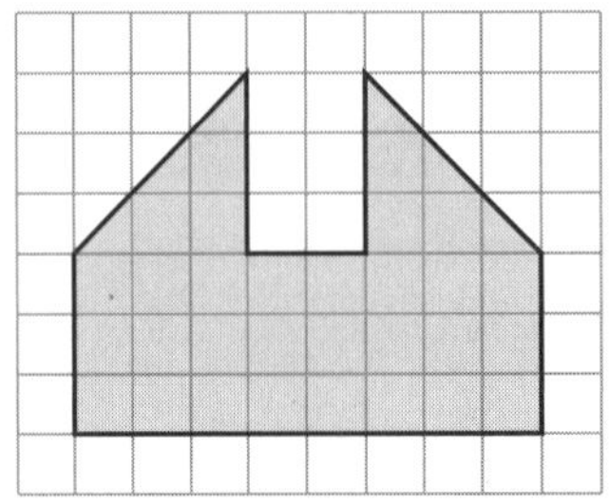

3. 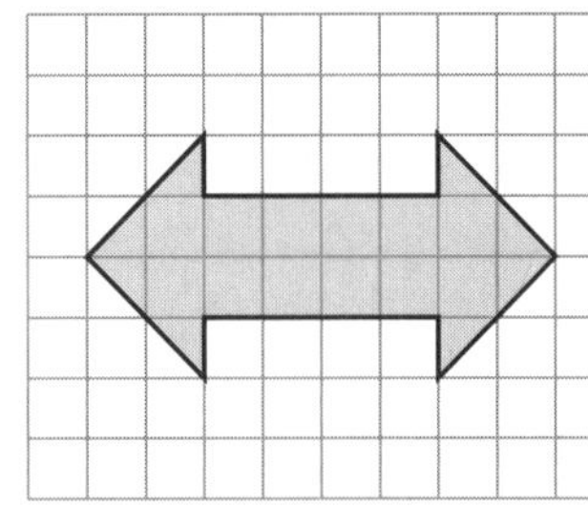

Find the area of the figure.

4. 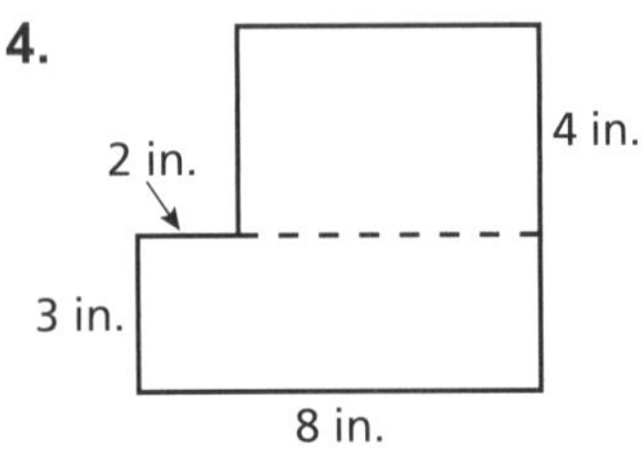

5. 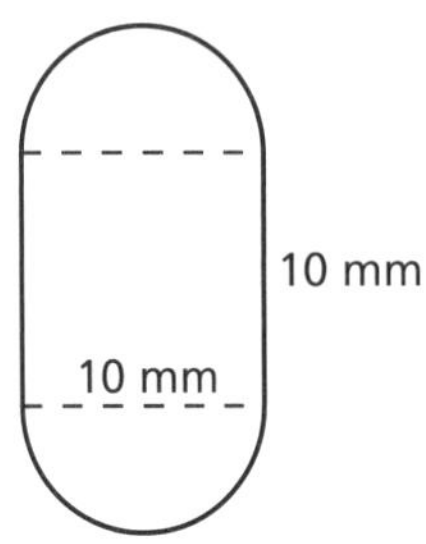

6. 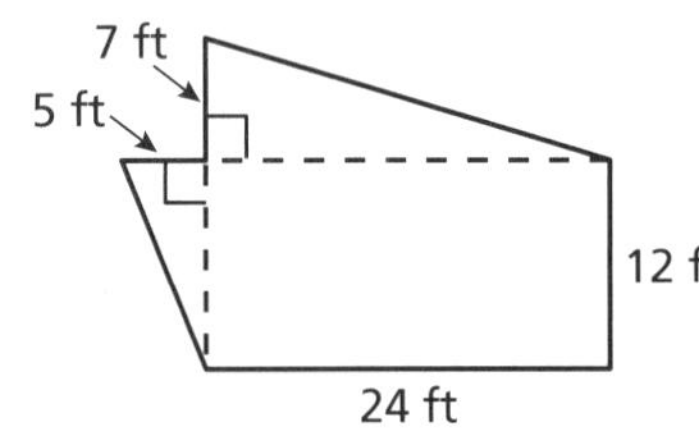

7. 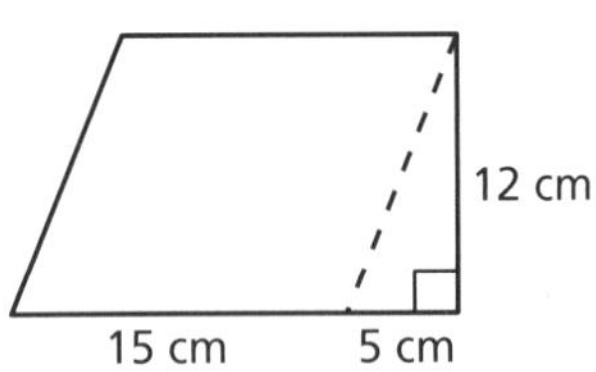

8. 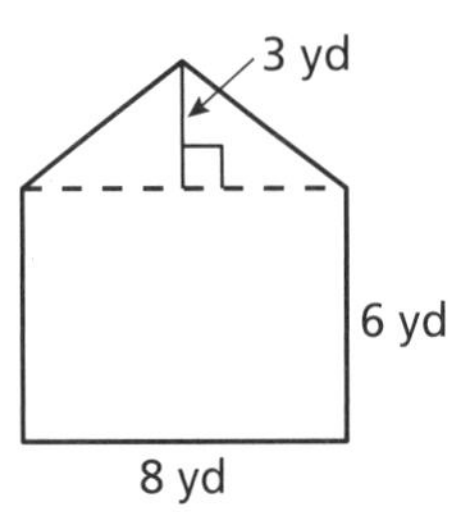

9. 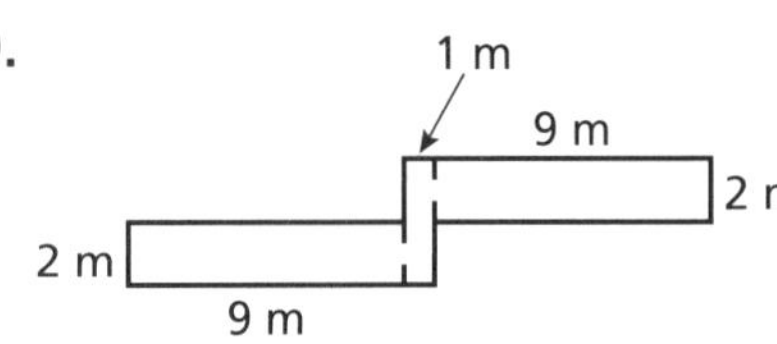

10. A garden is made up of two squares and a quarter circle. What are the perimeter and area of the garden?

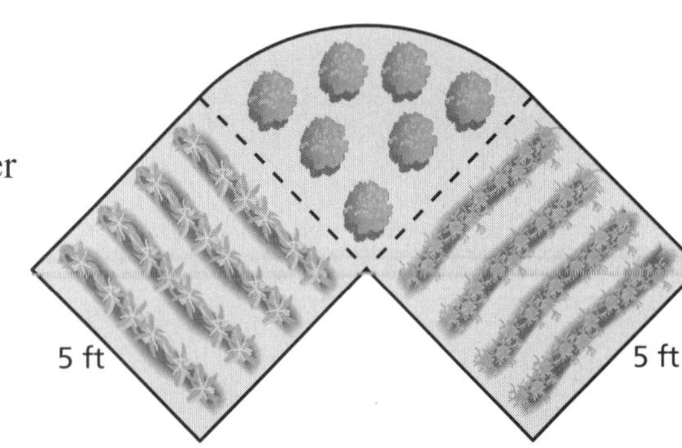

11. A pinwheel design for a quilt is shown at the right. Each square of the design has a side length of 3 inches.

a. Find the total area of the design.

b. Find the total area of the four large triangles.

c. The quilt will be made of 30 pinwheel designs. The material to make the large triangles costs $0.08 per square inch. How much will it cost to purchase the material to make the triangles for the quilt?

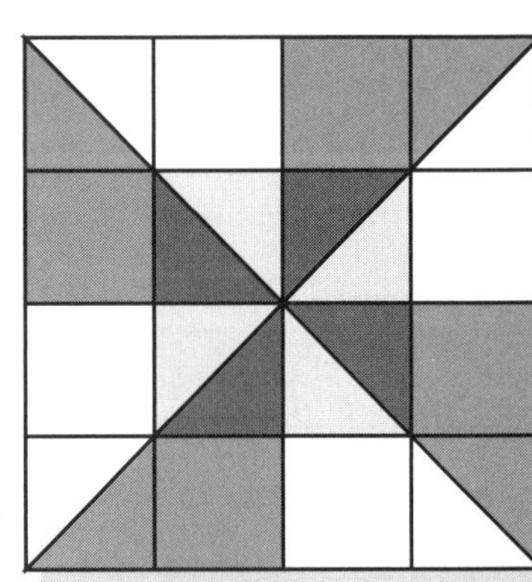

Name______________________________ Date__________

6.4 Practice B

Find the perimeter and area of the figure.

1.

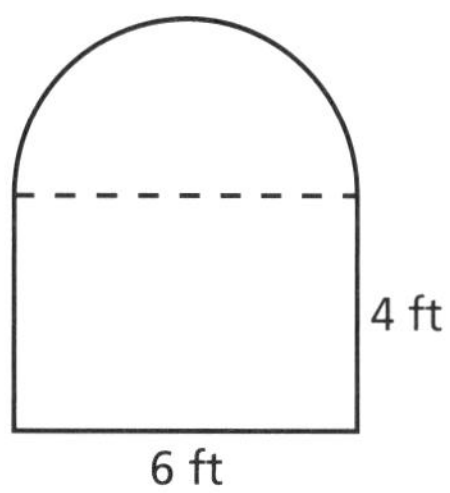

2.

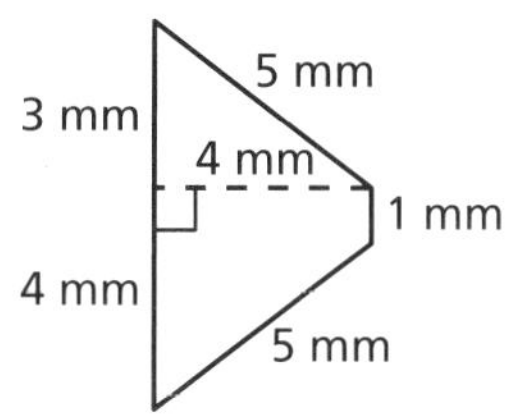

3.

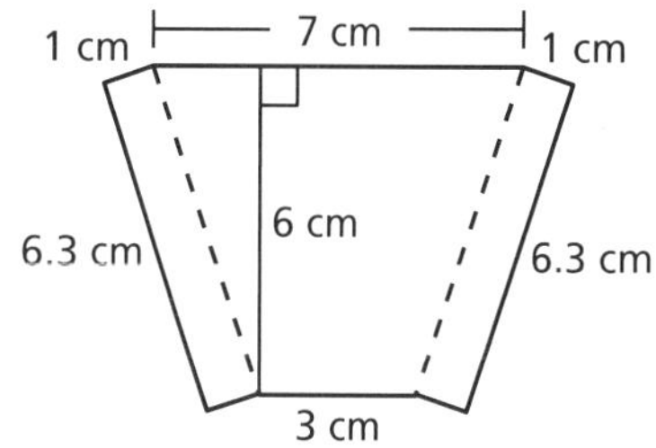

Find the area of the shaded region of the figure.

4.

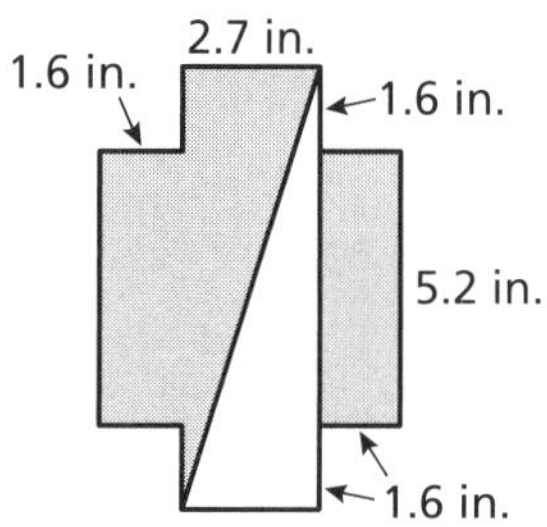

5.

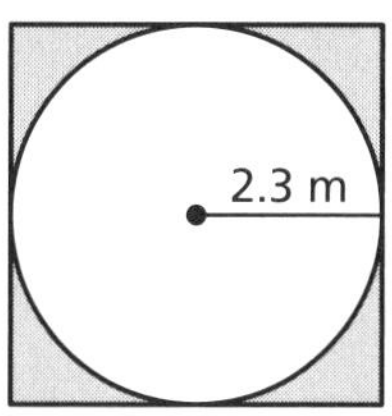

6.

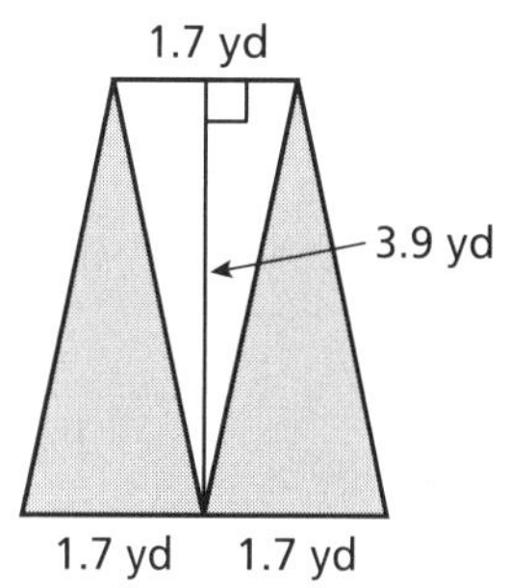

7. Describe and correct the error in finding the area of the shaded region of the figure.

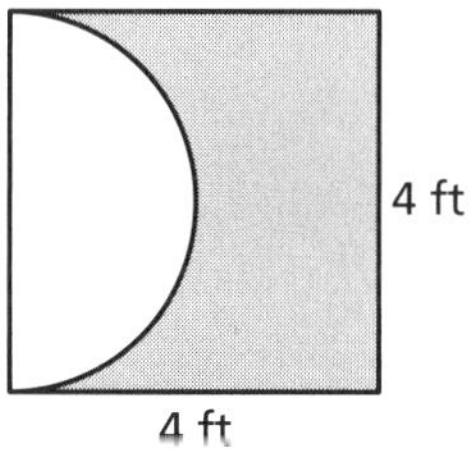

✗ Shaded area $\approx (4 \bullet 4) - (3.14 \bullet 2^2)$

$\approx 16 - 12.56$

$\approx 3.44 \text{ ft}^2$

8. Jackson Well Middle School's logo is shown at the right.

a. Find the perimeter of the logo.

b. Find the area of the logo.

c. Will the logo fit on a notebook cover that is 11 inches long and 8.5 inches wide? Explain.

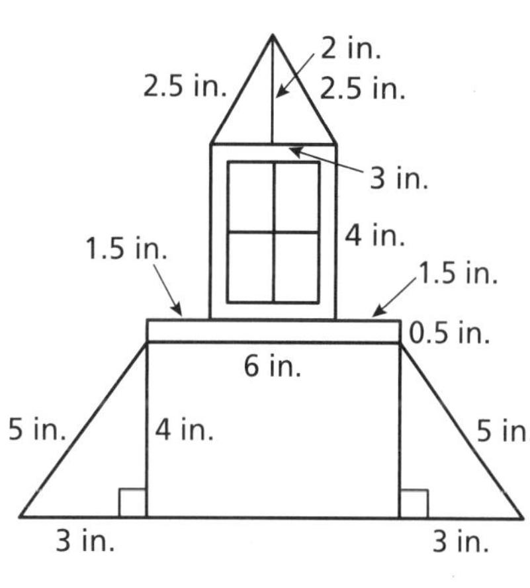

Jackson Well Middle School

Name ______________________ Date __________

6.4 Enrichment and Extension

Composite Letters

The letters consist of rectangles, semicircles, and circles. Find the area of the letter.

1.

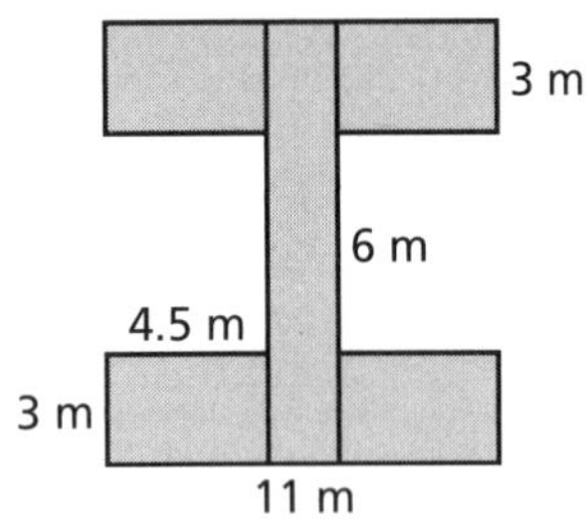

2.

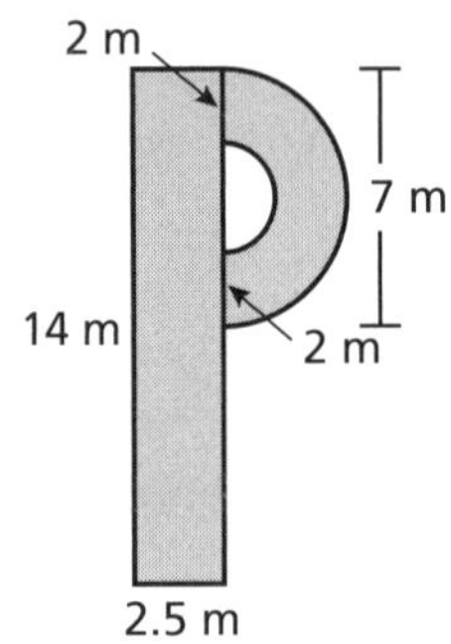

3.

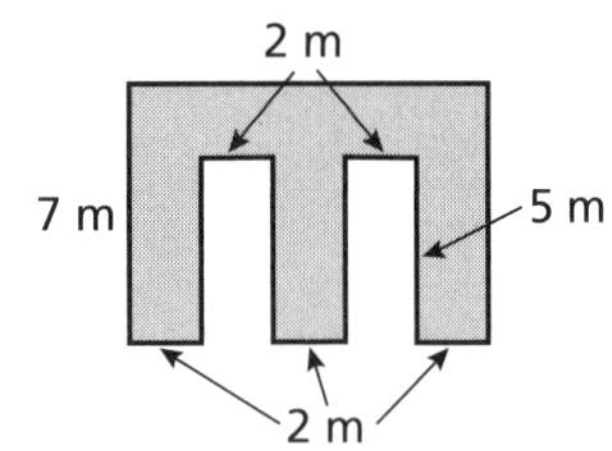

4.

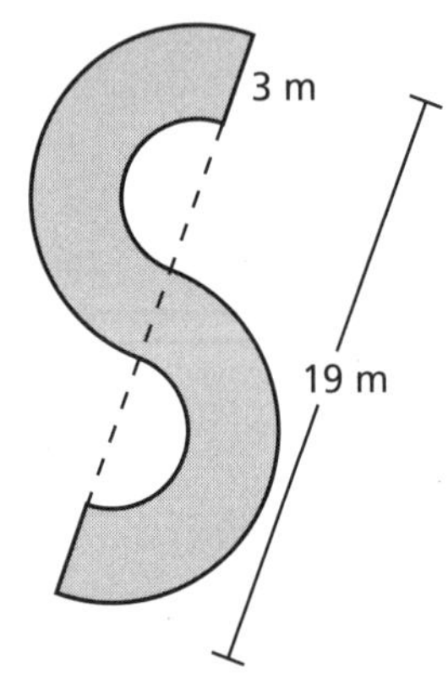

5.

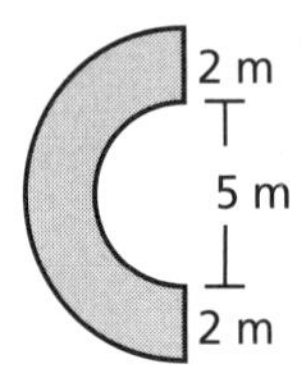

6.

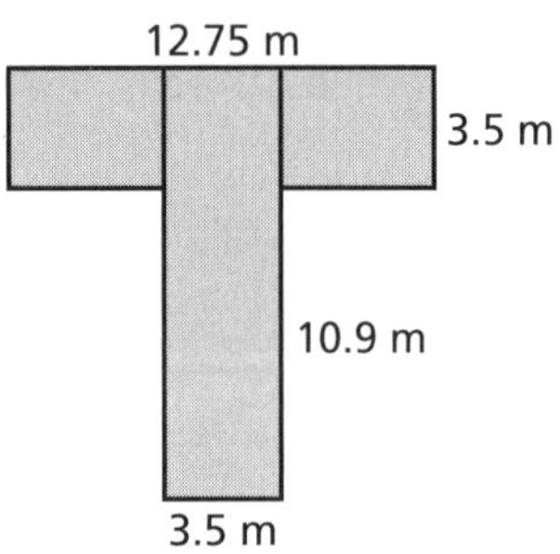

7.

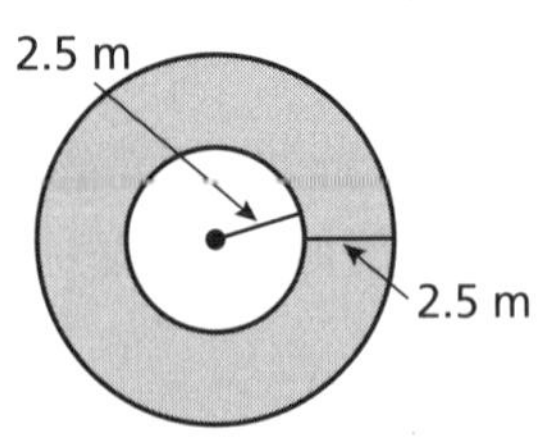

8.

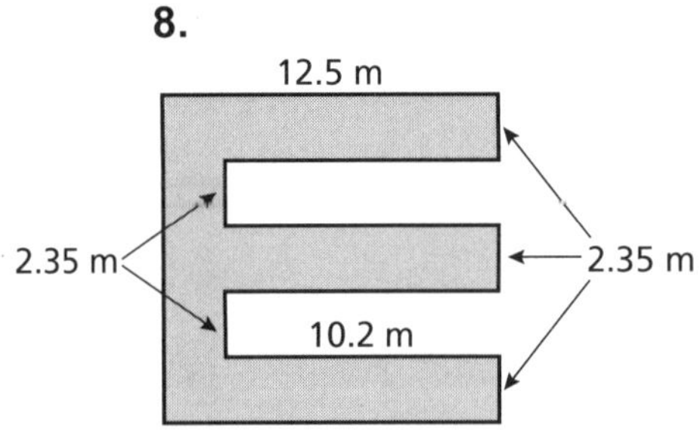

9.

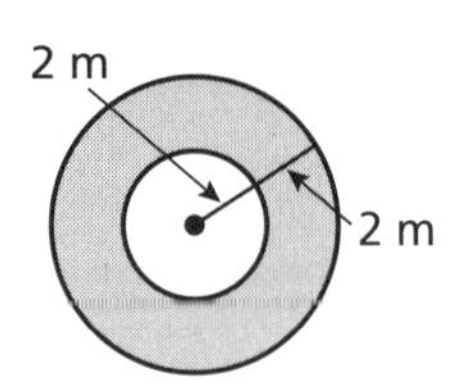

10. Order the letters from least to greatest area. What do they spell?

Name______________________________ Date__________

6.4 Puzzle Time

What Did One Flea Say To Another Flea?

A	B	C	D	E	F
G	H				

Complete each exercise. Find the answer in the answer column. Write the word under the answer in the box containing the exercise letter.

61.12 cm^2 **HOP**
126 cm^2 **A**
46 cm^2 **DO**
100.48 cm^2 **DOG**

Find the area of the shaded region of the figure.

A.

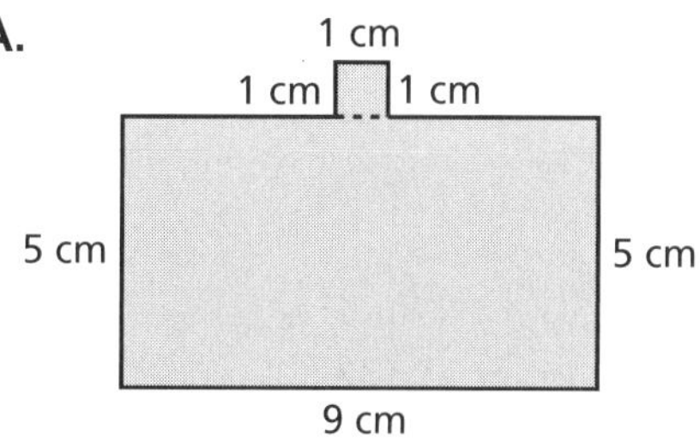

B.

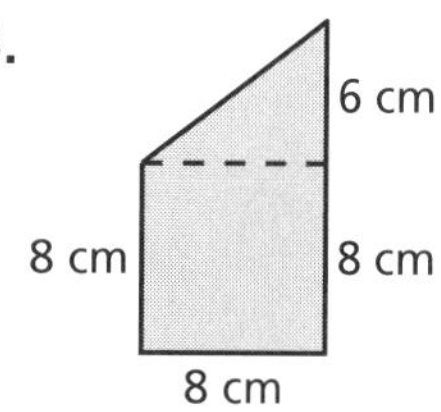

C.

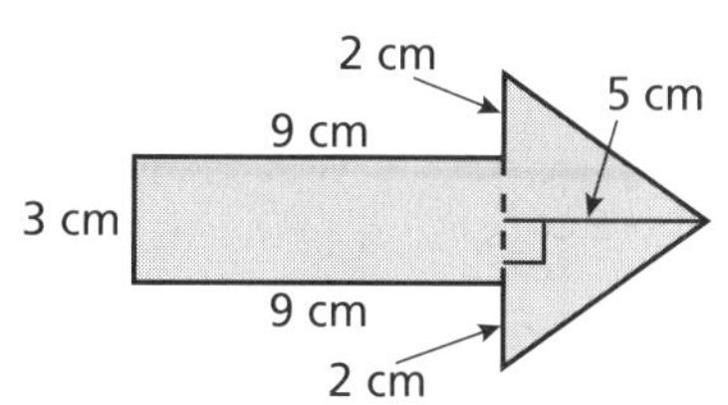

D.

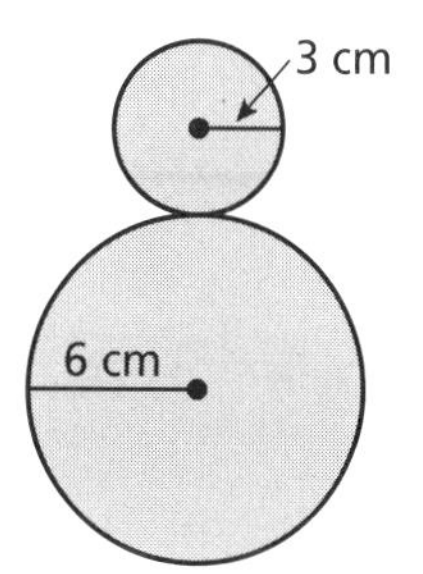

E.

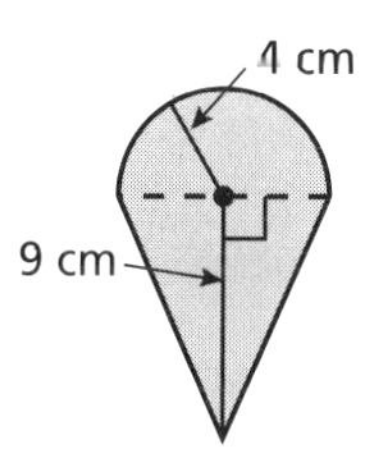

F.

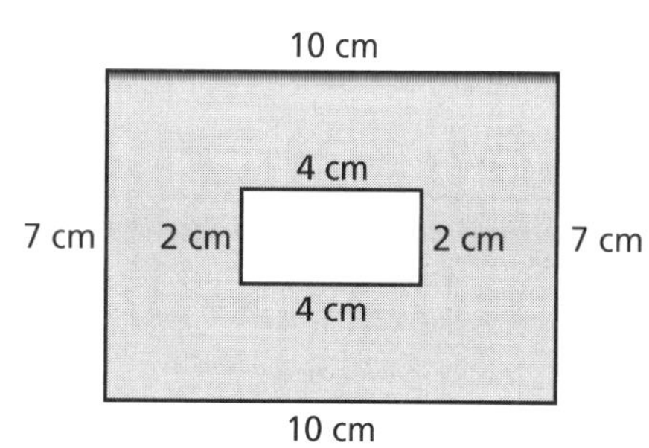

G.

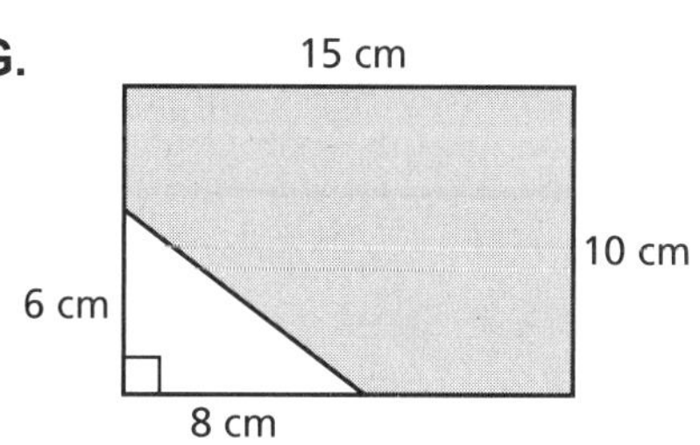

H.

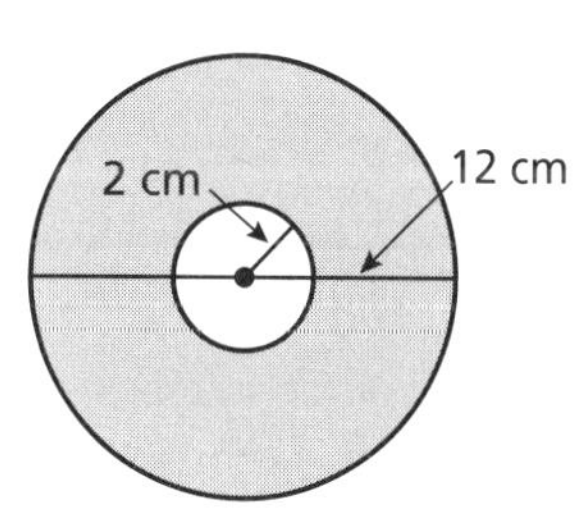

44.5 cm^2 **WALK**
141.3 cm^2 **OR**
88 cm^2 **WE**
62 cm^2 **ON**

Name ______________________________ Date ________

Chapter 6

School-to-Work

For use after Section 6.4

Garden Designer

A garden designer needs to understand plants, climates, and different types of soil. What plants will grow in the garden? How will the plants survive in the shade, sun, or wind? How can the yard or park be more attractive?

A garden designer also needs to consider a client's needs, budget, and ability to maintain the garden. He or she must also be able to calculate how much edging or fencing will be needed for the plant borders, and what areas will be covered in stone, grass, topsoil, mulch, gravel, and so on.

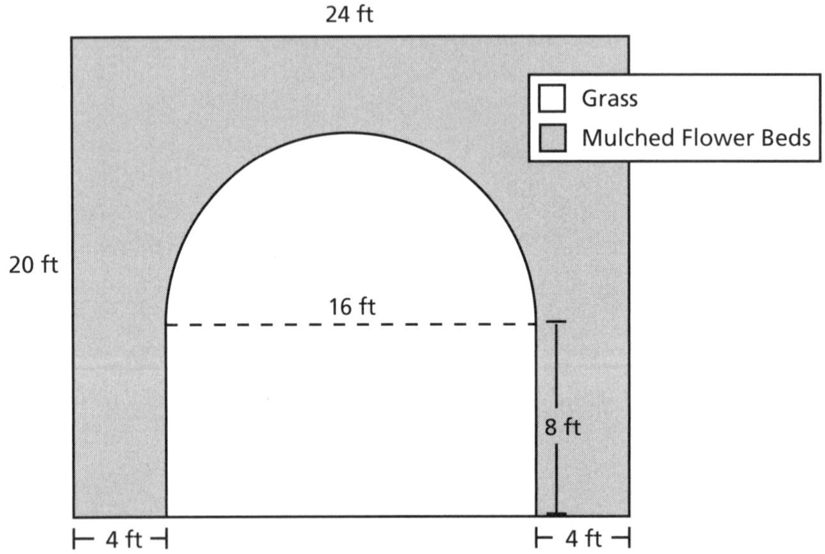

A garden designer created the small yard shown above.

1. How long is the fence around the outside of the yard?

2. Edging will keep the grass out of the flower beds. How many feet of edging are needed? Round up to the nearest foot.

3. How many square feet of grass does the yard contain?

4. How many square feet of flower beds are there? If one cubic foot of mulch covers 4 square feet of flower bed, how much mulch will be needed?

5. Design a patio in the grassy area. How much edging do you need around the patio? How many 1-foot square pavers do you need to make the patio?

6. Design your own plan for a yard or park. Make a poster of your design. Label the dimensions of each section. Include the amount of edging, pavers, and other materials you will need. Research the cost of these items and include prices on your poster. Present your poster as if you were presenting it to a client.

Name______________________________ Date__________

Chapter 6 Study Help

You can use a **word magnet** to organize formulas or phrases that are associated with a vocabulary word or term.

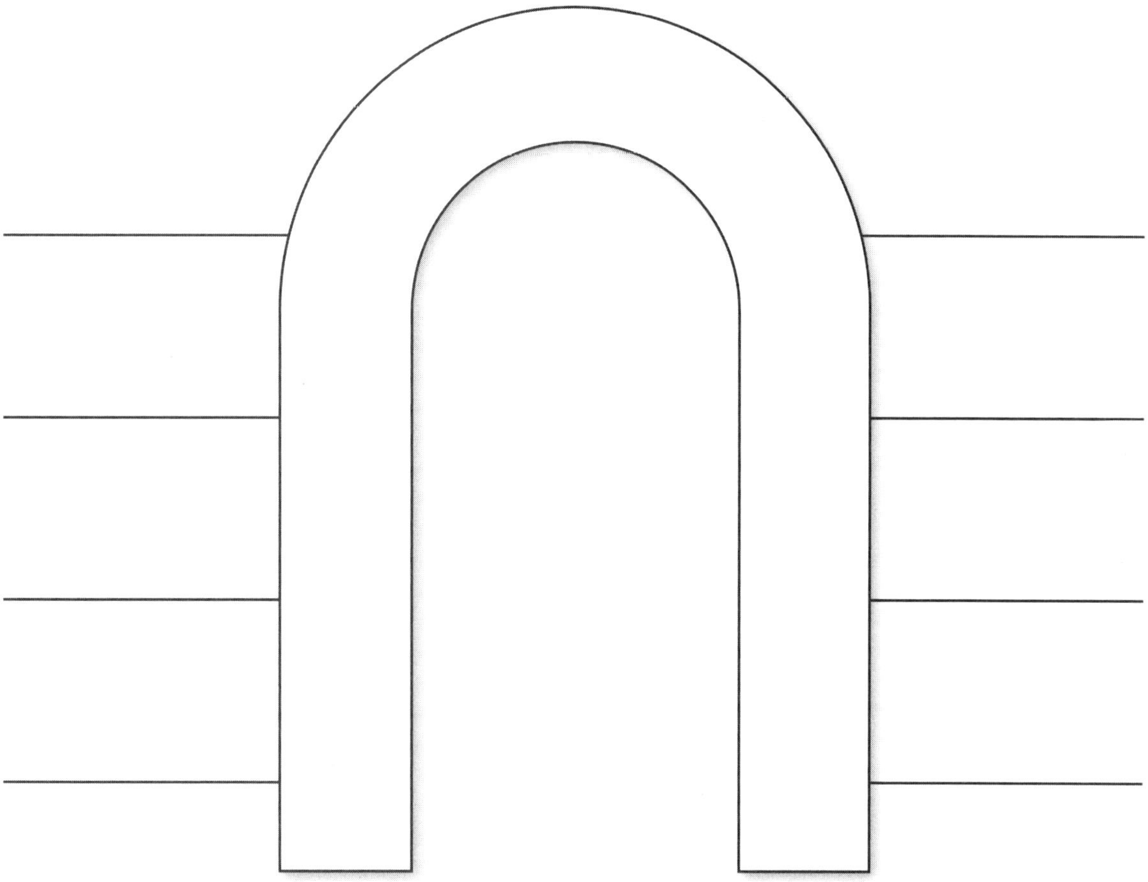

On Your Own

Make a word magnet to help you study these topics.

1. semicircle
2. composite figure
3. perimeter

After you complete this chapter, make word magnets for the following topics.

4. area of a circle
5. area of a composite figure

Name ______________________________ Date __________

Chapter 6 Cumulative Practice

Evaluate the expression.

1. $\frac{1}{2}(8 \bullet 3)$ **2.** $16 + 22 + 18$ **3.** $6.2(1.4)$

4. $2(15) + 2(21)$ **5.** $\frac{1}{2}(12 + 13) \bullet 6$ **6.** $5\left(\frac{3}{5}\right)^2$

Find the indicated measurement. Use 3.14 or $\frac{22}{7}$ for π.

7. Perimeter

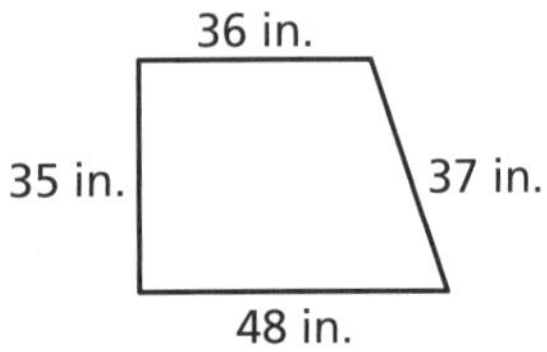

8. Area

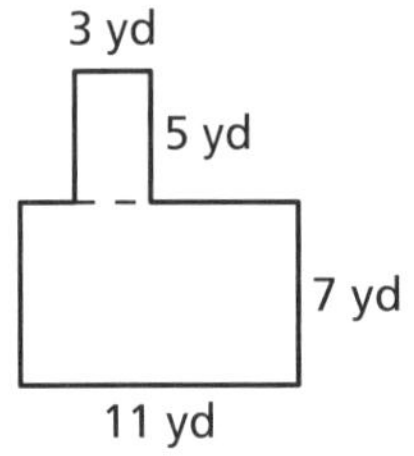

9. Circumference

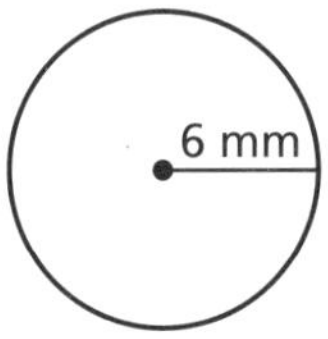

10. Area

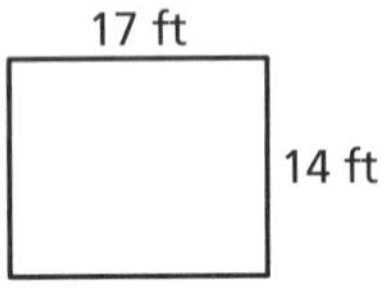

11. Area

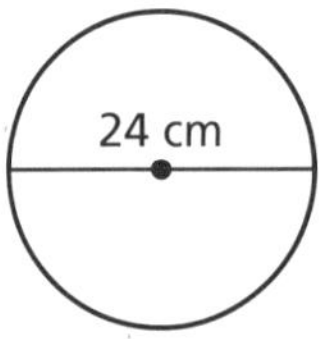

12. Perimeter

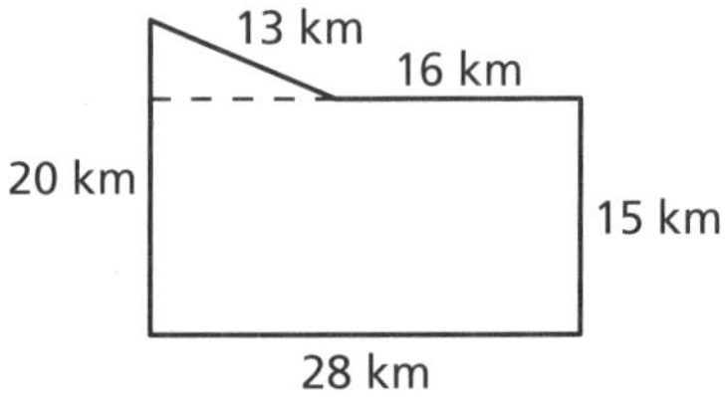

13. Area

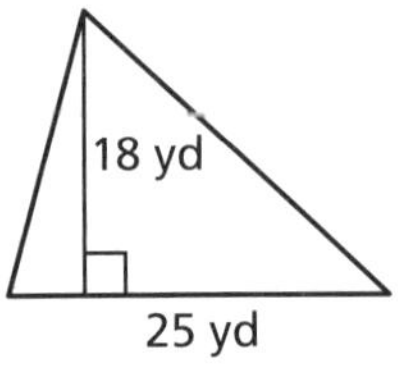

14. Perimeter

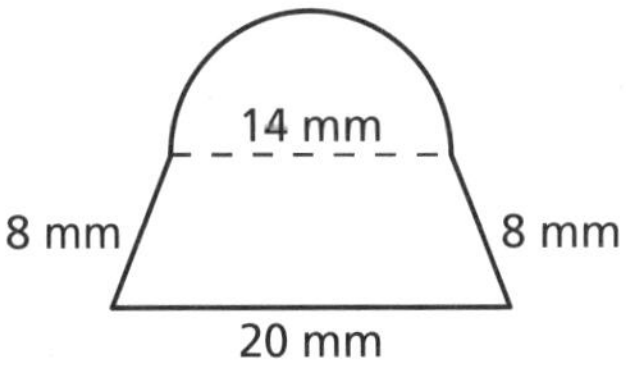

15. Area

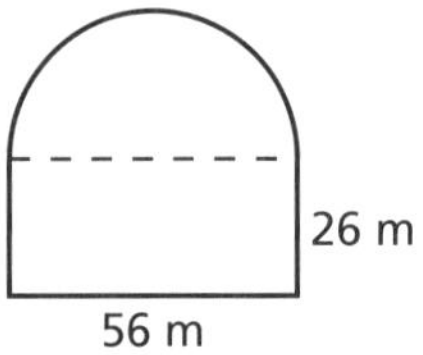

16. A calculator screen is in the shape of a rectangle. The screen is $2\frac{1}{2}$ inches long and $1\frac{3}{4}$ inches wide. What is the area of the screen?

17. A rainbow sticker is in the shape of a semicircle. The diameter is 4 centimeters long. What is the perimeter of the sticker?

18. The radius of the face of a nickel is about 0.4 inch. Find the circumference and area of the face of a nickel.

Name_________________________________ Date__________

Unit 3 Project: Composite Figures and Design

For use after Unit 3

Objective **Use straight lines and parts of circles in a design. Then apply the formulas from the chapter to find areas and perimeters and make a model of your design.**

Materials Models may need paper, cardboard, foam, poster board, markers, fabric, wood, etc.

Investigation Many designs use circles, parts of circles, and straight lines. Design a picture, mural, garden, apartment, park, swimming pool, or other object using straight lines, circles, and parts of circles. Make a sketch of your design.

Find at least 2 perimeters and 2 areas in your design. Explain how you use these calculations in the situation. (For example, to know how many tiles to buy for a bathroom, or how much fencing is needed to enclose a garden, etc.)

Decide whether you know how to find all the distances in your sketch. You may be able to ask your teacher for help with finding areas of unusual shapes.

Decide how to make your model or detailed presentation piece of your plan. What materials will you use?

If you have a more complex design, you may want to simplify.

Completing the Project Turn in your sketches, your model, and a report explaining what you created. Explain why you made the design choices you did. Then give the perimeters and areas you found, and explain why you might want to know them.

Name ______________________________ Date __________

Unit 3 Student Grading Rubric

For use after Unit 3

		Student Score	Teacher Score
Model 100 points			
	a. Name (5 points)	______	______
	b. Project Name (5 points)	______	______
	c. The sketch(es) are neat and well-labeled. (30 points)	______	______
	d. The model is neat and well made. (30 points)	______	______
	e. The model includes both straight lines and circles or parts of circles. (10 points)	______	______
	f. The model uses elements from the chapter. (20 points)	______	______
Report 100 points			
	a. Name (4 points)	______	______
	b. Class (2 points)	______	______
	c. Project Name (2 points)	______	______
	d. Due Date (2 points)	______	______
	e. Explanation of what the project represents. (20 points)	______	______
	f. Explanation of design—why various distances or shapes were used. (20 points)	______	______
	g. Perimeters are found correctly. (15 points)	______	______
	h. Explanation of why you would want to know each perimeter makes sense. (10 points)	______	______
	i. Areas are found correctly. (15 points)	______	______
	j. Explanation of why you would want to know each area makes sense. (10 points)	______	______
FINAL GRADE		_____	_____

Unit 3

Teacher's Project Notes

For use after Unit 3

Materials The materials students use for their models will depend on the type of item they designed. Graph paper and rulers may be helpful for all students. For a garden or building, students may need cardboard for the base and structure, pompoms for bushes, pipe cleaners to mark borders, sugar cubes to build a fountain, etc. Encourage students to be creative.

Alternatives Students could create a full-scale project at school or at home or this could be a class project: design a maze, have the class choose one to model, determine how much material they need to lay it out (e.g. rope on the athletic field, masking tape on the floor of the gym), and then build the full-scale version.

A student could find a real-life example in a reference book or online—e.g. a garden, a circular building, a corn maze. He or she could then determine the perimeters and areas.

Common Errors Watch for students who may be over-ambitious and create a design that is too complex. Requiring each student to turn in a rough sketch before they complete the project will help you control this. Does he or she know how to find the area of an octagon? Are the shapes or design too complicated to model? Will the calculations take too long?

Suggestions You may want to introduce the project with a slide show or images showing numerous designs to help spark ideas. For example, photos of gardens with straight or curved walls and flower beds, a floor plan for a house, mosaics, geometric sculptures, etc.

Discuss how to limit the project to make it doable: A studio apartment with a rectangular common room, quarter-circle kitchen island, and square bathroom is about the right size; their 12-bedroom dream house is too complex. A mural should feature interesting geometric shapes they can work with, not a realistic rendering of dolphins. (Though dolphins in one circle, manatees in a larger circle, and turtles in a third circle could work.)

Make clear what level of effort and detail you expect in the model.

For the report, students may be unsure about what explanations to include. You may want to give students some examples. A shortened example:

In my design of a pool, I included a lap pool, a diving pool, and a spa. I wanted to add something more exciting, so I put in the waterfall and made the pool at the top a half-circle...

I wanted the lap pool to be 50 feet long so I could easily figure out how many 100-yard laps I had done. The lap pool is 6 feet wide so two people can swim at once. I tried different sizes for the spa and 8 feet across seemed to fit best....

Unit 3

Grading Rubric

For use after Unit 3

Scoring Rubric	
A	179-200
B	159-178
C	139-158
D	119-138
F	118 or below

Model 100 points

- **a.** Name (5 points)
- **b.** Project Name (5 points)
- **c.** The sketches are neat and well-labeled. (30 points)
- **d.** The model is neat and well-made. (30 points)
- **e.** The model includes both straight lines and circles or parts of circles (10 points)
- **f.** The model uses elements from the chapter. (20 points)

Report 100 points

- **a.** Name (4 points)
- **b.** Class (2 points)
- **c.** Project name (2 points)
- **d.** Due date (2 points)
- **e.** Explanation of what the project represents. (20 points)
- **f.** Explanation of design—why various distances and shapes were used. (20 points)
- **g.** Perimeters are found correctly. (15 points)
- **h.** Explanation of why you would want to know each perimeter makes sense. (10 points)
- **i.** Areas are found correctly. (15 points)
- **j.** Explanation of why you would want to know each area makes sense. (10 points)

Chapter 7

Name ______________________________ Date __________

Chapter 7 Equations

Dear Family,

Have you ever had to plan a large party—perhaps a family reunion, a wedding, or a community fundraiser? Planning for a large event can be quite a challenge. Recruiting your student to help with the planning provides a great opportunity for them to use their math skills.

For example, you could ask your student to figure out the following.

- How much food is needed? Should you plan on just one portion per person, or multiple portions? Have your student write a rule (or equation) to determine the number of portions of food you need. Your student can write a rule even if you don't know how many people will be attending when you first start planning.
- Is the number of invitations needed equal to the number of people being invited? Have your student write a rule for the number of invitations you need and another rule for the cost of the postage.
- Each table can probably seat 8 or 10 people. You'll want to figure out how many tables you will need. Have your student write a math rule to determine this amount.
- How many tablecloths and table decorations will you need? If there will be serving tables, don't forget about decorating those as well.

Event planners often say that about two-thirds to three-quarter of invitees can be counted on to attend. Work with your student on a strategy to guess how many people you think will actually attend. Then have your student use the rules they wrote to estimate the number of portions, invitations, tables, and decorations that will be needed for the event.

Is your event a fundraiser? If so, figure out how much you will charge per person. Figure out how much you will spend on the whole event. Have your student write a rule to determine if you will make money for your cause.

You and your student can take satisfaction from your good planning—enjoy the event!

Nombre ______________________________ Fecha __________

Capítulo 7 Ecuaciones

Estimada Familia:

¿Alguna vez han tenido que planear una gran fiesta—quizás una reunión familiar, un matrimonio, o un evento de recaudación de fondos para la comunidad? Planear un evento grande puede ser un gran desafío. Incluir a su estudiante para ayudar con la planificación le ofrece una buena oportunidad de usar sus habilidades matemáticas.

Por ejemplo, puede pedir a su estudiante que averigüe lo siguiente:

- ¿Cuánta comida se necesita? ¿Debe planear solamente una porción por persona o múltiples porciones? Haga que su estudiante escriba una regla (o ecuación) para determinar el número de porciones de comida que necesita. Su estudiante puede escribir una regla aun si no se sabe cuántas personas asistirán cuando se empezó con la planificación.
- ¿El número de invitaciones requeridas es igual al número de personas invitadas? Haga que su estudiante escriba una regla para el número de invitaciones que necesita y otra regla para el costo del envío por correo.
- Probablemente cada mesa sea para 8 ó 10 personas. Querrá saber cuántas mesas necesita. Haga que su estudiante escriba una regla matemática para determinar esta cantidad.
- ¿Cuántos manteles y decoraciones de mesa necesitará? Si van a haber mesas para servirse, no olvide decorarlas también.

Los planificadores de eventos a menudo dicen que se puede calcular que alrededor de dos tercios a tres cuartos de los invitados asistirán. Trabaje con su estudiante en una estrategia para adivinar cuántas personas se cree que realmente asistirán. Luego haga que su estudiante use las reglas que escribieron para calcular el número porciones, invitaciones, mesas y decoraciones que se necesitarán para el evento.

¿Se trata de un evento de recaudación de fondos? De ser así, averigüe cuánto puede cobrar por persona. Averigüe cuánto gastará en todo el evento. Haga que su estudiante escriba una regla para determinar si va a obtener ganancias para su causa.

Usted y su estudiante pueden beneficiarse de su buena planificación—disfruten el evento!

Non ______________________________ Dat __________

Chapít 7 Ekwasyon

Chè Fanmi:

Èske sa janm rive ou pou ou gen pou planifye yon gwo fèt—petèt yon reyinyon familyal, yon maryaj, oswa yon evènman pou kolekte lajan nan kominote a? Planifye yon gwo evènman byen kapab yon defi. Rekrite elèv ou a pou ede avèk planifikasyon an ofri yon bèl opòtinite pou li sèvi ak aptitid matematik li.

Paregzan, ou ta kapab mande elèv ou a pou li jwenn bagay sila yo.

- Ki kantite manjeki nesesè? Èske ou ta dwe planifye sèlman yon pòsyon pa moun, oubyen plizyè pòsyon? Fè elèv ou a ekri yon règ (oswa ekwasyon) pou detèmine ki kantite pòsyon manje ou bezwen. Elèv ou a kapab ekri yon règ menm si ou pa konnen ki kantite moun k'ap vini lè ou fèk kòmanse planifye.
- Èske kantite envitasyon an bezwen egal ak kantite moun k'ap envite yo? Fè elèv ou a ekri yon règ pou kantite envitasyon ou bezwen an ak yon lòt règ pou kou tarif postal la.
- Chak tab kapab siman pran 8 oswa 10 moun. W'ap vle jwenn konbyen tab w'ap bezwen. Fè elèv ou a ekri yon règ matematik pou detèmine kantite sa a.
- Konbyen nap ak dekorasyon tab w'ap bezwen? Si ap gen tab sèvis, pa bliye dekore sa yo tou.

Òganizatè evènman di souvan ke ou kapab konte sou ant apeprè de tyè ak twa ka envite yo pou yo vini. Travay avèk elèv ou a sou yon estrateji pou devine konbyen moun nou panse ap vini toutbon. Apre sa fè elèv ou a sèvi ak règ li ekri yo pou estime kantite pòsyon, envitasyon, tab, ak dekorasyon k'ap nesesè pou evènman an.

Èske evènman ou an se pou kolekte lajan?Si se sa, jwenn konbyen w'ap fè chak moun peye. Jwenn konbyen w'ap depanse pou tout evènman an. Fè elèv ou a ekri yon règ pou detèmine si w'ap fè lajan pou kòz ou a.

Ou menm ak elèv ou a kapab jwenn satisfaksyon nan bon planifikasyon nou an—jwi fèt la!

Activity 7.1 Start Thinking!

For use before Activity 7.1

Look back at Chapter 6 and choose a word problem. Rewrite the problem in your own words using fewer words.

Exchange problems with a classmate and solve your classmate's problem. Did your classmate provide all of the necessary information for you to solve the problem?

List some advantages and disadvantages of rewriting a word problem in your own words.

Activity 7.1 Warm Up

For use before Activity 7.1

Solve the problem.

1. A car travels at an average rate of 30 miles per hour for 2 hours. How far has the car traveled?

2. A picture frame is shaped like a rectangle with a length of 10 inches and a width of 8 inches. What is the area of the picture frame?

3. You are walking 1 mile home from school. When you are halfway home, you realize that you forgot a book in your locker and have to go back and get it. Once you finally get home, how far have you walked?

Lesson 7.1 Start Thinking!
For use before Lesson 7.1

Describe some instances of when you have used two expressions that are equal to each other in your daily life.

Lesson 7.1 Warm Up
For use before Lesson 7.1

Rewrite the problem until it is easy to read. Then solve the problem.

1. You are walking at a steady pace of 15 minutes per mile on a treadmill at the gym. While on the treadmill, you watch one 30-minute long news program. How far did you walk during the news program?

2. In one game, $\frac{2}{3}$ of a pitcher's pitches are strikes. The other $\frac{1}{3}$ of the pitches are balls. If the pitcher throws 99 pitches in the game, how many of the pitches are strikes?

Name____ Date__________

7.1 Practice A

Rewrite the problem until it is easy to read. Then solve the problem.

1. The handwritten rough draft of your book report is 4 pages long. Typing at a steady rate of 20 words per minute, it takes you 48 minutes to enter the report onto a computer. In typed form, the report is about 3 pages long. How many words long is your report?

2. You and a friend are riding your bikes home from a park. You ride at a steady rate. The park is 6 miles from your home. About halfway home, the wind starts blowing at about 5 miles per hour from your left. You keep riding at the same rate. You arrive home after one-half hour of riding. What was your riding rate?

Write the word sentence as an equation.

3. 4 is the quotient of 16 and a number n.
4. The sum of a number k and 5 is 11.
5. The difference of a number x and 2 is 7.
6. 7 times a number a is 42.
7. 6 is one-third of a number s.
8. A number t added to 13 equals 17.
9. 5 less than a number b is 16.
10. 30 is 6 multiplied by a number y.
11. A number z divided by 7 equals 4.
12. 5 more than a number m equals 8.
13. Describe the error in writing the sentence as an equation.

✗	24 is 8 multiplied by a number x. $24x = 8$

14. A crew is picking up chairs in the gymnasium after an event. After clearing 217 chairs from the gym floor, 36 chairs still remain. Write an equation you can use to find the number of chairs placed on the gym floor for the event.

15. You and three of your cousins shuck six dozen ears of corn for a picnic. Write an equation you can use to find the number of ears n for each person if each of you shucks the same number of ears.

16. You drop the rubber ball from height h. It bounces back to 60% of its original height. Write an equation you can use to find the height h.

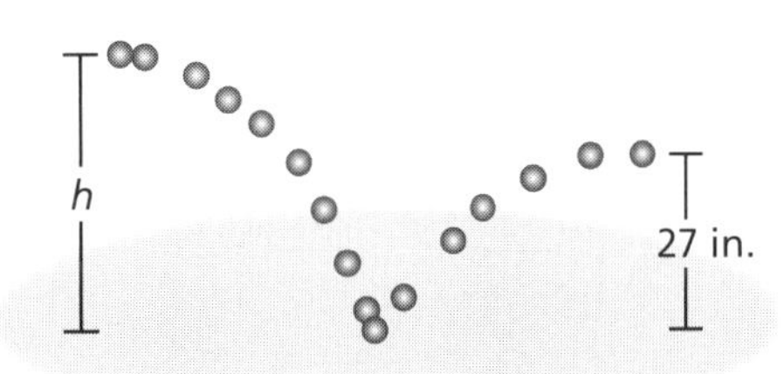

17. You earn 4070 points in level 1 of a video game and 2710 points in level 2. After level 3, your point total is 8180. Write an equation you can use to find the number of points p you scored in level 3.

Name ____________________ Date __________

7.1 Practice B

Write the word sentence as an equation.

1. 17 is n less than 41.
2. 168 divided by a number x equals 14.
3. The product of 17 and a number s is 153.
4. A number b increased by 37 is 112.
5. 23 is the quotient of a number t and 61.
6. 114 is 37 more than a number g.
7. 46 less than a is 33.
8. One-fifth of a number d is 22.

Write an equation that can be used to find the value of *x*.

9. Circumference of circle: 36π ft

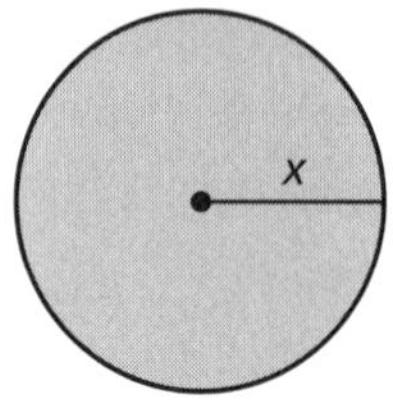

10. Area of triangle: 108 cm^2

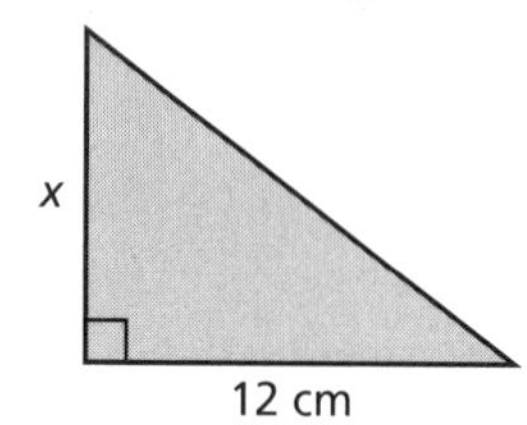

11. You want to put 520 quarters in coin wrappers. You need one wrapper for every $10 in quarters. Write an equation you can use to find how many wrappers w you need.

12. You use a metal detector at the beach. You find 2 quarters, 12 dimes, and 23 pennies. Write an equation you can use to find how many more pennies p you need to find in order to have a total of $2.00.

13. In one minute, you climb halfway up a rock wall. In another minute, you are 24 feet above the ground after covering half of the remaining height. Write an equation you can use to find the total height h of the rock wall.

14. A golf driving range has small buckets of golf balls for $6 each and medium buckets of golf balls for $8 each. One day, golfers use 27 small buckets and some medium buckets for a total cost of $626. Write an equation you can use to find the number m of medium buckets used.

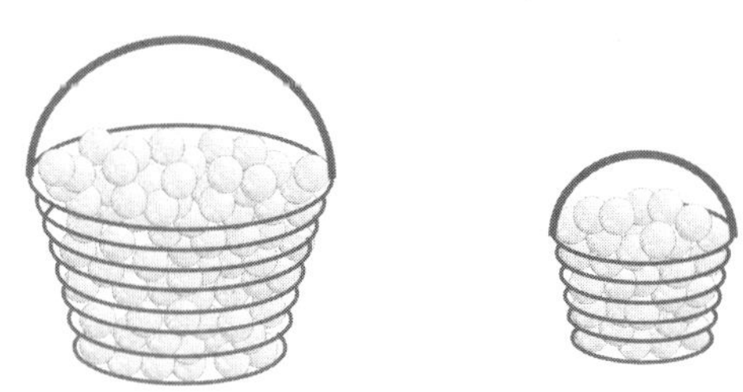

15. A silkworm winds its cocoon out of one long silk fiber. To make silk thread, 3 to 10 of these silk fibers are unwound from their cocoons and combined into a single thread with a typical length of 300 yards.

    a. Explain why you cannot write an equation to find the exact total length of the silk fibers used in a 300-yard silk thread.

    b. Choose a reasonable number of silk fibers in a thread. Write an equation to find the total length of the silk fibers used in 300 yards of the thread.

Name______________________________ Date__________

7.1 Enrichment and Extension

Writing Other Types of Equations

Write the word sentence as an equation.

1. A number *n* squared plus 7 is 23.

2. Ten decreased by a number *p* squared is 9.

3. A number *x* minus another number *y* is equal to 8.

4. A number *x* squared plus another number *y* squared equals 4.

Maria and Sam have been asked to write the following word sentence as an equation:

A number x plus 2 squared equals 9.

Their answers are shown.

Maria: $x + 2^2 = 9$ **Sam:** $(x + 2)^2 = 9$

5. Who has the correct answer?

6. Explain the error in the other solution.

7. How could you word the sentence for the other equation so there is no confusion?

8. Use your answer to Exercise 7 to write a word sentence for the equation $(3 - x)^2 = 4$.

Name ______________________________ Date __________

7.1 Puzzle Time

Why Did The Sea Monster Eat Six Ships That Were Carrying Potatoes?

A	B	C	D	E	F
G	H	I	J		

Complete each exercise. Find the answer in the answer column. Write the word under the answer in the box containing the exercise letter.

$\frac{x}{3} = 12$ JUST
$550 + x = 1250$ SHIP
$x + 5 = 14$ IT
$x - 13 = 15$ ONE
$24 = 4 + x$ CAN

Write the word sentence as an equation.

A. The sum of a number x and 5 equals 14.

B. A number x decreased by 6 is 5.

C. 7 times a number x is 42.

D. A number x divided by 8 equals 11.

E. 24 equals 4 more than a number x.

F. 9 is one-third of a number x.

G. 12 is the quotient of a number x and 3.

H. 13 less than a number x equals 15.

I. You throw a football 20 yards. Your friend throws the same football x yards. The football was thrown a total distance of 50 yards. Write an equation you can use to find the distance x that your friend threw the football.

J. Students raised \$550 by having a car wash. They need \$1250. Write an equation you can use to find the amount x that the students still need to raise.

$\frac{x}{8} = 11$ ONE
$20 + x = 50$ POTATO
$7x = 42$ NO
$9 = \frac{1}{3}x$ EAT
$x - 6 = 5$ SEEMS

Activity 7.2 Start Thinking!

For use before Activity 7.2

If you are trying to weigh a pet, it is often easier to stand on a scale holding the pet to find your combined weight.

How is the situation related to solving an equation?

Hint: Use the weight of the pet as the variable in the equation.

Activity 7.2 Warm Up

For use before Activity 7.2

Evaluate the expression.

1. $3 + x$ when $x = 2$

2. $a + 16$ when $a = 4$

3. $10 + z$ when $z = 5$

4. $17 - c$ when $c = 9$

5. $8 - w$ when $w = 8$

6. $s - 20$ when $s = 34$

Lesson 7.2 Start Thinking!
For use before Lesson 7.2

Two operations are **inverse operations** if one operation "undoes" the other.

Of the four operations (addition, subtraction, multiplication, division), which are inverse operations? Why?

This idea can be applied to real-life situations. What are the inverses of the following actions?

1. Unbutton

2. Ascend

3. Deposit

4. Turn off

Describe another pair of real-life inverse actions.

Lesson 7.2 Warm Up
For use before Lesson 7.2

Use a scale to model and solve the equation.

1. $n + 9 = 12$

2. $r + 4 = 8$

3. $p + 1 = 4$

4. $t - 2 = 4$

5. $d - 5 = 2$

6. $x - 3 = 1$

Name______________________________ Date__________

7.2 Practice A

Tell whether the given value is a solution of the equation.

1. $x + 16 = 20;\ x = 4$

2. $p - 4 = 28;\ p = 32$

3. $4w = 44;\ w = 10$

4. $\dfrac{y}{6} = 6;\ y = 24$

Solve the equation. Check your solution.

5. $x - 5 = 9$

6. $y - 12 = 0$

7. $q + 8 = 25$

8. $f - 22 = 14$

9. $8 + s = 10$

10. $r - 3.2 = 1.7$

11. $8.9 = v + 7.3$

12. $\dfrac{1}{3} + n = \dfrac{2}{3}$

13. $\dfrac{2}{3} = \dfrac{1}{4} + g$

Describe and correct the error in solving the equation.

14. ✗

$$\begin{aligned} 13 + m &= 56 \\ \underline{+13} \quad & \quad \underline{+13} \\ m &= 69 \end{aligned}$$

15. ✗

$$\begin{aligned} 27 &= n - 15 \\ \underline{-15} \quad & \quad \underline{+15} \\ 12 &= n \end{aligned}$$

Write the word sentence as an equation. Then solve the equation.

16. 20 equals 8 more than a number y.

17. The sum of a number x and 12 equals 15.

18. 4 less than a number g equals 9.

19. A number w decreased by 10 is 3.

20. The height of a desk is 11 inches shorter than the height of a chair. Write and solve an equation to find the height of the desk.

21. The Florida Panther has an average height of 30 inches. It is 20 inches taller than the Northern Mockingbird. Write and solve an equation to find the average height of the Northern Mockingbird.

22. The advertised price of a cell phone is \$149 after a \$50 mail-in rebate. Write and solve an equation to find the price of the cell phone before the rebate is applied.

Name ______________________________ Date __________

7.2 Practice B

Tell whether the given value is a solution of the equation.

1. $2.5w = 12.5;\ w = 5$

2. $\frac{y}{8} = 7;\ y = 64$

3. $39 = 3.9t;\ t = 10$

4. $\frac{1}{4} = \frac{1}{8}m;\ m = 2$

Write the word sentence as an equation. Then solve the equation.

5. A number a decreased by 13.4 is 2.6.

6. 27 less than a number h equals 3.5.

7. 46 equals 2.5 more than a number z.

8. The sum of a number b and 4.7 equals 10.9.

Solve the equation. Check your solution.

9. $x - 72 = 136$

10. $251 = 148 + j$

11. $\frac{4}{5} + a = 1$

12. $n - 10 = 13 + 5$

13. $v + 17 - 11 = 65$

14. $47 - 15 + c = 79$

15. $7 + 57 = 3 + y$

16. $30 + 12 = e - 42$

17. $21 - 16 + \ell = 14 - 4$

Write and solve an addition equation to find *x*.

18. Perimeter = 30 in.

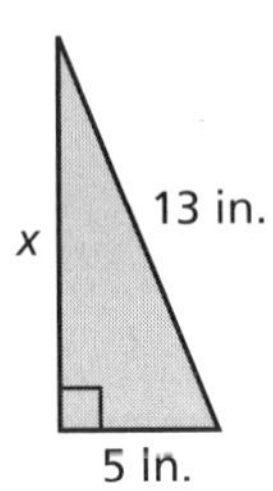

19. Perimeter = 43 m

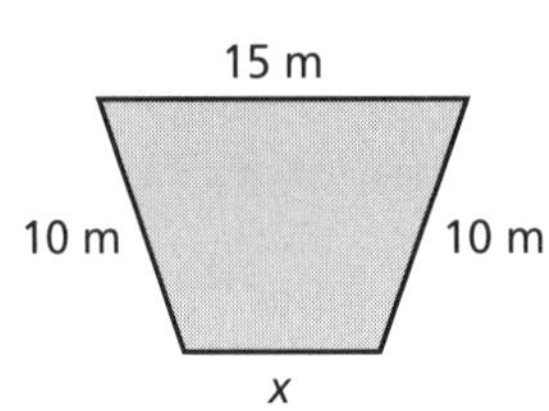

20. Perimeter = 16 ft

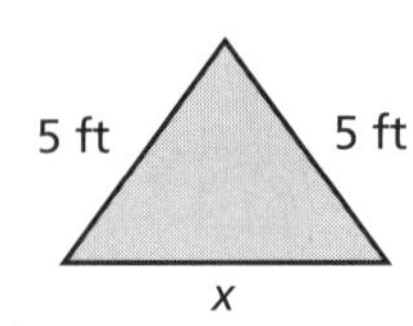

21. You are grocery shopping. You have $12.

a. Write and solve three equations to find the cost m of the milk, the cost c of the cereal, and the cost e of the eggs.

b. How much money do you have left if you purchase one of each item?

Grocery Items

Bread: $2.19
Milk: $1.56 more than bread
Cereal: $3.20 more than eggs
Eggs: $2.36 less than milk

22. A jacket is on sale for $10 off. You have a coupon worth $5.80 that brings the cost of the jacket down to $33.19. Write and solve an equation to find the original cost c of the jacket.

Name____ Date __________

7.2 Enrichment and Extension

Writing and Solving Different Types of Equations

Tell whether the given value is a solution of the equation.

1. $x^2 - 5 = 20; x = 5$

2. $x^2 + 11 = 45; x = 6$

3. $1 - x^2 = 0; x = 1$

4. $110 - x^2 = 19; x = 9$

5. $2x + x = 10; x = 4$

6. $5x - 3x = 6; x = 2$

7. $(x + 5)^2 = 144; x = 7$

8. $(x - 2)^2 + 1 = 2; x = 3$

9. Write two different addition equations that have 10 as the solution.

10. Write two different subtraction equations that have 7 as the solution.

11. Your softball team has 8, 10, 9, and 3 hits during four games. Write and solve an equation to find how many hits the team needs in the fifth game to have at total of 40 hits.

12. Write and solve an equation to find how many miles you must hike on Friday to hike a total of 12 miles for the five days.

Day	Monday	Tuesday	Wednesday	Thursday	Friday
Miles Hiked	2.5	$1\frac{3}{4}$	$3\frac{1}{4}$	2	?

Name ______________________________ Date __________

7.2 Puzzle Time

What Do Kitty Cats Like To Eat For Breakfast?

Write the letter of each answer in the box containing the exercise number.

Solve the equation. Check your solution.

1. $p - 8 = 4$
2. $k - 2 = 12$
3. $9 = h - 15$
4. $y + 4 = 7$
5. $z + 5 = 21$
6. $63 = r + 31$
7. $x - 25 = 16$
8. $26 = m + 18$
9. $\frac{2}{3} = a - \frac{2}{3}$
10. $f + \frac{1}{4} = \frac{7}{8}$
11. $2.3 = q - 3.6$
12. $j + 4.4 = 16.2$

Answers

K. 16

I. $\frac{5}{8}$

E. 24

S. 14

R. 5.9

C. 41

I. 32

P. 12

S. 8

M. 3

E. 11.8

I. $1\frac{1}{3}$

4	10	7	12		5	11	9	2	1	6	3	8

Activity 7.3

Start Thinking!

For use before Activity 7.3

Write a problem that can be modeled with the equation $4x = 28$. Be sure to provide all the necessary information.

Activity 7.3

Warm Up

For use before Activity 7.3

Solve the equation. Check your solution.

1. $x + 7 = 10$

2. $y + 13 = 21$

3. $z + 3 = 27$

4. $a - 12 = 16$

5. $b - 9 = 4$

6. $15 = c - 1$

Lesson 7.3 Start Thinking!

For use before Lesson 7.3

Explain how to use the distance formula $(d = rt)$ to find your average rate getting to school in the morning. Be sure to include units in your explanation.

Lesson 7.3 Warm Up

For use before Lesson 7.3

Solve the equation. Check your solution.

1. $5x = 205$

2. $16 = 8w$

3. $77 = 7b$

4. $7 = v \div 6$

5. $30 = p \div 2$

6. $k \div 4 = 26$

Name ______________________________ Date __________

7.3 Practice A

Solve the equation. Check your solution.

1. $\frac{x}{2} = 9$

2. $4 = \frac{t}{4}$

3. $\frac{w}{10} = 8$

4. $s \div 7 = 6$

5. $5a = 15$

6. $8 \bullet d = 40$

7. $60 = 20m$

8. $7g = 14$

9. $9y = 72$

10. $3 \bullet n = 63$

11. $4 = \frac{v}{11}$

12. $\frac{c}{7} = 5$

13. $\frac{b}{5} = 2.2$

14. $h \div 3 = 50$

15. $24k = 60$

16. $210 = 7r$

Describe and correct the error in solving the equation.

17.

X

$$\frac{x}{9} = 3$$

$$9 \bullet \frac{x}{9} = 3$$

$$x = 3$$

18.

X

$$4 \bullet z = 32$$

$$\frac{4 \bullet z}{4} = 4 \bullet 32$$

$$z = 128$$

19. A teacher tells 36 students to form 4 equal groups. Write and solve a multiplication equation to find how many students s there should be in each group.

20. You have been saving \$12 each week for many weeks. One day, you decide to count your savings and find that you have \$384. Write and solve a multiplication equation to find how many weeks w you have been saving.

21. A 14-inch pepperoni pizza is sliced into 8 equal pieces. One slice contains 352 calories. Write and solve a division equation to find how many calories c there are in the whole pizza.

22. In a cheerleading competition, each team receives scores in eight categories. One team has a mean score of 7.25 in the eight categories. Write and solve an equation to find the team's total score s.

23. Write and solve an equation to find the width w of the rectangle. Explain how you know what units to use with the answer.

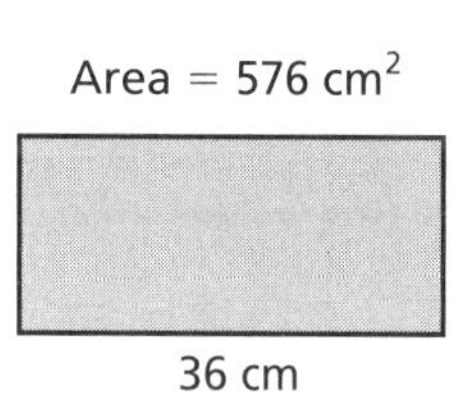

Name ______________________________ Date __________

7.3 Practice B

Solve the equation. Check your solution.

1. $6a = 78$

2. $7 = \frac{s}{20}$

3. $96 = 24p$

4. $g \div 18 = 6$

5. $\frac{v}{27} = 4$

6. $13 \bullet x = 84.5$

7. $\frac{c}{16.2} = 3.8$

8. $12.4b = 117.8$

Solve for *x*. Check your answer.

9. Rectangle

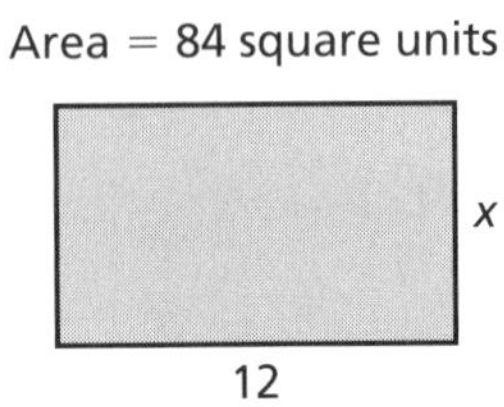

10. Triangle

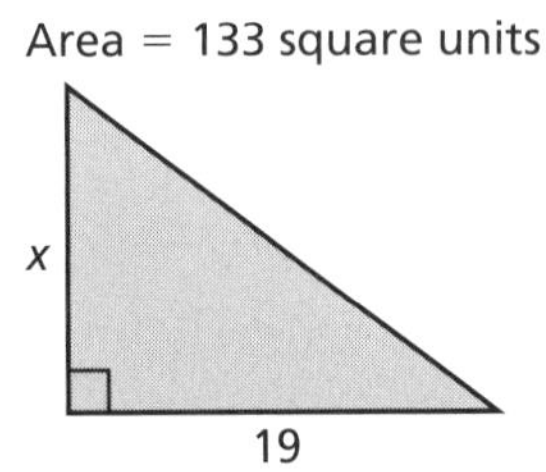

11. Circle

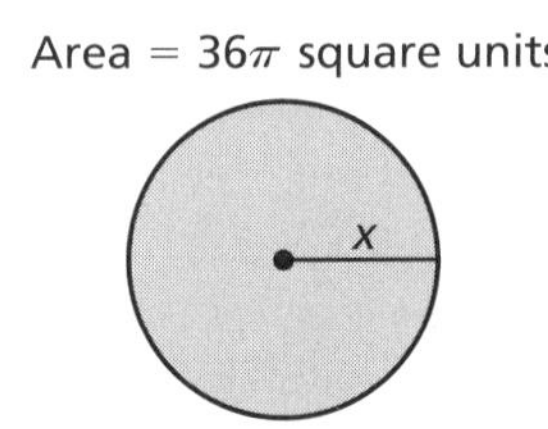

12. You want to know the diameter of the circular swimming pool. Explain how you can use the circumference. Then find the diameter to the nearest foot.

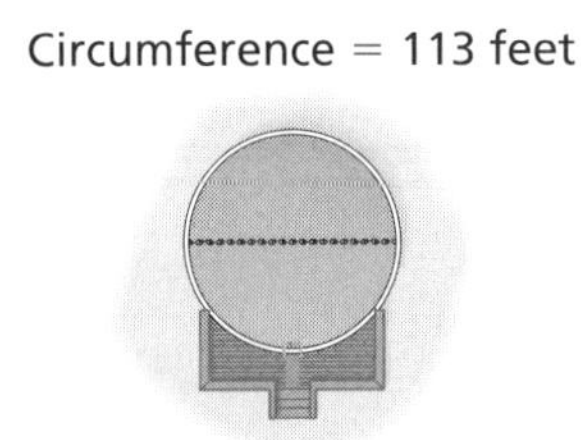

13. Your family has a car with a 15-gallon gas tank. Find the price of one gallon of gas in your area. Then write and solve a division equation to find the price p to fill the car's gas tank when it is empty.

14. Using a special discount, you download 15 songs for \$10.68. The regular price of each song is \$0.89. What is the percent of the discount?

15. In a cranberry punch recipe, four cups of cranberry juice and one quart of pineapple juice make up half of the punch. How many cups of punch does the recipe make?

16. You buy dog biscuits at the bulk rate price of \$2.25 per pound. The scale shows the weight of 25 biscuits. Find the cost of 40 biscuits. Explain how you found your answer.

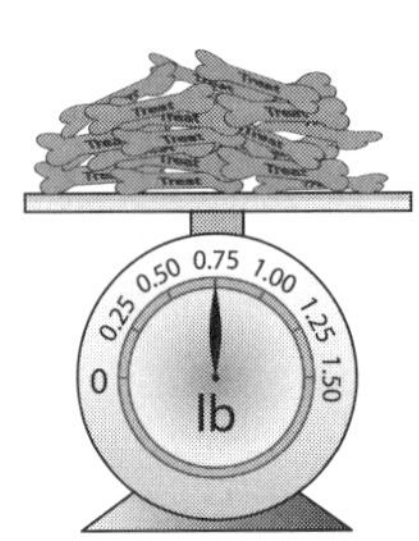

Name____ Date__________

7.3 Enrichment and Extension

Writing Repeating Decimals as Fractions

Example: Write $0.\overline{21}$ as a fraction.

Step 1: Let $x = 0.\overline{21}$ or 0.212121….

Step 2: Because the decimal has 2 repeating digits, multiply each side of $x = 0.212121...$ by 10^2, or 100.

$x = 0.212121\ldots$ Write equation.

$100x = 21.212121\ldots$ Multiply each side by 100.

Step 3: Subtract x from $100x$.

$100x = 21.212121\ldots$ Write equation.

$-x = 0.212121\ldots$ Subtract x.

$99x = 21$ Simplify.

Step 4: Solve for x.

$99x = 21$ Write equation.

$\frac{99x}{99} = \frac{21}{99}$ Divide each side by 99.

$x = \frac{21}{99}$, or $\frac{7}{33}$ Simplify.

So, $0.\overline{21}$ is equivalent to $\frac{7}{33}$.

Write the decimal as a fraction or mixed number.

1. $0.\overline{7}$ **2.** $0.\overline{6}$ **3.** $0.\overline{36}$

4. $0.\overline{51}$ **5.** $0.\overline{43}$ **6.** $0.\overline{09}$

7. $4.\overline{89}$ **8.** $0.\overline{735}$ **9.** $0.\overline{029}$

10. How can you use the method shown above to write the decimal $0.2\overline{3}$ as a fraction? Explain.

Name ______________________________ Date __________

7.3 Puzzle Time

What Did The Dirt Say When It Began To Rain?

A	B	C	D	E	F
G	H	I	J	K	L
M	N	O	P		

Complete each exercise. Find the answer in the answer column. Write the word under the answer in the box containing the exercise letter.

7 NAME
56 WILL
54 IF
9 MUD
10 MY
13 WILL
2 CALLED
40 HEAVY

Solve the equation. Check your solution.

A. $\frac{a}{9} = 6$

B. $7 = \frac{z}{6}$

C. $y \div 4 = 10$

D. $25 = \frac{k}{5}$

E. $2s = 16$

F. $8 \bullet t = 96$

G. $50 = 5x$

H. $56 = 8k$

I. $4b = 52$

J. $39 = 6 \bullet c$

K. $14 = n \div 5$

L. $10 = v \div 6$

M. $x \div 16 = 3.5$

N. $\frac{w}{25} = 4.4$

O. $11.5 \bullet d = 23$

P. $4.5v = 40.5$

12 UP
6.5 CHANGE
125 RAIN
110 BE
42 THIS
8 KEEPS
60 I
70 AND

Activity 7.4 Start Thinking!

For use before Activity 7.4

Choose a number. Write it down. Then follow these steps.

Step 1 Add 3 to your number.

Step 2 Multiply the result by 2.

Step 3 Subtract 6 from the result.

Step 4 Divide the result by 2.

What is your final result? Try a different number. Do you still get the same result?

What does this game have to do with inverse operations?

Activity 7.4 Warm Up

For use before Activity 7.4

Describe how you can "undo" the operation described.

1. Add 3.

2. Subtract 5.

3. Multiply by 6.

4. Divide by 2.

5. Add 4.57.

6. Multiply by 2.5.

Lesson 7.4

Start Thinking!

For use before Lesson 7.4

How are two-step equations related to the cost of purchasing prints from an online photo website?

Lesson 7.4

Warm Up

For use before Lesson 7.4

Solve the equation. Check your solution.

1. $4 + \frac{x}{7} = 12$

2. $\frac{a}{3} - 1 = 5$

3. $\frac{y}{5} + 2 = 6$

4. $2p - 3 = 11$

5. $6t + 14 = 32$

6. $5w - 2 = 53$

Name____ Date__________

7.4 Practice A

Solve the equation. Check your solution.

1. $2j + 1 = 5$

2. $\frac{b}{3} - 7 = 2$

3. $6 = 4a - 10$

4. $60 = 4m + 12$

5. $2z - 8 = 20$

6. $\frac{r}{5} + 9 = 10$

7. $5t + 10 = 35$

8. $18 = \frac{f}{7} + 11$

9. $\frac{n}{10} + 13 = 27$

10. $7z + z = 72$

11. $3j + 2j = 15$

12. $9e - 3e = 48$

13. $x - 24 = 36 - 6$

14. $1 = \frac{s}{15} - 2$

15. $15 = 10 + y + 2$

Describe and correct the error in solving the equation.

16. ✗

$$\frac{d}{2} - 5 = 8$$
$$d - 5 = 16$$
$$d = 21$$

17. ✗

$$16 = 4g - 4$$
$$12 = 4g$$
$$3 = g$$

18. A gym charges $12 per swimming class and a one-time registration fee of $10. A student paid a total of $82. Write and solve an equation to find the number of swimming classes the student took.

19. You spend $75 to go surfing. It costs $10 per day to rent a surfboard and $15 per day to rent a wetsuit. Write and solve an equation to find the number of days you rent the surfboard and wetsuit.

20. You and 9 friends purchase tickets to an amusement park. The total price of the tickets is $650. Each person uses a discount coupon. The total price with the discount is $530. Write and solve an equation to find the discount each person received.

21. You have 47 marbles. Your friend says that you have 15 more than 4 times the number of marbles that he has. You say that you have 7 more than 5 times as many marbles as he has.

a. Explain how both of you can be right.

b. How many marbles does your friend have?

Name ____________________ Date ________

7.4 Practice B

Solve the equation. Check your solution.

1. $m + 6 = 17 + 2$

2. $\frac{c}{10} + 32 = 41$

3. $4x - 2 = 30$

4. $29 = 6y - 25$

5. $\frac{n}{11} - 3 = 27$

6. $14 = \frac{z}{4} - 8$

7. $\frac{w}{7} + 3 = 43$

8. $81 = 6k + 3k$

9. $9 + 4 = f + 3$

10. $25 = \frac{b}{12} + 21$

11. $85 = 72 + \frac{t}{3}$

12. $18r - 12r = 192$

13. $24 = 2(5 + y)$

14. $7(p - 3) = 21$

15. $\frac{b - 9}{3} = 12$

Write and solve an equation to find *x*.

16. Perimeter = 16 inches

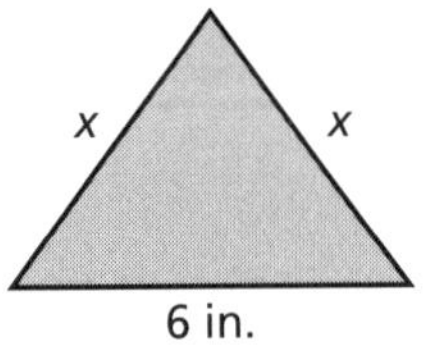

17. Perimeter = 36 meters

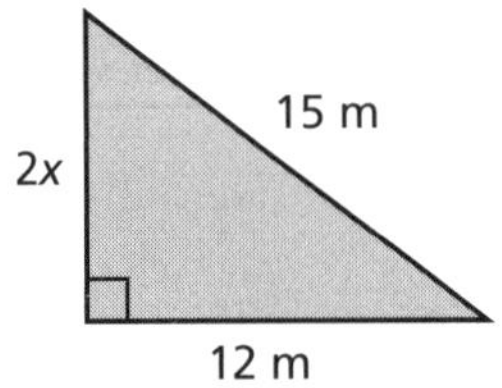

18. Perimeter = 13 feet

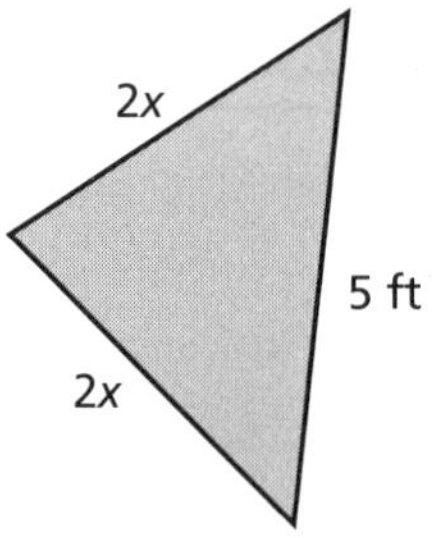

19. You and a friend need to deliver 776 ounces of cookies. You deliver 45 cookie boxes and your friend delivers 52 cookie boxes. Write and solve an equation to find the weight of each cookie box.

20. You and four friends have \$350 to spend on fishing licenses and fishing poles. It costs \$100 for five licenses. Write and solve an equation to find the maximum amount that each person can spend on a fishing pole.

21. After six 100-point tests, you have 552 points.

a. How many points do you need to score on your next 100-point test to have a mean score of 91 points?

b. Would a mean score of 91 points after 7 tests be *greater than* or *less than* your mean score after 6 tests?

c. Is it possible that your mean score does not change after the seventh test? Explain.

Name____ Date__________

7.4 Enrichment and Extension

Consecutive Whole Numbers

Consecutive whole numbers **are whole numbers that are "next to each other," such as 3 and 4 or 57 and 58. To find consecutive whole numbers, choose a whole number and add 1 to it.**

An equation can be used to find two consecutive whole numbers whose sum is given.

Example: Write and solve an equation to find two consecutive whole numbers whose sum is 15.

Let x be the first number and $x + 1$ be the second number.

first number → x; second number → $x + 1$

$x + x + 1 = 15$	Write equation.
$2x + 1 = 15$	Combine like terms.
$-1 = -1$	Subtract 1 from each side.
$2x = 14$	Simplify.
$\frac{2x}{2} = \frac{14}{2}$	Divide each side by 2.
$x = 7$	Simplify.

So, the numbers are 7 and $7 + 1 = 8$. You can check your answer by finding the sum, which is $7 + 8 = 15$.

Write and solve an equation to find two consecutive whole numbers that have the given sum.

1. 11 **2.** 39 **3.** 141 **4.** 689

5. How would the equation in the example above change if you wanted to find three consecutive whole numbers whose sum is 15? What are the numbers?

6. Use your answer in Exercise 5 to write and solve an equation to find three consecutive whole numbers whose sum is 36.

7. Consecutive even numbers are numbers such as 2 and 4, or 28 and 30. Write and solve an equation to find two consecutive even numbers whose sum is 50.

Name ______________________________ Date __________

7.4 Puzzle Time

Why Did The Cookie Go To The Doctor?

Write the letter of each answer in the box containing the exercise number.

Solve the equation. Check your solution.

1. $7 + \frac{x}{3} = 13$

2. $\frac{b}{4} - 10 = 13$

3. $3w - 6 = 18$

4. $7 + \frac{z}{6} = 32$

5. $8a - 48 = 0$

6. $11c - 31 = 68$

7. $\frac{t}{15} - 10 = 14$

8. $116 = 5s + 16$

9. $51 + \frac{y}{3} = 63$

10. $3.6x - 9 = 30.6$

11. $\frac{r}{5} + 2.1 = 4.8$

12. $5w - 20.3 = 25.2$

13. $2z + 4z = 18$

14. $21 = 9b - 6b$

15. $8.3t - 4.3t = 48$

16. $42y + 12 - 5 = 49$

17. $16 + \frac{a}{6} - 3 = 22$

18. $\frac{x - 7}{12} = 12$

Answers

M. 6
E. 11
W. 20
I. 150
N. 12
U. 18
C. 9.1
E. 151
S. 1
F. 9
Y. 92
L. 36
R. 13.5
G. 360
A. 54
M. 8
E. 7
H. 3

13	18		8	17	16		6	14	10	9	4	15	7		12	11	1	5	3	2

Activity 7.5 Start Thinking!
For use before Activity 7.5

Describe how you can use a formula to find the diameter of a telephone pole if you know its circumference.

Activity 7.5 Warm Up
For use before Activity 7.5

Find the area of the figure.

1. 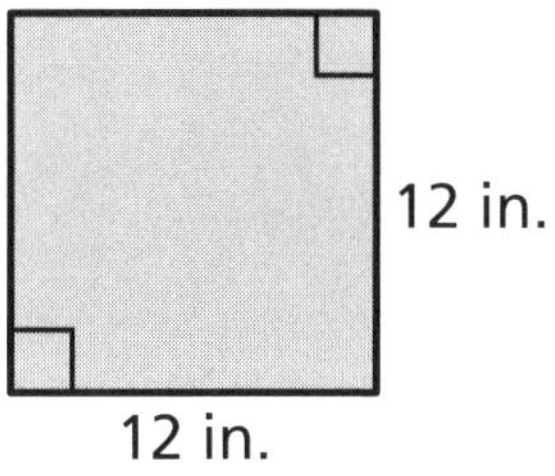

2. 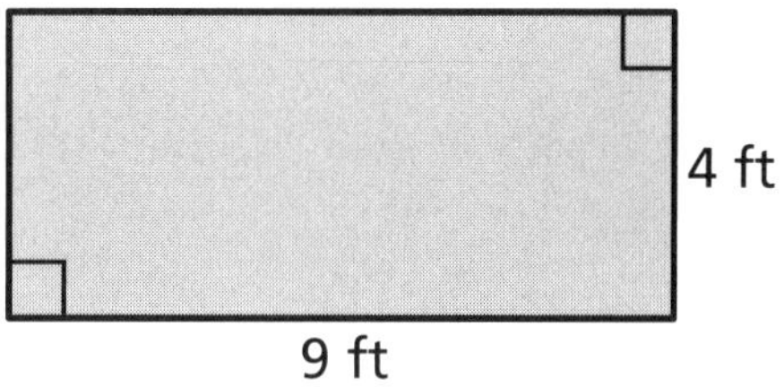

3. 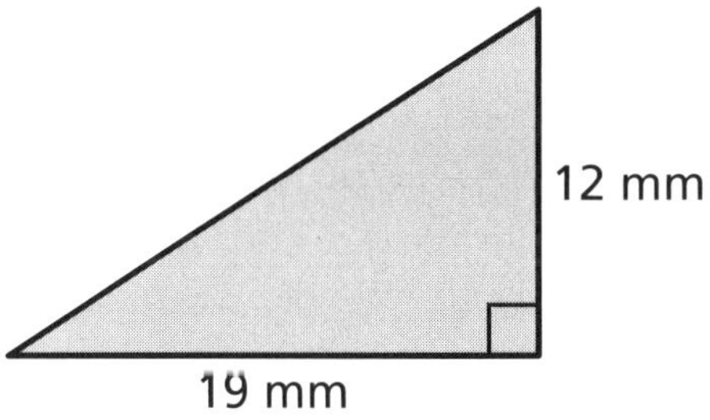

4. 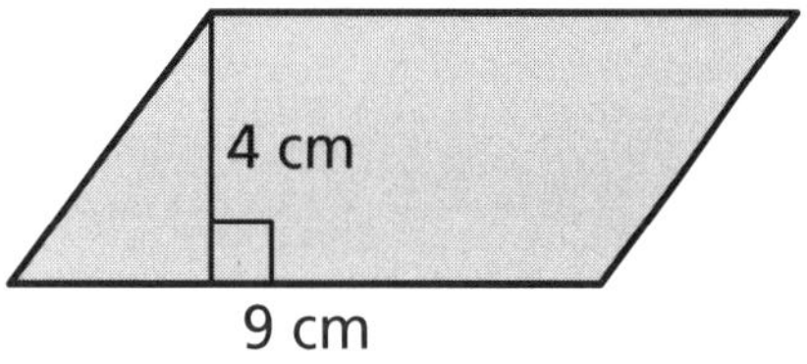

5. 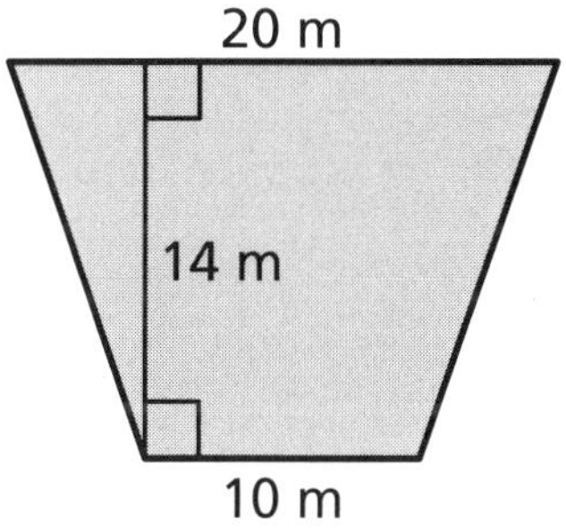

6. 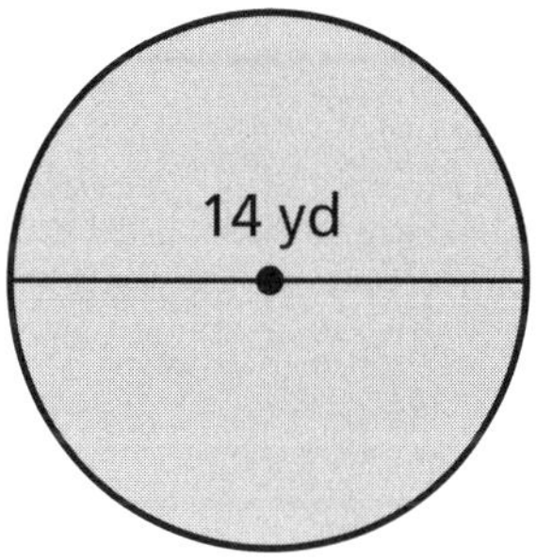

Lesson 7.5 Start Thinking!

For use before Lesson 7.5

One television screen has a height of 19 inches and an area of 665 square inches.

Another television screen has a height of 19 inches and a perimeter of 90 inches.

Are the television screens the same size?

Lesson 7.5 Warm Up

For use before Lesson 7.5

Write and solve an equation to find the missing dimension of the figure. Check your solution.

1. Area $= 24\text{ mm}^2$

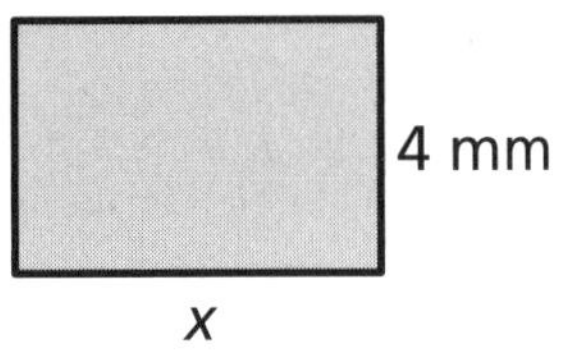

2. Area $= 5\text{ in.}^2$

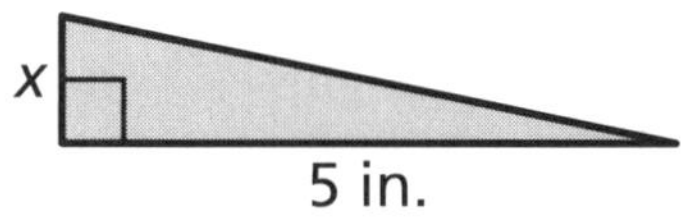

3. Perimeter $= 18$ ft

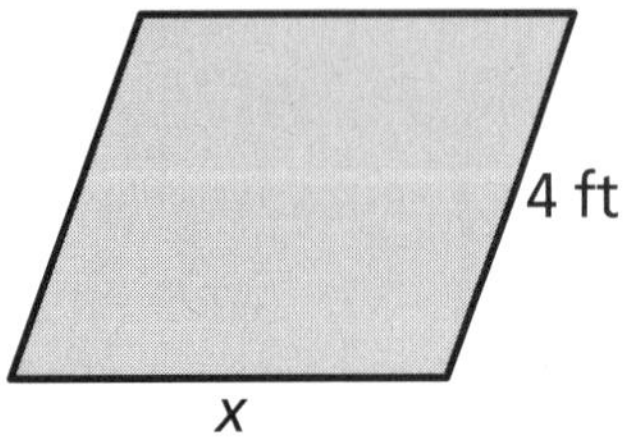

4. Circumference $= 314$ m

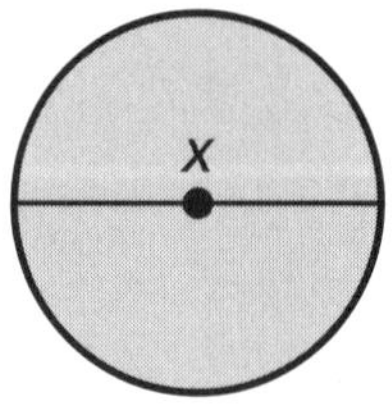

Name____ Date __________

7.5 Practice A

Write and solve an equation to find *x*. Check your solution.

1. Area = 15 in.^{2}

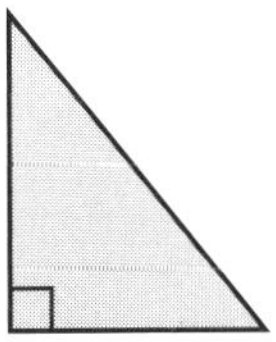

2. Circumference = 44 ft

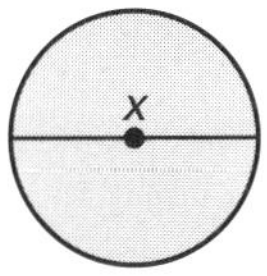

3. Perimeter = 46 yd

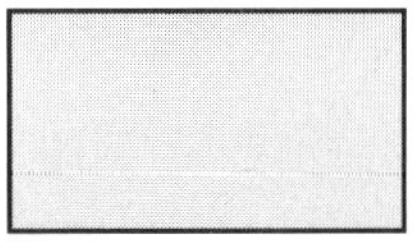

4. Area = 60 cm^2

5. Area = 54 km^2

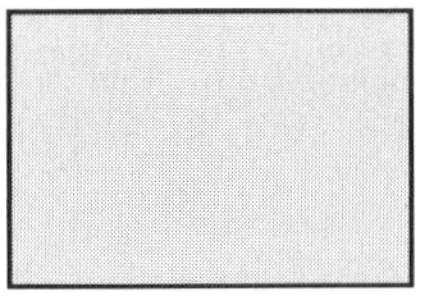

6. Perimeter = 24 m

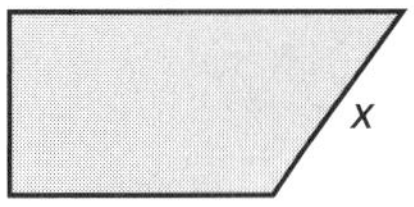

7. Perimeter = 56 m

8. Area = 50 in.^{2}

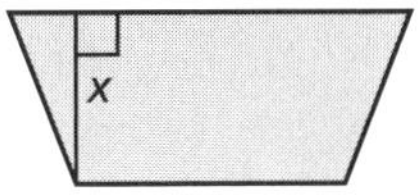

9. Circumference = 25.12 ft

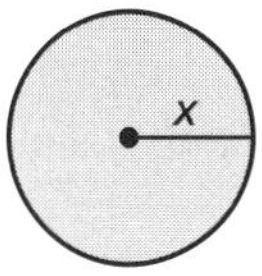

10. Describe and correct the error in finding the diameter x of the circle.

✗

$$C = 2\pi x$$
$$157 = 3(3.14)x$$
$$157 = 6.28x$$
$$\frac{157}{6.28} = \frac{6.28x}{6.28}$$
$$25 = x$$

Circumference = 157 cm

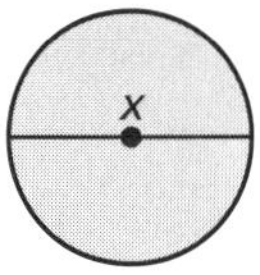

11. The perimeter of the stop sign is 261.6 centimeters. What is the side length x?

12. When a new apartment building is constructed, each bedroom must have at least one rectangular window large enough to escape from in the event of a fire. The escape window must have an area of at least 820 square inches. What is the shortest height of an escape window that is 20 inches wide?

Name ______________________________ Date __________

7.5 Practice B

Write and solve an equation to find *x*. Check your solution.

1. Area $= 308\text{ m}^2$

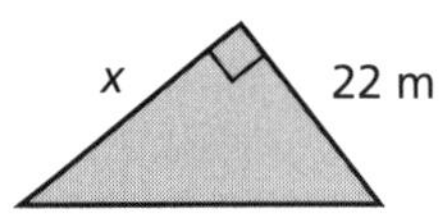

2. Circumference $= 62.8$ ft

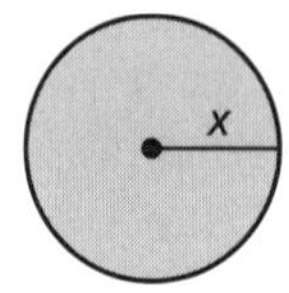

3. Area $= 496\text{ in.}^2$

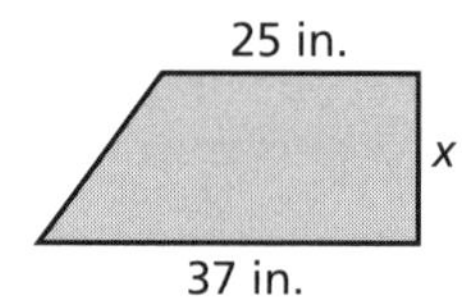

4. Perimeter $= 118$ mm

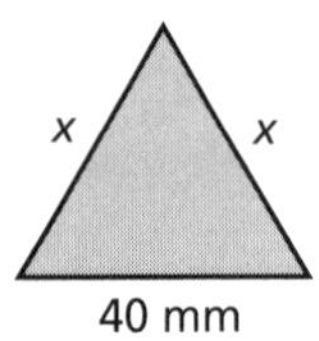

5. Area $= 285\text{ cm}^2$

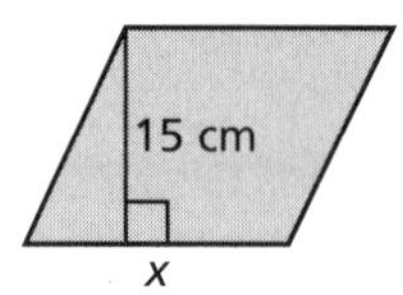

6. Perimeter $= 228.5$ ft

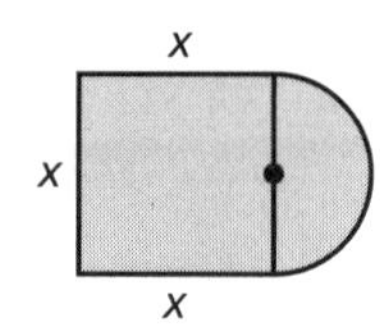

7. A container for suntan lotion is made by heat sealing one end of a plastic tube. The length of the straight seam formed is half the circumference of the tube. What is the diameter of the tube before it is sealed?

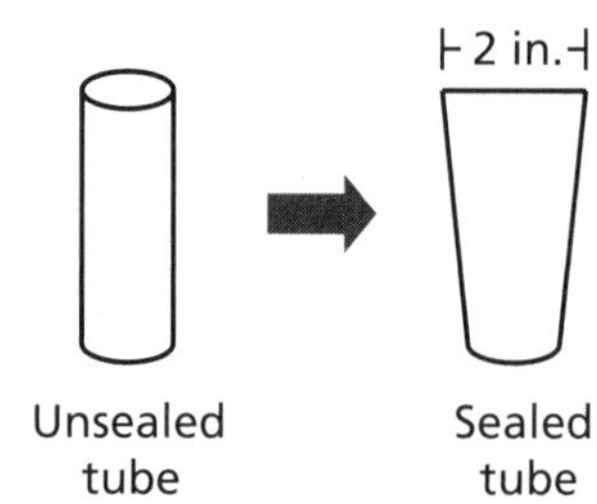

8. You bend an 11-foot-long piece of sheet metal into a circle to form a fire pit. What area is enclosed by the fire pit? Use $\frac{22}{7}$ for π.

9. The shaded rectangular region of the basketball court is called "the paint" and has an area of 228 square feet. What is the area of the semicircular region outside the paint?

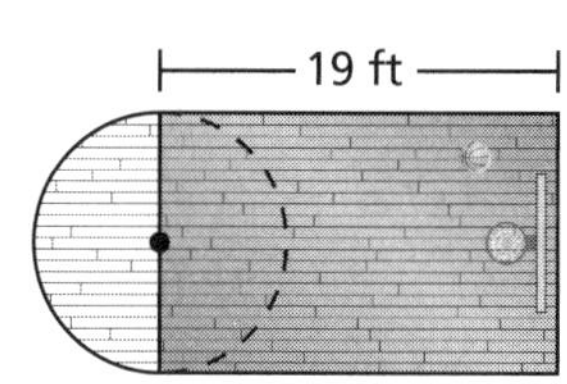

10. You are painting the side of the building. It has an area of 429 square feet. What is the highest point you will have to reach?

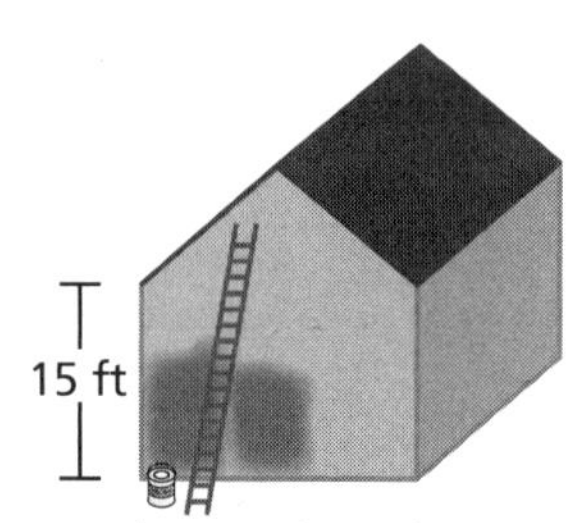

Name______________________________ Date__________

7.5 Enrichment and Extension

Using Mental Math and Guess, Check, and Revise

Example: The area of a circle is 200.96 square meters. What is the radius of the circle?

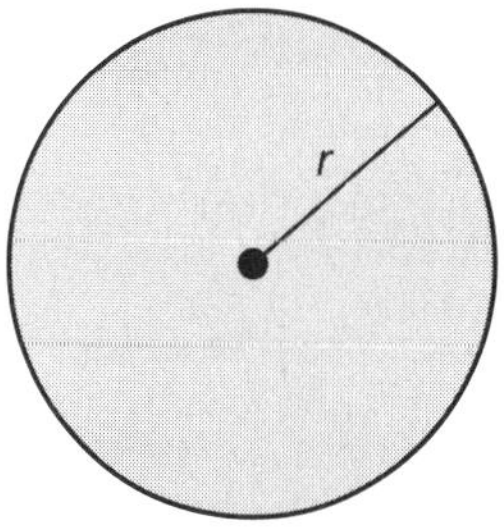

$A = \pi r^2$ Write formula for area of a circle.

$200.96 = 3.14r^2$ Substitute 200.96 for *A* and 3.14 for π.

$\frac{200.96}{3.14} = \frac{3.14r^2}{3.14}$ Divide each side by 3.14.

$64 = r^2$ Simplify.

You can solve the equation $r^2 = 64$ using mental math.

$r^2 = 64$ Ask yourself "What number times itself is equal to 64?"

$(8)^2 = 64$ You know that $8 \bullet 8 = 64$.

$64 = 64$

So, the radius of the circle is 8 meters.

Write and solve an equation to find the missing dimension of the figure.

1.

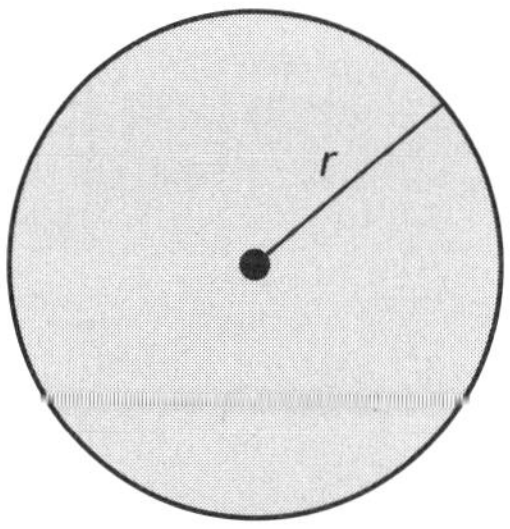

Area $= 28.26$ in.^{2}

2.

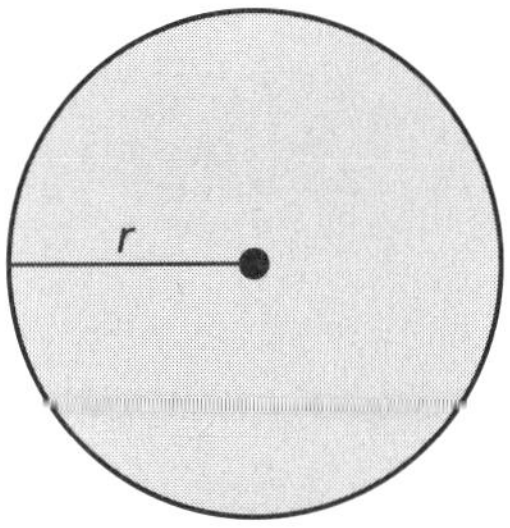

Area $= 153.86$ cm^2

3.

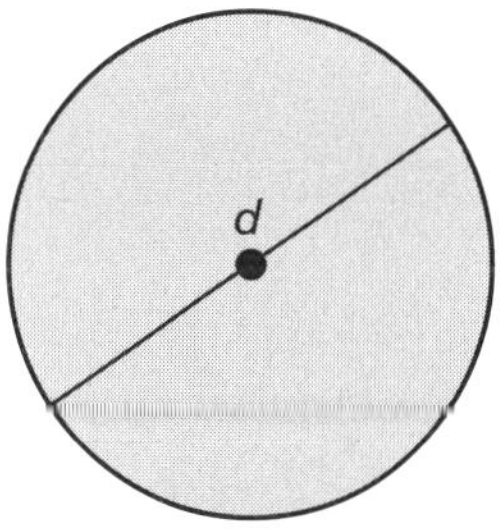

Area $= 706.5$ ft^2

4. The area of a circle is 176.625 square millimeters. Use the guess, check, and revise strategy to find the radius of the circle.

5. The area of a circle is 346.185 square yards. Use the guess, check, and revise strategy to find the radius of the circle.

Name ______________________________ Date __________

7.5 Puzzle Time

What Word Starts With *E*, Ends With *E*, And Only Has One Letter In It?

Write the letter of each answer in the box containing the exercise number.

Write and solve an equation to find the missing dimension of the figure. Check your solution.

1. Area = 42 ft^2

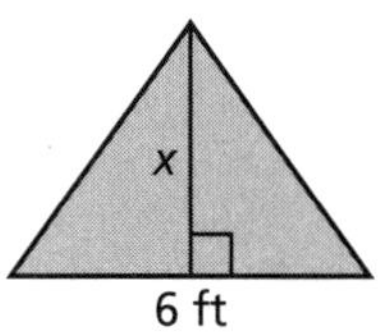

2. Perimeter = 72 ft

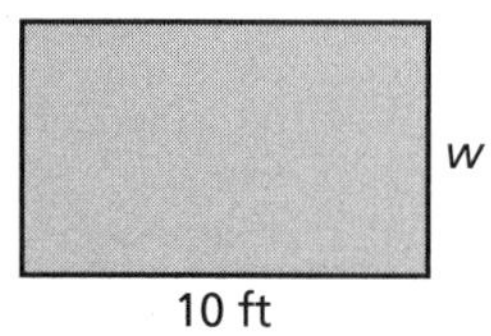

3. Area = 81π ft^2

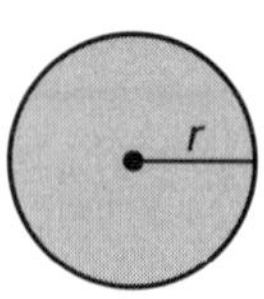

4. Perimeter = 56 ft

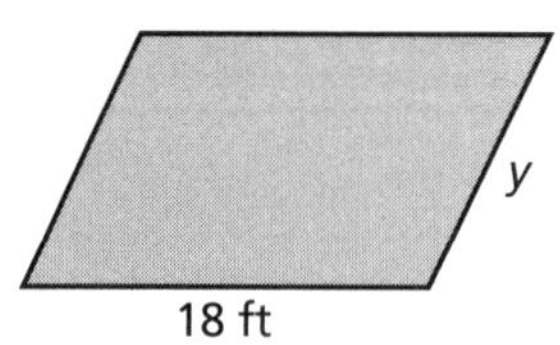

5. Area = 28 ft^2

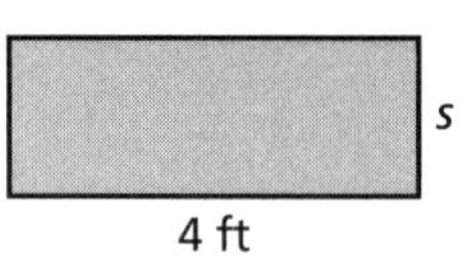

6. Area = 36 ft^2

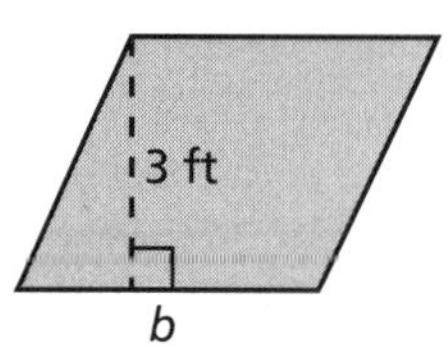

7. Circumference = 15π ft

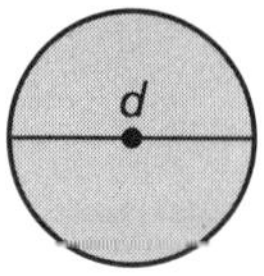

8. The side of a staircase door is made up of a rectangle and a triangle. The area of the door is 42 square feet. What is the height *h* of the door?

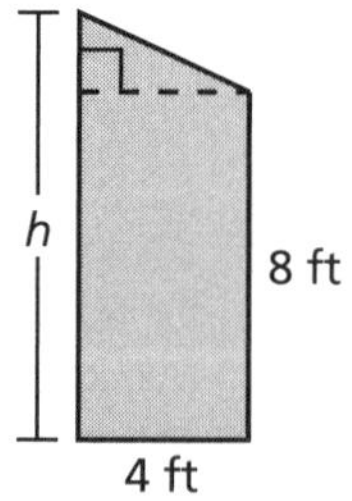

Answers

N. 13 ft

E. 7 ft

O. 12 ft

P. 15 ft

L. 14 ft

E. 26 ft

E. 9 ft

V. 10 ft

2	8	4	5	1	6	7	3

Activity 7.6 Start Thinking!

For use before Activity 7.6

When you hear the word "volume," what do you think of?

Give an example of when you might want to know the volume of an object.

What kind of units are used to measure volume?

Activity 7.6 Warm Up

For use before Activity 7.6

Solve the equation. Check your solution.

1. $2x = 10$

2. $14x = 7$

3. $7x = 63$

4. $25x = 300$

5. $9x = 162$

6. $16x = 24$

Lesson 7.6 Start Thinking!

For use before Lesson 7.6

Tell whether *volume* applies to the situation.

A Wrapping paper used to wrap a gift

B Milk poured into a glass

C Air needed to blow up a ball

D Fabric for making a quilt

E How much will fit in a suitcase

Describe another situation in which volume would apply.

Lesson 7.6 Warm Up

For use before Lesson 7.6

Write and solve an equation to find the missing dimension of the rectangular prism.

1. Volume $= 30\text{ m}^3$

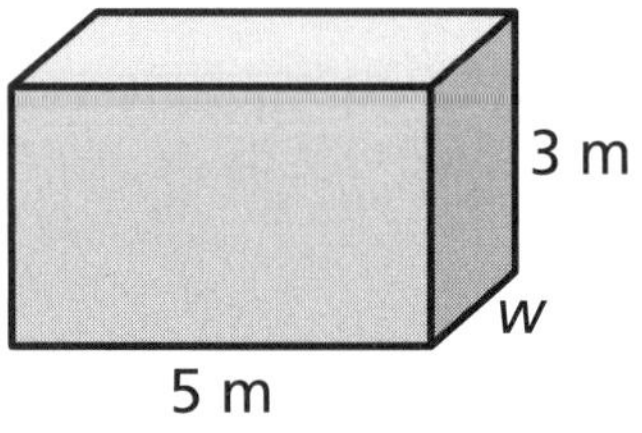

2. Volume $= 5600\text{ ft}^3$

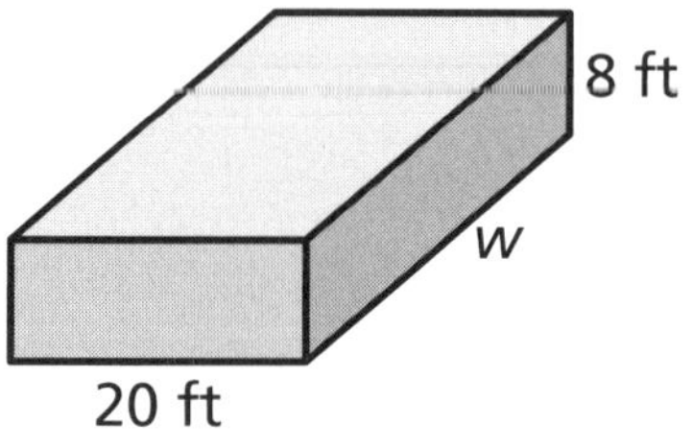

3. Volume $= 64\text{ in.}^3$

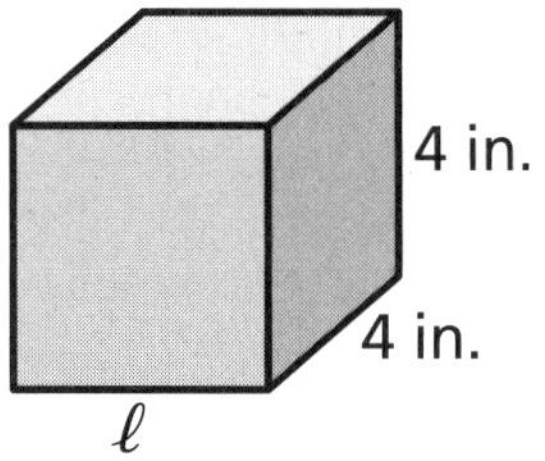

4. Volume $= 100\text{ cm}^3$

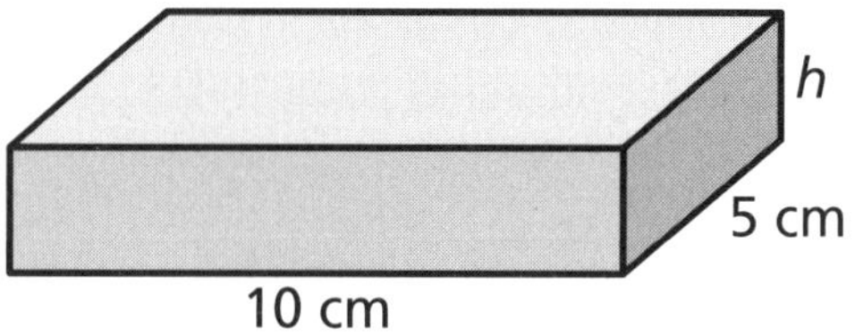

Name ________________________________ Date ____________

7.6 Practice A

Find the number of cubes it takes to fill the box.

1. 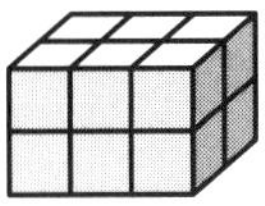

2. 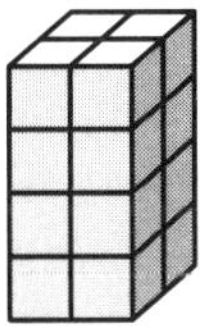

3. 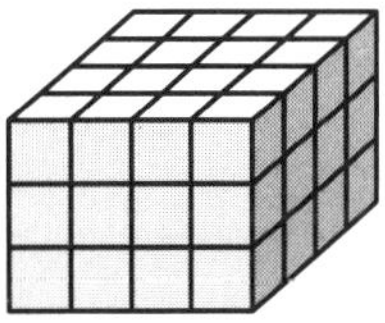

Find the volume of the rectangular prism.

4. 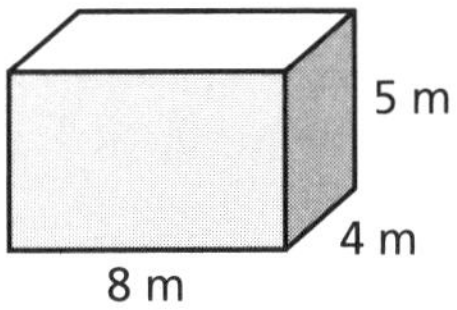

5. 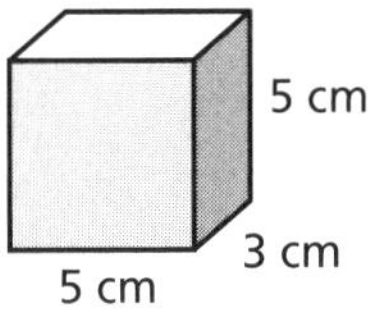

6. 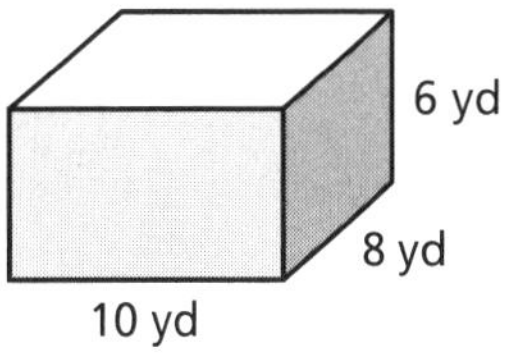

Write and solve an equation to find the missing dimension of the rectangular prism.

7. Volume $= 231 \text{ in.}^3$

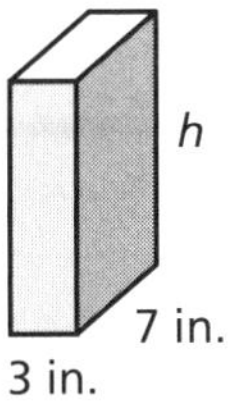

8. Volume $= 6400 \text{ m}^3$

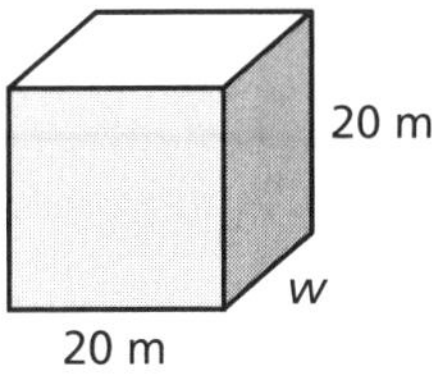

9. Volume $= 864 \text{ yd}^3$

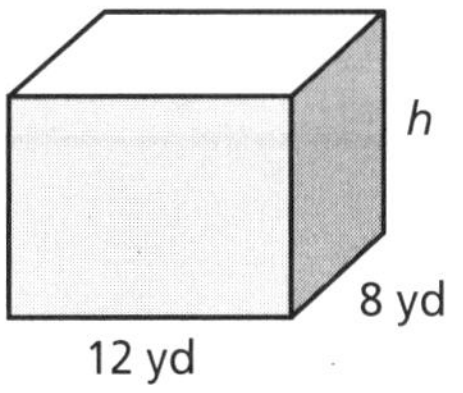

10. Volume $= 18{,}000 \text{ cm}^3$

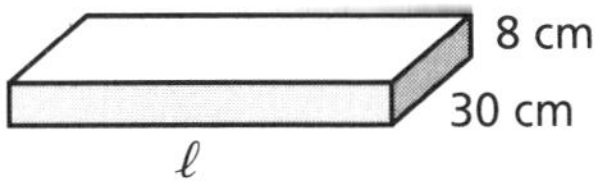

11. Volume $= 16 \text{ ft}^3$

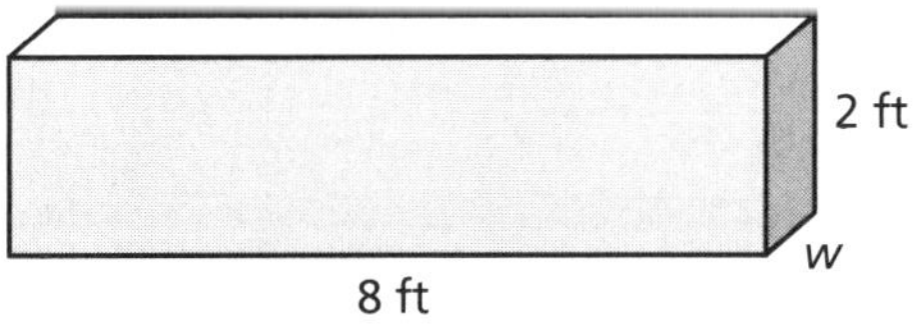

12. An office cubicle measures seven feet by eight feet with a five foot wall. What is the volume of the cubicle?

13. A cube with side length 4 centimeters is 25% full of sand. What is the volume of the sand?

14. A cube has sides of length 2 meters. Explain what happens to the volume of the cube if the length of the sides is doubled.

Name ____ Date __________

7.6 Practice B

1. A storage trunk is 3 feet long and 1 foot wide. The height of the trunk is 1.25 feet. What is the volume of the trunk?

2. A filing cabinet is 35 centimeters by 70 centimeters by 115 centimeters. The bottom drawer is 30% of the volume of the filing cabinet. What is the volume of the bottom drawer?

Write and solve an equation to find the missing dimension of the rectangular prism.

3. Volume $= 1 \text{ km}^3$

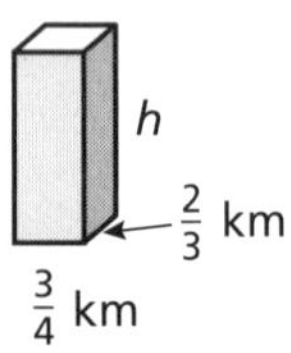

4. Volume $= 23.04 \text{ ft}^3$

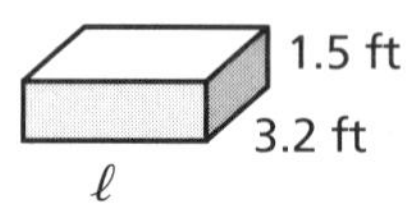

5. Volume $= 0.072 \text{ mi}^3$

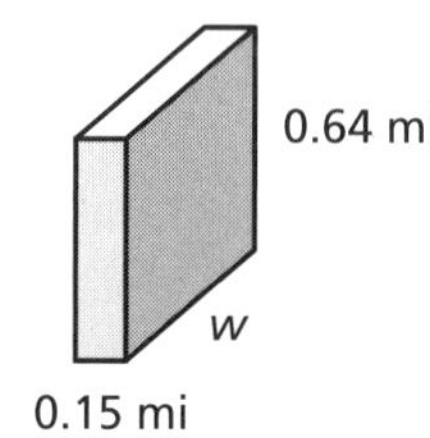

6. Volume $= 220.8 \text{ mm}^3$

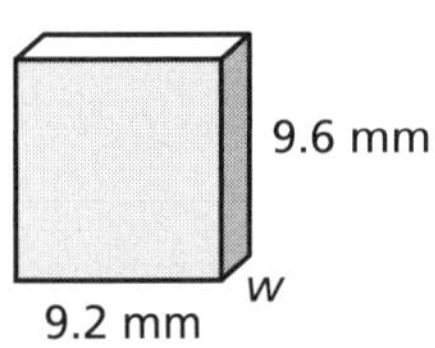

7. Volume $= 4\frac{1}{2} \text{ in.}^3$

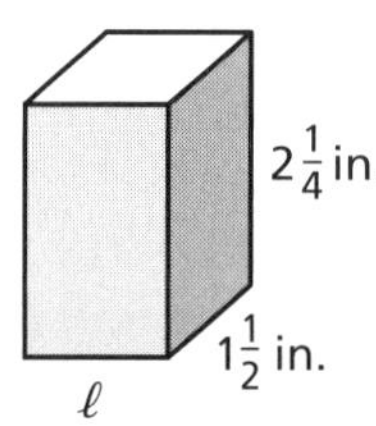

8. Volume $= 5 \text{ cm}^3$

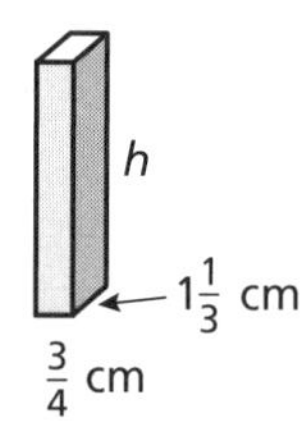

9. A cellular phone is in the shape of a rectangular prism. The height of the phone is 6 millimeters and the width is 50 millimeters. The volume is 22,500 cubic millimeters. What is the length of the cellular phone?

10. A cube is made up of a group of smaller, identical cubes. The cube has a side length of three inches. What is the volume of one of the smaller cubes?

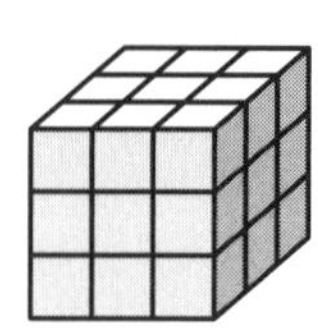

11. A calendar that has one page per day is 7 inches long, 7 inches wide, and 2 inches high. What is the volume of one page of the calendar? Round your answer to the nearest thousandth.

12. What happens to the volume of a rectangular prism if you multiply the length, the width, and the height by 2?

Name______________________________ Date__________

7.6 Enrichment and Extension

Displacement

1. A rectangular tank is $\frac{1}{2}$ filled with water. When the metal prism is placed in the tank, it is $\frac{3}{5}$ full. Find the height h of the metal prism.

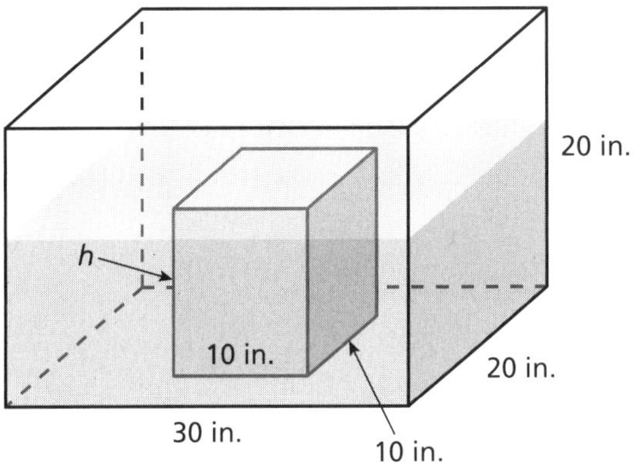

2. A rectangular tank is $\frac{3}{4}$ filled with water. When the metal prism is placed in the tank, it is $\frac{79}{100}$ full. Find the width w of the metal prism.

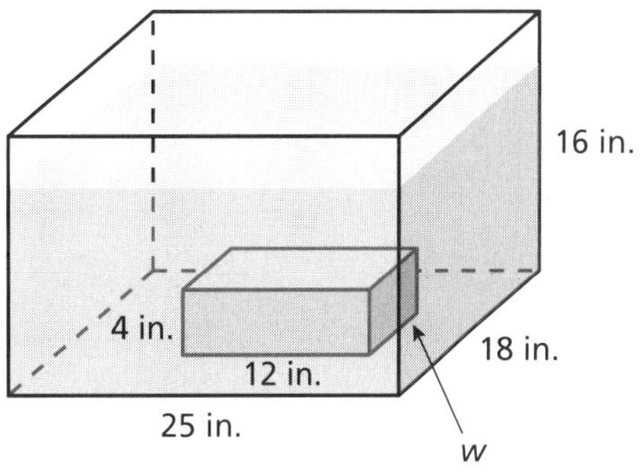

3. A tank is $\frac{5}{8}$ filled with water. When the metal cube is placed in the tank, it is $\frac{3}{4}$ full. Use the guess, check, and revise strategy to find the dimensions of the cube.

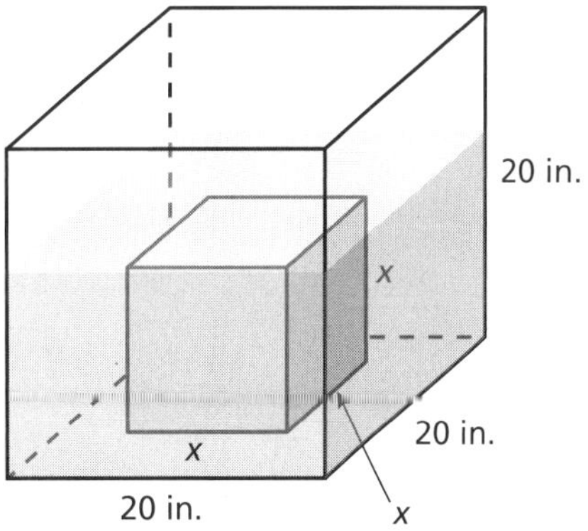

4. A tank is $\frac{1}{2}$ filled with water. The water level rises when four 5-inch metal cubes are placed in the tank. Find the height of the water level before and after the cubes are placed in the tank.

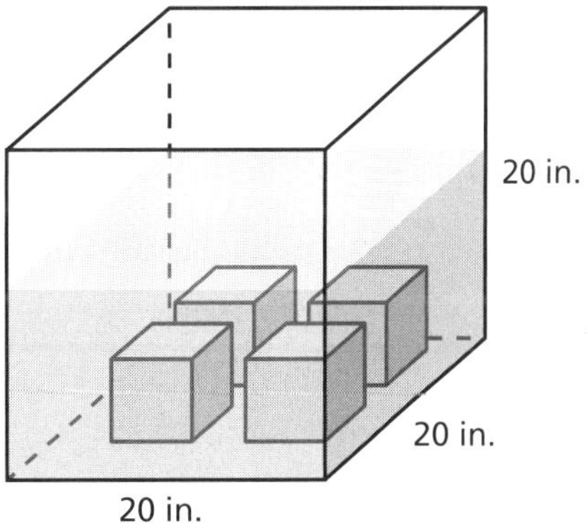

Name ______________________________ Date __________

7.6 Puzzle Time

What Did The Necktie Say To The Hat?

A	B	C	D	E	F
G	H				

Complete each exercise. Find the answer in the answer column. Write the word under the answer in the box containing the exercise letter.

3 in. WILL
5 in. AROUND
360 in.^3 YOU
385 in.^3 I

Find the volume of the rectangular prism.

A.

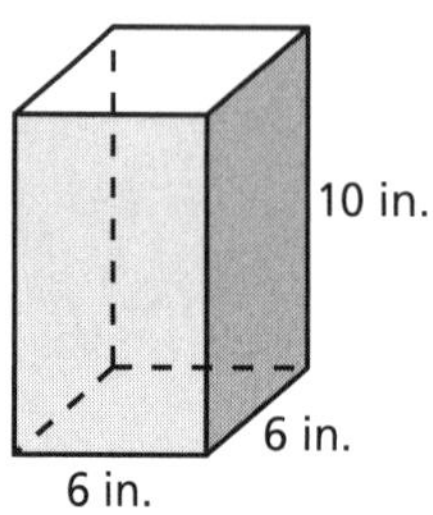

B.

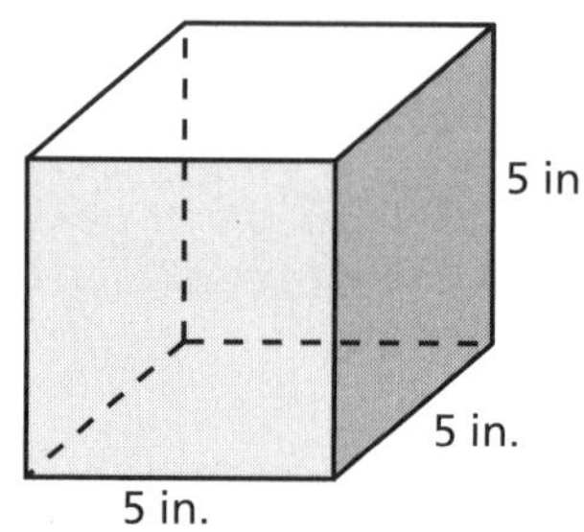

C.

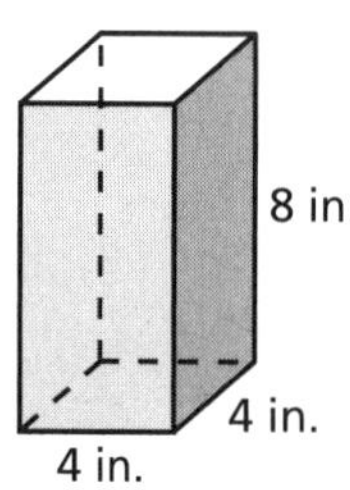

D.

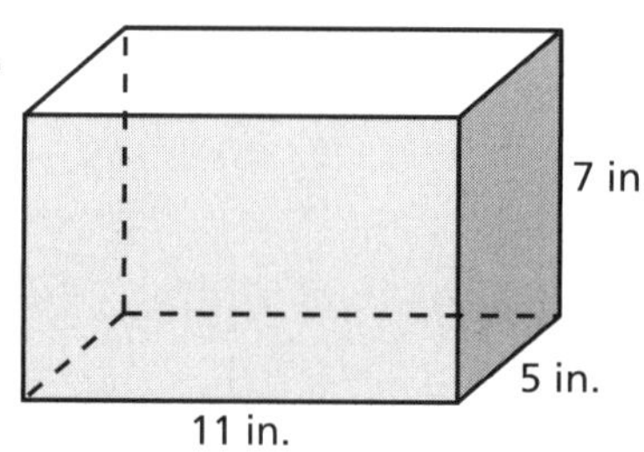

Write and solve an equation to find the missing dimension of the rectangular prism.

E. $V = 27 \text{ in.}^3$

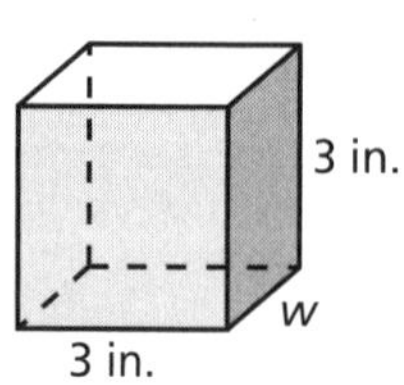

F. $V = 352 \text{ in.}^3$

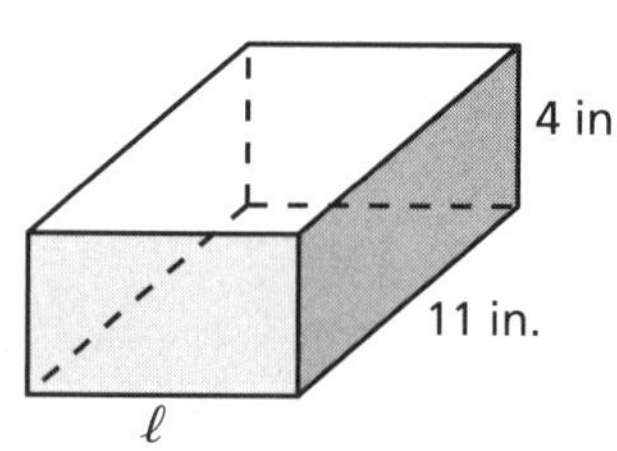

G. $V = 624 \text{ in.}^3$

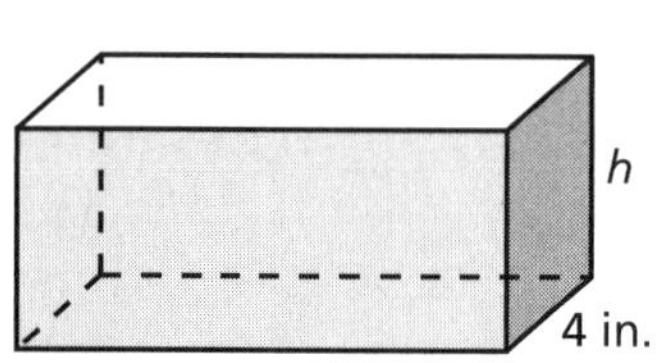

H. $V = 60 \text{ in.}^3$

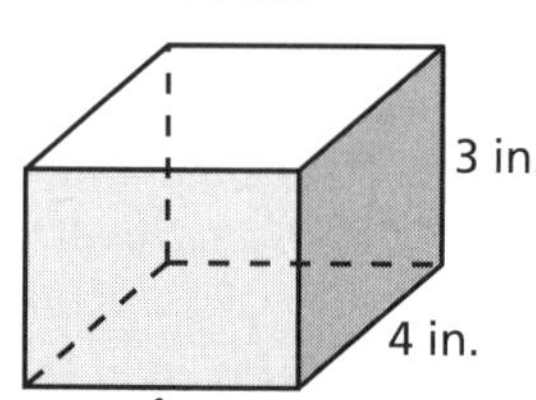

125 in.^3 GO
13 in. HANG
8 in. JUST
128 in.^3 AHEAD

Name____ Date__________

Chapter 7

School-to-Work

For use after Section 7.3

Electrician

An electrician installs, repairs, and maintains electrical systems. He or she must be able to read blueprints, diagnose and repair electrical problems, and solve basic mathematical equations.

The sum of the electrical currents entering a connection is equal to the sum of the currents leaving the connection. This is known as *Kirchhoff's Law.*

1. A 6-milliampere current enters a connection, and one of the two wires leaving the connection carries 2 milliamperes of current. How much current does the second wire carry? Write and solve an addition equation.

2. A connection has two incoming wires and one outgoing wire. The outgoing wire carries 22.5 amperes of current. If one incoming wire carries 12 amperes of current, how much does the other incoming wire carry?

Electrical wires are run through a metal or plastic tube called a *conduit.* An electrical conduit protects and guides the wires.

3. A wiring job calls for 38 pieces of $\frac{1}{2}$-inch conduit, each $8\frac{3}{4}$ feet long. How many feet of conduit is this?

4. You use $41\frac{1}{2}$ feet of conduit from a 100-foot roll. How much remains?

5. You have 12 feet of conduit. How many $\frac{3}{4}$-foot sections can you cut? Write and solve a multiplication equation.

6. To reach the correct height to connect with other components, a motor must be raised $5\frac{3}{4}$ inches. You have a block that has a height of $2\frac{1}{2}$ inches. What height should the second block be?

7. The voltage, current, and resistance in an electrical circuit are related by the formula $\text{current} \times \text{resistance} = \text{voltage}$. (Current is measured in amperes, resistance in ohms, and voltage in volts.) What is the resistance in a circuit with 60 amperes of current and 120 volts?

8. One meter of cable weighs 0.008 kilogram. A roll of cable weighs 6.24 kilograms. How many meters of cable are in the roll?

Name ______________________________ Date __________

Chapter 7 Study Help

You can use an **example and non-example chart** to list examples and non examples of a vocabulary word or term.

Examples	Non-Examples

On Your Own

Make an example and non-example chart to help you study these topics.

1. solutions of equations
2. inverse operations
3. equations solved using addition or subtraction
4. equations solved using multiplication or division
5. two-step equations
6. like terms

After you complete this chapter, make example and non-example charts for the following topics.

7. plane figures
8. rectangular prisms
9. volume

Name ______________________________ Date __________

Chapter 7

Financial Literacy

For use after Section 7.3

Energy Efficiency

How much does it cost to run your refrigerator for one year? a clothes washer? an air conditioner? All of these appliances use electricity. If your family can use less electricity—by using the appliances less, or by using more energy-efficient appliances—you will reduce your utility bills.

Energy is measured in kilowatt-hours. Every electrical item in your household uses a certain amount of power, measured in watts or kilowatts. (One kilowatt equals 1000 watts.) The energy used by an item is the amount of power it needs to operate, multiplied by the amount of time it needs the power.

Choose three different electrical items in your home. (Examples might include the refrigerator, microwave, fan, air conditioner, light, or radio.)

1. Find out how much power each uses. Here are some suggestions:
   - Look for a label on the appliance that gives the amount of power needed to run the item, in watts. Convert this to kilowatts.
   - Some labels list current and voltage but not power. To find the power (in watts), multiply the current (in amperes) times the voltage (in volts).
   - If you can't find labels, look up reasonable estimates for each appliance online. The Department of Energy website is a good source.
2. For each item, estimate how many hours each day it is operating at full power. A refrigerator is usually operating at full power about one third of the time it is on, or 8 hours a day. An air conditioner that turns on and off might only operate at full power one third of the time. A radio playing quietly needs less power than one at full volume.
3. Multiply the values from Exercises 1 and 2 to estimate the kilowatt-hours used by each appliance in a day. Then estimate the kilowatt-hours used in a month (30 days) and a year (12 months).
4. Find the cost of electricity in your area. It should be listed on a utility bill. Use this to estimate how much it costs to operate each item for a month or a year.
5. If each appliance used 75% of the energy it uses now, how much money would you save?
6. A clothes dryer uses 3.5 kilowatts of energy. Over the course of a month it costs \$13.44 to run the dryer. Electricity costs \$0.12 per kilowatt-hour. Explain how you can determine how many hours each month the drier runs.

Name ______________________________ Date __________

Chapter 7 Technology Connection

For use after Section 7.4

Solving Multi-Step Equations

You may find it helpful to use a calculator to solve multi-step equations with large numbers or decimals.

EXAMPLE 1 Solve $19.39 + 4.8x = 34.51$.

SOLUTION

Step 1 Enter 34.51 [−] 19.39 to undo addition.

Step 2 Most calculators will store the answer, 15.12, so that you don't need to write it down. Just enter [÷] 4.8 to undo the multiplication.

ANSWER 3.15

You can apply this method to solve equations that require more steps.

EXAMPLE 2 Solve $\frac{x}{1.2} + 1.8 = 9 - 1.45$.

SOLUTION

Step 1 Enter 9 [−] 1.45 to simplify the right side of the equation.

Step 2 Enter [−] 1.8 to undo addition.

Step 3 Enter [×] 1.2 to undo division.

ANSWER 11.22

Use a calculator to solve the equation.

1. $106y + 317 = 5299$
2. $0.93x - 0.904 = 0.2864$
3. $325.85 = 0.14m - 40.67$
4. $\frac{x}{52} - 73 = 36$
5. $\frac{w}{27} + 372 = 431$
6. $203 = \frac{m}{67} + 189$
7. $212x - 5216 = 9310 + 2646$
8. $4.838 - 1.202 = 1.85z - 3.209$
9. $41 + \frac{a}{109} = 22.3 + 21.5$
10. $0.65 + 9.85 = \frac{c}{0.65} + 2.6$

Name________________________________ Date__________

Chapter 7 Cumulative Practice

Evaluate the expression when $a = 3$, $b = 4$, and $c = 5$.

1. $a + \frac{b}{2}$　　**2.** $c - b$　　**3.** $2ac$

4. Six friends have lunch in a restaurant and share the cost equally. The total cost including the tip is d dollars.

a. Write an expression for the amount of money each person pays.

b. Evaluate the expression when $d = 58.50$ to find each person's cost when the total cost is \$58.50.

Solve the equation. Check your solution.

5. $\frac{n}{3} = 17$　　**6.** $d - 8 = 2$　　**7.** $1.3w + 2.5 = 10.3$

8. $7s - 3 = 81$　　**9.** $4j = 28$　　**10.** $a - \frac{2}{5} = \frac{3}{5}$

11. $\frac{u}{4} + 12 = 13$　　**12.** $45 = 2x + 4x$　　**13.** $1.6 = y + 0.7$

14. $q \div 0.5 = 22.5$　　**15.** $58 = c - 32$　　**16.** $\frac{z}{8} - 2 = 4$

17. You spend \$39 for a large planter and six plants. The planter costs \$18. Write and solve an equation to find the cost of each plant.

Write and solve an equation to find the missing dimension of the figure.

18. Volume $= 120 \text{ in.}^3$

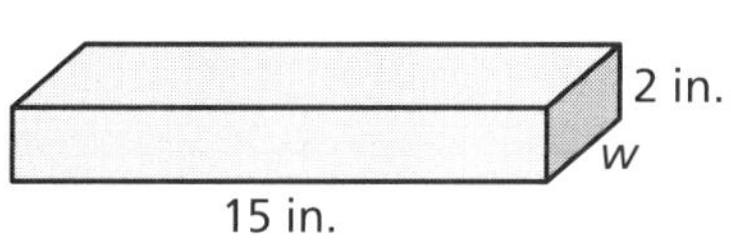

19. Perimeter $= 50$ cm

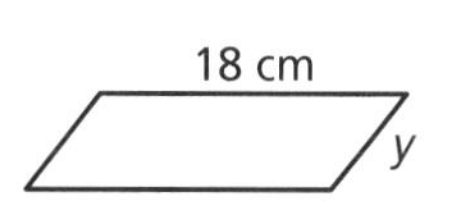

20. Area $= 60 \text{ m}^2$

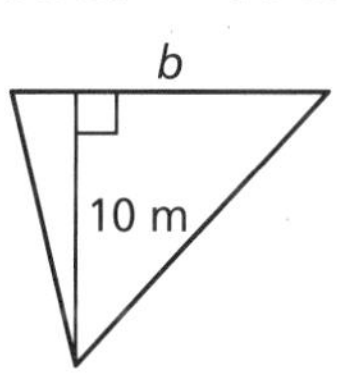

21. Volume $= 216 \text{ ft}^3$

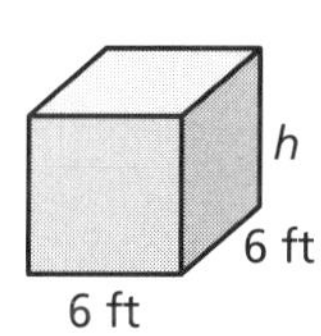

22. A lunch tray is in the shape of a trapezoid with a height of 7 inches and one base length of 14 inches. The area is 77 square inches. What is length of the other base?

Chapter 8

Name____ Date__________

Chapter 8 Inequalities

Dear Family,

Gardeners are familiar with uncertainty. Will there be enough sun? Will there be enough rain? Did I use too much fertilizer? Planning a garden can be a challenge, whether in a small container or over several acres.

You might work with your student to plan and create a small potted garden. Make sure you plant more seeds than you need—some will not germinate and some will produce weak plants. Make sure the plants get enough sun but not too much heat. Have your student write an *inequality* to represent each of these situations.

As your garden grows, ask your student to keep track of the growing conditions. Track your garden's basic needs and have your student write an inequality to represent these situations:

- How tall are the plants likely to get?
- Research how much water the plants need. Check the soil's moisture content every day—plants need water to survive. However, too much water can be just as bad as too little.
- Make sure the recommended amount of sunshine is available. The seed packet will usually tell you the minimum amount required.
- In a potted garden, your plants will probably need some fertilizer to stay healthy. Keep an eye out for signs of overfeeding, however.

Not all problems in mathematics involve a single answer. Many problems have answers that fall into a range. Your plants need at least enough fertilizer to grow, but you must limit the amount of fertilizer to what the plant can safely use. You must make sure the water stays in the right range.

It's hard to beat the satisfaction of growing a successful garden—and the fruits of your labor are beautiful to behold!

Nombre ________________________________ Fecha ________

Capítulo 8 Desigualdades

Estimada Familia:

Los jardineros están familiarizados con la incertidumbre. ¿Habrá suficiente sol? ¿Habrá suficiente lluvia? ¿Usé demasiado fertilizante? Planear un jardín puede ser un desafío, ya sea en un contenedor pequeño o a lo largo de muchos acres.

Puede trabajar con su estudiante para planear y crear un pequeño jardín en maceta. Asegúrese de plantar más semillas de las que necesita—algunas no germinarán y otras producirán plantas débiles. Asegúrese que las plantas tengan suficiente sol, pero no demasiado calor. Haga que su estudiante escriba una *desigualdad* para representar cada una de estas situaciones.

A medida que su jardín crece, pida a su estudiante que registre las condiciones de crecimiento. Registre las necesidades básicas de su jardín y haga que su estudiante escriba una desigualdad para representar estas situaciones:

- ¿Qué tan altas van a crecer las plantas?
- Investiguen la cantidad de agua que necesitan las plantas. Revisen la humedad del suelo todos los días—las plantas necesitan agua para sobrevivir. sin embargo, demasiada agua puede ser tan malo como demasiado poco.
- Asegúrese que la cantidad de luz solar recomendada esté disponible. Normalmente, el paquete de semillas indicará la cantidad mínima requerida.
- En un jardín en maceta, probablemente sus plantas necesitarán algo de fertilizante para estar sanas. No obstante, revise que no esté sobrealimentándolas.

No todos los problemas en matemáticas implican una respuesta única. Muchos problemas tienen respuestas que caen dentro de un rango. Sus plantas necesitan al menos fertilizante para crecer, pero debe limitar la cantidad de fertilizante que la planta puede usar con seguridad. Debe asegurarse que el agua se mantenga dentro del rango correcto.

Es difícil tener una mayor satisfacción que la de hacer crecer un jardín exitoso—¡y el fruto de su labor es algo bello para contemplar!

Non ____ Dat __________

Chapít 8 Inegalite

Chè Fanmi:

Jadinye abitye avèk ensètitid: èske ap gen ase solèy? Èske ap gen ase lapli? Èske mwen sèvi ak twòp angrè? Planifye yon jaden kapab yon defi, ke se nan yon ti po oswa sou plizyè ak.

Ou gen dwa travay ak elèv ou a pou planifye ak kreye yon ti jaden po. Asire ou ke ou plante plis semans pase sa ou bezwen—genyen ki pap donnen epi genyen k'ap bay plant fèb. Asire ou ke plant yo jwenn ase solèy men pa twòp chalè. Fè elèv ou a ekri yon *inegalite* pou reprezante chak sitiyasyon sa yo.

Amezi jaden ou an ap grandi, mande elèv ou a pran nòt kondisyon kwasans yo. Pran nòt bezwen esansyèl jaden ou an epi fè elèv ou a ekri yon inegalite pou reprezante sitiyasyon sila yo:

- Ki wotè plant yo siman ap rive?
- Fè rechèch sou ki kantite dlo plant yo bezwen. Tcheke imidite sòl la chak jou—plant bezwen dlo pou yo siviv. Sepandan, twòp dlo gen dwa fè menm mal ak pa ase.
- Asire ou ke kantite solèy ki rekòmande a disponib. Pake semans la ap abityèlman di ou kantite minimòm ki nesesè a.
- Nan yon jaden po, plant ou yo ap siman bezwen yon ti angrè pou yo rete an sante. Siveye siy twòp angrè, sepandan.

Se pa tout pwoblèm nan matematik ki gen yon sèl repons. Anpil pwoblèm gen repons ki tonbe nan yon entèval. Plant ou yo bezwen omwen ase angrè pou yo grandi, men ou dwe limite kantite angrè a sou sa plant la kapab itilize san danje. Ou dwe asire ou ke dlo a rete nan bon entèval la.

Sa difisil pou depase satisfaksyon reyisi kiltive yon jaden—epi fwi travay ou bèl pou wè!

Activity 8.1

Start Thinking!

For use before Activity 8.1

$1 < 2$ and $2 > 1$ are inequalities.

Think about the word *inequality*.

What does *equality* mean?

What does the prefix *in-* mean?

What are some other words in the English language that use the prefix *in-* in the same way?

Activity 8.1

Warm Up

For use before Activity 8.1

Complete the statement with <, >, or =.

1. $-5 \; \underline{\;\;?\;\;} \; -2$
2. $4 \; \underline{\;\;?\;\;} \; -6$
3. $7 \; \underline{\;\;?\;\;} \; 9$
4. $-8 \; \underline{\;\;?\;\;} \; -12$
5. $-1 \; \underline{\;\;?\;\;} \; -1$
6. $-15 \; \underline{\;\;?\;\;} \; 15$

Lesson 8.1 Start Thinking!

For use before Lesson 8.1

The speed limit in a school zone is 15 miles per hour.

How are speed limits related to inequalities?

Lesson 8.1 Warm Up

For use before Lesson 8.1

Match the inequality with its graph.

1. $x > -1$ **2.** $x < 1$ **3.** $x > 1$ **4.** $x \leq -1$

A.

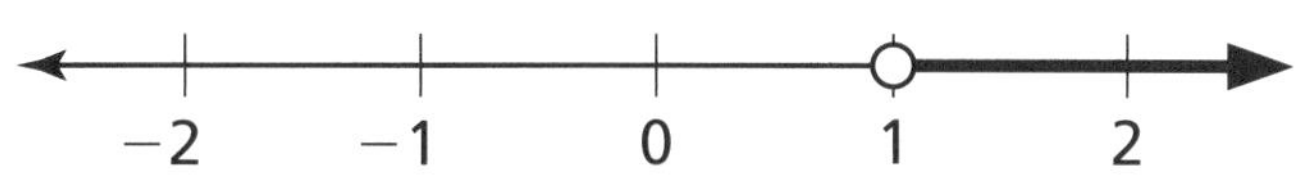

B.

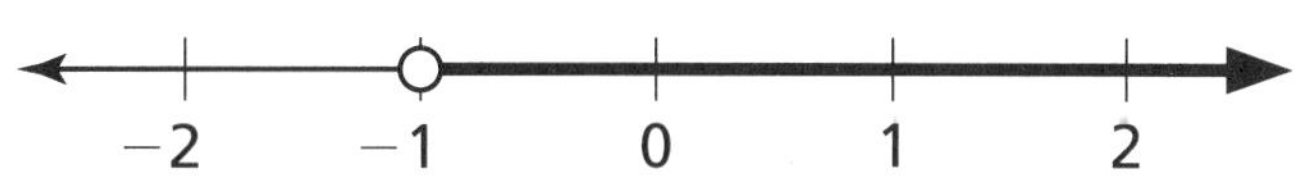

C.

D.

Name ______________________________ Date __________

8.1 Practice A

Write the word sentence as an inequality.

1. 2 is more than a number v.
2. A number h is at most 15.
3. A number p is less than $\frac{1}{2}$.
4. 12 is fewer than a number n.
5. $\frac{3}{5}$ is no less than a number a.
6. A number b is no more than 17.
7. 6 plus a number x is at least 12.
8. A number k minus 7 is greater than 10.

Tell whether the given value is a solution of the inequality.

9. $y \leq 11;\ y = 8$
10. $q + 1 \geq 7;\ q = 3$
11. $4 < u - 9;\ u = 13$
12. $5m < 72;\ m = 15$
13. $\frac{7}{8} \leq 2c;\ c = 3$
14. $10 \geq g + 3;\ g = 7$

Graph the inequality on a number line.

15. $k > 1$
16. $w \leq 5$
17. $r < 0$
18. $t \geq \frac{1}{3}$
19. $s > 6$
20. $z \leq -\frac{3}{2}$
21. $2.5 < n$
22. $-\frac{2}{3} < x$
23. $3 \geq a$

24. A lifeboat can carry up to 24 people. Write an inequality to represent this situation.

25. A USB flash drive costs \$16. You have \$50.

    a. Write an inequality to represent the number of USB flash drives you can buy.

    b. Can you buy 4 USB flash drives? Explain.

26. A produce box can hold no more than 25 pounds of potatoes.

    a. Write and graph an inequality to represent this situation.

    b. Is 9.8 a solution of the inequality?

    c. Name a number that is *not* a solution of the inequality and explain your answer.

Name____ Date__________

8.1 Practice B

Write the word sentence as an inequality.

1. A number b times 3.5 is no less than 21.

2. The quotient of a number y and 9 is greater than 4.

3. The difference between a number h and $\frac{1}{4}$ is at most 0.

4. The sum of a number w and 2.56 is at least 10.24.

5. The product of 6 and number c is less than 12.

Tell whether the given value is a solution of the inequality.

6. $\frac{x}{6} \le 1;\ x = 5$

7. $1.4v > 16;\ v = 11$

8. $t + 9.8 \ge 21.01;\ t = 11.1$

9. $2\frac{1}{2} < \frac{m}{30};\ m = 90$

10. $\frac{1}{2} > 3p;\ p = \frac{1}{6}$

11. $2.16 \ge 3z - 0.5;\ z = 0.6$

Graph the inequality on a number line.

12. $r \le 2$

13. $k \ge -4$

14. $6.5 > a$

15. $\frac{5}{3} < g$

16. $u \le -8$

17. $y > 0$

18. $-18 \ge s$

19. $w < -56.8$

20. $n > \frac{7}{5}$

21. A highway passes under a road. The clearance height is 7.75 feet. Write and graph an inequality that represents the height of a vehicle that can travel on the highway.

22. You must be at least 48 inches tall to go on the Scrambler at an amusement park. You must be at most 48 inches tall to go on the Busy Boats at the park.

    a. Write an inequality that represents the height you must be for each of the rides.

    b. You are allowed to ride both rides. What is your height?

23. Describe the numbers that are solutions of both of the following inequalities:

$$n \le 8 \text{ and } n > 2.$$

Name ______________________________ Date ________

8.1 Enrichment and Extension

Compound Inequalities

A *compound inequality* is a mathematical sentence that uses more than one inequality symbol. It is the result of combining two inequalities. For example, the two inequalities $x > 2$ and $x < 5$ can be combined to form the compound inequality $2 < x < 5$. It is read as "*x is greater than 2 and x is less than 5.*"

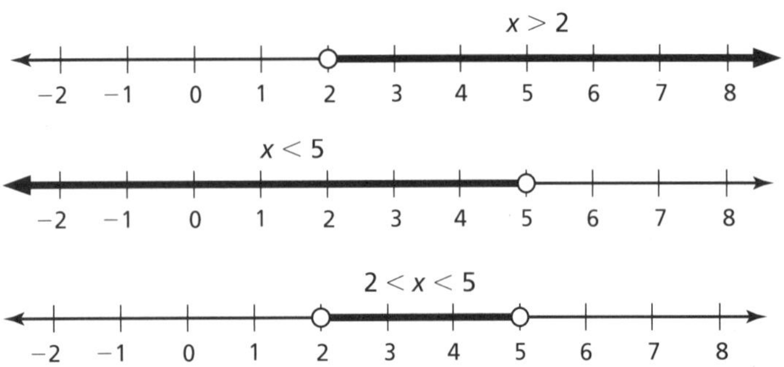

Graph $x > 2$.

Graph $x < 5$.

Graph the common solutions.

Identify each integer that is a solution of the compound inequality.

1. $0 < x < 6$

2. $3 \leq x \leq 8$

3. $4 \leq x < 11$

4. $-2 < x \leq 2$

Match each inequality with its graph.

5. $1 \leq x < 7$

6. $1 < x < 7$

7. $1 \leq x \leq 7$

8. $1 < x \leq 7$

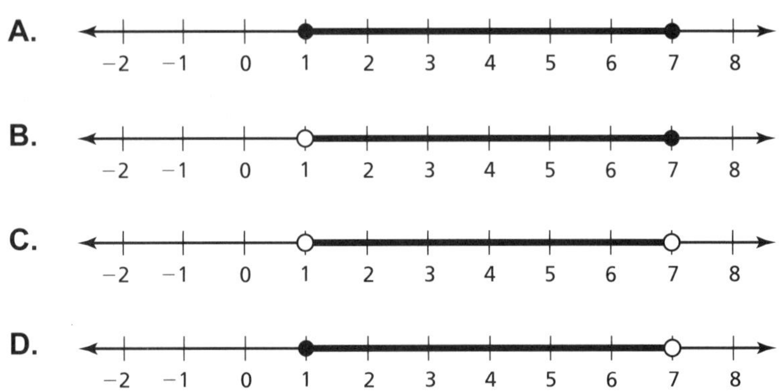

Graph the compound inequality on a number line.

9. $0 \leq x < 3$

10. $2 < x < 9$

11. $5 \leq x \leq 11$

12. $-1 < x \leq 6$

13. $-5 < x < -3$

14. $10 \leq x < 70$

15. A compound inequality can be made up of two inequalities using the word "and" (as shown above) or the word "or." Graph the compound inequality $x < 0$ *or* $x > 2$. Can this be written as a single inequality? Explain.

Name______________________________ Date__________

8.1 Puzzle Time

What Kind Of Cheese Comes With A House?

Write the letter of each answer in the box containing the exercise number.

Write the word sentence as an inequality.

1. A number x is more than 15.
2. A number b is less than 23.
3. A number y is at most 8.
4. Three plus a number a is greater than or equal to 19.

Tell whether the given value is a solution of the inequality.

5. $\frac{a}{4} > 5;\ a = 28$
6. $z + 4.5 \leq 13;\ z = 9.5$

Write an inequality that represents the graph.

7. 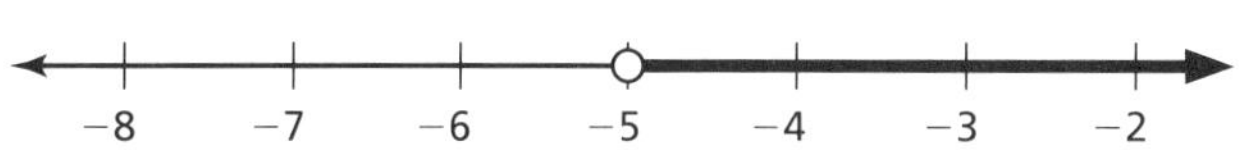

8. 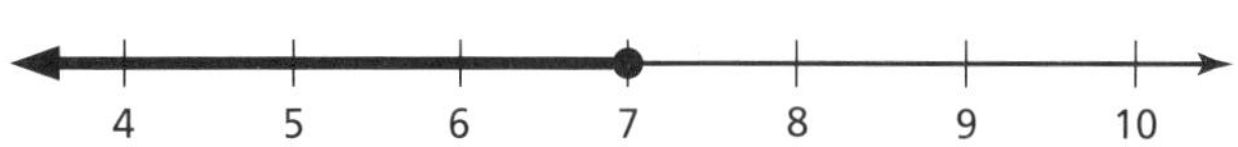

9. 

Match each inequality with its graph.

10. $x < \frac{4}{5}$
11. $a \geq -3$
12. $p \leq 2.6$
13. $y > -\frac{2}{3}$

Answers

E. $x > -5$

C. $y \leq 8$

T. $x \geq -3$

H. 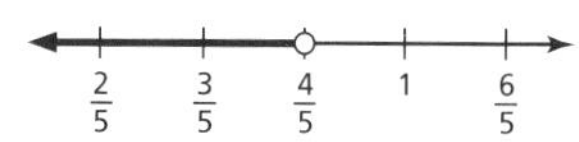

E. $x > 15$

A. no

G. 

E. $3 + a \geq 19$

S. 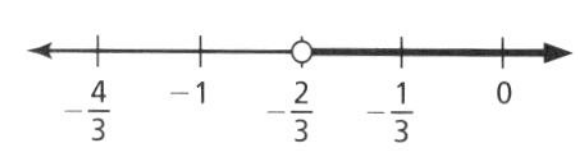

O. $b < 23$

C. $x \leq 7$

T. 

E. yes

8	2	12	9	6	11	5		3	10	1	7	13	4

Activity 8.2 Start Thinking!

For use before Activity 8.2

Describe a real-life situation that can be represented by the inequality $x + 5 > 15$.

Activity 8.2 Warm Up

For use before Activity 8.2

Graph the inequality on a number line.

1. $a < 2$

2. $b > 1$

3. $c \leq 0$

4. $d \geq 2.5$

5. $f > -2$

6. $g \leq -5$

Lesson 8.2 Start Thinking!

For use before Lesson 8.2

A classroom should have no more than 22 students.

A math classroom meets this requirement. After two students transferred into the class, it still meets the requirement.

Write an inequality that represents this situation. Solve it to find the possible numbers of students in the class before the two students transferred.

Lesson 8.2 Warm Up

For use before Lesson 8.2

Solve the inequality. Graph the solution.

1. $x + 2 < 6$

2. $2 + h > 11$

3. $y - 3 \leq 2$

4. $p - 15 \geq 0$

5. $k + 3 \leq 14$

6. $b - 3 > 3$

Name ______________________________ Date __________

8.2 Practice A

Solve the inequality. Graph the solution.

1. $n - 9 \geq 2$

2. $v + 10 \leq 14$

3. $p + \frac{1}{4} < \frac{5}{4}$

4. $x - 3 > 8$

5. $20 < k + 15$

6. $\frac{4}{5} \leq m - \frac{1}{5}$

7. $12 \geq h - 8$

8. $4.4 > 2.4 + b$

9. $w - 36 \leq 64$

10. $a + 16 \geq 25$

11. $r + \frac{2}{3} > \frac{8}{3}$

12. $y - 19 < 51$

Write the word sentence as an inequality. Then solve the inequality.

13. 6 more than a number is at most 10.

14. Four less than a number is more than 3.

15. 0.6 is no less than 2.4 subtracted from a number.

16. The sum of a number and 14 is at least 18.

Describe and correct the error in solving the inequality.

17. X

$$\begin{aligned} 3 &> g - 4 \\ \underline{+4} & \quad \underline{+4} \\ 7 &< g \end{aligned}$$

18. X

$$\begin{aligned} x + 5 &\geq 11 \\ \underline{-5} & \quad \underline{+5} \\ x &\geq 16 \end{aligned}$$

19. You can spend at most \$10 at the mall. You want to buy a book that costs \$6.75 and a cold drink. Write and solve an inequality to represent the amount of money you can spend on your cold drink.

20. An order from an online bookstore takes at least four weeks to arrive. You ordered some books online nine days ago. Write and solve an inequality to represent the possible number of days it will take for your books to arrive.

21. The school auditorium can hold at most 480 people. There were 185 advance tickets sold for the school play. Write and solve an inequality to represent the number of people who can attend the play if all the people who bought advance tickets attend the play.

Name______________________________ Date__________

8.2 Practice B

Solve the inequality. Graph the solution.

1. $x + 12 \geq 15$
2. $c - 8 < 4$
3. $q - 9.8 > 1.2$
4. $t + 54 \leq 85.6$
5. $9\frac{1}{4} < z - 4\frac{1}{8}$
6. $\frac{6}{5} \geq g + \frac{3}{5}$
7. $5.76 \leq k + 5.76$
8. $6 > \frac{1}{3} + x$
9. $u + \frac{1}{3} \leq \frac{1}{2}$
10. $y - 58 < 74$
11. $d + \frac{5}{8} \leq 1$
12. $h - 3.1 > 1.3$

Write the word sentence as an inequality. Then solve the inequality.

13. The difference between a number and 4.5 is greater than 8.
14. The total of $\frac{3}{4}$ and a number is no more than 2.
15. 9.1 less than a number is less than 4.6.
16. A number minus $\frac{7}{12}$ is at least $3\frac{1}{2}$.
17. $5\frac{1}{4}$ is greater than a number plus $2\frac{1}{10}$.

Solve the inequality. Graph the solution.

18. $a - \frac{1}{3} - \frac{1}{6} < \frac{1}{2}$
19. $n + 1.25 - 0.75 > 6$
20. $12.4 + 6.07 \leq v - 8.13$
21. You are cooking a turkey. The turkey must reach a temperature of at least 165 degrees to be fully cooked. The temperature is 135 degrees. Write and solve an inequality to represent the number of degrees the temperature must increase for the turkey to be done.
22. The memory card on a cell phone can store up to 50 phone numbers. There are 21 phone numbers stored on the cell phone. Write and solve an inequality to represent the number of additional phone numbers you can store on the cell phone.
23. The possible values of x are given by $x + 2.1 \leq 6.5$. What is the greatest possible value of $\frac{x}{4}$?

Name ______________________________ Date __________

8.2 Enrichment and Extension

Solving a Compound Inequality

Example: Solve $3 < x + 1 < 6$. Graph the solution.

$3 < x + 1 < 6$	Write inequality.
$-1 \quad\;\; -1 \quad -1$	Subtract 1 from each part of the inequality.
$2 < x < 5$	Simplify.

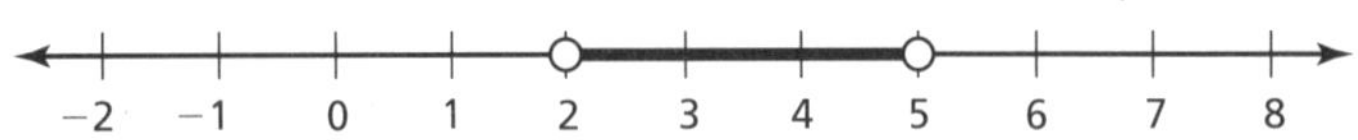

Example: Solve $x + 2 \geq 7$ *or* $x - 1 \leq 1$. Graph the solution.

Solve each inequality separately.

$x + 2 \geq 7$	Write inequality.	$x - 1 \leq 1$	Write inequality.
$-2 \quad -2$	Subtract 2 from each side.	$+1 \quad +1$	Add 1 to each side.
$x \geq 5$	Simplify.	$x \leq 2$	Simplify.

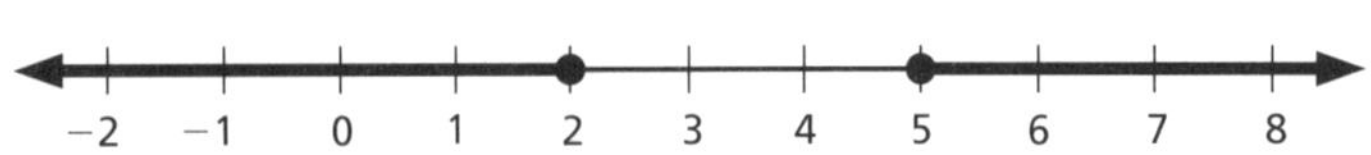

Solve the inequality. Graph the solution.

1. $1 < x - 2 < 4$
2. $3 \leq x + 3 \leq 7$
3. $2 < x - 5 \leq 6$
4. $8 \leq x + 4 < 12$
5. $x > 6$ *or* $x + 10 < 14$
6. $x + 1 \leq 4$ *or* $x - 3 \geq 2$
7. $x - 5 \geq 2$ *or* $x + 8 < 8$
8. $x + 7 < 10$ *or* $9 + x \geq 13$
9. What do you notice about the solution of the compound inequality $x + 3 > 3$ *or* $x + 3 \leq 3$?

Name____________________ Date__________

8.2 Puzzle Time

A Man Went To the Rocket Station And Asked For A Ticket To The Moon...

A	B	C	D	E	F
G	H	I	J	K	L

Complete each exercise. Find the answer in the answer column. Write the word under the answer in the box containing the exercise letter.

$x < 11$

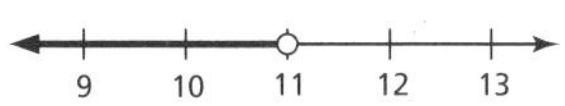

I'M

$x \leq 22.4$

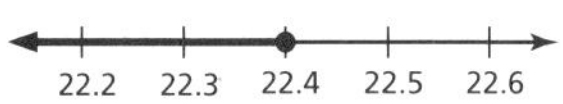

FULL

$x \leq 17$

ATTENDANT

$x > \frac{1}{6}$

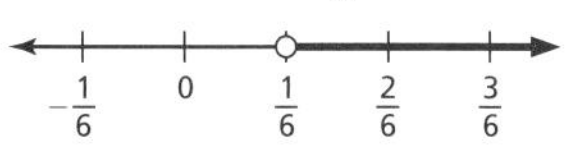

NOW

$x < 110$

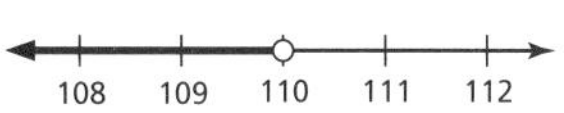

THE

$x \geq 9$

SAID

Solve the inequality. Graph the solution.

A. $x - 5 < 6$

B. $7 + x > 9$

C. $5 \geq x - 7$

D. $12 \leq x + 3$

E. $20 > 14 + x$

F. $39 + x \leq 56$

G. $x - 23 < 87$

H. $x - 19 \geq 19$

I. $8.4 < x + 4.2$

J. $14.9 \geq x - 7.5$

K. $\frac{7}{8} \leq \frac{1}{4} + x$

L. $\frac{1}{2} + x > \frac{2}{3}$

$x \geq \frac{5}{8}$

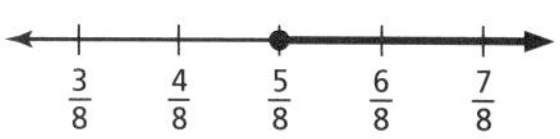

RIGHT

$x \leq 12$

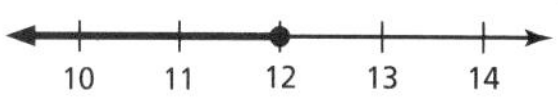

SIR

$x < 6$

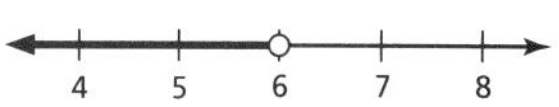

THE

$x > 38$

MOON

$x > 2$

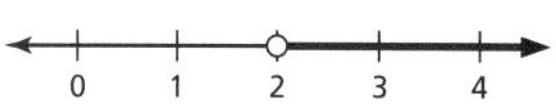

SORRY

$x > 4.2$

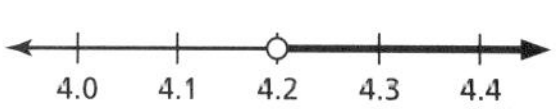

IS

Activity 8.3 Start Thinking!

For use before Activity 8.3

How are the inequalities $5x < 20$ and $5x > 20$ related to the equation $5x = 20$?

Describe how to solve the equation $5x = 20$.

Can you use the same method to solve the inequalities? Use the inequalities $5x < 20$ and $5x > 20$ to support your answer.

Activity 8.3 Warm Up

For use before Activity 8.3

Solve the inequality.

1. $x + 9 < 11$

2. $10 < y - 14$

3. $14 \geq 1 + r$

4. $21 \leq c + 8$

5. $7 + s < 12$

6. $x - 4 \leq 1$

Lesson 8.3 Start Thinking!

For use before Lesson 8.3

On some cell phone plans, text messages cost $0.20 per message.

Write a word problem involving the cost of text messages that can be solved with an inequality using division.

Exchange problems with a classmate and solve your classmate's problem.

Lesson 8.3 Warm Up

For use before Lesson 8.3

Solve the inequality. Graph the solution.

1. $\frac{x}{4} > 2$

2. $\frac{t}{2} \geq 1$

3. $\frac{n}{3} \leq 3$

4. $6f < 18$

5. $15b > 90$

6. $12y \leq 0$

Name ______________________________ Date ________

8.3 Practice A

Solve the inequality. Graph the solution.

1. $\frac{b}{3} > 6$ **2.** $2w < 16$ **3.** $7p \le 70$ **4.** $n \div 3 \ge 12$

5. $200 < 10x$ **6.** $55 > 11a$ **7.** $21 \ge \frac{g}{4}$ **8.** $17 \le \frac{k}{9}$

9. $h \div 4 \le 0$ **10.** $5v \ge 125$ **11.** $8t > 24$ **12.** $\frac{m}{35} < 1$

13. $13y \ge 104$ **14.** $\frac{r}{4} \le 15$ **15.** $\frac{c}{6} < 13$ **16.** $4s > 28$

Write the word sentence as an inequality. Then solve the inequality.

17. A number q divided by 14 is no less than 4.

18. The product of 21 and a number u is at most 126.

19. The quotient of a number z and 7 is at least 12.

20. A number a divided by 22 is greater than 5.

21. A number g multiplied by 13 is no more than 117.

Describe and correct the error in solving the inequality.

22. ✗

$$5x < 10$$
$$\frac{5x}{5} < \frac{10}{10}$$
$$x < 1$$

23. ✗

$$2n \ge 28$$
$$2n - 2 \ge 28 - 2$$
$$n \ge 26$$

24. A wheelbarrow can carry up to 300 pounds of weight. A bag of soil weighs 20 pounds. Write and solve an inequality to represent the number of bags of soil the wheelbarrow can carry.

25. A table is 3 feet wide. The length of the table can be adjusted as needed. You need at least 24 square feet of space on the table. Write and solve an inequality to represent the minimum length the table should have.

26. A step on a stairway is at least 8 inches high. Write and solve an inequality to represent the minimum number of steps between floors that are 10 feet apart.

Name____ Date__________

8.3 Practice B

Solve the inequality. Graph the solution.

1. $\frac{3}{4}m \leq 6$

2. $6.4v \geq 32$

3. $8b > 2$

4. $\frac{2}{5}x < 3$

5. $42h \geq 105$

6. $\frac{u}{9} \leq 14$

7. $\frac{r}{4} < 3.1$

8. $9.15a > 21.96$

9. $7.2p > 64.8$

10. $\frac{k}{2} < \frac{7}{20}$

11. $\frac{y}{15} \leq 2.7$

12. $5.4g \geq 2.7$

13. $\frac{6}{5}n < 9$

14. $1.05c > 8.4$

15. $6.76s \geq 16.9$

16. $\frac{w}{45} \leq 0.08$

Write the word sentence as an inequality. Then solve the inequality.

17. 10 times a number q is at least 2.01.

18. The quotient of a number z and 18 is at most $\frac{2}{3}$.

19. A number d divided by 8 is no more than 3.43.

20. The product of 5 and a number t is greater than 12.

21. The quotient of a number b and 3 is less than $\frac{1}{12}$.

Graph the numbers that are solutions to both inequalities.

22. $y - 2 > 5$ and $6y \leq 60$

23. $p + 6 \geq 11$ and $3p < 18$

24. $3m \geq 12$ and $\frac{m}{2} > 3$

25. $x - 10 < 0$ and $\frac{x}{4} \leq 1$

26. Three friends decide to share the cost to rent an apartment equally. The apartments that they are considering cost at least \$1200 per month. Write and solve an inequality to represent each person's share of the rental cost.

27. A printer can print up to 80 pages a minute. Write and solve an inequality to represent the number of minutes required to print a report of 380 pages.

28. A highway has at least one rest stop every 25 miles. Write and solve an inequality to represent the number of rest stops in the first 200 miles of the highway.

Name ______________________________ Date __________

8.3 Enrichment and Extension

Solving a Compound Inequality

Example: Solve $4 < 2x < 8$. Graph the solution.

$4 < 2x < 8$	Write inequality.
$\frac{4}{2} < \frac{2x}{2} < \frac{8}{2}$	Divide each part of the inequality by 2.
$2 < x < 4$	Simplify.

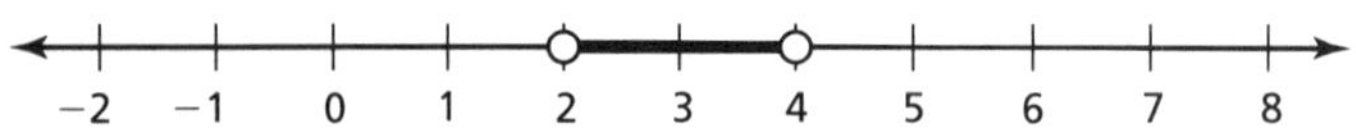

Example: Solve $\frac{x}{3} \geq 2$ *or* $4x \leq 12$. Graph the solution.

Solve each inequality separately.

$\frac{x}{3} \geq 2$	Write inequality.	$4x \leq 12$	Write inequality.
$\frac{x}{3} \bullet 3 \geq 2 \bullet 3$	Multiply each side by 3.	$\frac{4x}{4} \leq \frac{12}{4}$	Multiply each side by 4.
$x \geq 6$	Simplify.	$x \leq 3$	Simplify.

Solve the inequality. Graph the solution.

1. $1 < \frac{x}{2} < 3$

2. $9 \leq 3x \leq 15$

3. $3 \leq \frac{x}{4} < 6$

4. $30 < 5x \leq 40$

5. $2x > 10$ *or* $x - 3 < 1$

6. $\frac{x}{3} \leq 1$ *or* $7x \geq 42$

7. $\frac{x}{2} > 5$ *or* $10x \leq 40$

8. $x + 9 < 11$ *or* $\frac{x}{12} \geq \frac{1}{4}$

9. How are the graphs of the solutions for Exercises 1–4 different from the graphs of the solutions for Exercises 5–8? Explain.

Name______________________________ Date__________

8.3 Puzzle Time

How Many Paws Does A Bear Have?

Write the letter of each answer in the box containing the exercise number.

Solve the inequality. Graph the solution.

1. $\frac{x}{9} < 5$

2. $\frac{x}{7} > 3$

3. $\frac{x}{4} \geq 16$

4. $11x < 99$

5. $6x \geq 30$

6. $8x \leq 64$

7. $\frac{x}{5} > 14$

8. $3x < 48$

9. $9x \geq 63$

10. $\frac{x}{2} \leq 55$

11. $12x > 60$

12. $\frac{x}{6} \leq 16$

Answers

M. $x \leq 8$ 

N. $x \geq 64$ 

A. $x \leq 96$ 

W. $x < 16$ 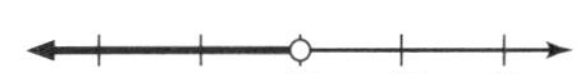

A. $x < 45$ 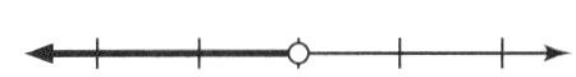

O. $x \leq 110$ 

W. $x < 9$ 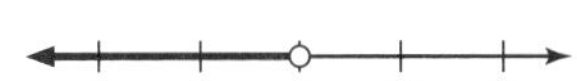

D. $x > 70$ 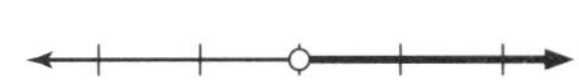

N. $x > 5$ 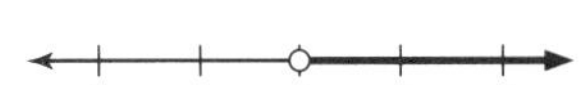

P. $x > 21$ 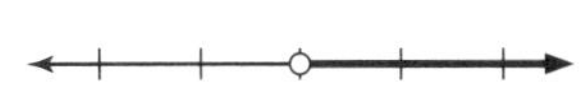

A. $x \geq 7$ 

E. $x \geq 5$ 

10	3	5		2	12	8		9	11	7		6	1	4

Activity 8.4 Start Thinking!

For use before Activity 8.4

Copy the graph and label the parts of the graph with each of the following age categories: preschooler, infant, school-aged, tween, toddler

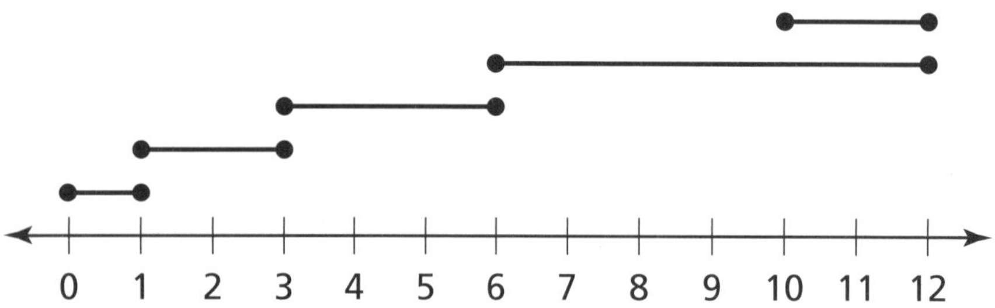

Are all tweens considered school-aged?

Are all school-aged children considered tweens?

Activity 8.4 Warm Up

For use before Activity 8.4

Tell whether the given value is a solution of the inequality.

1. $x + 1 > 8; x = 6$

2. $y - 8 \leq 9; y = 8$

3. $3r \geq 9; r = 3$

4. $\frac{s}{6} < 10; s = 66$

5. $5 \geq z - 1; z = -1$

6. $4 \geq p + 2; p = -4$

Lesson 8.4

Start Thinking!

For use before Lesson 8.4

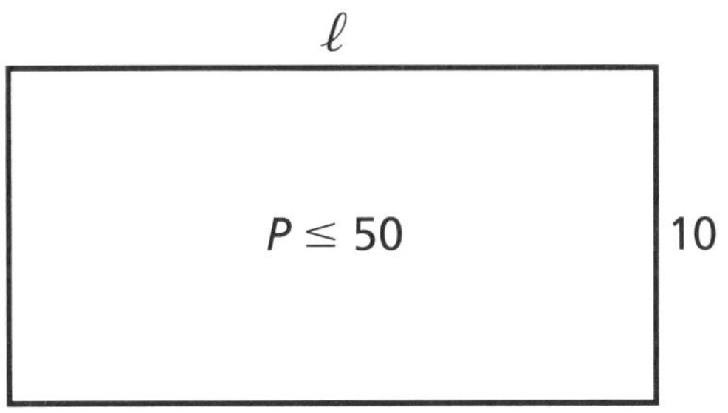

The perimeter of a rectangle must be less than or equal to 50 units. If the width of the rectangle is 10 units, what are all of the possible values for the length of the rectangle?

Describe a real-world situation with the same set-up. Draw a detailed sketch of the situation.

Lesson 8.4

Warm Up

For use before Lesson 8.4

Solve the inequality. Graph the solution.

1. $x - 1 \leq 19$

2. $8x > 12$

3. $\frac{x}{3} < 2$

4. $5 + x \geq 12$

5. $12x \leq 6$

6. $x + 3.2 > 4$

Name ____________________ Date __________

8.4 Practice A

Solve the inequality. Graph the solution.

1. $10s - 6 \leq 14$

2. $4k + 9 < 33$

3. $\frac{n}{3} + 11 > 24$

4. $\frac{x}{7} - 1 \geq 2$

5. $68 < 8g + 12$

6. $17 > 2b - 13$

7. $24 \leq 11p - 42$

8. $2 \geq \frac{c}{5} - 6$

9. $\frac{v}{9} + 24 > 32$

10. $\frac{h}{4} + 2 < 2$

11. $\frac{y}{4} - 12 \leq 3$

12. $0.2a + 7 \geq 8$

Describe and correct the error in solving the inequality.

13. ✗

$$3q - 9 \geq 18$$
$$3q \geq 9$$
$$q \geq 3$$

14. ✗

$$7d + 14 < 98$$
$$7d < 7$$
$$d < 1$$

15. The total volume of a box is 1440 cubic inches. There are 380 cubic inches of paper in the box. The volume of one file is at least 50 cubic inches. Write and solve an inequality to represent the number of files that can be packed in the box.

16. You have 550 minutes of music stored on your MP3 player. One song is at most 6 minutes in length. Your MP3 player can store 5800 minutes of songs. Write and solve an inequality that represents the number of songs you can store on your MP3 player.

17. You want to withdraw money from your savings account to buy two tickets to a dance. The tickets cost \$6.50 each. Your bank requires that you have at least \$50 in your account to avoid a fee. Write and solve an inequality to represent the amount of money you must have in your account to avoid paying a fee.

18. A trumpet player is performing in a 2-hour concert. She has practiced for 10 hours so far and plans to practice for 1.5 hours each day before the concert. She wants to practice at least 40 hours before the concert. Write and solve an inequality to represent the number of days she must practice to meet her goal.

Name____ Date__________

8.4 Practice B

Solve the inequality. Graph the solution.

1. $91d + 36 < 309$

2. $10 \geq \frac{r}{5} + 5$

3. $\frac{a}{6} + \frac{2}{3} > 1\frac{1}{2}$

4. $8.1k - 6.28 > 5.87$

5. $8u - 3 \leq 1.8$

6. $\frac{n}{12} - 4.68 \geq 7.08$

7. $15 < 15 + \frac{z}{3}$

8. $27 > 0.3x - 15$

9. $3.14r + 6 \leq 27.98$

10. $0.14c + 2.48 > 9.06$

11. $1.1m - 0.06 < 2.8$

12. $7y - 1\frac{1}{4} \geq 4$

The area of the figure is at most 600 square centimeters. Write and solve a two-step inequality to represent the possible values of *x*. Use 3.14 for π.

13.

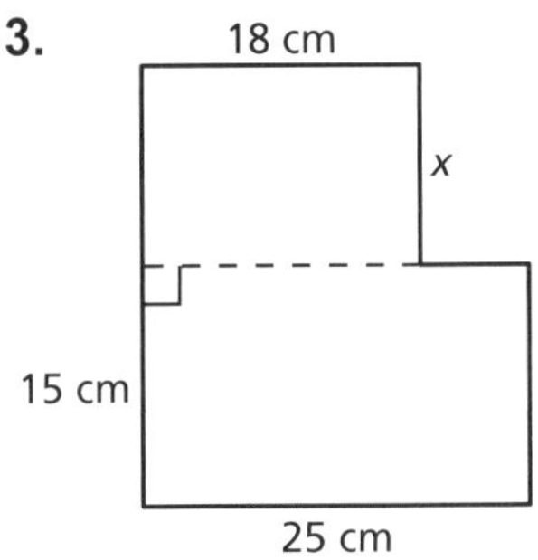

14.

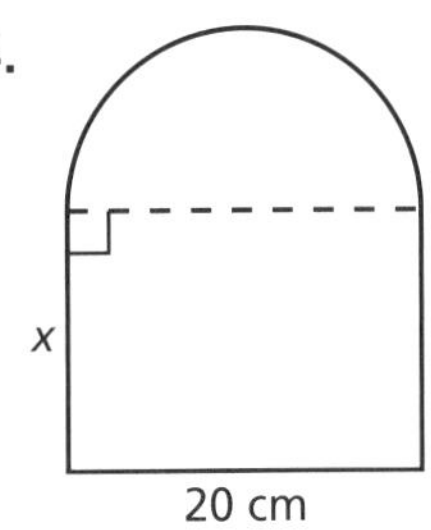

15.

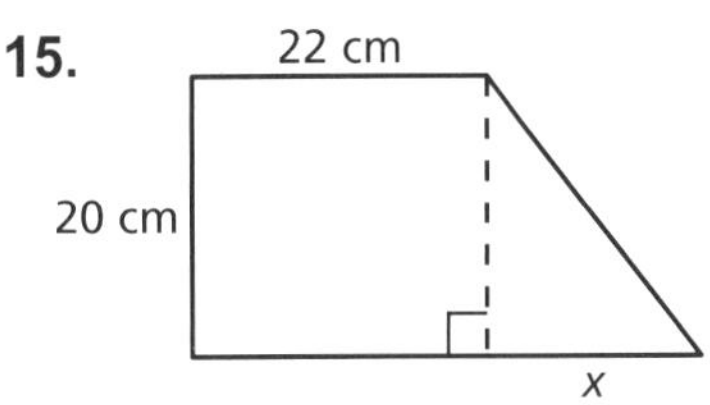

16. Admission to a town carnival is \$3. Tickets for games and rides cost \$0.50 each. You want to spend less than \$15 at the carnival. Write and solve an inequality to represent the number of tickets you can buy without going over your budget. What is the greatest number of tickets you can buy?

17. Your scores for the first 5 levels of a video game are 250, 190, 380, 420, and 310 points. Write and solve an inequality to represent the score you need on the sixth level to have a mean score of at least 350 points for the first 6 levels.

18. You are on a 500-mile road trip. You want to reach your destination in no more than 9 hours. You travel at an average speed of 62 miles an hour for the first 4 hours of the trip. Write and solve an inequality to represent your average speed for the remainder of the trip.

19. You have a \$60 gift card to spend at a department store. You want to buy some \$15 T-shirts and a hat. The hat costs \$14.75. Write and solve an inequality to represent the possible number of T-shirts you can buy.

Name ______________________________ Date __________

8.4 Enrichment and Extension

Solving Multl-Step Inequalities

Solve the inequality. Graph the solution.

1. $4(x + 1) < 6$

2. $2(x - 3) \geq 10$

3. $\frac{1}{2}(x + 8) \leq 11$

4. $\frac{x + 5}{6} > 2$

5. $\frac{2x - 1}{4} \geq 1$

6. $\frac{3x + 2}{5} \leq 7$

7. $8 < \frac{4 + 10 + x}{2}$

8. $5(2x + 6) \geq 40$

9. $3(7 + 10x) < 21$

10. You get scores of 85 and 91 on two history tests. Write and solve an inequality to find the scores you can get on your next history test to have an average of at least 90. Explain your answer.

11. You and a group of friends wait $\frac{1}{2}$ hour to ride an amusement park ride. You go on the ride a second time and wait $\frac{1}{3}$ hour. You want to go on a third time. Write and solve an inequality to find how many minutes you can wait for your average waiting time to be at most $\frac{1}{3}$ hour. (*Hint:* Convert the waiting times to minutes.)

12. Write and solve an inequality to find the possible values of x so that the rectangle has an area of more than 130 square units. Explain your answer.

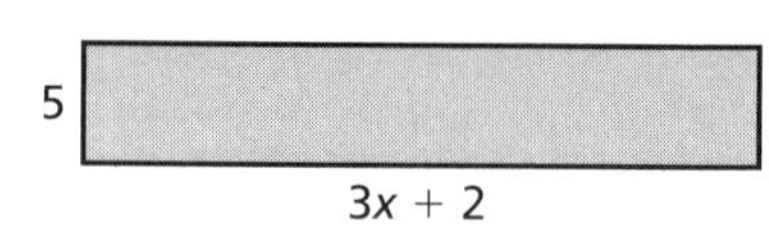

Name______________________________ Date__________

8.4 Puzzle Time

Why Did The Man Bring His Dog To The Railroad Station?

Write the letter of each answer in the box containing the exercise number.

Solve the inequality. Graph the solution.

1. $\frac{x}{4} + 3 \le 6$
2. $6x - 4 > 8$
3. $9 \le 9 + 5x$
4. $7 > \frac{x}{5} - 7$
5. $4x - 18 \ge 2$
6. $30 + 5x < 80$
7. $16 \le \frac{x}{7} + 11$
8. $0.3 + 3.7x \le 7.7$
9. $4.4 > \frac{x}{6} - 2.6$
10. $8x + 11.1 > 27.9$

Answers

O. $x \ge 0$ 

T. $x \le 2$ 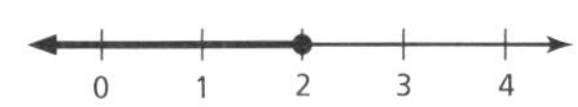

I. $x \ge 5$ 

M. $x > 2.1$ 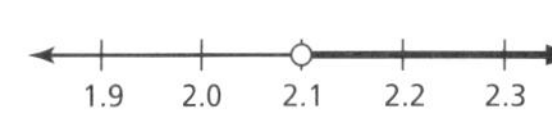

H. $x < 10$ 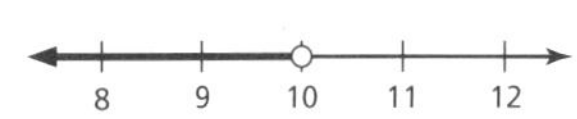

R. $x < 70$ 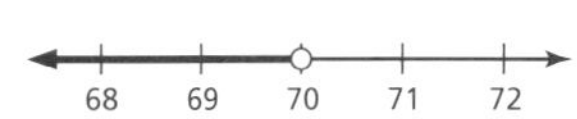

I. $x \ge 35$ 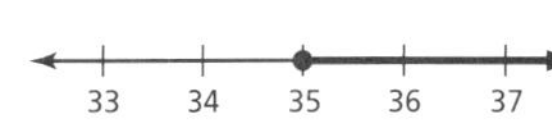

N. $x < 42$ 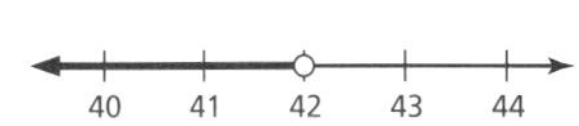

T. $x > 2$ 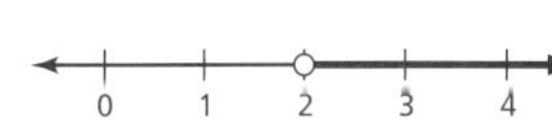

A. $x \le 12$ 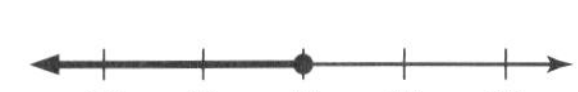

8	3		2	4	1	7	9		6	5	10

Name ____ Date __________

Chapter 8

School-to-Work

For use after Section 8.3

Veterinary Assistant

Veterinary assistants help to care for animals. They assist the veterinarian with exams, care for ill animals and animals recovering from surgery, clean cages and exam rooms, administer tests, draw blood, prepare medications, and help to run the office, including billing customers. Veterinary assistants may work in veterinarian offices, zoos, animal hospitals, or research laboratories.

1. A medication is designed for cats and kittens 9 pounds and under.

a. Write an inequality for the situation.

b. A cat weighs 7 pounds 9 ounces. Should it be given this medication, or another one?

2. A client brings in a 12-week old puppy adopted from a shelter. The puppy weighs 28 pounds. According to your chart, what will you tell the client to expect?

Breed	Weight w at 12 weeks
Small breed	$w \leq 7.5$ lb
Medium breed	$7.5 \text{ lb} < w \leq 27$ lb
Large breed	$27 \text{ lb} < w \leq 30$ lb
Very large breed	$w > 30$ lb

A veterinary assistant at the zoo is giving advice to help raise an orphaned *sugar glider*. A sugar glider is a small furry marsupial originally from Australia.

3. A baby sugar glider requires a temperature of 86 to 93 degrees Fahrenheit.

a. Write two inequalities for the situation.

b. Give an example of an acceptable temperature.

c. Give an example of an unacceptable temperature. Is it too hot or too cold?

4. At this age, the sugar glider's weight should increase by at least 60% over the course of 10 days.

a. Write an inequality for the situation. Use f for the first weight and s for the second weight.

b. The sugar glider weighed 15 grams when brought in and now weighs 26 grams. Is it growing as expected?

5. The assistant estimates that the sugar glider was 65 days old when it arrived at the zoo. It should start eating solids when it is 100 days old and be weaned from milk by the time it is 130 days old. Write and solve an expression to find the number of days after the sugar glider arrives for it to be weaned from milk.

Name____ Date__________

Chapter 8 Study Help

You can use a **Y Chart** to compare two topics. List differences in the branches and similarities in the base of the Y.

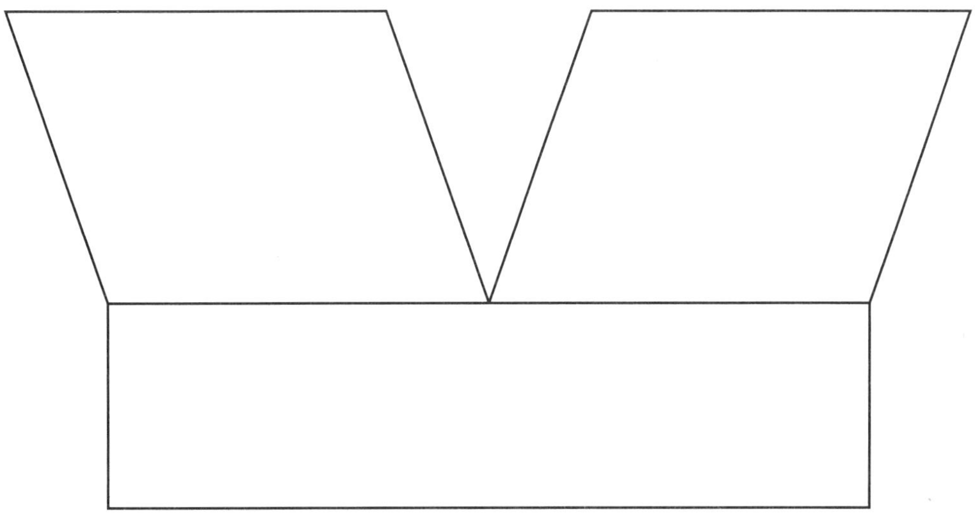

On Your Own

Make a Y chart to help you study and compare these topics.

1. writing equations and writing inequalities
2. graphing the solution of an equation and graphing the solution of an inequality
3. graphing inequalities that use $>$ and graphing inequalities that use $<$
4. graphing inequalities that use $>$ or $<$ and graphing inequalities that use $\geq$ or $\leq$
5. solving inequalities using addition and solving inequalities using subtraction

After you complete this chapter, make Y charts for the following topics.

6. solving inequalities using multiplication and solving inequalities using division
7. solving two-step equations and solving two-step inequalities
8. Pick two other topics that you studied earlier in this course and make a Y chart to compare them.

Name ______________________________ Date __________

Chapter 8 Cumulative Practice

Write the phrase or word sentence as an expression, equation, or inequality.

1. 9 is greater than a number h.

2. 6 less than a number v is 12.

3. The sum of 8 and a number k

4. A number b is at least 13.

5. 24 times a number x

6. A number w divided by 40 is greater than 3.

7. The difference of 4 and a number a

8. 4 plus a number c is less than 15.

Write the expression, equation, or inequality as a phrase or word sentence.

9. $c \div 6$

10. $m - 9 = 12$

11. $4u \geq 28$

Write an inequality and a word sentence that represent the graph.

12. 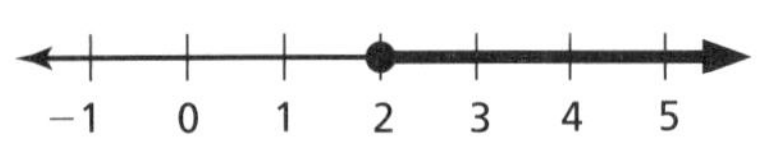

13. 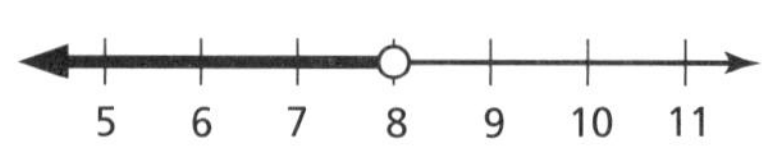

14. A theater can hold 500 people. 135 people are in the theater. Write an inequality that represents the number of people p that can enter the theater.

Graph the inequality on a number line.

15. $d \geq 2$

16. $t \leq 1.5$

17. $g > \frac{3}{4}$

18. A restaurant has a menu for children ages 10 and younger. Write and graph an inequality that represents the age of a child who can order off the children's menu.

Solve the equation or inequality. Graph the solution of each inequality.

19. $y + 5 = 13$

20. $2b + 3 = 27$

21. $\frac{p}{5} \geq 1$

22. $u - \frac{1}{8} > \frac{7}{8}$

23. $8c - 12 \geq 36$

24. $2z = 6$

25. $95 = 24m + 53$

26. $38 \leq \frac{k}{4} - 10$

27. $198 > 11v$

28. You are packing books in a duffel bag. Each book weighs 3 pounds. The total weight of the duffel bag must be less than 50 pounds. The empty duffel bag weighs 5 pounds. Write and solve an inequality that represents the number of books you can pack in the duffel bag.

Chapter 9

Name ____ Date __________

Chapter 9 Tables, Graphs, and Functions

Dear Family,

Keeping to a regular program of exercise can be a chore. Choosing an activity that you enjoy is a good start. However, we often need extra motivation. Setting goals and keeping track of your progress can be a great way to keep you going.

You and your student can use a table to keep track of your exercise program. Each day, you can record what you did. Choose the type of information you want to record. Decide whether you are counting the number of repetitions, the distance, the length of time, or the speed.

My Exercise Progress		
Day	**Miles Run**	**Sit-ups**
1	1.1	15
2	1.1	16
3	1.1	16
4	1.2	15
5	1.2	17
6	1.2	16
7	1.2	18

Help your student make a graph of the data in your table. Ask your student how the graph helps you visualize your progress. You may ask your student the following questions:

- "If we are counting repetitions, do we want the graph to go up or down?" Your student will likely say "up."
- "How about if we are counting the length of time, do we want the graph to go up or down?" Your student will likely say "down."
- "Judging from the graph, are we going to make our goals?"

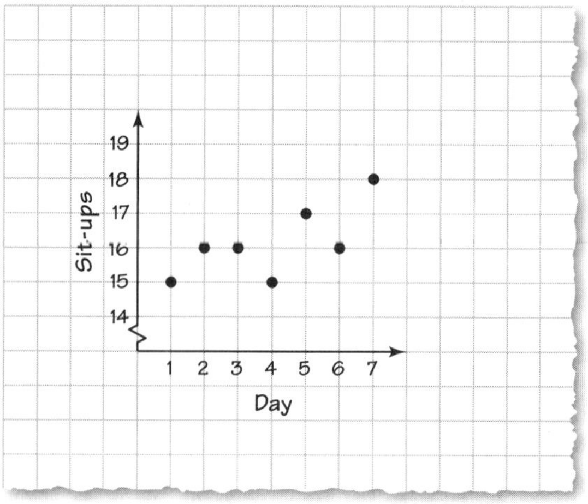

You and your student may choose to compare your data. Talk about a strategy to do this. It may not be fair to directly compare the data—talk about a way to compare the graphs. How fast are you both approaching your goal?

Here's hoping you reach your goals together!

Nombre ______________________________ Fecha ________

Capítulo 9 Tablas, Gráficos y Funciones

Estimada Familia:

Apegarse a un programa regular de ejercicios puede ser toda una faena. Elegir una actividad que disfrute hacer es un buen comienzo. Sin embargo, con frecuencia necesitamos una motivación adicional. Establecer reglas y hacer un seguimiento de su progreso puede ser una buena manera de mantenerse activo.

Usted y su estudiante pueden usar una tabla para hacer un seguimiento de su programa de ejercicios. Cada día, pueden registrar lo que hayan hecho. Elijan el tipo de información que desean registrar. Decidan si están contando el número de repeticiones, la distancia, la duración de tiempo o la velocidad.

Mi Progreso de Ejercicios		
Día	**Millas Recorridas**	**Abdominales**
1	1.1	15
2	1.1	16
3	1.1	16
4	1.2	15
5	1.2	17
6	1.2	16
7	1.2	18

Ayude a su estudiante a hacer un gráfico de datos en su tabla. Pregunte a su estudiante cómo ayuda el gráfico a visualizar su progreso. Puede hacer a su estudiante las siguientes preguntas:

- "Si estamos contando repeticiones, ¿queremos que el gráfico vaya para arriba o abajo?" Su estudiante probablemente dirá "para arriba."
- "Y si estamos contando la duración de tiempo, ¿queremos que el gráfico vaya para arriba o abajo?" Su estudiante probablemente dirá "para abajo."
- "A partir del gráfico, ¿vamos a alcanzar nuestras metas?"

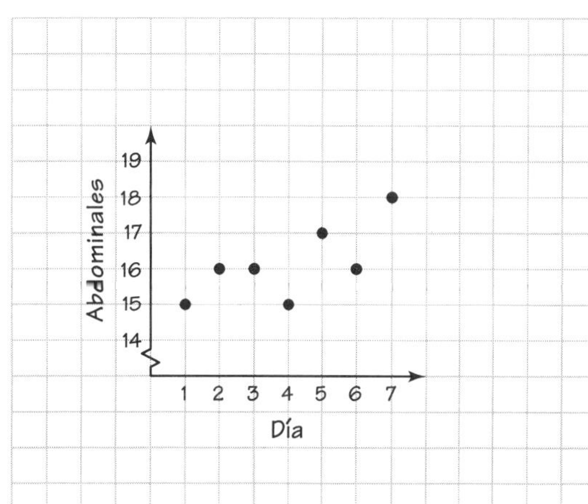

Usted y su estudiante pueden elegir comparar sus datos. Conversen acerca de una estrategia para lograr esto. Puede no ser justo comparar directamente los datos—conversen acerca de una manera para comparar los gráficos. ¿Qué tan rápido se están acercando a su meta?

¡Esperamos que alcancen sus metas juntos!

Non ______________________________ Dat __________

Chapít 9 Tablo, Graf, ak Fonksyon

Chè Fanmi:

Kenbe yon pwogram egzèsis regilye kapab yon kòve. Chwazi yon aktivite ou renmen se yon bon kòmansman. Sepandan, nou souvan bezwen motivasyon anplis. Tabli bi epi pran nòt pwogrè ou kapab yon bon jan pou fè ou kontinye.

Ou menm ak elèv ou a kapab sèvi ak yon tablo pou swiv fil pwogram egzèsis ou a. Chak jou, ou kapab note sa ou fè. Chwazi tip enfòmasyon ou vle note yo. Deside si w'ap konte kantite repetisyon, distans, kantite tan, oswa vitès la.

Pwogrè Egzèsis Mwen		
Jou	Mil Mwen Kouri	Redrèsman Chita
1	1.1	15
2	1.1	16
3	1.1	16
4	1.2	15
5	1.2	17
6	1.2	16
7	1.2	18

Ede elèv ou a fè yon graf done yo nan tablo ou a.

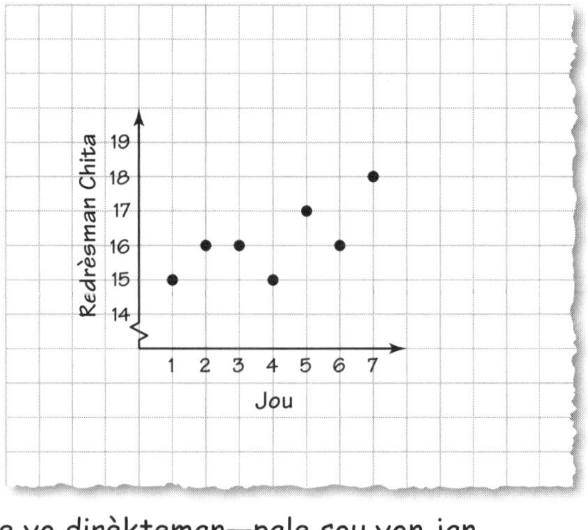

Mande elèv ou a kijan graf la ede ou vizyalize pwogrè ou. Ou gen dwa poze elèv ou a keksyon sila yo:

- "Si n'ap konte repetisyon, èske nouvle pou graf la monte oubyen desann?" Elèv ou a ap siman di "monte".
- "E si n'ap konte kantite tan an, èske nou vle pou graf la monte oubyen desann?" Elèv ou a ap siman reponn "desann".
- "Dapre graf la, èske n'ap atenn bi nou?"

Ou menm ak elèv ou a gen dwa chwazi konpare done nou yo. Pale sou estrateji pou fè sa. Sa gen dwa pa jis pou konpare done yo dirèkteman—pale sou yon jan pou konpare graf yo. Ak ki vitès nou tou de ap pwoche bi nou?

Nou swete nou atenn bi nou yo ansanm!

Activity 9.1 Start Thinking!

For use before Activity 9.1

What is the cost of a container of milk at your school cafeteria? Complete the diagram below, using the cost of a container of milk at your school cafeteria.

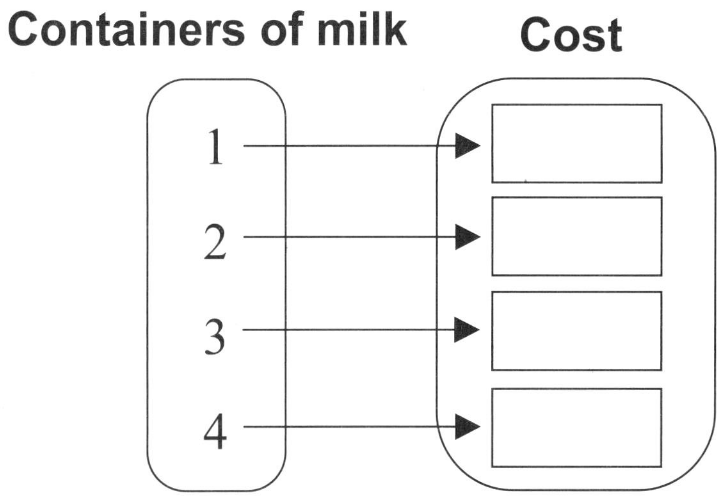

This type of diagram is called a *mapping diagram*. Why do you think it is called that?

Activity 9.1 Warm Up

For use before Activity 9.1

Find the volume of the rectangular prism.

1.

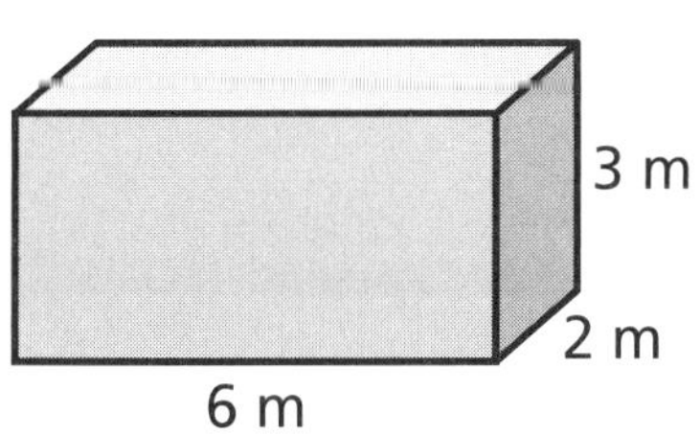

2.

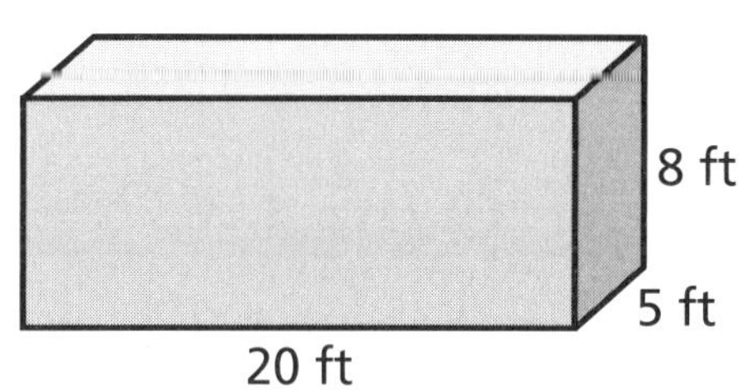

3.

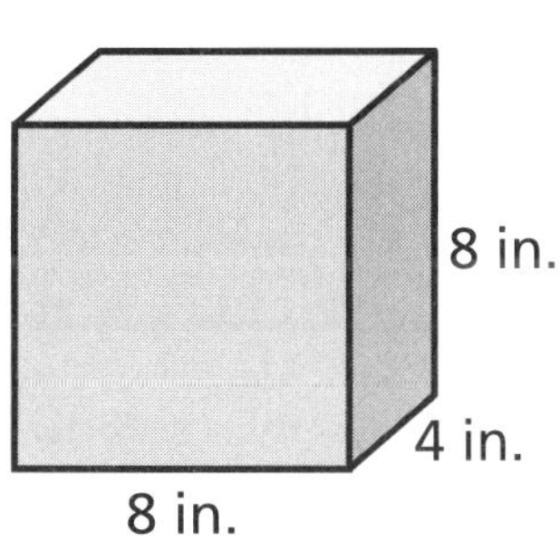

4.

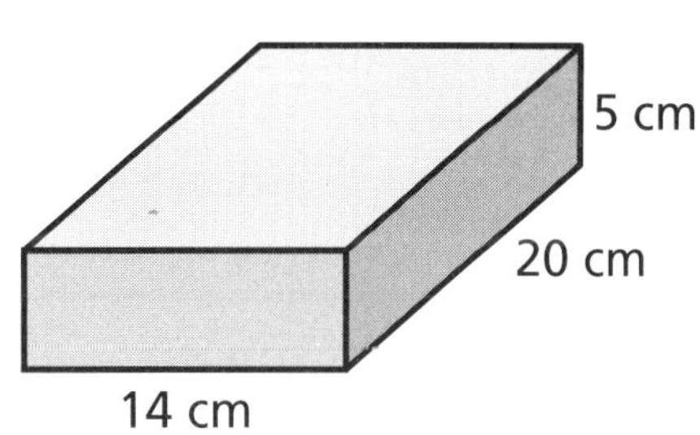

Lesson 9.1 Start Thinking!

For use before Lesson 9.1

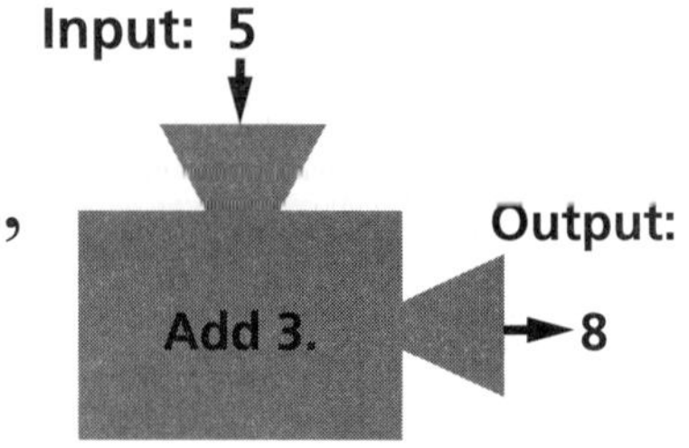

The function machine shown is for the rule "Add 3." Using this function machine, what is the output for an input of "12"?

Guess the Function Game (2 players)

1. Decide who will be Player A and Player B.
2. Player A makes up a function rule. Do NOT tell Player B!
3. Player B gives an input number.
4. Player A uses the rule and mental math to calculate the output. Tell Player B the output.
5. Repeat Steps 3 and 4 until Player B can correctly guess the function rule.
6. Switch roles and play again.

Lesson 9.1 Warm Up

For use before Lesson 9.1

Describe the pattern in the mapping diagram. Copy and complete the diagram.

1. Input Output

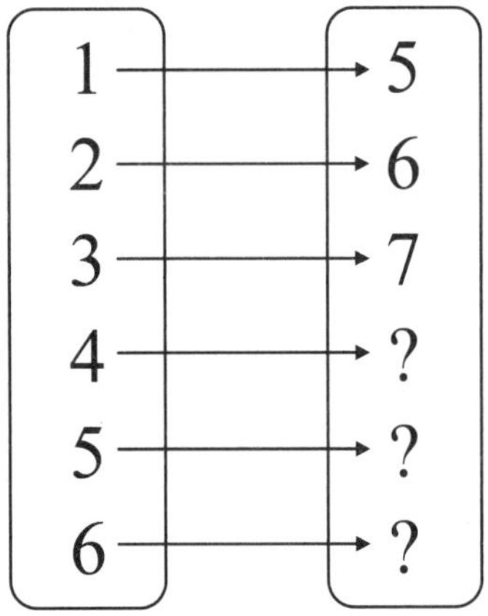

2. Input Output

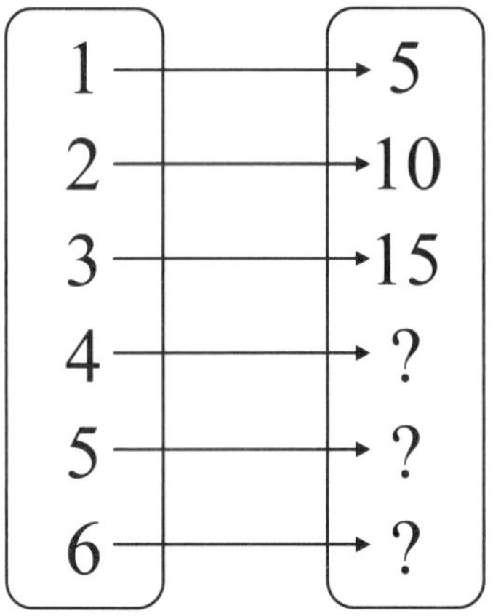

Name________________________________ Date__________

9.1 Practice A

Describe the pattern in the mapping diagram. Copy and complete the diagram.

1. Input Output

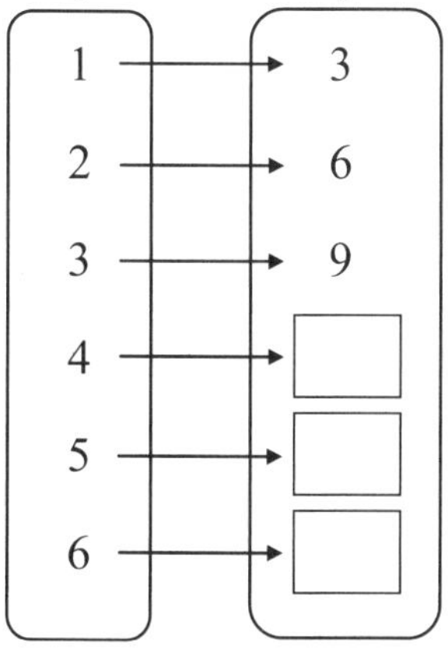

2. Input Output

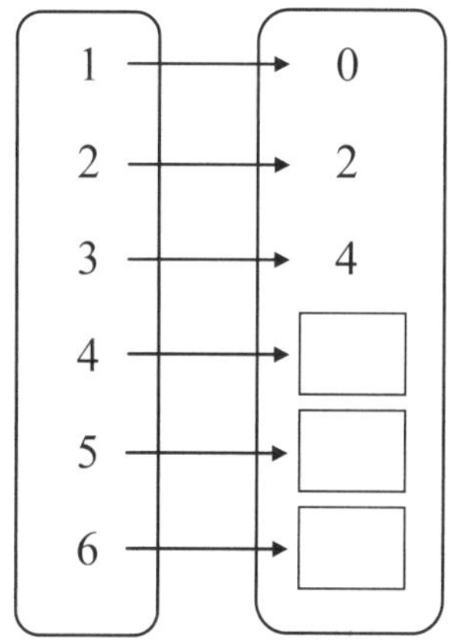

List the ordered pairs shown in the mapping diagram.

3. Input Output

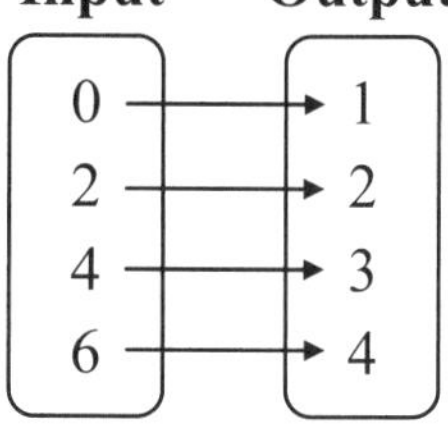

4. Input Output

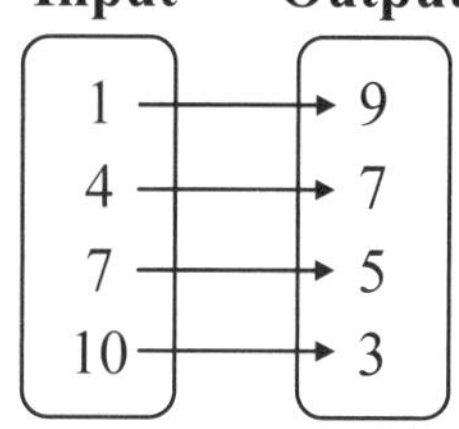

Draw a mapping diagram of the set of ordered pairs.

5. (1, 2), (3, 5), (6, 9), (10, 12)

6. (0, 7), (2, 5), (5, 8), (4, 9)

7. The table shows the number of tickets purchased and the total cost.

Tickets	Total Cost
1	$14
2	$24
3	$30
4	$32

a. Use the table to draw a mapping diagram.

b. Describe the pattern. How does the cost per ticket change as you buy more tickets?

c. Based on this pattern, how much would you expect to pay for 5 tickets?

d. Compare the costs for 3 tickets and 5 tickets. What can you suggest?

e. Explain why this pattern could not continue for up to 8 tickets.

Name ______________________ Date __________

9.1 Practice B

List the ordered pairs shown in the mapping diagram.

1. Input Output

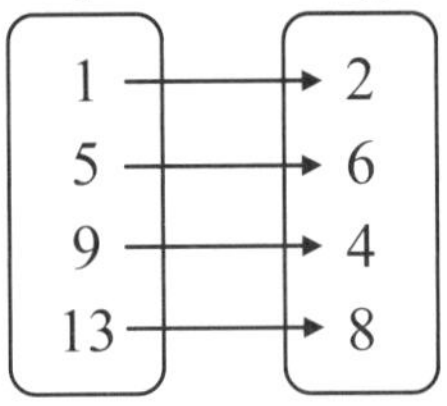

2. Input Output

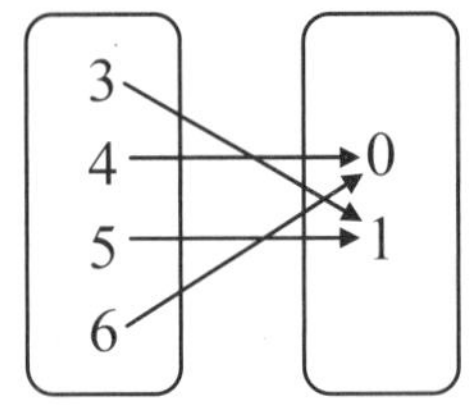

Draw a mapping diagram of the set of ordered pairs.

3. (0, 5), (4, 12), (6, 13), (7, 9)

4. (1, 0), (3, 0), (5, 4), (7, 4), (9, 4)

Draw a mapping diagram for the graph. Then describe the pattern of inputs and outputs.

5.

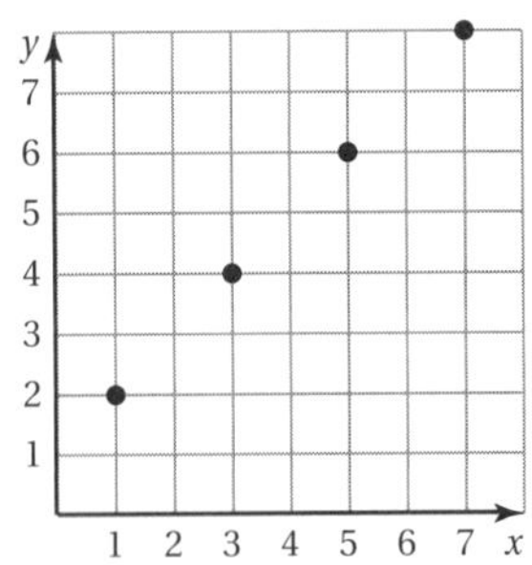

6.

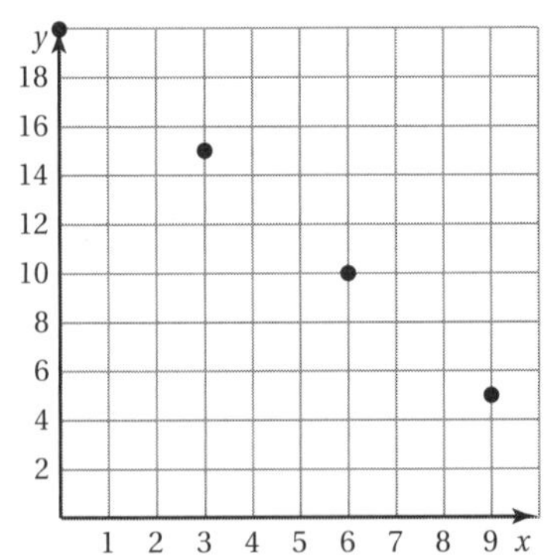

7. The table shows the cost of a collect call.

Minutes	1	2	3	4	5	6	7
Cost	$3	$3.25	$3.50	?	?	?	?

a. Complete the table.

b. Draw a mapping diagram for the table.

c. List the ordered pairs.

d. Graph the ordered pairs in a coordinate plane.

e. Describe the pattern. How does the cost change as the number of minutes increases?

Name______________________________ Date__________

9.1 Enrichment and Extension

Identifying Functions

You learned that a *function* pairs each input with exactly one output. Tell whether the mapping diagram represents a function. Explain.

1. Input Output

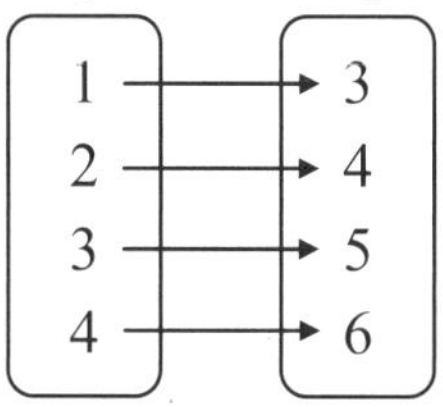

2. Input Output

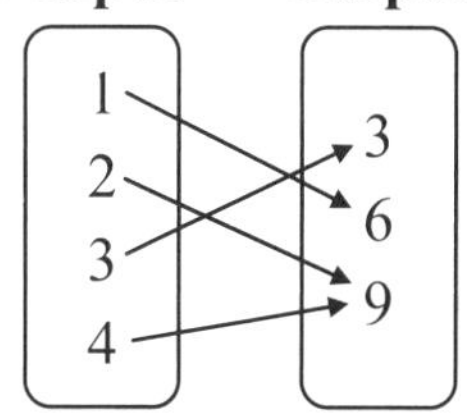

3. Input Output

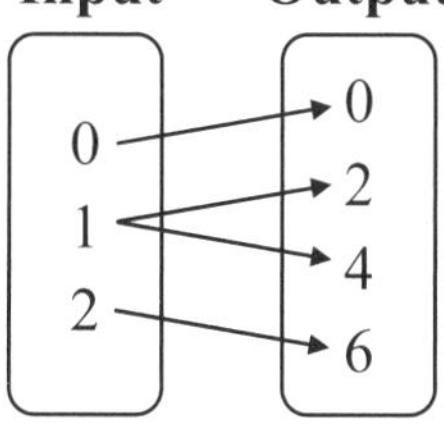

4. Input Output

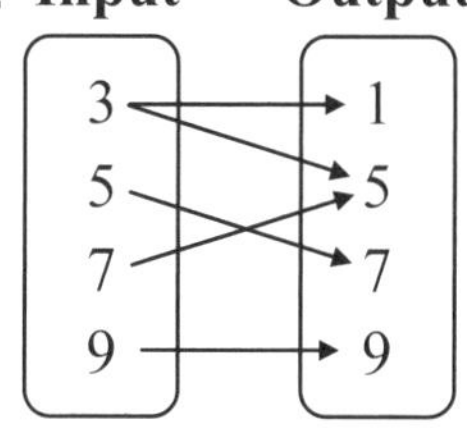

(a) Draw arrows to complete a mapping diagram of a function. (b) Draw one more arrow so that the mapping diagram does not represent a function.

5. Input Output

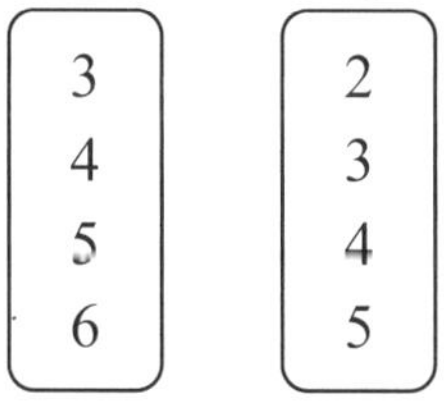

6. Input Output

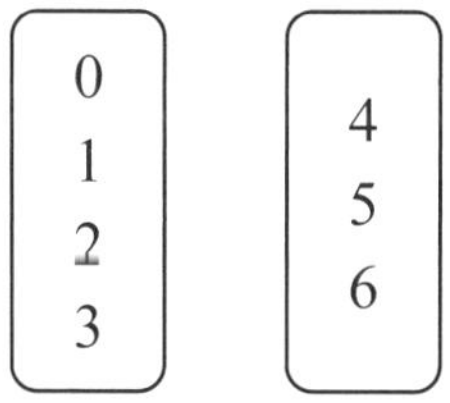

(a) Find another point (x, y) so that the set of points represents a function.
(b) Find a point so that the set of points does not represent a function.

7. $(0, 0), (1, 1), (2, 2), (x, y)$

8. $(0, 9), (2, 8), (4, 7), (6, 6), (8, 5), (x, y)$

9. $(1, 3), (2, 3), (3, 4), (4, 4), (5, 5), (6, 5)\ (x, y)$

Name ______________________________ Date __________

9.1 Puzzle Time

What Flowers Grow Between Your Nose And Your Chin?

Write the letter of each answer in the box containing the exercise number.

List the ordered pairs shown in the mapping diagram.

1. Input Output

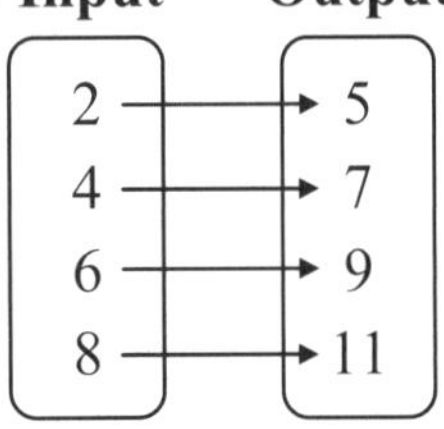

2. Input Output

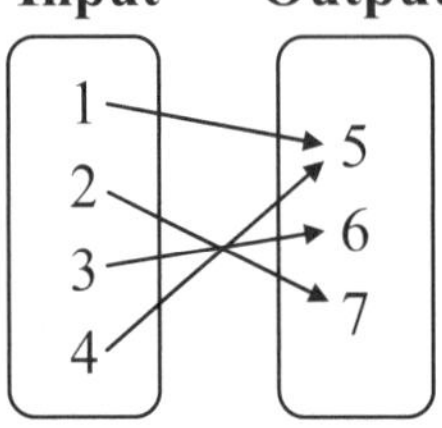

Draw a mapping diagram of the ordered pairs.

3. (2, 4), (6, 8), (9, 11), (15, 17)

4. (2, 1), (3, 1), (5, 3), (6, 3)

Describe the pattern of inputs and outputs.

5.

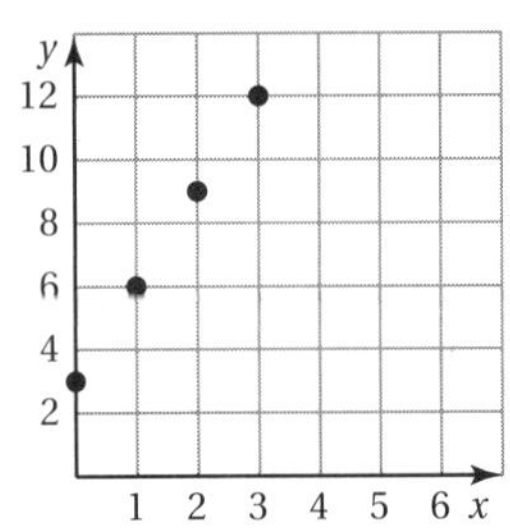

6.

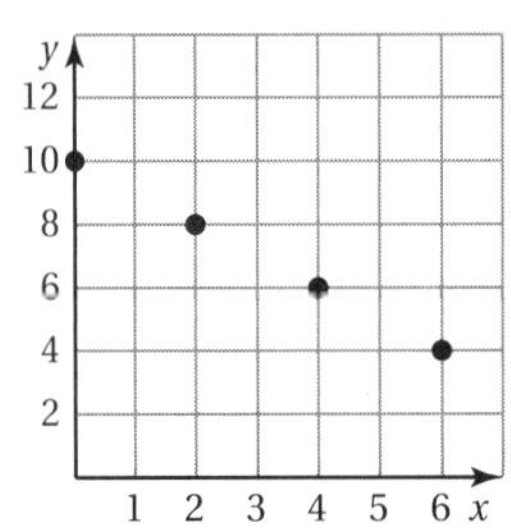

Answers

S. As each input increases by 2, the output decreases by 2.

L. (2, 5), (4, 7), (6, 9), (8, 11)

P. Input Output

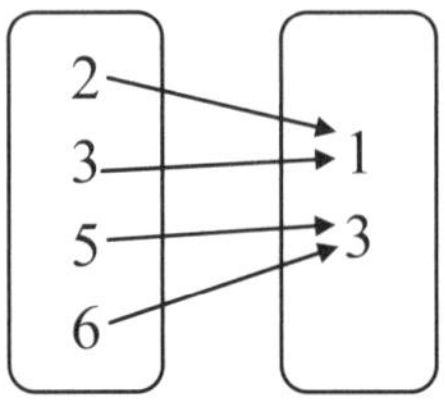

M. As each input decreases by 2, the output increases by 2.

R. (2, 11), (4, 9), (6, 7), (8, 5)

I. Input Output

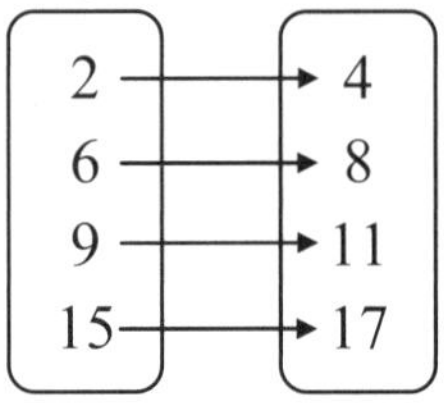

U. (1, 5), (2, 7), (3, 6), (4, 5)

T. As each input increases by 1, the output increases by 3.

A. (1, 5), (2, 6), (3, 7), (4, 5)

D. As each input increases by 1, the output decreases by 3.

5	2	1	3	4	6

Activity 9.2 Start Thinking!

For use before Activity 9.2

Think about the function rule "Multiply by 7."

How can you write an equation for this function?

Explain how this equation can be used to describe the area of a rectangle with a length of 7 meters.

Activity 9.2 Warm Up

For use before Activity 9.2

1. What is the perimeter of a square with a side length of 5 inches?
2. What is the area of a triangle with a base of 15 feet and a height of 2 feet?
3. What is the circumference of a circle with a radius of 12 feet? Use 3.14 for π.
4. What is the surface area of a cube with an edge length of 5 centimeters?
5. What is the perimeter of a rectangle with a length of 9 meters and a height of 7 meters?
6. What is the area of a circle with a diameter of 10 inches? Use 3.14 for π.

Lesson 9.2 Start Thinking!

For use before Lesson 9.2

Using two variables, write an equation for a function that describes a real-life situation. Explain the situation and what each of the variables represents.
Which variable represents input values?
Which variable represents output values?
Then make a mapping diagram that includes four different input values.

Lesson 9.2 Warm Up

For use before Lesson 9.2

Write an equation that describes the function.

1.

Input	Output
0	0
1	10
2	20
3	30

2.

Input	Output
1	17
2	18
3	19
4	20

3.

Input	Output
2	1
4	2
6	3
8	4

4.

Input	Output
20	10
40	30
60	50
80	70

Name____ Date__________

9.2 Practice A

Write an equation that describes the function.

1.

Input	Output
0	10
1	11
2	12
3	13

2.

Input	Output
0	0
1	3
2	6
3	9

Write an equation that describes the function.

3. The output is eight less than the input.

4. The output is double the input.

Find the value of *y* for the given value of *x*.

5. $y = x - 1; x = 5$

6. $y = 8x; x = 3$

7. $y = 4x - 1; x = 10$

8. $y = \frac{x}{2} + 5; x = 2$

Tell whether the ordered pair is a solution of the equation.

9. $y = x + 5; (2, 7)$

10. $y = 9x; (3, 12)$

11. $y = 2x + 3; (4, 12)$

12. $y = \frac{x}{2} - 4; (8, 0)$

13. You are running at a rate of 6 miles per hour.

a. Write an equation that relates the distance d traveled in h hours.

b. How many miles do you run in 2 hours?

14. The cost of admission for a student is $4 less than the cost of admission for an adult.

a. Write an equation that relates the cost of admission for a student s with the cost of admission for an adult a.

b. What is the cost of admission for a student when the cost of admission for an adult is $7.50?

c. What is the cost of admission for an adult when the cost of admission for a student is $2?

Name ______________________________ Date __________

9.2 Practice B

Write an equation that describes the function.

1.

Input	Output
10	14
20	24
30	34
40	44

2.

Input	Output
10	5
12	6
14	7
16	8

Write an equation that describes the function.

3. The output is five times the input.

4. The output is two less than the input.

Find the value of *y* for the given value of *x*.

5. $y = 3x - 4;\ x = 2$

6. $y = \frac{x}{3} - 1;\ x = 6$

Tell whether the ordered pair is a solution of the equation.

7. $y = \frac{x}{3} - 4;\ (24, 5)$

8. $y = 2x + 7;\ (5, 17)$

9. You are traveling on a turnpike at a rate of 70 miles per hour.

a. Write an equation that relates the distance d traveled in h hours.

b. How many miles do you travel in 3.5 hours?

Find the value of *x* for the given value of *y*.

10. $y = 6x - 4;\ y = 20$

11. $y = \frac{x}{2} + 3;\ y = 8$

12. Your school club is selling popcorn at the football game. The cost of making the popcorn is \$90. You charge \$1.50 for each bag of popcorn.

a. Write an equation you can use to find the profit P for selling b bags of popcorn.

b. You will *break even* if the cost of making the popcorn equals your income. How many bags of popcorn must you sell to break even?

13. Write a function for the area A of a square given the perimeter P of the square.

Name____ Date__________

9.2 Enrichment and Extension

Function Notation

In Lesson 9.2, you wrote functions as equations in two variables, *x* and *y*, with *x* as the inputs and *y* as the outputs. Another way to write a function is to use *function notation*.

Function in *x* and *y*		Function notation
$y = 4x$	⟶	$f(x) = 4x$
$y = 3x + 5$	⟶	$f(x) = 3x + 5$

$f(x)$ is the output. The symbol $f(x)$ is read as "*f* of *x*."

Example: Find $f(4)$ for $f(x) = x + 3$

$f(x) = x + 3$	Write function.
$f(4) = 4 + 3$	Substitute 4 for *x*.
$f(4) = 7$	Add.

So, *f* of 4 is equal to 7.

Rewrite the function using function notation.

1. $y = 2x$ **2.** $y = x - 1$ **3.** $y = 4x + 7$ **4.** $\frac{3}{4}x + 5 = y$

Find $f(0)$, $f(2)$, and $f(4)$ for the function.

5. $f(x) = 8x$ **6.** $f(x) = 7 + x$ **7.** $f(x) = 5 - x$

8. $f(x) = 3x + 8$ **9.** $f(x) = 9 - 2x$ **10.** $f(x) = \frac{1}{2}x$

11. $f(x) = \frac{1}{4}x + 6$ **12.** $f(x) = 3 + 1.6x$ **13.** $f(x) = x^2 + 1$

14. Critical Thinking Give an example of a function where $f(0) = 2$ and $f(1) - 2$.

Name ______________________________ Date __________

9.2 Puzzle Time

Does It Take Longer To Run From First Base To Second, Or From Second To Third?

A	B	C	D	E	F
G	H	I	J	K	L

Complete each exercise. Find the answer in the answer column. Write the word under the answer in the box containing the exercise letter.

44 A
$y = x + 5$ TO
$y = 3x$ FROM
No THE
$y = 8x$ THIRD
9 SHORTSTOP

Write an equation that describes the function.

A.

Input	Output
0	0
1	3
2	6
3	9
4	12

B.

Input	Output
15	5
25	15
35	25
45	35
55	45

C. The output is five more than the input.

D. The output is eight times the input.

E. The output is one-third the input.

F. The output is thirteen more than four times the input.

Find the value of *y* for the given value of *x*.

G. $y = x + 7;\ x = 5$

H. $y = 6x - 4;\ x = 8$

I. $y = 2x + 4;\ x = 2.5$

J. $y = 9x - 3;\ x = 3$

Tell whether the ordered pair is a solution of the equation.

K. $y = 7x + 1;\ (0, 2)$

L. $y = \frac{x}{2} + 8;\ (24, 20)$

Yes MIDDLE
$y = \frac{1}{3}x$ BECAUSE
24 IN
12 IS
$y = x - 10$ SECOND
$y = 4x + 13$ THERE

Activity 9.3 Start Thinking!
For use before Activity 9.3

When have you seen or used a chart or table outside of math class?

What does a table help you do with information?

Activity 9.3 Warm Up
For use before Activity 9.3

Find the percent of the number.

1. 5% of 60

2. 10% of 2

3. 12% of 25

4. 35% of 50

5. 75% of 16

6. 125% of 24

Lesson 9.3 Start Thinking!

For use before Lesson 9.3

Describe something in real life that has an input and an output.

Lesson 9.3 Warm Up

For use before Lesson 9.3

Copy and complete the input-output table for the function.

1. $y = x - 3$

Input, *x*	10	11	12	13
Output, *y*				

2. $y = 8x$

Input, *x*	1	2	3	4
Output, *y*				

3. $y = x + 6$

Input, *x*	0	1	2	3
Output, *y*				

4. $y = \frac{x}{2}$

Input, *x*	0	2	4	6
Output, *y*				

Name____ Date __________

9.3 Practice A

Copy and complete the input-output table for the function.

1. $y = x + 3$

Input, x	1	2	3	4
Output, y				

2. $y = 5x$

Input, x	1	3	5	7
Output, y				

Write an equation for the function. Then copy and complete the table.

3. The output is 6 more than the input.

Input, x	0	1	2	3
Output, y				

4. The output is 3 times the input.

Input, x	0	2	4	6
Output, y				

Write an equation for the function shown by the table.

5.

Input, x	1	2	3	4
Output, y	6	7	8	9

6.

Input, x	1	3	5	7
Output, y	4	12	20	28

7. Copy and complete the table.

For each output, multiply the input by 2, then add 1.

Input, x	0	2	4	5		
Output, y	1	5	9	11	15	19

8. You are saving money to purchase a new cell phone. Your parents have agreed to pay you \$3 for each household job that you complete each week and give you an additional \$5 for the week.

a. Copy and complete the input-output table.

Number of jobs completed in a week, x	1	4	8	10	12
Money earned, y	8				

b. How is the output related to the input?

c. Write a function rule in which x is the input and y is the output.

Name ____ Date __________

9.3 Practice B

Write an equation for the function. Then copy and complete the table.

1. The output is 7 more than the input.

Input, x	1	3	5	7
Output, y				

2. The output is 10 times the input.

Input, x	0	2	4	6
Output, y				

Write an equation for the function shown by the table.

3.

Input, x	0	4	8	12
Output, y	0	1	2	3

4.

Input, x	4	6	8	10
Output, y	1	3	5	7

5. Copy and complete the table.

For each output, multiply the input by 5, then add 3.

Input, x	0	1	3	4		
Output, y	3	8	18	23	28	43

6. You live in the Eastern Time Zone.

a. Your cousin lives in the Central Time Zone. The table shows some times in the two time zones. Copy and complete the input-output table.

Eastern Time Zone, x	3	5	6	8	10	11
Central Time Zone, y	2	4	5	7		

b. Write a function rule in which x is the input and y is the output.

c. Your grandparents live in the Mountain Time Zone. The table shows some times in the two cities. Copy and complete the input-output table.

Eastern Time Zone, x	3	5	6	8	10	11
Mountain Time Zone, y	1	3	4	6		

d. Write a function rule in which x is the input and y is the output.

e. If it is 12 P.M. in the Central Time Zone, what time is it in the Mountain Time Zone?

Name____ Date__________

9.3 Enrichment and Extension

Playing a Game

Use the following rules to play the game.

- **Choose a number from 1 to 4.**
- **Use this number as the input of the function described to find the output.**
- **If the output is odd, move forward that many spaces.**
- **If the output is even, move backward that many spaces.**
- **Continue until you reach or go beyond the finish.**

	Start $y = x$	$y = 4x$	The output is 3 times the input.		
			The input is 1 less than the output.	$y = 2x - 1$	The output is 1 more than twice the input.
Finish	$y = 4x - 3$				$y = x + 2$
	$y = 2x$	The output is 2 more than 4 times the input.	$y = x^2$	The output is the difference of 6 and the input.	The output is 2 less than 3 times the input.

1. How many turns did it take you to finish?

2. Make an input-output table for the function described in each space. Which spaces do you want to avoid? Explain.

3. What numbers will get you to the finish the fastest?

Name ______________________________ Date __________

9.3 Puzzle Time

What Is The Difference Between One Yard And Two Yards?

Write the letter of each answer in the box containing the exercise number.

Write an equation for the function shown by the table.

1.

Input, *x*	0	1	2	3
Output, *y*	4	5	6	7

2.

Input, *x*	0	1	2	3
Output, *y*	0	9	18	27

3.

Input, *x*	1	2	3	4
Output, *y*	11	12	13	14

4.

Input, *x*	5	7	9	11
Output, *y*	1	3	5	7

5.

Input, *x*	4	8	12	16
Output, *y*	2	4	6	8

6.

Input, *x*	0	2	4	6
Output, *y*	0	16	32	48

Answers

E. $y = x - 4$

C. $y = 9x$

N. $y = 8x$

F. $y = x + 10$

A. $y = \frac{1}{2}x$

E. $y = x + 4$

5		3	1	6	2	4

Activity 9.4

Start Thinking!

For use before Activity 9.4

Explain what the table and graph below represent. Which form is easier to use to predict a given temperature? Predict the Celsius temperature for a temperature of 75 degrees Fahrenheit.

Celsius	Fahrenheit
0°	32°
20°	68°
30°	86°
100°	212°

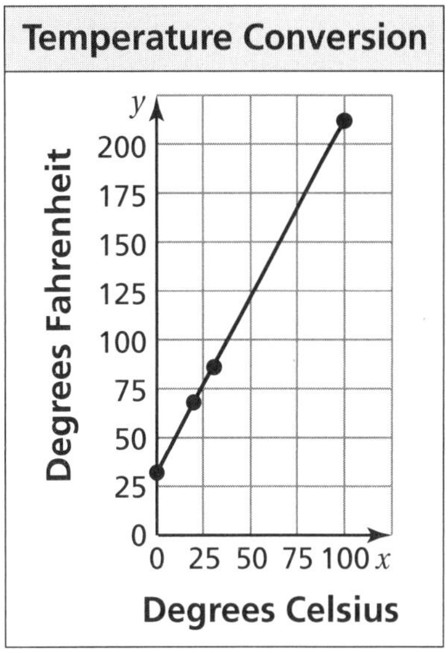

Activity 9.4

Warm Up

For use before Activity 9.4

Name the ordered pair.

1.

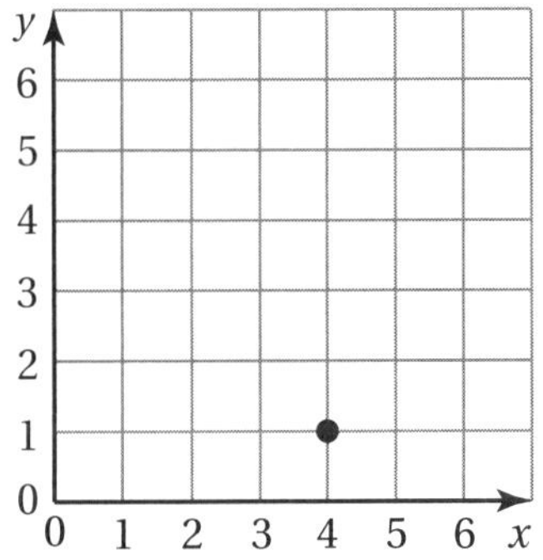

2.

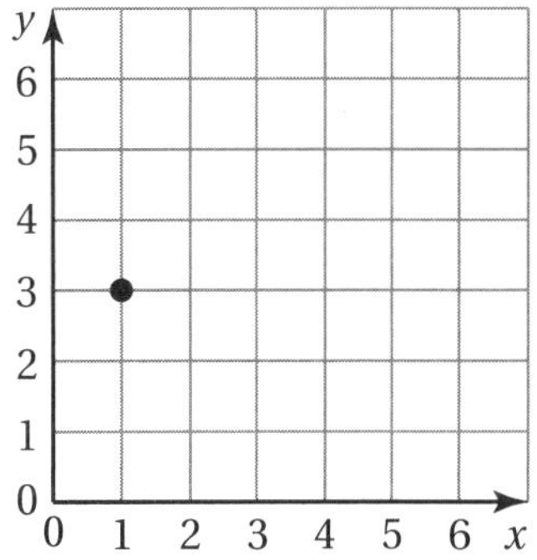

Lesson 9.4 Start Thinking!

For use before Lesson 9.4

Do you think that there is a relationship between a person's height and shoe size? Explain.

Divide the class into two groups—boys and girls.

In each group, make a table that includes the height and shoe size of each member.

Height	Shoe size
55 in.	5
60 in.	5.5

Make a graph of the data.

Using the graph, can you see any relationship between height and shoe size? Explain.

Lesson 9.4 Warm Up

For use before Lesson 9.4

Graph the data.

1.

Input, *x*	1	2	3	4
Output, *y*	2	4	6	8

2.

Input, *x*	0	2	4	6
Output, *y*	1	2	3	4

3.

Input, *x*	1	2	3	4
Output, *y*	3	5	7	9

Name______________________________ Date__________

9.4 Practice A

Draw a graph of the data.

1.

Input, *x*	1	3	5	7
Output, *y*	6	5	4	3

2.

Input, *x*	0	1	2	3
Output, *y*	0	3	6	9

Graph the function.

3. $y = x + 2$

4. $y = 5x$

5. $y = 3x + 1$

6. $y = \frac{x}{2}$

7. Use the function $q = 4d$ to find the number of quarters q that you can exchange for d dollars. Graph the function.

Match each graph with a function.

8.

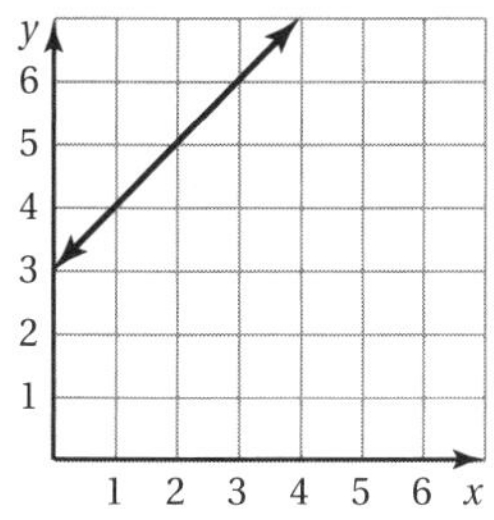

9.

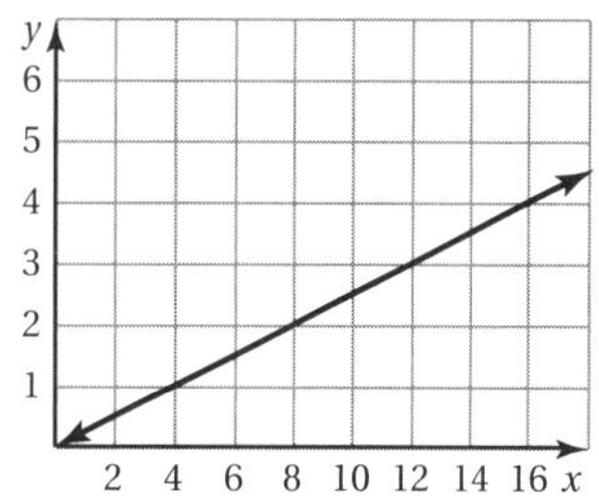

A. $y = \frac{x}{4}$

B. $y = x + 3$

10. A shoe store is having a sale in which every pair of sandals is 30% off.

a. Write a function that you can use to find the amount of discount d on an item with a regular price p.

b. Graph the function using the inputs 10, 20, 30, 40, and 50 for p.

c. You buy a pair of sandals that has a regular price of \$25. Find the discount and the sale price of the sandals.

d. You buy a pair of sandals and get a discount of \$4.50. What is the regular price of these sandals?

Name ______________________________ Date __________

9.4 Practice B

Draw a graph of the data.

1.

Input, *x*	2	4	6	8
Output, *y*	0	3	6	9

2.

Input, *x*	1	3	5	7
Output, *y*	0	2.5	5	7.5

Graph the function.

3. $y = 2x + 3$

4. $y = \frac{x}{2} + 3$

5. $y = 0.75x$

6. $y = \frac{2}{3}x + 2$

7. Use the function $c = 1.5s$ to find the cost c to download s songs. Graph the function.

8. The table shows the number of miles that you cycle in x hours.

Hours, *x*	1	2	3	4	5
Miles, *y*	8	16	24	32	40

a. Graph the data.

b. Find a function to describe the data.

c. How many miles can you cycle in 3.5 hours?

d. How many hours will it take you to cycle 60 miles?

9. The graph of a function is a straight line that goes through the points $(4, 2)$, $(7, 6)$, and $(10, y)$. What is the value of y?

10. Describe and correct the error in graphing the line from the input-output table.

Input, *x*	2	4	6	8
Output, *y*	4	3	2	1

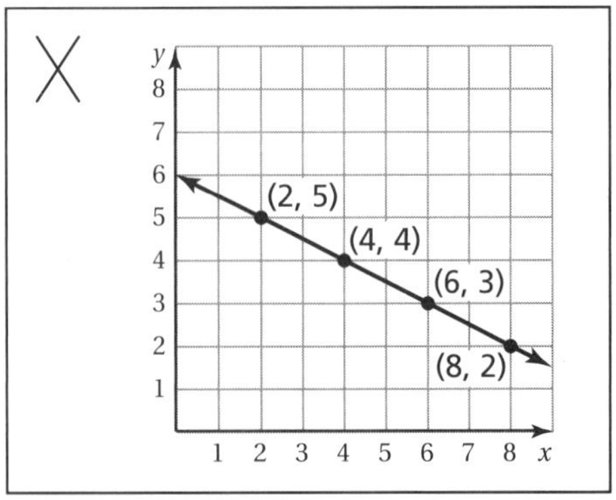

Name______________________________ Date__________

9.4 Enrichment and Extension

Graphing Functions

1. A graph represents all of the solutions of an equation. In the graph at the right, which point is a solution of each equation? Explain.

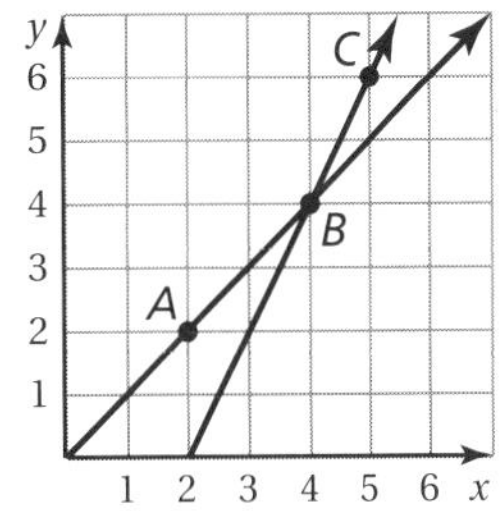

Use the graph of the function to find the point that is a solution of each equation.

2. $y = x + 1$

   $y = \frac{1}{2}x + 3$

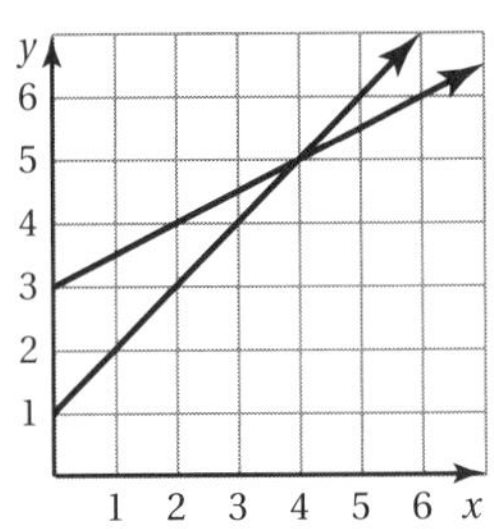

3. $y = x + 1$

   $y = 2x$

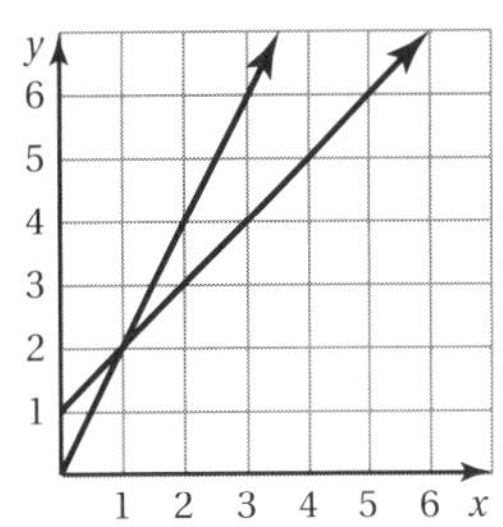

Graph the functions in the same coordinate plane. Find the point that is a solution of each equation.

4. $y = x$

   $y = 0.5x + 3$

5. $y = 2x + 1$

   $y = x + 3$

6. $y = x$

   $y = 2x - 4$

   $y = \frac{1}{2}x + 2$

7. $y = x - 1$

   $y = 0.25x + 2$

   $y = \frac{1}{2}x + 1$

Name ____ Date __________

9.4 Puzzle Time

Why Did The Girl Go Outside With Her Purse Open?

A	B	C	D	E	F
G	H	I	J		

Complete each exercise. Find the answer in the answer column. Write the word under the answer in the box containing the exercise letter.

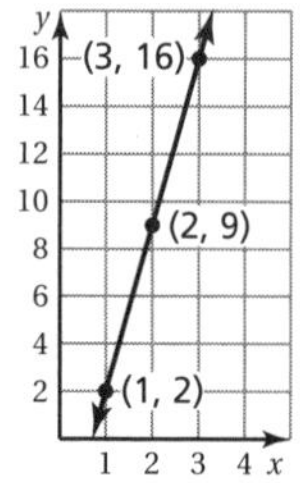

CHANGE

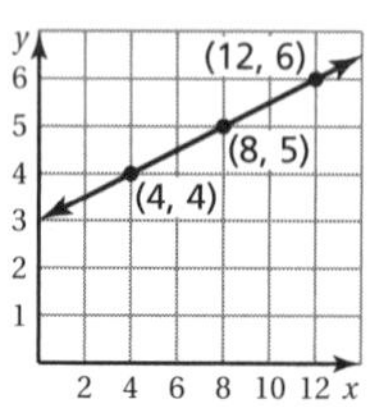

THE

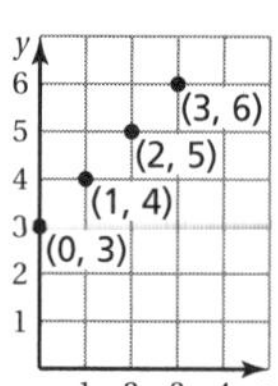

TO

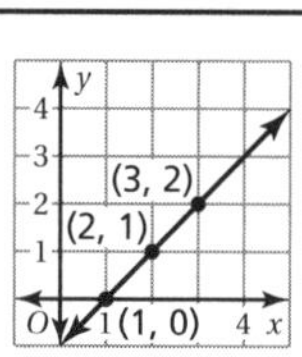

IN

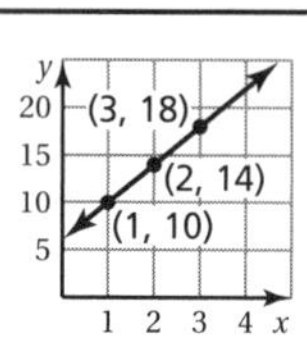

WAS

Graph the data or function.

A.

Input, x	0	1	2	3
Output, y	3	4	5	6

B.

Input, x	0	2	4	6
Output, y	0	2	4	6

C. $y = x + 5$

D. $y = 3x$

E. $y = 4x + 6$

F. $y = \frac{x}{2}$

G. $y = 7x - 5$

H. $y = x - 1$

I. $y = \frac{1}{4}x + 3$

J. $y = 1.5x + 2$

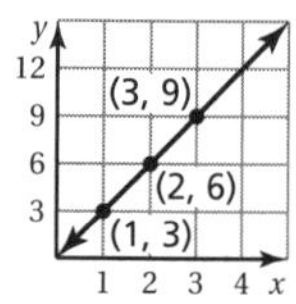

THERE

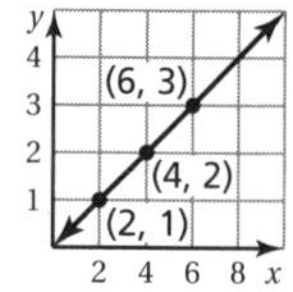

ANY

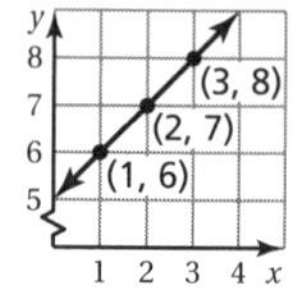

IF

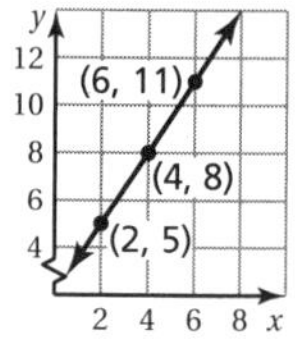

WEATHER

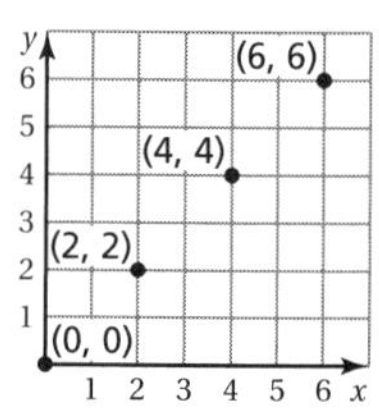

SEE

Activity 9.5 Start Thinking!

For use before Activity 9.5

The table below shows the population (in millions) of Florida from 2000 to 2007.

Year	2000	2001	2002	2003
Population	16.0	16.3	16.7	17.0
Year	2004	2005	2006	2007
Population	17.3	17.7	18.1	18.3

Make a graph of the data. Then describe the characteristics of the graph.

Do you think you can use the graph to predict the current population of Florida?

If so, make a prediction and check the prediction at www.census.gov.

Activity 9.5 Warm Up

For use before Activity 9.5

Plot the ordered pair in a coordinate plane.

1. $(4, 1)$ **2.** $(1, 3)$ **3.** $(5, 2)$

4. $(3, 0)$ **5.** $(2, 2)$ **6.** $(0, 5)$

Lesson 9.5 Start Thinking!
For use before Lesson 9.5

When you hear the term *linear*, what do you think of?

Explain what you think that the graph of a *linear function* looks like. Make a graph to illustrate your answer.

Lesson 9.5 Warm Up
For use before Lesson 9.5

Copy and complete the table. Then make a graph of the data. Write an equation for the function.

1. Find the cost of a cell phone plan for m months at $50 per month.

Months, *m*	1	2	3	4
Cost, *C*				

2. Find the area of the triangle.

Height, *h*	1	2	3	4
Area, *A*				

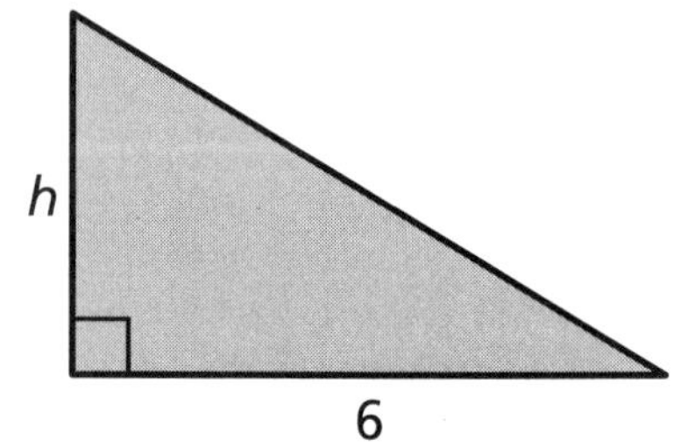

Name______________________________ Date__________

9.5 Practice A

Copy and complete the table. Then make a graph of the data. Write an equation for the function.

1. Find the cost of purchasing t tickets at \$7 per ticket.

Tickets, *t*	1	2	3	4
Cost, *c*				

Does the graph represent a linear function? Explain.

2.

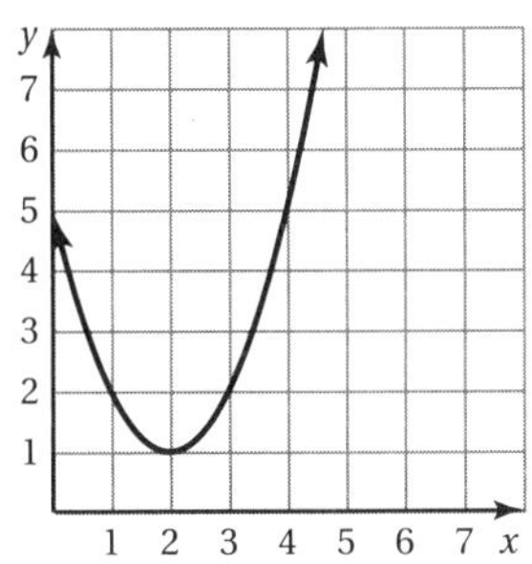

3.

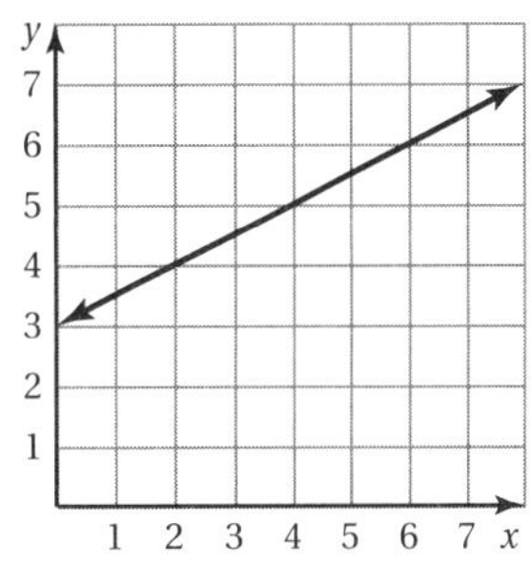

Does the input-output table represent a linear function? Explain.

4.

Input, *x*	1	2	3	4
Output, *y*	2	5	8	11

5.

Input, *x*	1	3	5	7
Output, *y*	1	4	3	6

6. The graphs show the number of miles driven by you and by your friend.

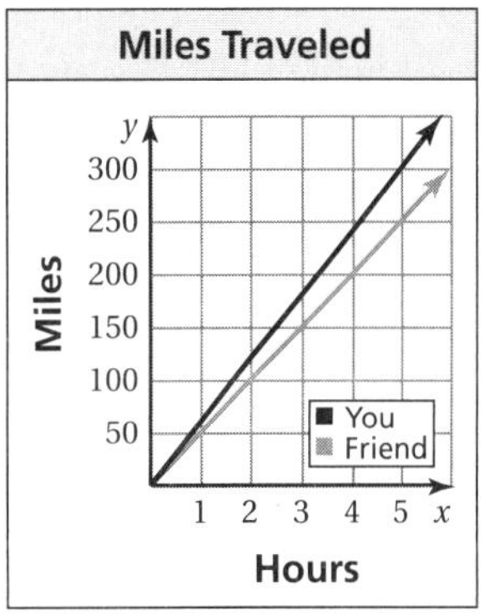

a. Do you or your friend drive more miles in an hour?

b. How long will it take you to drive 300 miles?

c. How long will it take your friend to drive 300 miles?

d. If neither of you stops, will your friend catch up with you? Explain.

e. If your destination is 300 miles away, how long will you wait for your friend?

Name ______________________________ Date __________

9.5 Practice B

Copy and complete the table. Then make a graph of the data. Write an equation for the function.

1. Find the cost of purchasing b bracelets at \$5.50 per bracelet.

Bracelets, *b*	1	2	3	4
Cost, *c*				

Does the graph represent a linear function? Explain.

2.

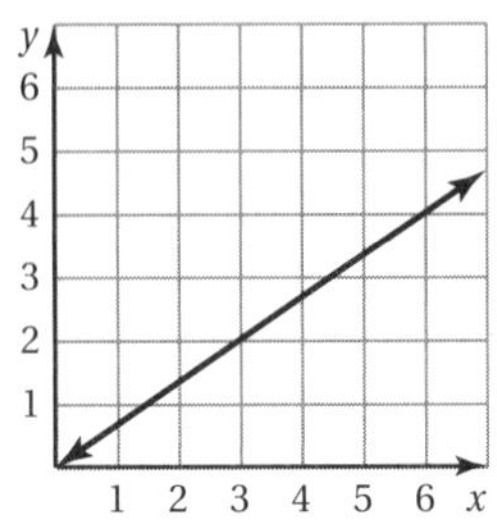

3.

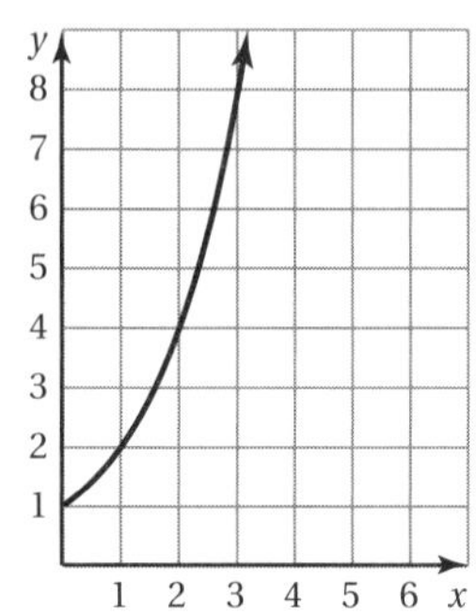

Does the input-output table represent a linear function? Explain.

4.

Input, *x*	1	2	3	4
Output, *y*	1	4	9	16

5.

Input, *x*	0	2	4	6
Output, *y*	1	2	3	4

Graph each linear function. Which graph is steeper? Explain.

6. $y = 4x$ and $y = \frac{1}{4}x$

7. $y = x$ and $y = 3x + 1$

8. Copy and complete the table using the graph. Then find the differences. What do you notice about the differences?

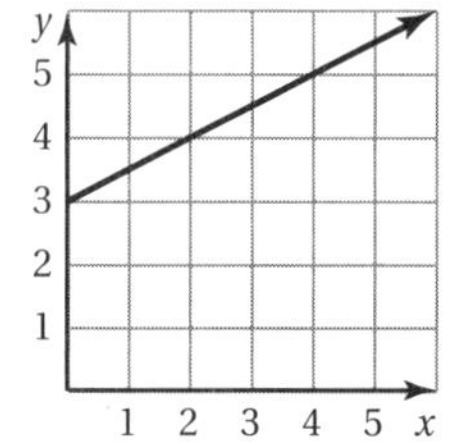

x	0	1	2	3	4	5
y	3	3.5				

Differences: 0.5 ? ? ? ?

Name______________________________ Date__________

9.5 Enrichment and Extension

Comparing Graphs Involving x^2

Copy and complete the input-output table for the function. Then draw a graph of the function.

1. $y = x^2$

Input, x	0	1	2	3
Output, y				

2. $y = x^2 + 1$

Input, x	0	1	2	3
Output, y				

3. $y = x^2 + 4$

Input, x	0	1	2	3
Output, y				

4. $y = x^2 + 8$

Input, x	0	1	2	3
Output, y				

5. How are the graphs in Exercises 1–4 similar? How are they different?

Use an input-output table to graph the function. Use the values 0, 1, 2, and 4 for x.

6. $y = 2x^2$ **7.** $y = 4x^2$ **8.** $y = \frac{1}{2}x^2$ **9.** $y = \frac{1}{4}x^2$

10. Using the results in Exercises 6–9, what happens to the graph of $y = ax^2$ as a increases? What happens as a gets closer to 0?

11. The graphs of $y = x^2$ and $y = 2^x$ are shown. Looking at the graphs, which function appears to increase more rapidly?

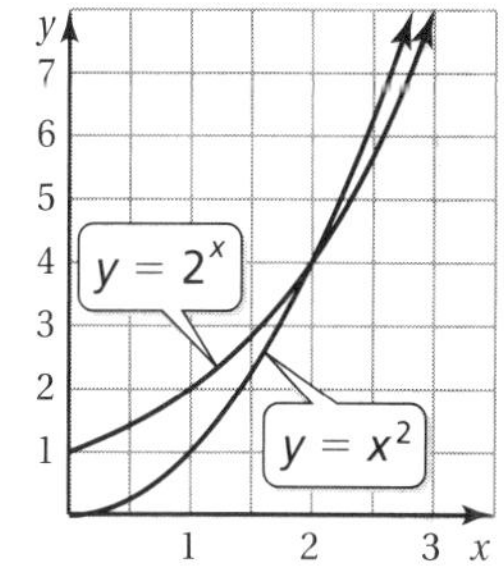

12. Use an input-output table to continue to graph the functions $y = x^2$ and $y = 2^x$ for input values greater than 3. What do you notice about the graphs? Do they share any common points? Explain.

13. Using your results from Exercise 12, how do you know the graphs do not share any more common points as x increases? Explain.

Name ______________________________ Date __________

9.5 Puzzle Time

What Animal Should You Never Play Cards With?

Write the letter of each answer in the box containing the exercise number.

Does the graph or table represent a linear function?

1.

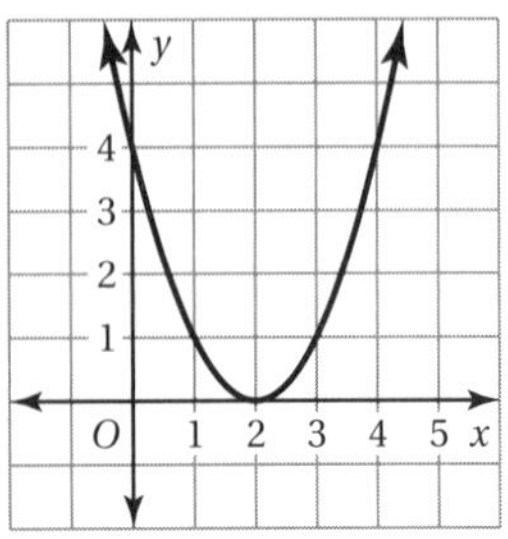

2.

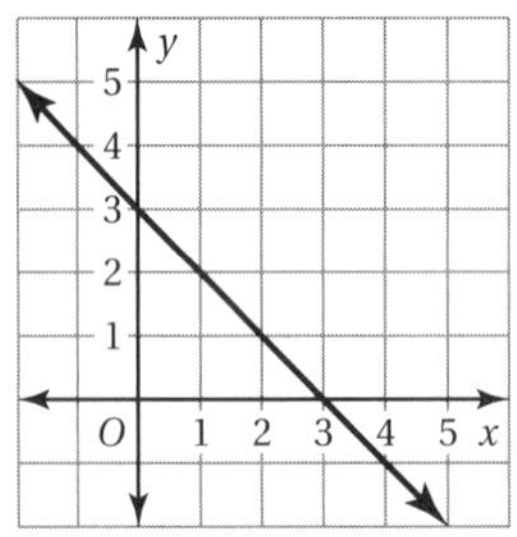

3.

Input, *x*	0	1	2	3
Output, *y*	4	5	6	7

4.

Input, *x*	0	1	2	3
Output, *y*	0	1	4	9

Answers

M. $y = \frac{1}{3}x$

H. no

S. $y = 2x$

E. yes

T. linear

A. $y = 4x + 2$

C. not linear

A. $y = 3x$

Which graph is steeper?

5. $y = 3x;\ y = \frac{1}{3}x$

6. $y = 2x;\ y = 4x + 2$

Is the function relating the figure number, *x*, to the number of dots, *y*, linear or not linear?

7. Figure 1 Figure 2 Figure 3 Figure 4

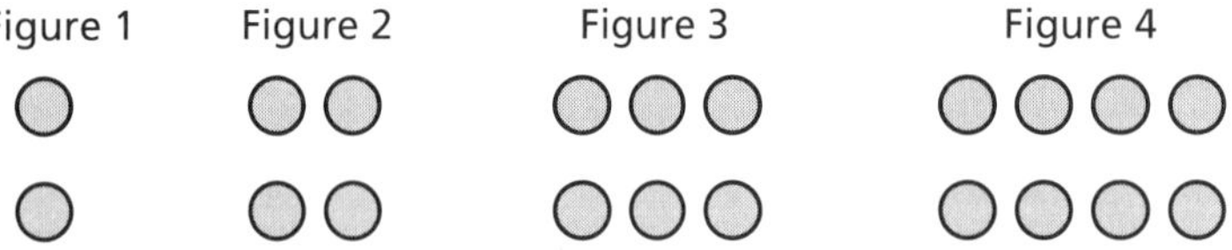

8. Figure 1 Figure 2 Figure 3 Figure 4

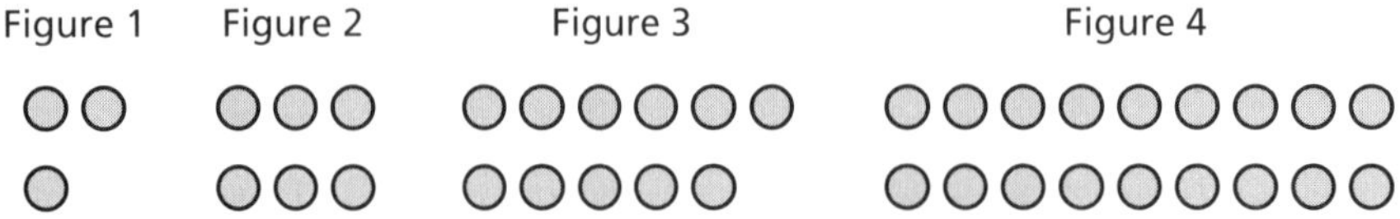

6		8	4	3	2	7	5	1

Name ____ Date __________

Chapter 9

School-to-Work

For use after Section 9.3

Building Contractor

A building contractor creates and maintains structures. They work with wood, metal, plastic, and concrete. Accurately measuring is crucial. A builder needs to be able to estimate the cost of jobs, determine the right type and amount of material needed for a job, and correctly measure and cut materials.

You are framing a roof for a house that is 28 feet wide. The roof should rise one foot for every two feet of horizontal space.

Distance from outer wall (feet)	0	2	4	6	8
Height (feet)	18	19	20	21	22

1. Write an equation for the function shown in the table.

2. How far from the outer wall is the center of the roof?

3. How high above the ground is the roof peak?

4. The garage will be 20 feet wide and the roof will slope in the same way it does on the house. At the outer wall the roof of the garage is 9 feet high. Where is the center of the garage, and how high is the roof at that point?

You are installing kitchen cabinets. Each cabinet requires 1.5 feet along the wall of the kitchen.

5. Create an input-output table. The input is the number of cabinets, and the output is the width. Use values from 0 to 6 cabinets.

6. Write a function for the amount of space remaining along a 12-foot long wall as each cabinet is installed. (For example, 12 feet for 0 cabinets, 10.5 feet for 1 cabinet, and so on.)

7. The countertop will be flush to the wall at one end, and will extend 1 inch beyond the end of the row of cabinets at the other end. Write a function for the length (in inches) of the countertop needed given x cabinets.

You estimate that a job will take you four 8-hour days to complete.

8. The job will require \$900 worth of materials. To earn \$20 per hour, how much should you charge?

9. You estimate that if you work with an apprentice who earns \$10 per hour, you can complete the project in 2 and a half days. How much should you charge now to cover labor and materials? (*Hint:* Remember that you must pay yourself and the apprentice.)

Name ____ Date __________

Chapter 9 Study Help

You can use a **four square** to organize information about a topic. Each of the four squares can have various categories, such as *definition*, *vocabulary*, *example*, *non-example*, *words*, *algebra*, *table*, *numbers*, *visual*, *graph*, or *equation*.

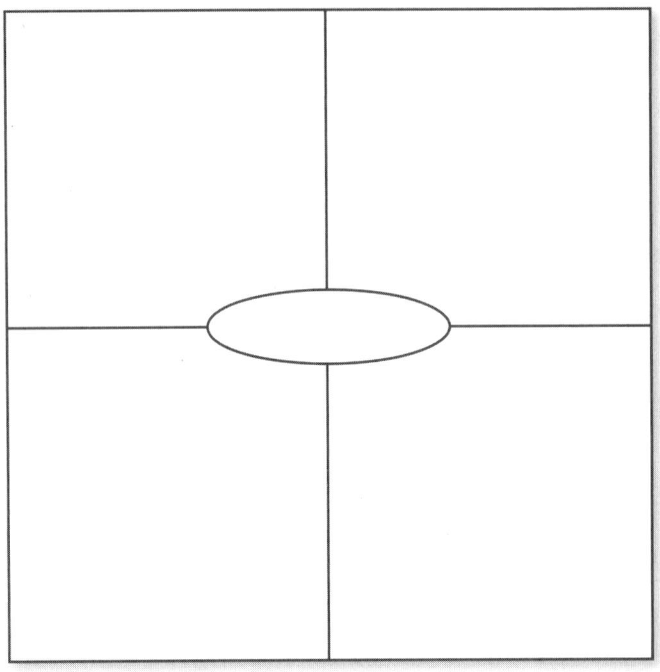

On Your Own

Make a four square to help you study these topics.

1. function

2. mapping diagram

3. input-output table

After you complete this chapter, make four squares for the following topics.

4. graph of a function

5. linear function

6. Pick three other topics that you studied earlier in this course. Make a four square for each topic.

Name____ Date __________

Chapter 9 Financial Literacy

For use after Section 9.5

Understanding Price and Demand

Picture someone describing a small business—a coffee shop, a company that does yard work, a graphics design firm, or any other business you find interesting. Imagine someone saying the following:

- The number of employees we can hire next quarter is a function of how much new business we generate this quarter.
- Daily sales are a function of the weather—when the weather is good, sales are good.
- Business is a function of word of mouth—if people aren't talking about us, our sales go down.

In each case someone is describing how their business depends on some factor.

1. Explain how the use of *function* in the three examples relates to the mathematical meaning of *function*.
2. In each example, what would cause the output to go up?
3. Write your own business example using the phrase *is a function of*, based on the three examples.

Demand for a product or service depends on the price. The *demand function* is often modeled as a linear function. These questions consider the price that a local band charges for a compact disc recording.

4. When the price of a product goes up, do you expect the demand for the product to go up or down?
5. The band's manager determines that the demand function is $D = 60 - 2P$, where D is the number of compact discs the band sells at price P.
   a. How many discs do they sell when the price is \$10?
   b. At what price do they sell 50 discs?
   c. The graph of the function includes the point (0, 60). Explain what this point means in the situation.
   d. At what price do they sell 0 discs?
6. Suppose the band has a few hit songs and becomes more popular. Would demand for discs go up or down? How would this affect the graph of the demand function?

Name ______________________________ Date __________

Chapter 9 Cumulative Practice

Plot the ordered pairs in the same coordinate plane.

1. (1, 2) **2.** (4, 2) **3.** (3, 5) **4.** (2, 0)

5. (0, 5) **6.** (7, 1) **7.** (5, 3) **8.** (5, 4)

Does the graph represent a linear function? Explain.

9.

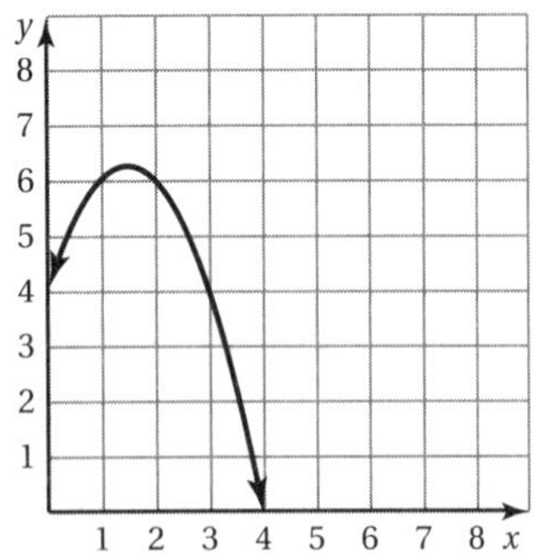

10.

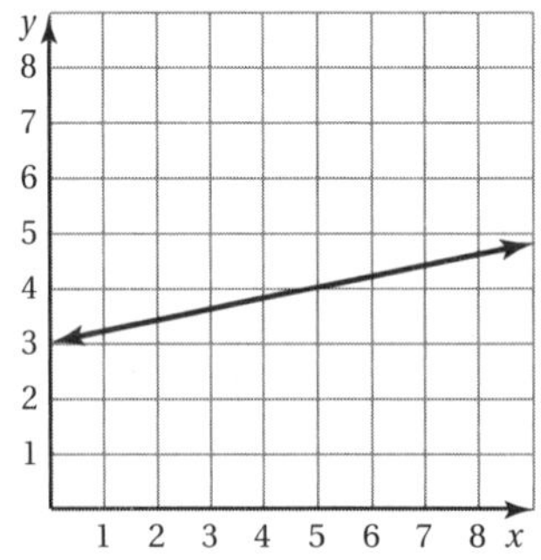

11.

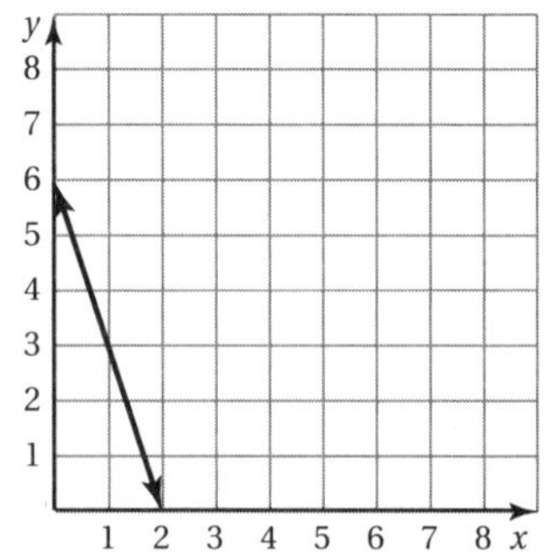

Write an equation for the function. Then copy and complete the table.

12. The output is the quotient of 12 and the input.

Input, x	1	2	3	4
Output, y				

13. The output is the sum of the input and 15.

Input, x	1	2	3	4
Output, y				

Graph the function.

14. $y = 4x$ **15.** $y = 2x + 2$ **16.** $y = 2 + \frac{1}{2}x$

17. A store is having a clearance sale where everything is 60% off.

a. Write a function that you can use to find the amount of discount d on an item with a regular price p.

b. Graph the function using the inputs 20, 40, 60, 80, and 100 for p.

c. You buy a sweater that has a regular price of \$30. What is the sale price of the sweater?

18. The table shows the amount you receive for recycling aluminum cans.

Pounds of aluminum cans	2	4	6	8
Dollars received	0.80	1.60	2.40	3.20

a. Use the table to draw a mapping diagram.

b. Describe the pattern. How does the amount you receive increase as you recycle more pounds of aluminum cans?

Name____ Date__________

Unit 4 Project: Tying Knots

For use after Unit 4

Objective **Analyze data from an experiment.**

Materials Rope or clothesline (90 centimeters); meter stick

Investigation Work in a group of three. Choose one person to tie knots in the rope, one person to measure the rope, and one person to record the results.

Measure the length of the rope in centimeters. This is the length of the rope when the number of knots is zero. Record the length in the table below.

Number of knots	0	1	2	3	4	5	6	7	8
Length of rope (cm)									

Tie a knot in the rope. Measure the length of the rope after the knot is tied. Record this length in the table. Tie another knot with the same tightness in the rope. Measure the length of the rope and record this length in the table. Repeat this process until you complete the table.

Data Analysis

- What happens to the length of the rope as the number of knots increases?
- By how many centimeters does the length of the rope change from zero knots to one knot? One knot to two knots? Is the change relatively consistent?
- Let the input be the number of knots and the output be the length of the rope. Make a mapping diagram of your data. Write ordered pairs to represent the data and then graph the ordered pairs. What does the y-intercept represent in this situation?
- Write an equation that gives the length of the rope in terms of the number of knots tied. Use your equation to estimate the length of the rope after 10 knots have been tied. Tie two more knots in the rope. How does the length of the rope compare with the estimate you found using your equation?
- Compare your equation and graph with those of other groups in your class. Do you have the same slopes? y-intercepts? equations? Explain.
- Write three inequalities to show how your slope compares to the slopes of three other groups in your class.
- What effect, if any, do each of these things have on the data: the tightness of each knot; the space between the knots; the thickness of the rope.

Written Summary Write a report in which you describe your data and analyze the results. Include comparisons of your data to the class data. Include all equations, mapping diagrams, and graphs. Be sure to include the answers to the questions above.

Name ______________________________ Date __________

Unit 4 Student Grading Rubric

For use after Unit 4

		Student Score	Teacher Score
Cover Page	10 points		
	a. Name (4 points)	______	______
	b. Class (2 points)	______	______
	c. Project Name (2 points)	______	______
	d. Due Date (2 points)	______	______
Investigation	40 points		
	a. Table is complete. (30 points)	______	______
	b. Data is carefully measured and recorded. (10 points)	______	______
Data Analysis	130 points		
	a. All questions are answered completely. (30 points)	______	______
	b. Mapping diagram and ordered pairs are correct. (30 points)	______	______
	c. Equation is correctly written. (15 points)	______	______
	d. Graph is clearly labeled and correct. (20 points)	______	______
	e. Meaning of graph is well-explained. (10 points)	______	______
	f. Inequalities are correctly written. (10 points)	______	______
	g. Effect of knot tightness, space between knots, and thickness of rope is accurately addressed. (15 points)	______	______
Written Summary	30 points		
	a. Method of investigation is described. (15 points)	______	______
	b. Table and data analysis are clearly presented. (15 points)	______	______
FINAL GRADE		______	______

Unit 4 Teacher's Project Notes

For use after Unit 4

Materials A 90 centimeter rope, such as clothesline, and a meter stick for each group. Cut a piece of rope for every three students. You cannot reuse the rope from one class to another because of the knots, so be sure to cut enough rope for all the classes doing the project.

Alternatives For greater variety in the data, you can give each group a rope with a different thickness.

For a more guided version, this could be done as a class project, in which students take turns in front of the class tying knots in different types of ropes. The data the class gathers should be recorded in separate tables for each type of rope. Then the data can be compared.

Common Errors Students may measure carelessly. Make sure they hold the rope straight and measure carefully so that they get accurate data.

Make sure students tie their knots about the same tightness; they should not have some tight and some loose. This may make their measurements less consistent.

When students graph their ordered pairs, they may mistakenly connect the points. If they do this, ask students what an x-value of 2.5 represents in this situation. Guide students to see that it doesn't make sense to have 2.5 knots.

Suggestions If the rope is thinner than clothesline, students may have difficulty measuring precisely. Avoid giving students rope that is too thin or too thick.

Instead of a written summary, you may want students to do a poster to present all the data, graphics, and data analysis.

Unit 4 Grading Rubric

For use after Unit 4

Scoring Rubric	
A	179-200
B	159-178
C	139-158
D	119-138
F	118 or below

Cover Page 10 points

a. Name (4 points)

b. Class (2 points)

c. Project Name (2 points)

d. Due Date (2 points)

Investigation 40 points

a. Table is complete. (30 points)

b. Data is carefully measured and recorded. (10 points)

Data Anaylsis 130 points

a. All questions are answered completely. (30 points)

b. Mapping diagram and ordered pairs are correct. (30 points)

c. Equation is correctly written. (15 points)

d. Graph is clearly labeled and correct. (20 points)

e. Meaning of graph is well-explained. (10 points)

f. Inequalities are correctly written. (10 points)

g. Effect of knot tightness, space between knots, and thickness of rope is accurately addressed. (15 points)

Written summary 30 points

a. Method of investigation is described. (15 points)

b. Table and data analysis are clearly presented. (15 points)

Appendix B

Name ______________________ Date ________

Appendix B Exploring the Number Line

Dear Family,

Running errands can be a time-consuming process. Organizing our errands decreases the time it takes to finish the task. You and your student can work on strategies to organize the day's errands.

One way to organize your route is to list the times you have to be at each place. By placing the errands on a timeline, you may be able to group the times to avoid making extra trips.

Another way is to find the distances to each place. If all of the locations lie in a line, you can simply put them in order to ensure that you will travel the shortest distance overall.

Organizing a trip when the destinations are not in a straight line can be more difficult. How can you keep the overall trip short? One way is to plot each place on a *coordinate plane*—a map is a great coordinate plane to use.

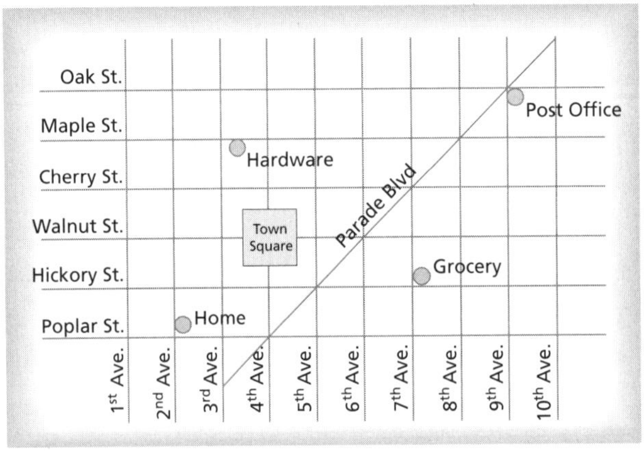

From your starting point, choose the nearest errand. Then choose the point nearest to that one, and so on. This will give you one of the shortest routes to all your destinations.

Next time you are running errands, work out a plan with your student to find the shortest route using the method above. There may be other factors to consider. If one of your errands is to the grocery store, you may want to go there last to avoid having the food get too warm. Perhaps the post office will close unless you go there first. Talk about how this will affect your strategy for making the trip as short as possible.

Have a great trip!

Nombre ______________________________ Fecha __________

Apéndice B Explorando la Recta Numérica

Estimada Familia:

Hacer mandados puede ser un proceso que demanda mucho tiempo. Al organizar nuestros mandados, se reduce el tiempo que nos toma terminar la tarea. Usted y su estudiante pueden trabajar con estrategias para organizar los mandados del día.

Una forma de organizar su ruta es haciendo una lista de las veces que tienen que estar en cada lugar. Al colocar los mandados en una línea del tiempo, serán capaces de agrupar los tiempos para evitar hacer viajes extras.

Otro modo es buscando las distancias a cada lugar. Si todas las ubicaciones se encuentran en una línea, simplemente pueden ponerlas en orden para asegurarse de recorrer la distancia general más corta.

Organizar un viaje cuando los destinos no se encuentran en línea recta puede ser más difícil. ¿Cómo puede mantener el recorrido general corto? Una manera es representando gráficamente cada lugar en un *plano coordenado*—un mapa es un gran plano coordenado que se puede usar.

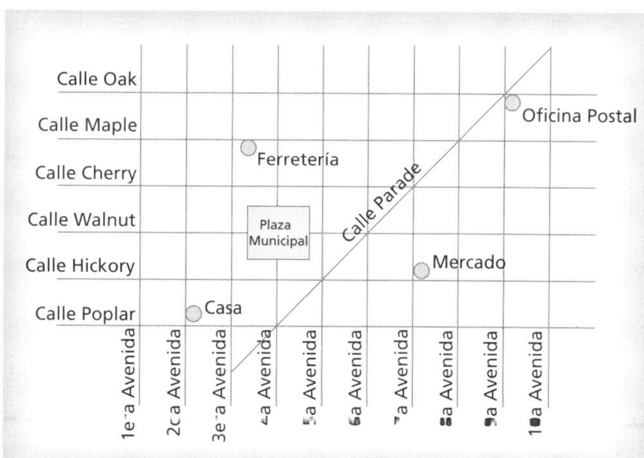

Desde su punto de partida, elijan el mandado más cercano. Luego escojan el punto más cerca de ése y así sucesivamente. Esto les dará una de las rutas más cortas a todos sus destinos.

La próxima vez que tenga que hacer mandados, haga un plan con su estudiante para encontrar la ruta más corta usando el método anterior. Puede haber otros factores para tomar en cuenta. Si uno de sus mandados es ir a la tienda de abarrotes, seguro desean ir allí al último, para evitar que la comida se caliente demasiado. Quizás la oficina de correos va a cerrar a menos que vayan allí primero. Converse acerca de cómo esto va a afectar su estrategia para hacer el viaje más corto posible.

¡Que tengan un buen viaje!

Non ______________________________ Dat __________

Apendis B Ann Esplore Liy Nimerik la

Chè Fanmi,

Fè komisyon kapab yon pwosesis ki pran tan. Lè ou òganize komisyon ou sa diminye tan sa pran pou fini tach la. Ou menm ak elèv ou a kapab travay sou estrateji pou òganize komisyon jounen an.

Yon jan pou òganize trajè ou se fè lis lè ou dwe pase chak kote. Lè ou mete komisyon yo sou yon echèl tan, ou gen dwa kapab gwoupe lè yo pou evite fè vwayaj anplis.

Yon lòt jan se jwenn distans pou al chak kote. Si tou kote yo sou yon liy, ou kapab senpleman mete yo nan lòd pou asire ou ke se distans ki pi piti a ke w'ap fè an total.

Òganize yon vwayaj lè destinasyon yo pa sou yon liy dwat kapab pi difisil. Kijan ou kapab kenbe vwayaj total la kout? Yon jan se plase chak kote sou yon *plan kowòdone*—yon kat se yon bon plan kowòdone pou itilize.

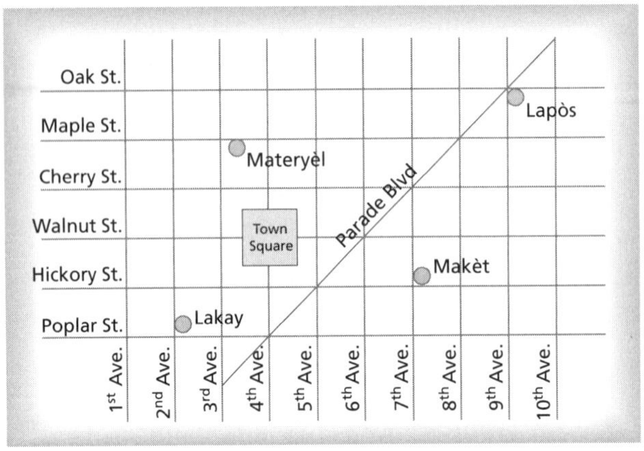

Apatide pwen depa ou, chwazi kote pou w'al fè komisyon ki pi pre a. Apre sa chwazi pwen ki pi pre li a, epi kontinye konsa. Sa ap ba ou youn nan wout ki pi kout pou al nan tout destinasyon ou yo.

Pwochen fwa ou pwal fè komisyon, fè yon plan avèk elèv ou a pou jwenn trajè ki pi kout la grasa metòd ki pi wo a. Gen dwa gen lòt faktè pou konsidere. Si youn nan komisyon ou yo se al nan makèt, ou gen dwa bezwen al la an dènye pou manje an pa vin twò cho. Petèt biwo lapòs la ap fèmen sòf si w'al la an premye. Pale sou kijan sa pwal afekte estrateji ou pou rann vwayaj la lepli kout posib.

Fè bon vwayaj!

Activity B.1 Start Thinking!

For use before Activity B.1

Copy the number line below. Choose four or five highlights from your day yesterday. For each highlight, use the start time to label the event on your number line.

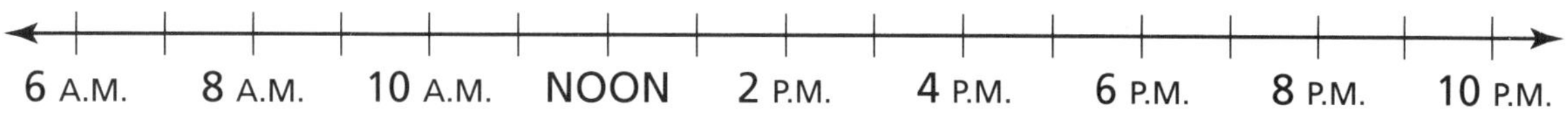

Activity B.1 Warm Up

For use before Activity B.1

Graph the integer on a number line.

1. -2 **2.** 5 **3.** -4

4. 3 **5.** 0 **6.** -1

Lesson B.1 Start Thinking!

For use before Lesson B.1

Use your number line from the Starting Thinking! for use before Activity B.1. Relabel the number line so that noon is at 0. Relabel the other times accordingly using positive and negative numbers.

Name one highlight that occurred after noon. What time did it occur? What number corresponds to that time?

Name one highlight that occurred before noon. What time did it occur? What number corresponds to that time?

Lesson B.1 Warm Up

For use before Lesson B.1

Write and graph the integer that represents the situation.

1. A football team gained 57 yards on one play.
2. The shores of the Dead Sea are 420 meters below sea level, the lowest dry land point on Earth.
3. The Kingda Ka roller coaster has a top speed of 128 miles per hour.
4. At its coldest, Mercury's surface temperature is 300 degrees Fahrenheit below zero.

Name____ Date __________

B.1 Practice A

Write and graph the integer that represents the situation.

1. You receive \$10 in a birthday card.

2. You lose a \$1 bill.

3. A fireman climbs 22 feet up a ladder.

4. You dig a hole 3 feet deep.

5. A bird's egg falls 16 feet from a nest.

6. You grew 2 inches last year.

Tell which integer is greater.

7. 11, 7

8. −5, 9

9. 8, −4

10. −2, −6

11. −21, −3

12. 17, −18

13. −33, 34

14. −14, −12

Find the absolute value of the integer.

15. 6

16. −7

17. −10

18. 11

19. −45

20. −21

21. 32

22. 61

Find the median of the data.

23. −2, 8, 11, −7, 13, −5, 1

24. 12, 0, 9, −31, −4, 16, −16, 8

25. You and a friend are collecting donations for a fundraiser. You collect more. You take \$8 out of your donations and add it to your friend's donations. Write integers showing how this changes your donations and how it changes your friend's donations.

26. An eagle soars 200 feet above sea level. A jellyfish swims 210 feet below sea level.

a. Write an integer for the position of each creature relative to sea level.

b. Find the absolute value of each integer.

c. Which creature is farther from sea level?

27. A soccer player weighs 91 pounds at the start of a game and 90 pounds at the end of the game.

a. Write an integer representing the weight change in pounds.

b. Write an integer representing the weight change in ounces.

28. One character in a science-fiction book travels 26 years into the future. Another character travels 28 years into the past.

a. Write integers representing the time traveled by thc two characters.

b. Which character traveled farthest in time? Explain.

Name _______________________________________ Date __________

B.1 Practice B

Write and graph the integer that represents the situation.

1. You gain 4 pounds in weight.
2. A barber shortens your hair by 7 centimeters.
3. A toy light saber extends 32 inches.
4. You climb 6 feet down a bunk bed ladder.
5. You remove 3 books from your locker.
6. A hot air balloon rises 142 feet.

Copy and complete the statement using <, >, or =.

7. $-123 \square -124$
8. $-17 \square 17$
9. $8 \square |-8|$
10. $|-12| \square |9|$

Order the values from least to greatest.

11. $5, -3, |-3|, 4, 0, 2, |1|$
12. $-6, -7, -8, |8|, |-7|, |-6|$
13. $0, 12, -11, |11|, |-9|, |-10|$
14. $-22, |-21|, |32|, -30, |24|, 28$

Simplify the expression.

15. $-|0|$
16. $-|-3|$
17. $-|-16|$
18. $-|9|$

19. A quarter falls from a hole in your pocket. Measure the number of inches from your pocket to the floor. Then write an integer that represents the change in the height of the quarter.

20. Two jets in flight are shown from the side.

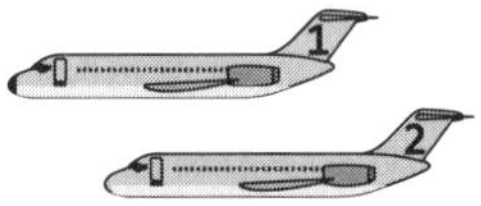

   a. Write an integer that represents a reasonable guess of the height (in feet) of jet 2 relative to jet 1.

   b. Write an integer the represents a reasonable guess of the forward progress (in feet) of jet 2 relative to jet 1.

   c. Jet 3 is 20 feet above jet 1 and 10 feet behind jet 2. Draw all three jets.

21. A student graphed her meal times for a day on the number line.

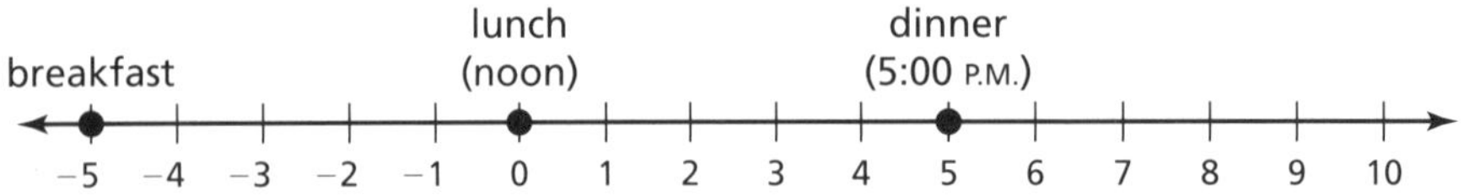

   a. What time was breakfast?

   b. The student also had snacks at 10:00 A.M. and at 8:00 P.M. Copy the graph and add her snack times.

   c. Other than lunch, what was the closest time to noon that the student ate?

Name______________________________ Date__________

B.1 Enrichment and Extension

Flags

Copy the flag shown. Assume $x > 0$ and $y < 0$. Tell whether the statement is *always, sometimes,* or *never* true. Use the table to color the portion of your flag that contains the exercise.

Answer	Color
Always	Black
Sometimes	Aqua
Never	Yellow

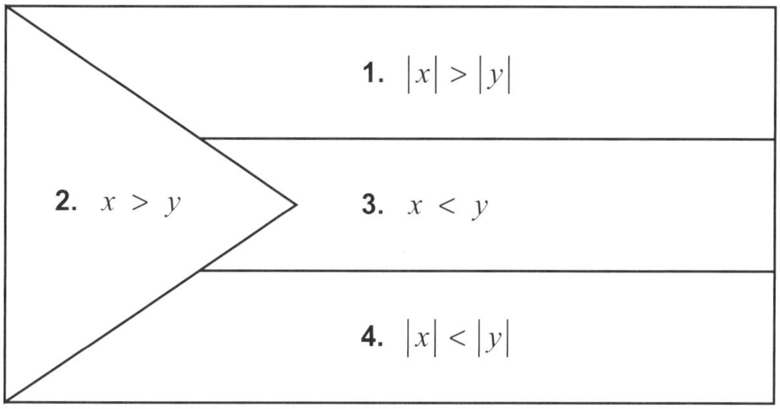

5. Which country does the flag represent?

Copy the flag shown. Tell whether the possible values of n are *all integers*, *all positive integers*, *all negative integers*, or *no integers*. Use the table to color the portion of your flag that contains the exercise.

Answer	Color
All integers	Black
All positive integers	Red
All negative integers	Yellow
No integers	Green

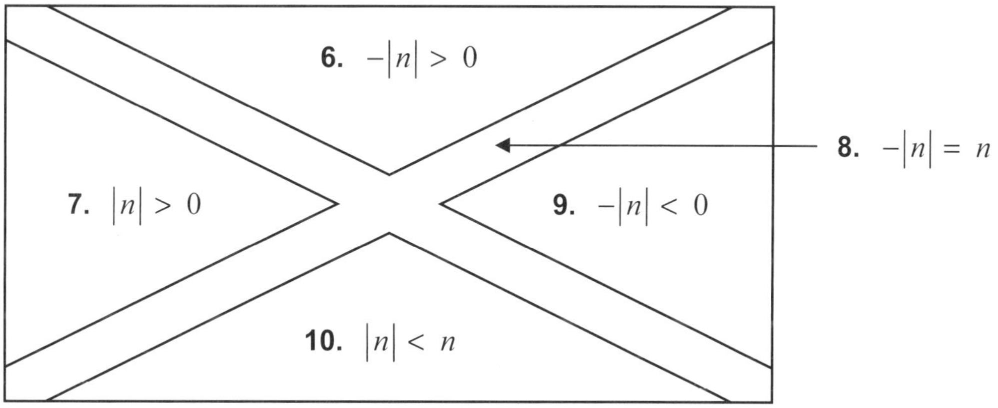

11. Which country does the flag represent?

Name ______________________________ Date __________

B.1 Puzzle Time

Why Do Clocks Seem So Shy?

A	B	C	D	E	F
G	H	I	J	K	L

Complete each exercise. Find the answer in the answer column. Write the word under the answer in the box containing the exercise letter.

-4 HAVE
-20 IN
27 THEIR
7 BECAUSE
5 RIGHT
18 FACES

Write the integer that represents the situation.

A. A football team gains 7 yards.

B. You miss 15 points on a test.

C. The price of gasoline increased 35 cents.

D. The temperature is 4 degrees below zero.

Tell which integer is greater.

E. $-3, -6$

F. $-5, 8$

G. $-9, 5$

H. $-25, -20$

Find the absolute value of the integer.

I. -4

J. 6

K. -27

L. 18

8 HANDS
-15 THEY
4 FRONT
-3 THEIR
6 OF
35 ALWAYS

Activity B.2 Start Thinking!

For use before Activity B.2

Temperatures can be positive or negative.

Use the following model to write and solve your own addition or subtraction problem involving a negative number.

At 3 A.M., it was –3°F outside. By 9 A.M., the temperature had risen 20°F. What was the temperature at 9 A.M.?

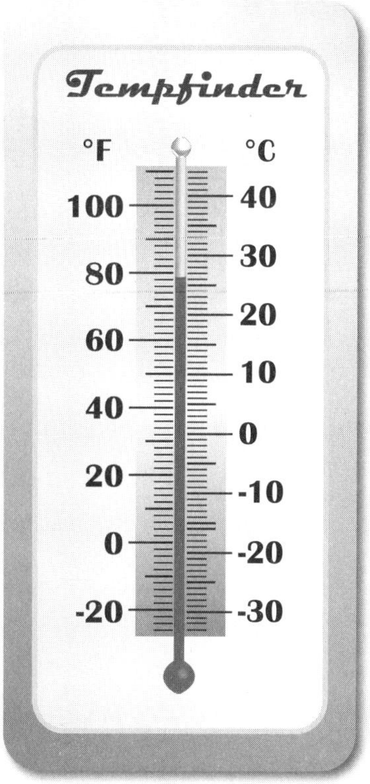

Activity B.2 Warm Up

For use before Activity B.2

Tell which integer is greater.

1. 2, 5

2. –1, 1

3. –7, 0

4. –5, –8

5. –10, –1

6. 6, –9

Lesson B.2 Start Thinking!

For use before Lesson B.2

An office building has an underground parking garage. The first through nineteenth floors have offices and there are 4 levels of parking. An elevator is used to access all of the floors.

Draw and label the floors of the building using integers. Make sure that you don't skip any integers and that the first floor of offices is labeled 1.

Write two word problems involving addition or subtraction. Both problems must involve one negative integer. Exchange problems with a classmate and solve your classmate's problems.

Lesson B.2 Warm Up

For use before Lesson B.2

Use a number line to find the sum or difference.

1. $-2 + 3$ **2.** $-8 + 2$ **3.** $-4 + 2$

4. $-2 - 5$ **5.** $-3 - 3$ **6.** $4 - 7$

Name______________________________ Date__________

B.2 Practice A

Use the number line to find the sum or difference.

1. $-3 + 5$

2. $2 - 6$

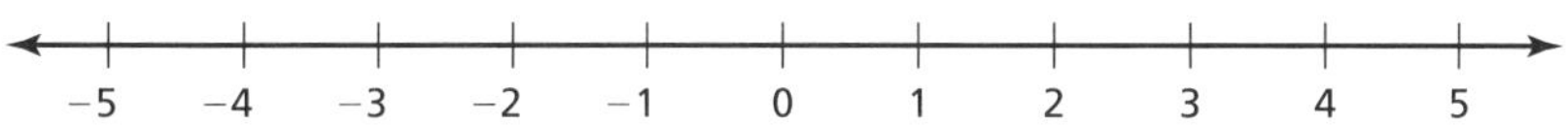

Use a number line to find the sum or difference.

3. $1 - 2$ **4.** $3 - 6$ **5.** $-3 + 2$ **6.** $-4 + 7$

7. $-3 - 1$ **8.** $-4 - 7$ **9.** $-6 + 5$ **10.** $-7 + 2$

11. $-8 - 7$ **12.** $-9 + 5$ **13.** $8 - 13$ **14.** $-1 + 9$

15. $-12 + 14$ **16.** $-11 - 5$ **17.** $-6 - 5$ **18.** $-12 + 17$

19. Describe and correct the error in finding $2 - 4$.

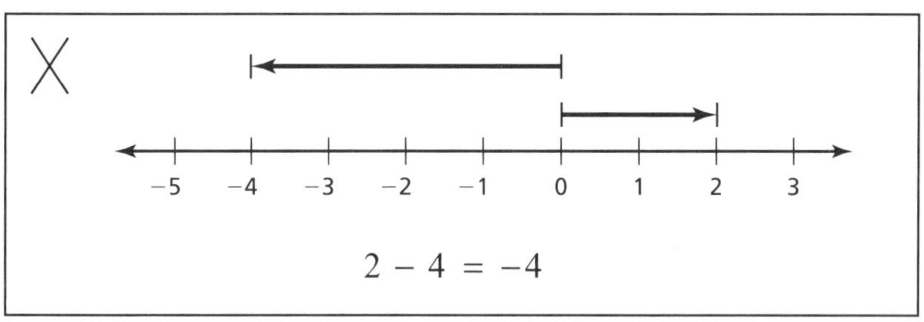

20. The daily low temperature is the sum of the daily "high" temperature and the "change" shown in the table.

a. Use a number line to find the daily low temperature in each city.

b. Order the low temperatures from least to greatest. Which city had the lowest temperature?

City	High (°C)	Change (°C)
Boston	−2	−5
Chicago	−1	−9
Miami	25	−10
New York	3	−5
Tampa	22	−16

21. While doing house work, you make a joke: "My efforts are all negative. Every time I take two steps forward, I fall three steps back."

a. Use a number line to solve $2 - 3$.

b. How is your solution related to the joke?

Name ______________________________ Date __________

B.2 Practice B

Use a number line to find the sum or difference.

1. $-3 + 5$ **2.** $-4 + 2$ **3.** $-3 - 2$ **4.** $-6 - 3$

5. $8 - 11$ **6.** $11 - 17$ **7.** $-12 + 15$ **8.** $-8 + 12$

9. $-19 + 6$ **10.** $-14 - 7$ **11.** $8 - 23$ **12.** $6 - 18$

Use a number line to evaluate the expression.

13. $-3 + 4 + 5$ **14.** $3 - 4 - 2$ **15.** $-1 + 2 - 6$ **16.** $-5 + 7 - 3$

17. $-8 - 2 + 6$ **18.** $-4 - 7 - 2$ **19.** $-6 + 3 - 5$ **20.** $7 - 2 - 11$

Use a number line to evaluate the expression when $a = 4$ and $b = -7$.

21. $|a + b|$ **22.** $|a| + |b|$ **23.** $|a - b|$ **24.** $|a| - |b|$

25. The number of shots used by a golfer on the first 8 holes is shown along with the "par" rating for each hole.

a. Copy and complete the table. Subtract par from the shots to find the scores against par for the first 8 holes.

b. Add to find the total score against par for the first 8 holes.

c. "Even par" means the total score against par is zero. How many shots does the golfer need on the last hole to have an "even par" round? Explain.

Hole	Shots	Par	Score against par
1	2	3	−1
2	5	4	1
3	3	3	
4	3	5	
5	5	4	
6	4	4	
7	6	5	
8	4	3	
9		4	

26. A dolphin is swimming 4 feet beneath the ocean's surface when it makes a series of leaps and dives. These moves change the dolphin's position relative to the surface by the following amounts (in feet):

7, −12, 6, −4, 11, −5, −4, 5

a. What is the dolphin's position relative to the surface after these moves?

b. Which changes represent *leaps*? Explain.

27. Which positive integers, when subtracted from 12, give a difference

a. less than 0? **b.** greater than 0? **c.** equal to 0?

Name________________________________ Date__________

B.2 Enrichment and Extension

Number Line Game

Use the following rules to play the game. Work with a partner.

- **Start at 0 on a number line.**
- **Roll a number cube and flip a coin.**
- **The number rolled is the number of units you move on the number line. If heads is flipped, move to the right on a number line. If tails is flipped, move to the left.**
- **Repeat this two more times. How close is your final number to 0?**

Example

Number cube	Coin flip	Action
Roll a 4.	Heads	Move 4 units right to 4.
Roll a 2.	Tails	Move 2 units left to 2.
Roll a 5.	Tails	Move 5 units left to end at **−3**. This is your final number.

Compare your final number to your friend's final number. The player that is farther from zero moves their climber up 1 unit on the mountain. Repeat until someone's climber goes over the side of the mountain. The player whose climber goes over loses. If both final numbers are the same, *both* climbers move up 1 unit on the mountain.

Player 1

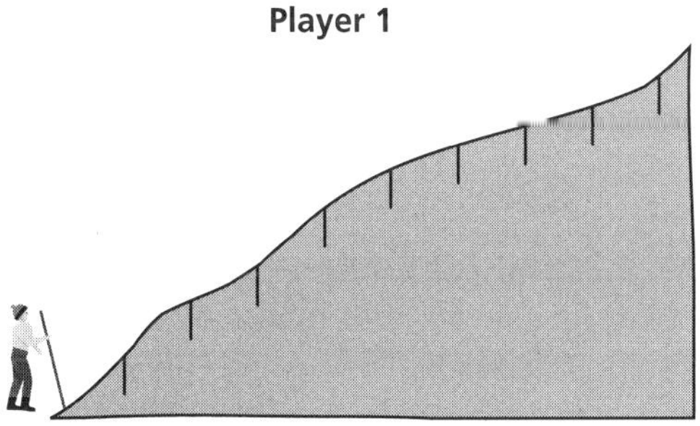

Player 2

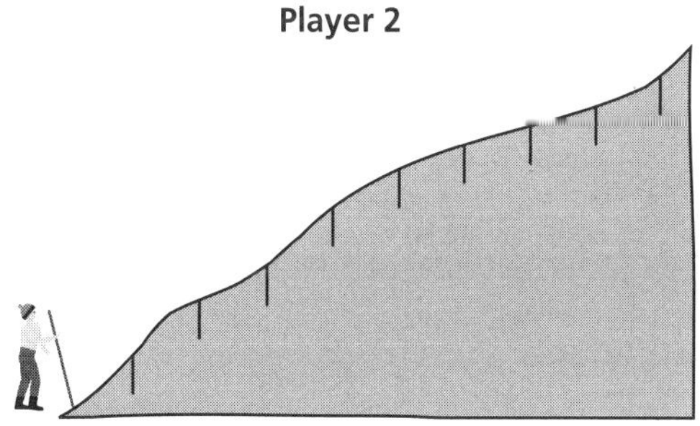

Tell whether the statement is *true* or *false*. If it is false, explain why.

1. A negative number minus a positive number is always negative.

2. A negative number plus a positive number is always positive.

3. A positive number minus a positive number is always positive.

Name ______________________________ Date __________

B.2 Puzzle Time

Where Does A Cat Go When It Loses Its Tail?

Write the letter of each answer in the box containing the exercise number.

Use a number line to find the sum or difference.

1. $-2 + 5$
2. $-9 + 6$
3. $-4 - 8$
4. $7 - 14$
5. $-8 - 9$
6. $-11 - 4$
7. $-12 + 6$
8. $-13 + 2$
9. $-3 - 5$
10. $-14 + 18$
11. $13 - 23$
12. $-1 + 9$
13. $-4 + 13$
14. $-6 + 6$

Answers

E. -6

T. -3

O. -10

R. -17

I. 3

A. 9

R. 0

S. -8

T. -15

L. -12

A. 4

O. -11

E. -7

T. 8

12	8		10		14	4	2	13	1	3		9	6	11	5	7

Activity B.3 Start Thinking!

For use before Activity B.3

Think of a positive fraction or mixed number that can be used to describe a situation.

Think of a negative fraction or mixed number that can be used to describe a situation.

Activity B.3 Warm Up

For use before Activity B.3

Find an integer that is between the two numbers.

1. 1 and 3

2. -1 and 5

3. -7 and -4

4. -8 and -5

5. -2 and 0

6. 0 and 5

Lesson B.3 Start Thinking!

For use before Lesson B.3

Write your age in years and months. Then convert the age to a mixed number in simplest form.

After everyone in the class has written their age as a mixed number, line up in order from youngest to oldest by comparing the mixed numbers.

Lesson B.3 Warm Up

For use before Lesson B.3

Find a fraction or a mixed number that is between the two numbers.

1.

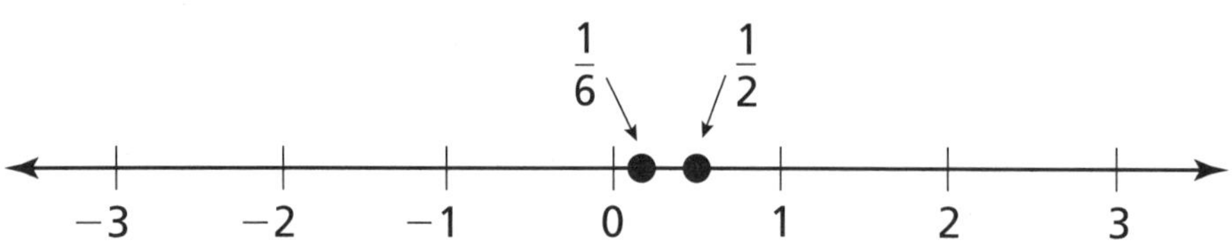

2.

3.

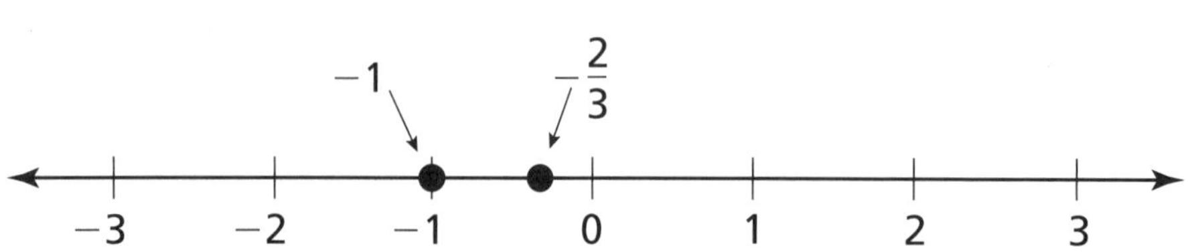

Name ______________________ Date __________

B.3 Practice A

Find a fraction or mixed number that is between the two numbers.

1.

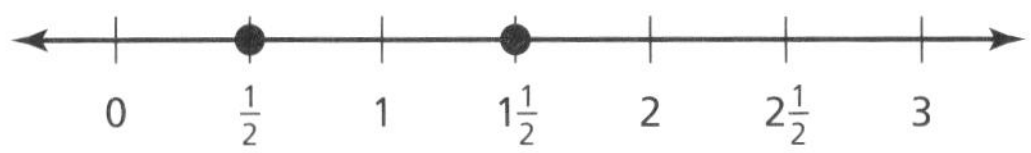

2.

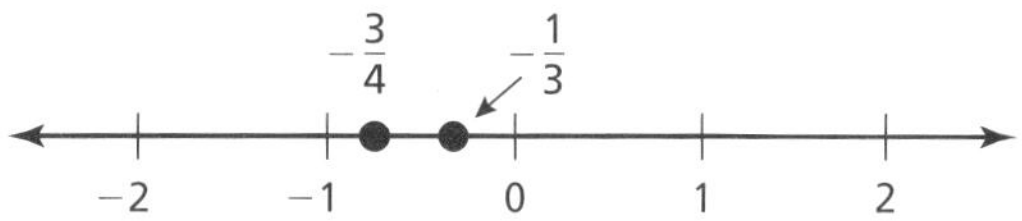

3.

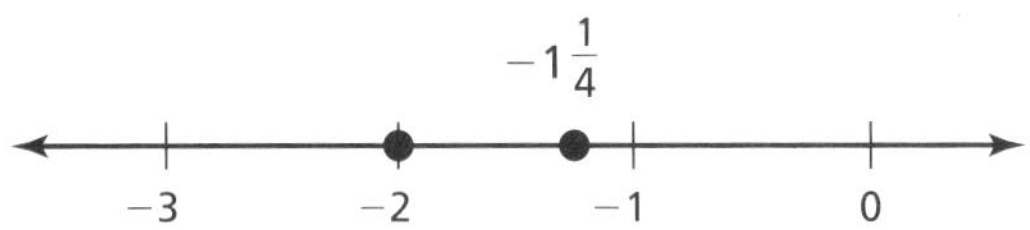

4.

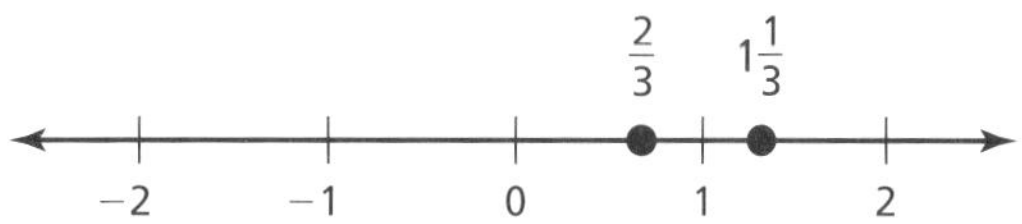

Which number is greater? Explain.

5. $-\frac{1}{5}, -\frac{3}{5}$

6. $3\frac{1}{8}, 3\frac{2}{9}$

7. $-\frac{10}{3}, -\frac{7}{4}$

8. $\frac{4}{5}, -1\frac{7}{8}$

9. $-\frac{7}{6}, -\frac{6}{7}$

10. $-2\frac{3}{4}, -2\frac{2}{3}$

11. $-\frac{7}{10}, -\frac{13}{10}$

12. $4\frac{5}{6}, 4\frac{1}{2}$

Find the median of the data.

13. $-\frac{3}{10}, -\frac{1}{10}, -\frac{2}{5}, -\frac{3}{10}, -\frac{1}{2}$

14. $1\frac{1}{12}, -\frac{3}{12}, 1\frac{5}{12}, \frac{7}{12}, -\frac{2}{12}$

15. $-\frac{2}{3}, -2, -1\frac{1}{3}, -1\frac{2}{3}, -2\frac{2}{3}$

16. $-\frac{5}{8}, 0, -\frac{1}{4}, 0, -\frac{3}{4}$

17. The position of a deep sea probe is $-2\frac{3}{4}$ fathoms relative to sea level. After finishing taking data, it moves to $-2\frac{5}{8}$ fathoms relative to sea level. Which was deeper, the first mission or the second?

18. The position of a camera lowered over a cliff changes $-\frac{1}{20}$ meter each minute. It starts at 0, the top of the cliff.

a. Use a number line to graph the position of the camera over 6 minutes.

b. After how many minutes does the camera reach a position of $-\frac{1}{10}$ meter from the edge of the cliff?

Name ____ Date __________

B.3 Practice B

Find a fraction or mixed number that is between the two numbers.

1.

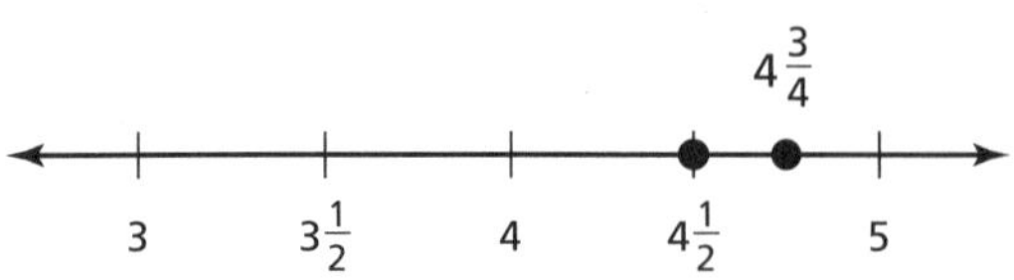

2.

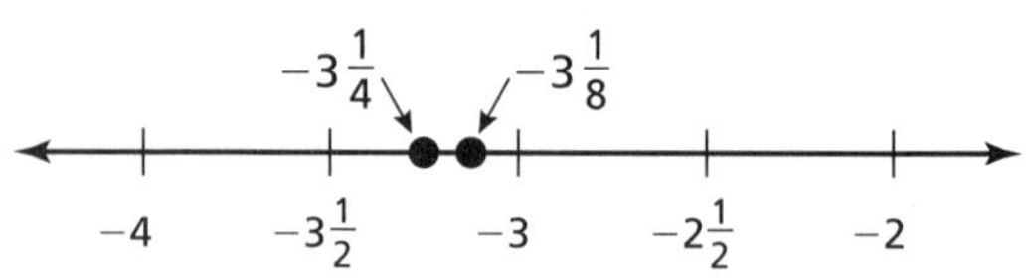

3.

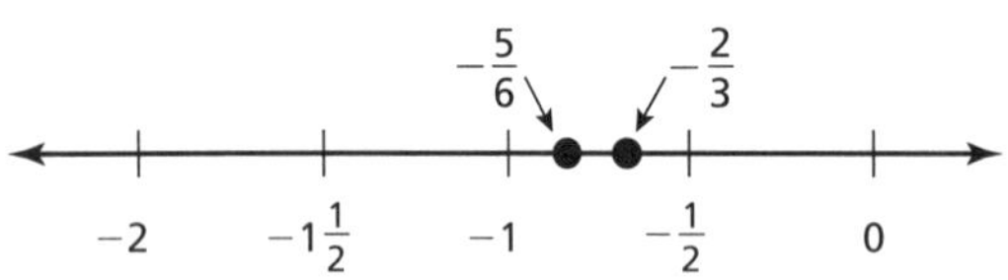

4.

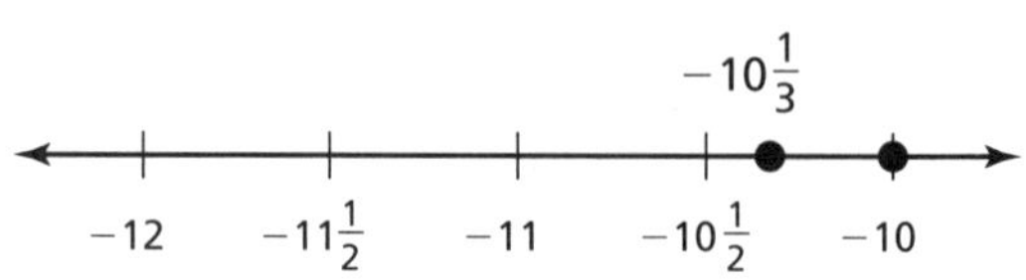

Which number is greater? Explain.

5. $-\frac{3}{8}, -\frac{4}{5}$

6. $-\frac{2}{9}, -\frac{1}{3}$

7. $\frac{3}{4}, -\frac{4}{3}$

8. $-\frac{13}{4}, -\frac{11}{3}$

9. $-1\frac{6}{10}, -1\frac{3}{10}$

10. $-\frac{2}{5}, -\frac{3}{10}$

11. $-\frac{10}{4}, -\frac{11}{5}$

12. $-1\frac{2}{3}, -1\frac{1}{2}$

Find the median of the data.

13. $-\frac{5}{6}, -\frac{2}{3}, -\frac{1}{2}, -\frac{7}{12}, -\frac{11}{12}$

14. $-\frac{5}{8}, -\frac{3}{4}, -1\frac{1}{8}, -\frac{3}{8}, -1\frac{1}{4}$

15. $-5\frac{1}{6}, -5\frac{3}{8}, -5\frac{7}{9}, -5\frac{6}{7}, -5\frac{3}{7}$

16. $-\frac{4}{25}, -\frac{9}{50}, -\frac{13}{100}, -\frac{3}{20}, -\frac{1}{5}$

17. Two runners slow down. One decelerates at $-\frac{5}{8}$ ft/sec^2, and the second at $-\frac{3}{5}$ ft/sec^2. Which runner slowed down more?

18. The level of water in a pool changes by $-2\frac{1}{4}$ inches each day. Use a number line to graph the position of the water below its starting point on 5 days if it starts at 0. On what day does it fall 4 inches below the starting level?

Name ______________________ Date __________

B.3 Enrichment and Extension

Fraction Game

Cut 10 index cards in half. Divide the cut index cards between you and a partner. Follow the rules below to play a game.

- **Write a fraction or mixed number on a piece of index card that is between the two numbers given.**
- **Compare your number with your friend's number. The person with the larger number gets both index cards. If the numbers are equivalent, each player writes a different fraction or mixed number. Repeat until there is not a tie.**
- **The person with the most index cards after the last set of numbers is the winner.**

1. $-\frac{1}{2}, \frac{5}{3}$

2. $\frac{1}{10}, -1\frac{6}{7}$

3. $-\frac{6}{5}, -\frac{3}{5}$

4. $-1\frac{3}{4}, -2\frac{1}{4}$

5. $-4\frac{7}{9}, -12\frac{3}{7}$

6. $-6\frac{4}{5}, -\frac{40}{7}$

7. $-\frac{37}{8}, -\frac{70}{13}$

8. $-\frac{21}{2}, -\frac{202}{21}$

9. $-20\frac{1}{2}, -15\frac{1}{3}$

10. $-62\frac{5}{9}, -70\frac{5}{11}$

Name ______________________________ Date __________

B.3 Puzzle Time

How Many Jelly Beans Can You Put In An Empty Jar?

A	B	C	D	E	F
G	H	I	J		

Complete each exercise. Find the answer in the answer column. Write the word under the answer in the box containing the exercise letter.

$1\frac{4}{5}$ BECAUSE
$-\frac{1}{2}$ ISN'T
$-2\frac{3}{4}$ THAT
$-3\frac{3}{8}$ ANYMORE
$\frac{2}{3}$ ONE

Which number is greater?

A. $-\frac{1}{2}, \frac{3}{5}$

B. $-\frac{2}{3}, -\frac{5}{6}$

C. $-3\frac{7}{8}, 1\frac{4}{5}$

D. $-5\frac{1}{4}, -5\frac{1}{2}$

E. $-2\frac{7}{8}, -2\frac{3}{4}$

F. $-\frac{13}{2}, -\frac{25}{4}$

Find the median of the data.

G. $-4\frac{3}{8}, -4\frac{1}{4}, -4\frac{1}{8}, -4, -3\frac{7}{8}$

H. $-\frac{1}{6}, -\frac{2}{3}, -\frac{1}{2}, -\frac{1}{3}, -\frac{5}{6}$

I. $-2\frac{3}{7}, -2\frac{6}{7}, -2\frac{5}{7}, -2\frac{4}{7}, -2\frac{2}{7}$

J. $-3\frac{1}{2}, -3\frac{1}{4}, 3, -3\frac{3}{4}, -3\frac{3}{8}$

$-\frac{25}{4}$ THE
$\frac{3}{5}$ ONLY
$-2\frac{4}{7}$ EMPTY
$-5\frac{1}{4}$ AFTER
$-4\frac{1}{8}$ JAR

Activity B.4 Start Thinking!
For use before Activity B.4

How are negative decimals and elevation related?

Activity B.4 Warm Up
For use before Activity B.4

Which number is greater?

1. $-\frac{5}{7}, \frac{1}{8}$

2. $-\frac{3}{4}, -\frac{2}{3}$

3. $-\frac{5}{2}, -\frac{15}{8}$

4. $-1\frac{1}{5}, -1\frac{1}{3}$

5. $2\frac{1}{2}, -2\frac{1}{2}$

6. $-3\frac{1}{2}, -3\frac{2}{3}$

Lesson B.4 Start Thinking!
For use before Lesson B.4

List some different methods of comparing positive decimals. Do these ways work for comparing negative decimals?

Lesson B.4 Warm Up
For use before Lesson B.4

Estimate the location of the point on the number line.

1. *A* **2.** *B* **3.** *C* **4.** *D*

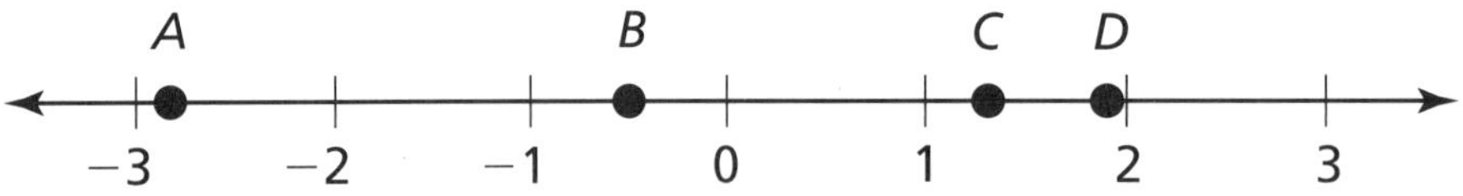

Name____ Date __________

B.4 Practice A

Estimate the location of the point on the number line.

1. E **2.** F **3.** G **4.** H

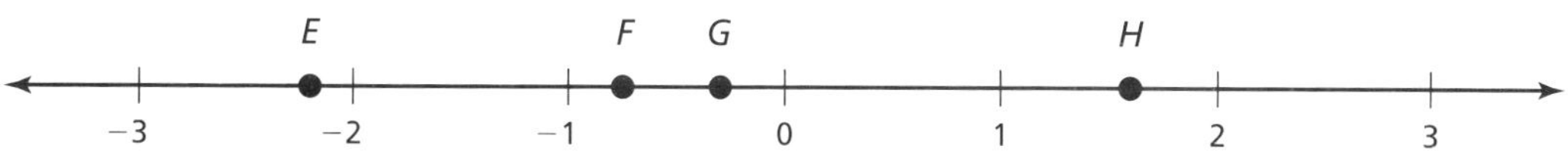

Which number is greater? Explain.

5. 2.1, −2.1 **6.** −8.4, 4.8 **7.** −0.08, −0.8 **8.** −3.08, −4.16

9. −3.95, 5.05 **10.** −4.82, −4.89 **11.** $-3.5, -3\frac{2}{3}$ **12.** $-4\frac{3}{4}, -4.70$

Use a number line to determine whether the number is between −1.3 and −2.4.

13. −1.8 **14.** −1.1 **15.** −2.6 **16.** −2

17. The table shows the percent change in sales compared to the previous month. Which month had the largest decrease? the largest increase?

Month	Jan.	Feb.	Mar.	Apr.	May	Jun.
Percent Change in Sales	−0.8	1.5	−1.9	0.5	−1.7	−0.6

18. An oceanographer takes readings at the following positions relative to sea level: −2.48 kilometers, −2.83 kilometers, −2.70 kilometers, and −2.15 kilometers. Order the positions from farthest from sea level to closest to sea level.

19. Describe and correct the error in determining which number is greater.

> ✗ $-5\frac{5}{6}$ is greater than −5.8 because
> $5\frac{5}{6}$ is greater than 5.8.

20. A biologist observing the level of a pond over a one week dry spell observes that the water level changes by −1.4 centimeters Monday, −1.8 centimeters Tuesday, and −1.6 centimeters Wednesday. On which day did the water level change the most?

21. A mapmaker models points east of town hall as positive and west as negative. The lighthouse is at −3.4 miles and the dock at −3.5 miles. Which is farther from the town hall?

Name ____ Date __________

B.4 Practice B

Estimate the location of the point on the number line.

1. W **2.** X **3.** Y **4.** Z

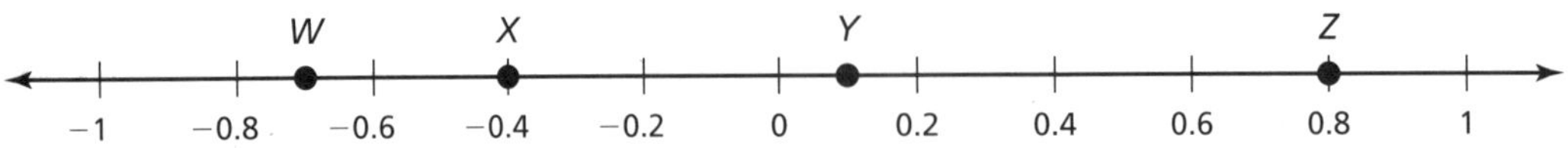

Which number is greater? Explain.

5. $5.38, -5.83$ **6.** $-6.3, -4.9$ **7.** $-0.11, -0.44$ **8.** $-2.05, -2.50$

9. $-8.7, 0.7$ **10.** $-4.9, -4.6$ **11.** $-\frac{1}{5}, -0.1$ **12.** $-3\frac{3}{8}, -3.8$

Use a number line to determine whether the number is between −11.7 and −13.8.

13. -11.5 **14.** -11.9 **15.** -14.0 **16.** -13.5

17. In physics, positive speeds denote upward motion and negative speeds denote downward motion. The table gives the speed of a ball thrown upward at a rate of 20.0 meters per second.

Time (seconds)	0	1	2	3	4
Speed (meters/second)	20.0	10.2	0.4	−9.4	−19.2

a. When was the speed greatest going upward?

b. When was the speed greatest going downward?

c. Between what two times was the speed zero? What does a speed of 0 mean?

18. The person in Exercise 17 stood at the edge of a cliff to throw the ball. Positive numbers indicate height above the edge of the cliff, and negative numbers below the edge.

Time (seconds)	0	1	2	3	4	5	6
Height (meters)	5	20.1	25.4	20.9	6.6	−17.5	−51.4

a. Is the ball higher at 1 second or at 3 seconds?

b. Is the ball lower at 5 seconds or 6 seconds? Does this make sense?

19. A stock lost value on both Monday and Tuesday. On Monday, it changed by -5.7 points and on Tuesday it changed by -3.8 points. On which day did it drop the least?

Name____ Date__________

B.4 Enrichment and Extension

Decimal Game

Find a partner. Follow the rules below to play a game.

- **For Rounds 1–12, each person rolls a number cube three times.**
- **Each person uses his or her numbers to write a negative decimal that is closest the given decimal without going over. If there is a tie, each player rolls again.**
- **The person who is closest to the given number wins the round. The person that wins the most rounds wins the game.**

1. -4.5

2. -3.6

3. -6.4

4. -4.1

5. -2.25

6. -3.45

7. -4.55

8. -5.01

9. -53.5

10. -24.0

11. -61.4

12. -32.6

If there is a tie, go to Round 13 for the tiebreaker.

13. -1.11

Name ______________________________ Date __________

B.4 Puzzle Time

What Did One Plate Say To The Other Plate?

Write the letter of each answer in the box containing the exercise number.

Which number is greater?

1. 4.8, −4.2
2. −2.7, 0.4
3. −2.1, −3.2
4. −21.5, −21.05
5. 5.2, −5.1
6. −6.4, 8.6
7. −3.07, −3.14
8. −6.75, −6.87
9. Use a number line to determine which number is between −6.9 and −7.9.

   A. −6.35 **B.** −7.5 **C.** −7.98

10. Use a number line to determine which number is between −4.4 and −5.8.

    A. −5.68 **B.** −4.14 **C.** −5.92

11. Use a number line to determine which number is between −2.61 and −5.49.

    A. −2.49 **B.** −5.51 **C.** −3.11

Answers

D. −5.1
O. 0.4
A. −4.2
N. −3.07
J. −6.4
H. B
E. −21.05
U. A
R. −2.7
S. 5.2
B. −3.2
C. −2.1
M. 8.6
S. −3.14
L. −6.75
T. −21.5
I. C
W. −6.87
N. 4.8

8	10	1	3	9		11	5		2	7		6	4

Activity B.5 Start Thinking!

For use before Activity B.5

Arrange two rulers as shown.

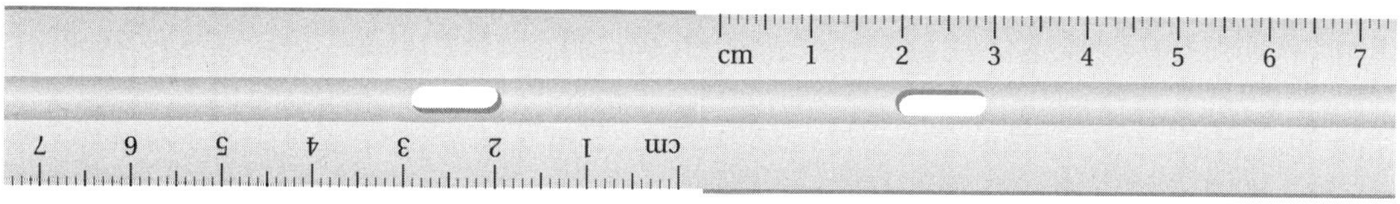

Explain how the rulers can be used to represent a number line. Then explain how you can use the rulers to help you solve the following problems.

$-2\frac{1}{4} + 4$ $\qquad$ $-3\frac{1}{2} + 3$

$3\frac{1}{12} - 4$ $\qquad$ $-6\frac{3}{4} - 2$

Activity B.5 Warm Up

For use before Activity B.5

Graph the fraction or decimal on a number line.

1. $-\frac{3}{4}$ **2.** $-\frac{2}{3}$ **3.** $\frac{3}{2}$

4. 5.2 **5.** -0.8 **6.** -1.25

Lesson B.5 Start Thinking!
For use before Lesson B.5

Write a word problem that can be solved using the expression $-5.5 + 3.1$. Then find the solution to your problem.

Lesson B.5 Warm Up
For use before Lesson B.5

Use a number line to find the sum or difference.

1. $8 - 4.2$

2. $-2\frac{1}{2} + 1$

3. $-4.75 + 3$

4. $-\frac{2}{3} - 1\frac{1}{3}$

5. $1 - 3.5$

6. $-\frac{1}{4} + 1\frac{1}{2}$

Name________________________________ Date__________

B.5 Practice A

Use a number line to find the sum or difference.

1. $3 - 2\frac{1}{3}$ **2.** $4 + 1\frac{2}{5}$ **3.** $\frac{3}{4} - 2$ **4.** $-1\frac{1}{2} + 2\frac{1}{2}$

5. $-2\frac{3}{5} - 2\frac{4}{5}$ **6.** $-\frac{5}{6} + 1\frac{1}{3}$ **7.** $-4\frac{1}{4} - \frac{1}{2}$ **8.** $\frac{13}{20} - \frac{17}{20}$

9. $12 - 4.9$ **10.** $-3.5 + 7$ **11.** $-2 - 3.9$ **12.** $0.09 + 0.16$

13. $-3.8 + 1.6$ **14.** $0.5 - 0.9$ **15.** $0 - 2.7$ **16.** $-1 + 2.6$

Use a number line to evaluate the expression.

17. $3 - 2\frac{1}{2} - 1\frac{1}{2}$ **18.** $-1\frac{1}{3} + \frac{2}{3} - 1\frac{1}{3}$ **19.** $-2 - \frac{1}{3} - \frac{1}{2}$

20. $3 - 5.5 + 1$ **21.** $-2.1 - 2.3 - 4.2$ **22.** $-2.5 + 6 - 0.8$

23. For the following rock climbing problems, assume that the top of the cliff is 0, positions below the edge are negative, and positions above the edge are positive.

a. A rock climber is at a position of -4.5 meters. He climbs up 2.8 meters. What is his position relative to the top of the cliff after this climb?

b. Another climber starts at the top edge of the cliff. She descends 5.0 meters, then climbs up 1.4 meters. Write and solve an expression to model her position. Explain each term in the expression.

c. A 20-meter rope is tied to a tree 1.5 meters above the top of the cliff. At what position is the bottom of the rope?

24. A football team starts on their own 17-yard line. On the next play they lose $11\frac{1}{2}$ yards. What yard line are they on now?

25. The deck of a boat is $4\frac{1}{3}$ feet above the water. Holding a 10-foot rope tied to the deck, Connor slides off the deck and swims down below the water, where he reaches the end of the rope. How deep is he?

26. You have \$20.00. You earn \$12.50 doing yard work and then spend \$13.85 on a new sweater. Write and solve an expression to find how much money you have left.

Name ______________________________ Date __________

B.5 Practice B

Use a number line to find the sum or difference.

1. $-2 + 1\frac{2}{3}$ **2.** $1 - 1\frac{3}{4}$ **3.** $1\frac{2}{5} - 3$ **4.** $-1\frac{1}{3} - 1\frac{1}{2}$

5. $-3\frac{2}{5} + 1\frac{3}{5}$ **6.** $-\frac{3}{10} + \frac{4}{5}$ **7.** $-1\frac{7}{8} - \frac{1}{2}$ **8.** $2\frac{3}{4} - 2\frac{7}{8}$

9. $2.3 - 4.1$ **10.** $-3.7 + 2.9$ **11.** $-4 - 0.8$ **12.** $5.15 - 7.35$

13. $-0.07 + 0.05$ **14.** $1.24 - 2$ **15.** $-1.8 - 4.4$ **16.** $-3.5 + 4.6$

Use a number line to evaluate the expression.

17. $\frac{4}{7} - 1\frac{1}{7} + \frac{5}{7}$ **18.** $\frac{2}{3} - \frac{5}{6} + \frac{1}{2}$ **19.** $-1 - \frac{1}{2} - \frac{1}{4}$

20. $8.9 - 11 - 3.6$ **21.** $-0.85 - 0.25 - 0.60$ **22.** $4.5 - 12 - 3.9$

23. A sightseer tosses a pebble into a sinkhole. The change in the pebble's position each half second is given in the table: positive change means it moves higher, and negative change means it moves lower. The pebble starts 3 feet above the edge of the sinkhole.

Time (seconds)	0.5	1	1.5
Change in Position (feet)	−2.75	−10.75	−18.75

a. What is the pebble's position relative to the edge of the sinkhole after 0.5 second? after 1 second?

b. The pebble hits the bottom of the sink hole after 1.5 seconds. How deep is the sinkhole?

24. Describe and correct the error in finding $-2.3 - 1$.

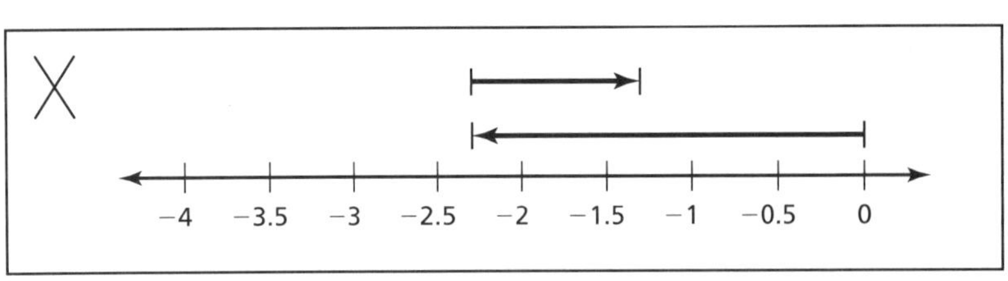

25. Your younger brother borrows \$6.50 to buy a toy. You represent this as your brother having -6.5 dollars. If your brother pays \$1.50 a week, write and solve an addition expression to find how much he has after 2 weeks.

Name______________________________ Date__________

B.5 Enrichment and Extension

Discovering Florida

2, 5	1, 1	1, 5	5, 8	4, 8	2, 6	2, 3
3, 3	1, 3	8, 9	6, 9	4, 9	2, 7	1, 9

Choose a set of numbers and fill in the boxes to make each equation true.

1. $-\square + \frac{6}{5} = -7\frac{\square}{5}$

2. $-\frac{3}{8} + \frac{3}{\square} = \frac{3}{\square}$

3. $-5\frac{2}{\square} + 8\frac{\square}{5} = 2\frac{14}{15}$

4. $\frac{7}{9} - \square = -6\frac{\square}{9}$

5. $1\frac{\square}{3} - 2\frac{\square}{9} = -\frac{8}{9}$

6. $-\frac{5}{\square} - 2\frac{1}{15} = -2\frac{\square}{10}$

7. $-\square.2 - 11.\square = -17.4$

8. $-9.\square + \square.8 = 0$

9. $-\square + 7.\square = 4.2$

10. $-\square.8 + 7.4\square = 5.61$

11. $-1\square.7 - 9.\square = -21.6$

12. $\square - 23.\square = -15.5$

13. The state of Florida was named by the Spanish explorer Juan Ponce de León. Two sets of numbers above are not used. Rearrange the two sets to find the year that his ships first landed on Florida's coast.

Name ______________________________ Date __________

B.5 Puzzle Time

What Did One Eye Say To The Other Eye?

A	B	C	D	E	F
G	H	I	J	K	L

Complete each exercise. Find the answer in the answer column. Write the word under the answer in the box containing the exercise letter.

$4\frac{1}{3}$ YOU
$-3\frac{1}{4}$ SMELL
-2.5 SOMETHING
-10 IT
$-9\frac{1}{2}$ TO
-11.75 SEEMS

Use a number line to find the sum or difference.

A. $-5 + 1\frac{1}{2}$

B. $-2\frac{1}{3} + 6\frac{2}{3}$

C. $-8 + 0.9$

D. $3 - 1\frac{1}{3}$

E. $-7\frac{1}{4} - 2\frac{3}{4}$

F. $-9 - 2.75$

G. $-3.4 - 0.5$

H. $6 - 8.5$

I. $-5\frac{1}{2} + 4$

J. $-4.75 + 1$

K. $-9\frac{1}{4} - \frac{1}{4}$

L. $-2\frac{1}{2} - \frac{3}{4}$

-3.9 THAT
$1\frac{2}{3}$ ME
-3.75 STARTING
$-3\frac{1}{2}$ BETWEEN
$-1\frac{1}{2}$ IS
-7.1 AND

Activity B.6 Start Thinking!

For use before Activity B.6

The graph shows temperature equivalents in Fahrenheit and Celsius.

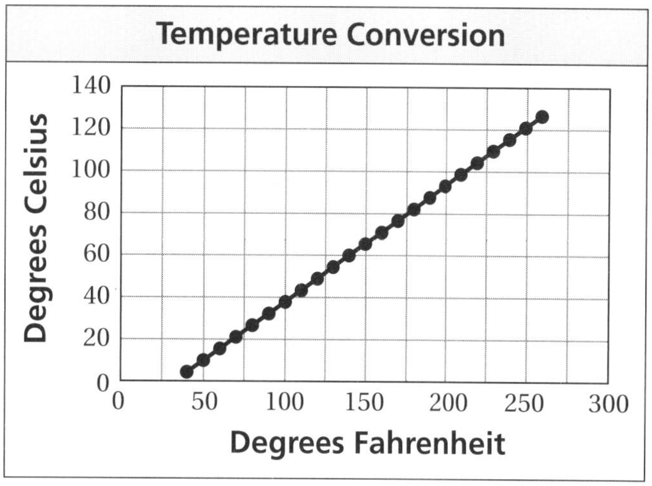

Copy the graph. To ensure accuracy, use a ruler and make sure that your graph includes the points $(50, 10)$ and $(140, 60)$.

Extend the axes so they include negative numbers. Use a ruler to extend the line into the new region.

How is your new graph different from the original? What does the new part of the graph represent? Use an example in your explanation.

Activity B.6 Warm Up

For use before Activity B.6

Plot the ordered pair in a coordinate plane.

1. $A(1, 3)$

2. $B(0, 4)$

3. $C(2, 2)$

4. $D(5, 0)$

5. $E(3, 1)$

6. $F(2, 5)$

Lesson B.6 Start Thinking!

For use before Lesson B.6

Use a coordinate plane to help you create directions to find a hidden treasure.

1. Choose an ordered pair for the location of the treasure.

2. Create a series of steps that will lead a treasure seeker to the treasure. The directions should begin at (0, 0) and each step should tell the treasure seeker to walk a certain number of "paces" north, south, east, or west. Be sure to make it tricky—hidden treasures should not be easy or quick to find! Make sure that the treasure seeker walks through all 4 quadrants of the map.

3. Switch directions with a classmate. Follow your classmate's directions and see if you both end up in the correct locations.

Lesson B.6 Warm Up

For use before Lesson B.6

Plot the ordered pair in a coordinate plane. Describe the location of the point.

1. $G(-1, -2)$ **2.** $H(2, -2)$ **3.** $I(5, 2)$

4. $J(0, -5)$ **5.** $K(-1, 0)$ **6.** $L(-3, 4)$

Name____ Date__________

B.6 Practice A

Write the ordered pair that corresponds to the point.

1. Point A

2. Point B

3. Point C

4. Point D

5. Point E

6. Point F

7. Point G

8. Point H

9. Point I

10. Point J

11. Point K

12. Point L

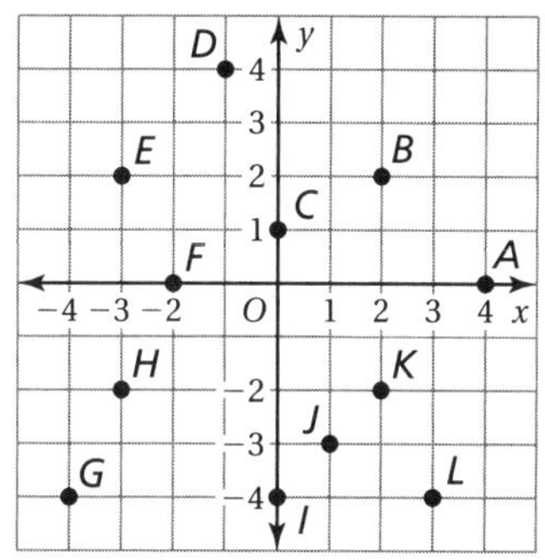

Plot the ordered pair in a coordinate plane. Describe the location of the point.

13. $M(-3, 3)$

14. $N(4, 1)$

15. $P(1, -1)$

16. $Q(-4, -2)$

17. $R(0, 2)$

18. $S(-4, 3)$

19. $T(-3, 0)$

20. $U(1, -2)$

21. $V(0, -1)$

22. $W(1, 2)$

23. $X(-2, -2)$

24. $Y(4, 0)$

25. Describe and correct the error in finding the ordered pair that corresponds to the point.

> ✗ The point is 6 units to the left of the y-axis and 3 units below the x-axis. The point is $(6, -3)$.

26. Plot and connect the points. What do you notice?
$(-3, -2)$, $(-1, -1)$, $(1, 0)$, $(3, 1)$

27. The coordinates of 3 vertices of a square are given. What are the coordinates of the missing vertex?
$(3, 2)$, $(-1, 2)$, $(-1, -2)$

28. Find the area of the triangle in the figure.

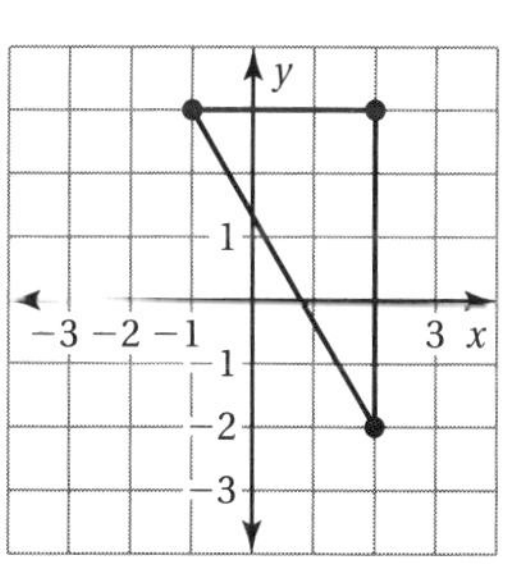

Name ______________________________ Date __________

B.6 Practice B

Write the ordered pair that corresponds to the point.

1. Point M **2.** Point N

3. Point P **4.** Point Q

5. Point R **6.** Point S

7. Point T **8.** Point U

9. Point V **10.** Point W

11. Point X **12.** Point Y

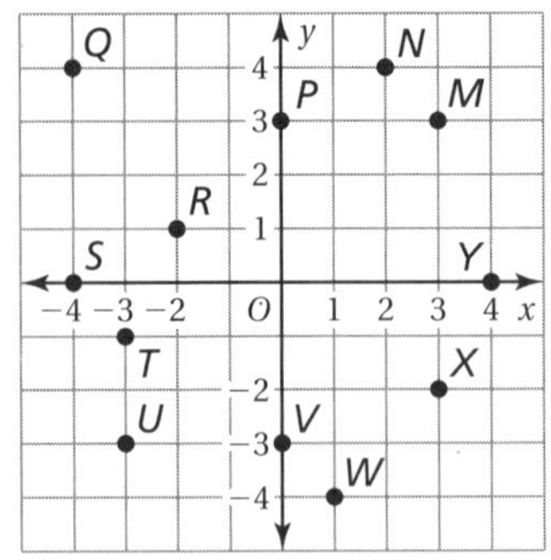

Plot the ordered pair in a coordinate plane. Describe the location of the point.

13. $A(0, 1)$ **14.** $B(4, 0)$ **15.** $C(1, 3)$ **16.** $D(-2, -1)$

17. $E(4, -3)$ **18.** $F(-1, 0)$ **19.** $G(0, -2)$ **20.** $H(-1, 1)$

21. $I(3, 3)$ **22.** $J(4, -1)$ **23.** $K(-3, -2)$ **24.** $L(-1, -4)$

25. Use the graph at the right.

a. The coordinates of three vertices of a rectangle are shown in the figure. Explain how you would find the coordinates of the fourth vertex.

b. What is the perimeter of the rectangle?

c. What is the area of the rectangle?

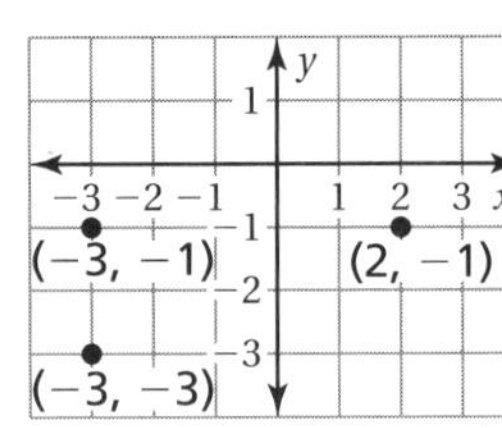

26. A wallpaper artist makes a design to repeat over a wall. Explain how you would move Figure A to get Figure B.

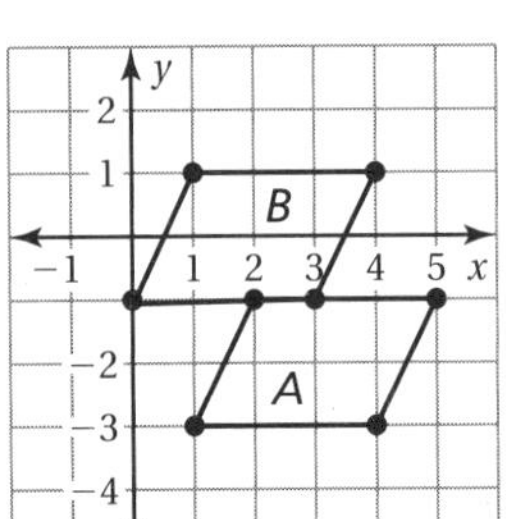

Name______________________________ Date __________

B.6 Enrichment and Extension

Plotting Fractions and Decimals in the Coordinate Plane

Copy the coordinate plane. Then plot the ordered pair in the coordinate plane. Describe the location of the point.

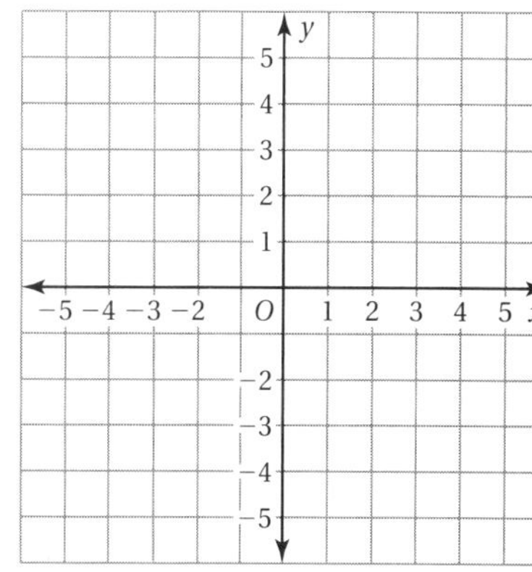

1. $A(2, 3.5)$ **2.** $B(0, -4.5)$ **3.** $C(-1.5, 1)$

4. $D(0.5, -2)$ **5.** $E(-3, 2.5)$ **6.** $F(-4, -0.5)$

7. $G\left(\frac{5}{2}, -\frac{5}{2}\right)$ **8.** $H\left(-1\frac{3}{4}, -5\right)$ **9.** $I\left(3\frac{1}{4}, 0\right)$

Write the ordered pair that corresponds to the given point in the coordinate plane.

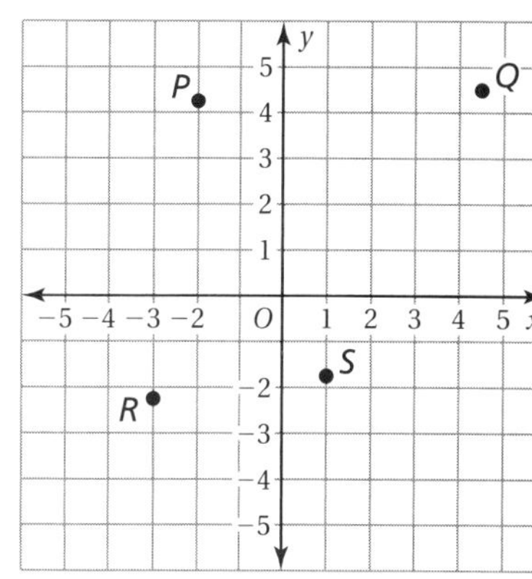

10. P **11.** Q

12. R **13.** S

14. Did you choose decimals or fractions to write your ordered pairs in Exercises 10–13? Explain your choice.

15. The coordinates of three vertices of a square are shown in the figure. What are the coordinates of the fourth vertex?

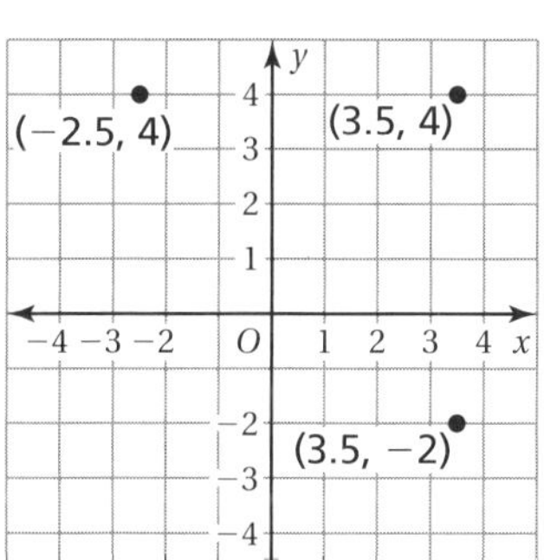

Name ______________________________ Date ________

B.6 Puzzle Time

What Does A Baker Put On His Bed?

Write the letter of each answer in the box containing the exercise number.

Write the ordered pair that corresponds to the point.

1. Point A

2. Point B

3. Point C

4. Point D

5. Point E

6. Point F

7. Point G

8. Point H

9. Point I

10. Point J

11. Point K

12. Point L

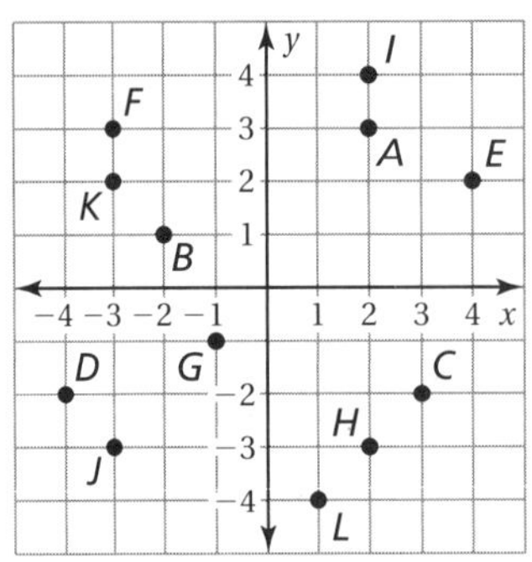

Answers

E. $(-4, -2)$

O. $(-3, 2)$

S. $(2, 3)$

I. $(2, -3)$

C. $(3, -2)$

K. $(-3, -3)$

E. $(-1, -1)$

T. $(4, 2)$

S. $(1, -4)$

O. $(-2, 1)$

H. $(-3, 3)$

E. $(2, 4)$

3	11	2	10	8	7		12	6	4	9	5	1

Activity B.7 Start Thinking!

For use before Activity B.7

For some real-life graphs, only positive values of x and y make sense. The graph below shows the distance a person can walk at an average rate of 4 miles per hour.

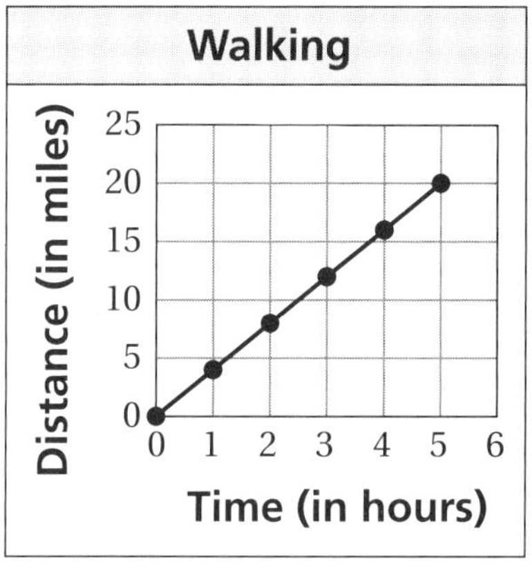

Why do only positive values make sense in this case?

Are there any real-life graphs in which negative numbers make sense too?

Activity B.7 Warm Up

For use before Activity B.7

Evaluate the expression.

1. $x + 3$ when $x = -2$

2. $x - 4$ when $x = 3$

3. $1 - x$ when $x = 5$

4. $4 - x$ when $x = 7$

5. $10 + x$ when $x = -6$

6. $x - 1$ when $x = -4$

Lesson B.7 Start Thinking!

For use before Lesson B.7

Choose a partner and follow the rules to play Guess the Equation.

Guess the Equation (2 players)

1. Each player writes an equation of the form:

   $y = \underline{\quad ? \quad} + x$ or $y = \underline{\quad ? \quad} - x$

   Do NOT tell the other player your equation.

2. Each player graphs their own equation.

3. The players exchange graphs.

4. The first player to correctly guess the equation from the graph wins the game.

Lesson B.7 Warm Up

For use before Lesson B.7

Make an input-output table for the function. Use the inputs −2, −1, 0, 1, and 2. Then draw the graph of the function.

1. $y = 6 - x$

2. $y = -2 - x$

3. $y = -x$

4. $y = x + 1$

5. $y = 2x$

6. $y = x + 4$

Name________________________________ Date__________

B.7 Practice A

Plot the data in the input-output table. Describe the pattern.

1.

Input, x	–1	0	1	2
Output, y	3	1	–1	–3

2.

Input, x	2	4	6	8
Output, y	–1	–1	–1	–1

3.

Input, x	–5	–3	–1	1
Output, y	1	3	5	7

4.

Input, x	–2	0	2	4
Output, y	5	2	1	2

Make an input-output table for the function. Use the inputs –4, –2, 0, 2, and 4. Then draw the graph of the function.

5. $y = 2x$ **6.** $y = x$ **7.** $y = 1 + 2x$ **8.** $y = x + 1$

9. $y = 3x + 1$ **10.** $y = -2 - x$ **11.** $y = 5x - 4$ **12.** $y = 2x + 4$

Match the function with its graph.

13. $y = 3$ **14.** $y = -x$ **15.** $y = 1 - 2x$

A.

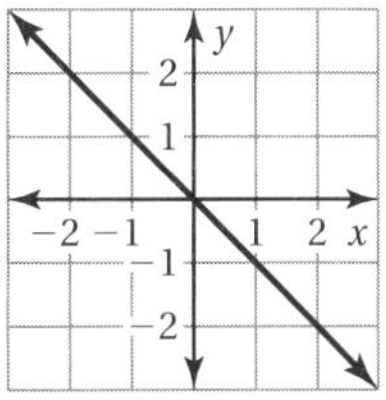

B.

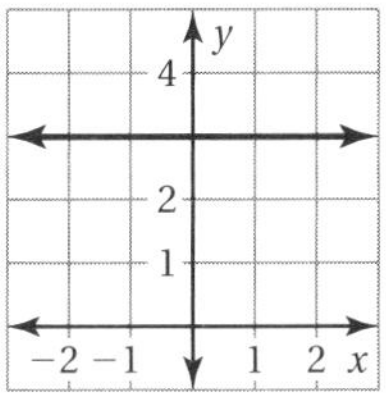

C.

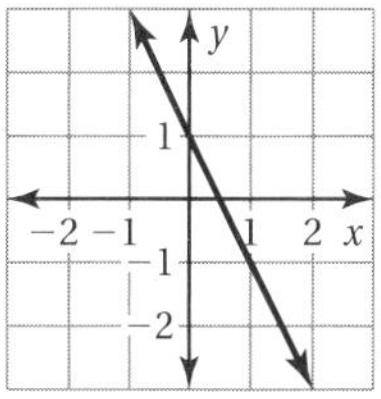

16. The table shows the relation between the voltage put into a circuit with 10 ohms resistance, and the current that flows through the circuit.

Voltage (volts)	0	10	20	30	40
Current (amperes)	0	1	2	3	4

a. Write an equation for the function shown in the table.

b. Suppose the voltage decreases by 20 volts. What happens to the current? Explain how you found your answer.

c. What current would be produced by 60 volts?

17. The temperature is 4° Fahrenheit at 3 P.M. Between 3 P.M. and 8 P.M. the temperature falls 2° each hour. Write an equation for the function, where h is the number of hours after 3 P.M. Then make an input-output table.

Name ______________________________ Date __________

B.7 Practice B

Make an input-output table for the function. Use the inputs −3, −1, 0, 1, and 3. Then draw the graph of the function.

1. $y = 2x - 3$ **2.** $y = 5 - x$ **3.** $y = 3 + 3x$ **4.** $y = -1 - x$

5. $y = 1$ **6.** $y = 4x - 10$ **7.** $y = -3 - x$ **8.** $y = x + 1$

Match the function with its graph.

9. $y = \frac{1}{2}x + 1$ **10.** $y = 4 - x$ **11.** $y = -3 + x$

A.

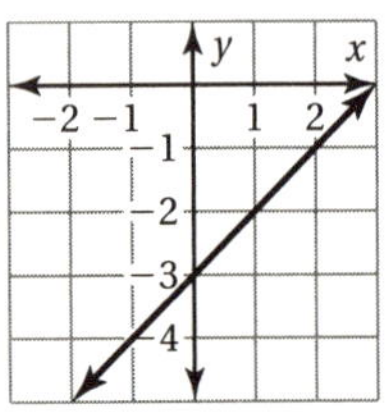

B.

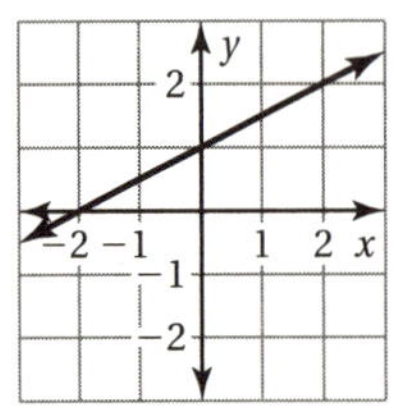

C.

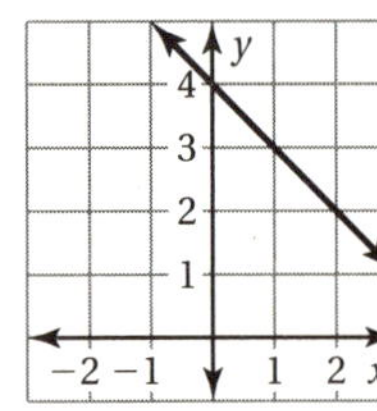

Write an equation for the function shown in the table.

12.

Input, x	−1	0	1	2
Output, y	1	0	−1	−2

13.

Input, x	−10	−5	0	5
Output, y	−50	−25	0	25

14. The table gives the cooking times for an unstuffed turkey.

Weight (pounds)	6	10	14	18
Cooking time (hours)	2	3	4	5

a. What happens to the cooking time when the weight of the turkey increases by 4 pounds?

b. How long would you expect to cook a 12-pound turkey? Explain how you can infer the cooking time even though it is not given in the table.

15. The graph shows the position relative to sea level d (in meters) of a diver returning to the surface. Time (in minutes) is represented by t.

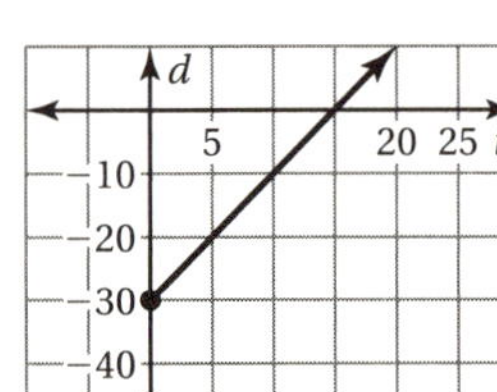

a. Write an equation for the function.

b. What is the diver's position at 10 minutes?

c. When will the diver reach the surface?

Name________________________ Date__________

B.7 Enrichment and Extension

Graphing Horizontal and Vertical Lines

1. You want to graph $y = 2.5$.

a. Copy and complete the table.

x	−2	−1	0	1	2
y					

b. Plot the ordered pairs represented by the table. Connect the points and describe the graph.

2. You want to graph $x = -3$.

a. Copy and complete the table.

x					
y	−2	−1	0	1	2

b. Plot the ordered pairs represented by the table. Connect the points and describe the graph.

Use what you learned in Exercises 1 and 2 to tell whether the statement is *true* or *false*. If it is false, change the statement to make it true.

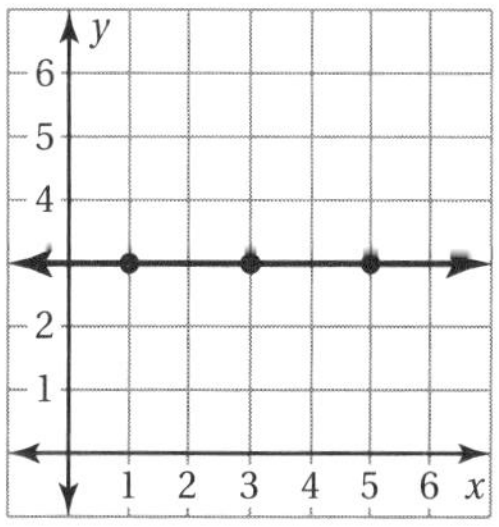

3. The graph of $y = a$ is a vertical line.

4. The graph of $x = b$ is a horizontal line.

5. The point (b, a) is a solution of both $y = a$ and $x = b$.

6. The equation of the line shown in the graph is $y = 3$.

7. $y = a$ is a function.

8. $x = b$ is a function.

Name ______________________________ Date __________

B.7 Puzzle Time

Which State Can You Wear?

Write the letter of each answer in the box containing the exercise number.

Describe the pattern in the input-output table.

1.

Input, x	−2	−1	0	1	2
Output, y	−1	0	1	2	3

2.

Input, x	−2	−1	0	1	2
Output, y	−3	−1	1	3	5

Write an equation for the function shown in the table.

3.

Input, x	−2	−1	0	1	2
Output, y	−4	−2	0	2	4

4.

Input, x	0	1	2	3	4
Output, y	4	2	0	−2	−4

5.

Input, x	−2	−1	0	1	2
Output, y	4	3	2	1	0

6.

Input, x	−2	−1	0	1	2
Output, y	−5	−3	−1	1	3

7.

Input, x	−2	−1	0	1	2
Output, y	0	1	2	3	4

8.

Input, x	−2	−1	0	1	2
Output, y	−8	−4	0	4	8

9.

Input, x	−2	−1	0	1	2
Output, y	0	−1	−2	−3	−4

Answers

M. The input values decrease by 1, and the output values increase by 1.

W. $y = 2x$

E. $y = 2x - 1$

Y. $y = -x - 2$

R. The input values increase by 1, and the output values increase by 1.

E. $y = x + 2$

N. $y = -x + 2$

J. $y = 4x$

S. The input values increase by 1, and the output values increase by 2.

E. $y = 4 - 2x$

A. The input values increase by 2, and the output values increase by 1.

5	7	3		8	6	1	2	4	9

Name____ Date__________

Appendix B

School-to-Work

For use after Section B.4

Bookkeeper

A bookkeeper maintains the financial records for a business. This includes recording daily expenses and incomes, tracking the balances in accounts, and preparing financial statements. Bookkeepers must be well-organized and careful: no one wants to be surprised at the amount of money in an account.

If you own a small business and don't want to hire a bookkeeper, you will need to learn simple bookkeeping yourself. Managing home finances, such as keeping a check registry and balancing your checkbook, are also considered bookkeeping.

Over the course of a morning, a salon received $45, $60, $35, $80, $20, and $75 for services, and spends $8 for coffee and $89 for a delivery of shampoo and conditioner.

1. Write these as positive or negative numbers, ordered from least to greatest. Income should be positive and expenses should be negative.

2. Which number is greatest?

3. Which number is least?

4. The salon started the day with $50. How much does it have at the end of the morning?

5. On another day, the salon takes in $891 for services, and spends $972. Which of these numbers has a greater magnitude? Did the salon take in more money or spend more?

A restaurant has $3000 in an account that pays suppliers. The table shows the amounts they pay out on Thursday.

Supplier	Payment
Abel's Farm	−$380
Seafood Market	−$760
Wholesale Baking	−$95
Merry Farm	−$480
Jo's Butcher	−$750

6. Which payment has the greatest magnitude? the smallest?

7. How much is left in the account at the end of the day?

8. Which costs more: the produce from *Merry Farm* or from *Abel's Farm*?

Name ______________________ Date __________

Appendix B Study Help

You can use an **information wheel** to organize information about a concept.

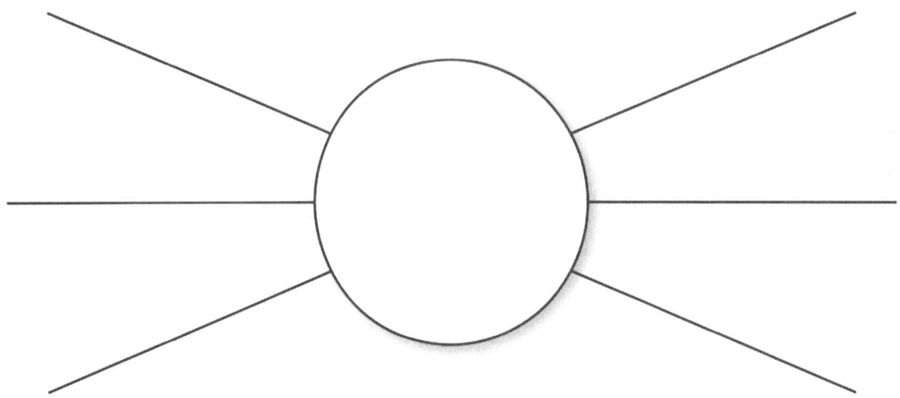

On Your Own

Make an information wheel to help you study these topics.

1. integers
2. adding on a number line
3. subtracting on a number line
4. fractions on the number line

After you complete this chapter, make information wheels for the following topics.

5. decimals on the number line
6. adding and subtracting fractions
7. adding and subtracting decimals
8. the coordinate plane
9. ordered pairs
10. quadrants
11. graphing

Name______________________________ Date__________

Appendix B Financial Literacy

For use after Section B.7

Running a Deficit

A **deficit** occurs when an organization or person spends more money than they take in. When this is expected and planned for, it can be part of normal business. For example, a garden supply company might earn a profit in the spring, summer, and fall, and run a deficit in winter. Or a manufacturing company with a big new order to fill might run at a deficit when they buy material to fill the order, and pay off the loan after filling the order.

A school event such as a fair or play expects to bring in money which will offset the cost of the production. In the exercises below, the treasurer for the Drama Club keeps track of the money needed to put on a play at the end of February.

September	Remaining balance from last year's play: \$600
October	Money from fundraiswers: \$1500
November	No income or expenses
December	Costume Fabric: –\$450, Lumber: –\$700, General Expenses: –\$225
January	Paint Etc: –\$375, Prop Rental: –\$200, General Expenses: –\$425
February	Light and Sound: –\$225, Donations from local businesses: \$525, General Expenses: –\$75, Promotion: –\$350

1. Create an input-output table using the months as the input and the total income or expense for the month as the output. Use numbers to represent the months. For example, September is 1, October is 2, and so on.

2. Plot the data in the table.

3. During what months do you have a positive cash flow? During what months do you have a negative cash flow?

4. Create another input-output table using the months as the inputs and the total balance in the account for that month, if all submitted bills were paid, as the output.

5. Plot the data in the table.

6. When does the balance in the account go below \$0?

7. Each year the school transfers some money from the general fund into the Drama Club account so that the bills can be paid. How much money is needed? When should it be transferred?

8. How much money does the Drama Club need to take in from the ticket sales to repay the general fund and leave at least \$500 in the Drama Club account for next year? At \$5 a ticket, how many tickets is this?

Name ____ Date __________

Appendix B Technology Connection

For use after Section B.5

Calculating with Negative Numbers

You can use a calculator to add and subtract positive and negative numbers. If you are not sure how your calculator enters or displays negative numbers, test a few simple problems entered in different ways.

EXAMPLE Find the sum $-3 + 1.8$ on a calculator with a change sign button.

SOLUTION

Enter 3 [+|–] to represent -3. Does your calculator display -3?

Next enter [+] 1.8 [=].

ANSWER -1.2

The negative sign may be to the far left of the display.

EXAMPLE Find the difference $-2.4 - 6$ on a calculator with a negative sign button.

SOLUTION

Use the [(–)] key to indicate the number you are about to enter is negative.

Enter [(–)] 2.4 [–] 6 [*ENTER*].

ANSWER -8.4

The negative sign is usually to the immediate left of the number.

Use a calculator to find the sum or difference.

1. $3 - 7.4$
2. $-8 + 13$
3. $14.5 - 107.3$
4. $-16.9 + 13.8$
5. $-0.78 - 0.42$
6. $8.1 - 16$
7. $0 - 0.18$
8. $-219 + 187$
9. $1 - 7 - 3$
10. $-6.2 + 1.3 - 5$
11. $20 - 18 - 11$
12. $-0.85 + 0.40 + 0.26$

Name________________________________ Date__________

Appendix B

Life Connections

For use after Section B.2

Football

In football, each team tries to move the ball into the other team's end zone to score. The team in possession of the football is the offense. They have 4 plays in which to try and retain possession of the ball while moving it forward at least 10 yards. If they succeed, a first down is called and they get another 4 plays to move another 10 yards. If they fail, the opposing team gets the ball.

The yardage on four plays is given. Yards advanced are positive, and yards lost are negative. Give the total yardage, and state whether the offense gets a first down.

1. 5 yards, –3 yards, 10 yards, –3 yards

2. 25 yards, –5 yards, –3 yards, –5 yards

3. 4 yards, 2 yards, 1 yard, 1 yard

4. –8 yards, 3 yards, 12 yards, –1 yard

5. 15 yards, 2 yards, –4 yards, 8 yards

6. 17 yards, 7 yards, 11 yards, 2 yards

7. A quarterback throws 2 completed passes, which gain 28 and 37 yards. He then throws an intercepted pass which loses 41 yards. Has his total passing yards *increased* or *decreased*? By how much?

A football field is not numbered in the same way as a number line. Instead, each team's territory starts at the zero yard line and goes to the 50 yard line in the middle of the field. State where the ball is on the field after each play.

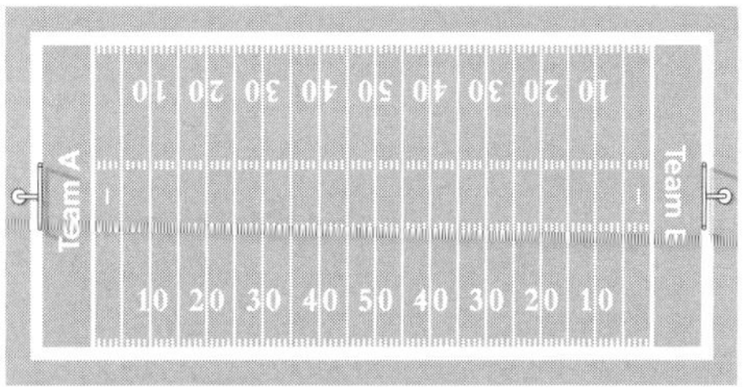

8. The ball is kicked at Team A's 30-yard line and caught by Team B's kick returner positioned 60 yards at the other end of the field.

9. Team B's kick returner runs the ball 21 yards forward before he is tackled.

10. On the next play, Team B's quarterback is tackled before he can throw, resulting in a loss of 8 yards.

11. On the next play, a successful pass from Team B earns them 39 yards.

12. On the next play, Team B's quarterback runs the ball 18 yards forward.

Name ______________________________ Date __________

Appendix B Cumulative Practice

Evaluate the expression.

1. $|-1|$

2. $-\frac{1}{2} + \frac{3}{4}$

3. $10 \div 2.5$

4. $\frac{7}{12} \bullet \frac{9}{14}$

5. $-8 - 11$

6. $6.28(2.5)$

7. $2\frac{2}{5} - 1\frac{3}{5}$

8. $|25|$

9. $6 - 9$

10. $-4.15 + 2.38$

11. $-9.3 - 1.7$

12. $\frac{15}{8} \div 6$

Evaluate the expression when $x = 6$, $y = -2$, and $z = -3$.

13. $\frac{x}{2} + z$

14. $y - x$

15. $x + y + z$

16. $|z|$

17. $z - y$

18. $|x + y|$

19. An elevator travels 40 feet up, 30 feet down, and 20 feet back up. Write and evaluate an expression that represents the total change in the height of the elevator.

20. A submarine is 250 feet below sea level. The submarine dives 100 feet deeper. Write and evaluate an expression that represents the position of the submarine after it finishes diving.

Which number is greater? Explain.

21. $-9, 4$

22. $-2.5, -2.05$

23. $-\frac{1}{6}, -\frac{1}{3}$

Plot the ordered pair in a coordinate plane. Describe the location of the point.

24. $P(-2, 0)$

25. $Q(3, -1)$

26. $R(0, 4)$

27. Make an input-output table for the function $y = 1 - x$. Use the inputs $-2, -1, 0, 1,$ and 2. Then draw the graph of the function.

28. A groundhog burrow has an elevation of -35 meters. Another groundhog burrow has an elevation of -32 meters. Which elevation is greater? How much greater?

Credits

9, 15, 21, 27, 47, 53, 59, 65, 71, 77, 83, 89, 106, 112, 118, 124, 130 Pellowski, Michael. *Wackiest Jokes in the World.* Sterling Publishing, March 1995; **148, 154** Walton, Rick. *Really, Really Bad School Jokes*. Candlewick, August 1995; **160, 166** Weitzman, Ilana, Eva Blank and Roxanne Green. *Jokelopedia: The Biggest, Best, Silliest, Dumbest, Joke Book Ever!*. Workman Publishing Company, October 2000; **172, 186, 192, 198, 204, 210** O'Donnell, Rosie. *Kids are Punny: Jokes Sent By Kids to The Rosie O'Donnell Show*. Warner Books, June 1997; **216** Hawthorn, Philip. The *Usborne Book of Silly Jokes*. Usborne Publishing, Ltd., 1991; **235, 241, 247, 253, 270, 276, 282, 288, 294, 300, 315, 321, 327, 333, 346, 352, 358, 364, 370, 388, 394, 400, 406, 412, 418, 424** O'Donnell, Rosie. *Kids are Punny: Jokes Sent By Kids to The Rosie O'Donnell Show*. Warner Books, June 1997

Answers

Chapter 1

1.1 Start Thinking!

For use before Activity 1.1

Sample answer: A word problem should be well written, concise, and include all the necessary details.

1.1 Warm Up

For use before Activity 1.1

1. 71 **2.** 114 **3.** 83

4. 90 **5.** 468 **6.** 50

1.1 Start Thinking!

For use before Lesson 1.1

Sample answer: You would map out the route, including stops along the way. You could find the distance and estimate the average speed to find how long it would take using the equation $d = rt$.

1.1 Warm Up

For use before Lesson 1.1

1. $2 \bullet 15$; \$30 **2.** $15 \div 5$; 3 minutes

3. $24 \div 4$; \$6 **4.** $200 - 70$; 130 miles

1.1 Practice A

1. 13 **2.** 6 **3.** 42 **4.** 5

5. 10 **6.** 9 **7.** 100 **8.** 6

9. 14 **10.** 6 **11.** 180 **12.** 10

13. 13 **14.** 17 **15.** 6

16. $3 \bullet 10 = 30$ should be done first. So, $30 - 6 = 24$.

17. \$26.00 **18.** 322 problems

19. a. \$60.00 **b.** Yes; 24 times \$5 is \$120.

20. a. 78 pages **b.** No; 2 times 52 is only 104 pages.

1.1 Practice B

1. 18 **2.** 24 **3.** 28

4. 6 **5.** 4 **6.** 42

7. 35 **8.** 4 **9.** 1

10. row 1: 40, 67, 103
row 2: 29, 68, 148
row 3: 65, 152, 324

11. 100 **12.** 0 **13.** 20

14. \$204.00 **15.** 28 cars

16. Your opponent won by one point, 59–58.

17. a. *Sample answer:* ten 4-legged and three 2-legged
b. Nine 2-legged and seven 4-legged
c. One 4-legged animal or two 2-legged animals

1.1 Enrichment and Extension

1. $2\frac{1}{2}$ **2.** 4 **3.** 36 **4.** 27

5. 25 **6.** 1 **7.** 40 **8.** 131

9. 2 **10.** 3 **11.** 59 **12.** 4

13.

First Number Cube (columns); Second Number Cube (rows)

	⚀	⚁	⚂	⚃	⚄	⚅
⚀	59	29	19	14	11	9
⚁	58	28	18	13	10	8
⚂	57	27	17	12	9	7
⚃	56	26	16	11	8	6
⚄	55	25	15	10	7	5
⚅	54	24	14	9	6	4

14. 9; it appears more often (3 times) in the table compared to the other values

1.1 Puzzle Time

THEY HAD SLEEPLESS KNIGHTS

1.2 Start Thinking!

For use before Activity 1.2

addition; *Sample answer:* heavier than (addition), shorter than (subtraction), twice (multiplication), split (division)

1.2 Warm Up

For use before Activity 1.2

1. 10 **2.** 6 **3.** 12

4. 8 **5.** 3 **6.** 4

Answers

1.2 Start Thinking!

For use before Lesson 1.2

Sample answer: It means that Susie lives 5 blocks closer to the school than Bill does. Bill walks 5 more blocks than Susie; to figure out how many blocks Susie walked, subtract 5 from the number of blocks Bill walked.

1.2 Warm Up

For use before Lesson 1.2

1. $12 + t$ **2.** $8h$ **3.** $7 - f$

4. $11 - w$ **5.** $n \div 2$ **6.** $2p$

1.2 Practice A

1. D **2.** E **3.** A

4. B **5.** F **6.** C

7. $4 + 6$ or $6 + 4$ **8.** $15 - 7$

9. $3x$ or $x \bullet 3$ **10.** $\frac{m}{4}$ **11.** $8p$ or $p \bullet 8$

12. $2x$ or $x \bullet 2$ **13.** $y - 10$ **14.** $k - 7$

15. $3 + c$ or $c + 3$ **16.** $11 - d$

17. Twelve less than a number means that you must subtract 12 from the number x; $x - 12$

18. 8 more than a number x

19. 20 decreased by the number n, or n less than 20

20. $5 - 2$

21. a. $3x + 2$ or $2 + 3x$

b. $3 \times 12 + 2$ or 38 years old

1.2 Practice B

1. D **2.** C **3.** A **4.** B

5. $8 + 5$ or $5 + 8$ **6.** $24 - 11$

7. $6y$ or $y \bullet 6$ **8.** $\frac{n}{7}$

9. $8 + e$ or $e + 8$ **10.** $2n + 6$

11. $x - 13$ **12.** $3m - 7$

13. $f + 3$ or $3 + f$ **14.** $25 - w$

15. $2 + 2 \bullet 6$ **16.** $16 - 3.20$

17. A; The next number is three more than the previous number.

18. a. 3 eggs **b.** $24x - 5$

c. 14 is the number of bags, 5 is the number of cookies in each bag, 2 is the number of cookies that are not needed to fill the bags, 3 is the number of batches, and 24 the number of cookies in each batch or recipe.

1.2 Enrichment and Extension

1. divided by; quotient of **2.** more than; difference of

3. less than; product of

Joke answer: "nice belt"

1.2 Puzzle Time

IT SNAPPED TO ATTENTION

1.3 Start Thinking!

For use before Activity 1.3

One sample method: Add $13 + 27 = 40$, $24 + 6 = 30$, and $19 + 11 = 30$. Then add $40 + 30 + 30 = 100$. Yes; addition is commutative. That is, the order in which you add does not matter.

1.3 Warm Up

For use before Activity 1.3

1. 22 **2.** 12 **3.** 15

4. 29 **5.** 22 **6.** 48

1.3 Start Thinking!

For use before Lesson 1.3

Sample answer: If you are finding the area of your bedroom that measures 10 feet by 12 feet, the area is $10(12) = 120$ square feet or $12(10) = 120$ square feet. The order in which you multiply doesn't affect the answer.

Sample answer: You are sitting between two friends at a movie theater. You talk to one friend, and then you talk to the other friend.

1.3 Warm Up

For use before Lesson 1.3

1. Commutative Property of Multiplication

2. Associative Property of Addition

3. Commutative Property of Addition

4. Associative Property of Multiplication

5. Commutative Property of Multiplication

6. Commutative Property of Addition

Answers

1.3 Practice A

1. C
2. F
3. E
4. A
5. B
6. D

7. $2 + (a + 8) = 2 + (8 + a)$ Commutative Property of Addition
   $= (2 + 8) + a$ Associative Property of Addition
   $= 10 + a$ Add 2 and 8.

8. $(2x) \bullet 5 = 5 \bullet (2x)$ Commutative Property of Multiplication
   $= (5 \bullet 2) \bullet x$ Associative Property of Multiplication
   $= 10x$ Multiply 5 and 2.

9. $7(4y) = (7 \bullet 4)y$ Associative Property of Multiplication
   $= 28y$ Multiply 7 and 4.

10. $(0 + c) + 12 = (c + 0) + 12$ Commutative Property of Addition
    $= c + (0 + 12)$ Commutative Property of Addition
    $= c + 12$ Addition Property of Zero

11. $(8e) \bullet 1 = 1 \bullet (8e)$ Commutative Property of Multiplication
    $= (1 \bullet 8) \bullet e$ Associative Property of Multiplication
    $= 8e$ Multiplication Property of One

12. $7 \bullet x \bullet 5 = 7 \bullet (x \bullet 5)$ Associative Property of Multiplication
    $= 7 \bullet (5 \bullet x)$ Commutative Property of Multiplication
    $= (7 \bullet 5) \bullet x$ Associative Property of Multiplication
    $= 35x$ Multiply 7 and 5.

13. $7 + a$
14. $5 \bullet 12$
15. $(6 + 9) + 7$
16. $(2 \bullet 5) \bullet 7$

17. $54 \bullet 1 = 54$; Any number times one equals itself.

18. First day: $14 + y$; Second day: $x + 6$;
    $14 + y = x + 6$
    $y + 14 = 6 + x$

1.3 Practice B

1. $16 + (f + 4) = 16 + (4 + f)$ Commutative Property of Addition
   $= (16 + 4) + f$ Associative Property of Addition
   $= 20 + f$ Add 16 and 4.

2. $(3y) \bullet 9 = 9 \bullet (3y)$ Commutative Property of Multiplication
   $= (9 \bullet 3)y$ Associative Property of Multiplication
   $= 27y$ Multiply 9 and 3.

3. $5(8y) = (5 \bullet 8)y$ Associative Property of Multiplication
   $= 40y$ Multiply 8 and 5.

4. $(0 + n) + 15 = (n + 0) + 15$ Commutative Property of Addition
   $= n + (0 + 15)$ Associative Property of Addition
   $= n + 15$ Addition Property of Zero

5. $(21 \bullet y) \bullet 1 = 1 \bullet (21 \bullet y)$ Commutative Property of Multiplication
   $= (1 \bullet 21)y$ Associative Property of Multiplication
   $= 21y$ Multiplication Property of One

6. $10 \bullet x \bullet 4 = 10 \bullet (x \bullet 4)$ Associative Property of Multiplication
   $= 10 \bullet (4 \bullet x)$ Commutative Property of Multiplication
   $= (10 \bullet 4) \bullet x$ Associative Property of Multiplication
   $40x$ Multiply 10 and 4.

7. $34 \bullet y \bullet 0 = 34 \bullet (y \bullet 0)$ Associative Property of Multiplication
   $= 34 \bullet 0$ Multiplication Property of Zero
   $= 0$ Multiplication Property of Zero

Answers

8. $35 + (p + 5) = 35 + (5 + p)$ Commutative Property of Addition

$= (35 + 5) + p$ Associative Property of Addition

$= 40 + p$ Add 35 and 5.

9. $11 + h$ **10.** $k \bullet 12$ **11.** $(21 + 9) + 8$

12. $(12 \bullet 5) \bullet 4$ **13.** $18 \bullet w$ **14.** $26 + c$

15. The change in grouping shows the Associative Property of Multiplication.

16. a. $21 + 19 + n$

b. $21 + 19 + n = (21 + 19) + n$ Associative Property of Addition

$= 40 + n$ Add 21 and 19.

17. $37 + t + 29 = 66 + t$

1.3 Enrichment and Extension

	Number of pairs	Sum of each pair	Sum
1. 1 through 9	9	10	45
2. 1 through 20	20	21	210
3. 1 through 50	50	51	1275
4. 1 through 199	199	200	19,900
5. 1 through 1000	1000	1001	500,500

6. $\frac{n(n + 1)}{2}$

1.3 Puzzle Time

IT WAS A COOKBOOK

1.4 Start Thinking!

For use before Activity 1.4

Yes; the product for both expressions is 640.

1.4 Warm Up

For use before Activity 1.4

1. 91 **2.** 207 **3.** 56

4. 205 **5.** 222 **6.** 126

1.4 Start Thinking!

For use before Lesson 1.4

Sample answer: If you are preparing goody bags, you distribute the candy evenly among the bags.

1.4 Warm Up

For use before Lesson 1.4

1. 116 **2.** 126 **3.** 364

4. 114 **5.** 240 **6.** 536

1.4 Practice A

1. 6, 7, 8 **2.** 7, 8, 7, 6 **3.** 6, 12, 12 **4.** 115

5. 150 **6.** 244 **7.** 336 **8.** 486

9. 413 **10.** $4x + 24$ **11.** $8c - 40$

12. $7y + 56$ **13.** $9e - 36$ **14.** $24 + 6n$

15. $21 + 7x + 28 = 7x + 49$

16. $9w + 54 + 4 = 9w + 58$

17. $15 + 5m - 7 = 5m + 8$

18. 5 must be multiplied by both x and 9; $5x + 45$

19. C; You need to add the two activities and then multiply the sum by 5.

20. $5(r + 3) = 5r + 15$

1.4 Practice B

1. 144 **2.** 243 **3.** 623 **4.** 624

5. 360 **6.** 396 **7.** $8a + 48$ **8.** $7p - 35$

9. $90 + 10x$ **10.** $12f - 60$

11. $12 + 6a + 54 = 6a + 66$

12. $4 + 3x + 15 = 3x + 19$

13. $8t + 40 + 15 = 8t + 55$

14. $4y + 44 - 10 = 4y + 34$

15. 6 must be multiplied by both x and 7. $6x + 42$

16. $7(c + 12) = 7c + 84$ **17.** $4(y + 3) = 4y + 12$

18. $8(x + 5)$ **19.** $9(4 + x)$ **20.** $2(10 - p)$

21. $6(1 + 3y + 4)$

22. $8 \bullet x + 10 + 2x + 4 \bullet (x + 8) - 6$

Answers

1.4 Enrichment and Extension

1.

$50(128 + 192 + 256)$
$= 50(128) + 50(192) + 50(256)$
$= 50(100 + 20 + 8) + 50(200 - 8) + 50(250 + 6)$
$= 5000 + 1000 + 400 + 10{,}000 - 400 + 12{,}500 + 300$
$= 28{,}800$

The company receives $28,800 for the in-ear model.

2. $100(96 + 192 + 160)$
$= 100(96) + 100(192) + 100(160)$
$= 100(100 - 4) + 100(200 - 8) + 100(150 + 10)$
$= 10{,}000 - 400 + 20{,}000 - 800 + 15{,}000 + 1000$
$= 44{,}800$

The company receives $44,800 for the over-the-ear model.

3. The company receives a total of $\$28{,}800 + \$44{,}800 = \$73{,}600$ for both models.

4. The price of the over-the-ear model is twice the price of the in-ear model. So, if each store orders twice as many in-ear models, the price they will pay for each model will be equal.

5. Commutative, Multiplication; Distributive; Commutative, Multiplication

1.4 Puzzle Time

ON ACCOUNT IT WAS KNOTTY

1.5 Start Thinking!

For use before Activity 1.5

Answers may vary. Check students' drawings.

1.5 Warm Up

For use before Activity 1.5

1. trapezoid **2.** rectangle **3.** square

4. right triangle **5.** parallelogram **6.** trapezoid

1.5 Start Thinking!

For use before Lesson 1.5

Sample answer: conversion formulas (length, temperature, capacity, mass, currency), $d = rt$

1.5 Warm Up

For use before Lesson 1.5

1. 40 in.^{2} **2.** 260 in.^{2} **3.** 50 cm^2

4. 36 ft^2 **5.** 21 mm^2 **6.** 54 m^2

1.5 Practice A

1. 206 **2.** 108 in.^{2}

3. 160 cubic units **4.** 48 in.^{2}

5. 42 in.^{2} **6.** 49.5 mm^2 **7.** 225 ft^2

8. 308 in.^{2} **9.** 200 cm^2 **10.** 159 **11.** 4.5

12. Divide the area by the length of the rectangle.

1.5 Practice B

1. 256 in.^{2} **2.** 102 cm^2 **3.** 768 ft^2 **4.** 273 mm^2

5. 322 cm^2 **6.** 294.5 cm^2 **7.** $8x - 25$

8. $7(w + 3) - w^2 = 7w + 21 - w^2$

9. $12(2y - 3) - y^2 = 24y - 36 - y^2$

10. 208

11. 6 by 20, 4 by 30, 8 by 15, 5 by 24, 3 by 40, 2 by 60, 1 by 120

1.5 Enrichment and Extension

1. $E = \dfrac{mv^2}{2}$

2. The car likely has the most kinetic energy because it has a much larger mass than the baseball or cyclist.

3. The 10-kg dog has 320 Joules of kinetic energy. The 8-kg dog has 400 Joules of kinetic energy. So, the 8-kg dog has more kinetic energy.

4. 290,400 Joules

5. No; at 11 meters per second, the car has 72,600 Joules of kinetic energy. So, the car's kinetic energy decreased by more than one half.

1.5 Puzzle Time

A SNOW BANK

School-to-Work

1. 195 ft^2; *Sample answer:* Break the cabinetry into two rectangles and a trapezoid. The area formulas that will be used are $A = bh$ and $A = h(b + B) \div 2$.

Answers

2. 13 hours
3. *Sample answer:* Answer should include research done to determine a competitive labor cost.
4. *Sample answer:* The cost of the tile and of any border needed at the walls.
5. *Sample answer:* Answer should include the labor cost, tile cost, and any other costs considered. The detailed explanation should show these costs broken up so the customer can see what they are charged for each part of the job.
6. \$390; Because the cost of the tile went up \$2 per square foot, the extra cost for the tile is 2(\$195).

Study Help

Available at *BigIdeasMath.com.*

Financial Literacy

1. *Sample answer:* The presentation should include the price you will charge, the location, supplies needed, income and expenses, and reasons why. The presentation should demonstrate the student's understanding of the chapter.

Cumulative Practice

1. $12 + 4c$ **2.** 0 **3.** $m + 17$

4. $6w$ **5.** 132 **6.** $3.2x$

7. $p + 20$ **8.** $5y + 50$ **9.** 9

10. 16 **11.** $22h$ **12.** $26 + 2a$

13. $9k + 117$ **14.** $g + 8.4$ **15.** $12v$

16. \$90 **17.** $25 + v$ **18.** 9

19. 16 **20.** 13 **21.** 34

22. 5 **23.** 2 **24.** 277.7

25. 20 **26.** 52 pounds **27.** $11 + n$

28. $7 - y$ **29.** $3x$ **30.** $20 \div 4$

31. $k - 13$ **32.** $c + 5$

33. $p \div 8$ **34.** $25g$

35. $a + 5$, where a represents your age

36. $3m - 6$; where m is your distance in miles from school; 6 miles

Chapter 2

2.1 Start Thinking!

For use before Activity 2.1

Sample answer: You could use number sense to estimate the number of red gumballs and the total number of gumballs. Then create a fraction with the numerator as the number of red gumballs and the denominator as the total number of gumballs.

2.1 Warm Up

For use before Activity 2.1

1. $\frac{3}{5}$ **2.** $\frac{5}{6}$ **3.** $\frac{1}{2}$

4. $\frac{2}{3}$ **5.** $\frac{5}{7}$ **6.** $\frac{1}{4}$

2.1 Start Thinking!

For use before Lesson 2.1

Sample answer:

Estimation is often quicker and easier than doing the full computation.

An estimate can often be done without paper or pencil.

Estimating is a good way to check the reasonableness of an exact answer.

Often when dealing with real-life problems, an estimate is all you need.

2.1 Warm Up

For use before Lesson 2.1

1. 1 **2.** $\frac{1}{2}$ **3.** $\frac{1}{4}$

4. $\frac{1}{2}$ **5.** $\frac{1}{4}$ **6.** 2

2.1 Practice A

1. Reasonable **2.** Reasonable

3. Not reasonable; A better estimate is 1.

4. Reasonable

5. Not reasonable; A better estimate is 35.

6. Reasonable

7. about $\frac{1}{2}$ **8.** about $\frac{1}{2}$

9. about 1 **10.** about $\frac{1}{4}$

11. $30 \times 3 = 90$ **12.** $48 \div 6 = 8$

Answers

13. $40 \times 5 = 200$

14. $140 \div 7 = 20$

15. $149 \times 1 = 149$

16. $72 \div 8 = 9$

17. $2\frac{2}{3}$ should be rounded to 3. $90 \div 3 = 30$

18. 20 sheets

19. You should over estimate to make sure everyone will have enough to drink.

20. $111 \div 5\frac{2}{3} \approx 120 \div 6 = 20$

2.1 Practice B

1. about 1

2. about $\frac{1}{2}$

3. $4 \times 10 = 40$

4. $40 \div 4 = 10$

5. $12 \times 4 = 48$

6. $22 \div 2 = 11$

7. $15 \times 2 = 30$

8. $68 \div 2 = 34$

9. $500 \div 5 = 100$

10. $16 \times 2 = 32$

11. $55 \div 5 = 11$

12. $60 \times 6 = 360$

13. $180 \div 9 = 20$

14. $100 \times \frac{1}{2} = 50$

15. $360 \div 12 = 30$

16. 80

17. 10

18. 6

19. Incorrect rounding was done.
$\frac{4}{11} \times \frac{11}{12} \approx \frac{1}{2} \times 1 = \frac{1}{2}$

20. 126 in.^{2}; Underestimate, by rounding $35\frac{5}{8}$ up to 36 and by rounding $7\frac{1}{5}$ down to 7.

21. Yes; An overestimate for the area of the deck is 280 square feet. Two cans of waterproofing sealer cover 300 square feet. So, you have enough waterproofing sealer.

2.1 Enrichment and Extension

1.

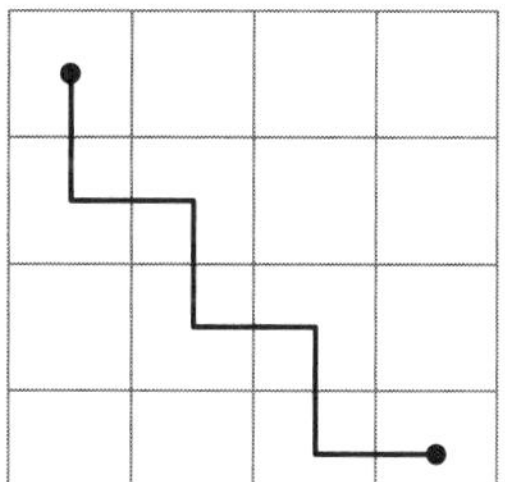

2.

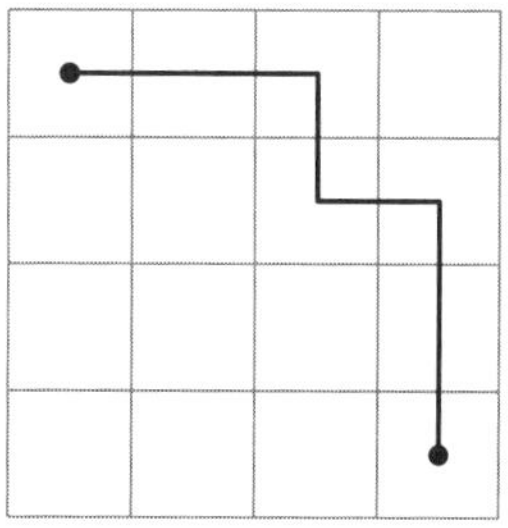

2.1 Puzzle Time

CENTRAL HARE CONDITIONING

2.2 Start Thinking!

For use before Activity 2.2

Sample answer: You can estimate that each person will eat $\frac{1}{4}$ of a pizza, and multiply by the number of people to figure out how many pizzas you need.

2.2 Warm Up

For use before Activity 2.2

1. $\frac{3}{5}$

2. $\frac{1}{2}$

3. $\frac{1}{2}$

4. 2

5. 3

6. $\frac{10}{3}$, or $3\frac{1}{3}$

2.2 Start Thinking!

For use before Lesson 2.2

Sample answer: The product of a whole number and a fraction is less than the whole number when the value of the fraction is less than 1. The product is greater than the whole number when the value of the fraction is greater than 1.

2.2 Warm Up

For use before Lesson 2.2

1. $\frac{9}{5}$, or $1\frac{4}{5}$

2. 6

3. $\frac{18}{7}$, or $2\frac{4}{7}$

4. $\frac{3}{4}$

5. 18

6. $\frac{26}{3}$, or $8\frac{2}{3}$

2.2 Practice A

1. $\frac{7}{8}$

2. $\frac{4}{5}$

3. $5\frac{1}{3}$

4. 8

5. $7\frac{1}{2}$

6. $4\frac{1}{2}$

7. 20

8. 28

9. $5\frac{1}{3}$

10. $2\frac{2}{3}$

11. 10

12. 25

Answers

13. The 8 should only be multiplied by the numerator, 2; $8 \times \frac{2}{5} = \frac{8 \times 2}{5} = \frac{16}{5} = 3\frac{1}{5}$

14. about 5 **15.** about 7 **16.** about 20

17. about 15 **18.** 12 t-shirts **19.** 24 months

20. Your friend spent $1 more.

21. 300 people

2.2 Practice B

1. $\frac{4}{5}$ **2.** $1\frac{1}{2}$ **3.** 9 **4.** 20

5. 15 **6.** $7\frac{1}{2}$ **7.** 28 **8.** 26

9. $22\frac{1}{2}$

10. Five was divided from the numerator, but 2 was divided from the denominator when simplifying.

$5 \times \frac{1}{10} = \frac{5}{10} = \frac{1}{2}$

11. about 5 **12.** about 16 **13.** about 21

14. 8 **15.** $6\frac{1}{4}$ **16.** 15 **17.** $1\frac{1}{2}$

18. 960 ft^2

19. 14 packages of white streamers

2.2 Enrichment and Extension

1. Alabama: 4320 km^2; Florida: 30,600 km^2; Georgia: 3850 km^2

2. Alabama: 2484 km^2; Florida: 12,240 km^2; Georgia: 2618 km^2

3. Georgia **4.** Alabama **5.** Georgia **6.** Florida

2.2 Puzzle Time

TO ROOT FOR THE HOME TEAM

2.3 Start Thinking!

For use before Activity 2.3

$\frac{1}{6}$

2.3 Warm Up

For use before Activity 2.3

1. $\frac{1}{8}$ **2.** $\frac{1}{4}$ **3.** $\frac{2}{3}$

4. $\frac{4}{9}$ **5.** $\frac{8}{19}$ **6.** $\frac{6}{7}$

2.3 Start Thinking!

For use before Lesson 2.3

$\frac{3}{4}$ and $\frac{5}{7}$; $\frac{15}{28}$; three-fourths of the height is shaded, five-seventh of the width is shaded, and 15 out of 28 squares are shaded.

2.3 Warm Up

For use before Lesson 2.3

1. $\frac{2}{15}$ **2.** $\frac{3}{10}$ **3.** $\frac{2}{9}$

4. $\frac{1}{24}$ **5.** $\frac{8}{27}$ **6.** $\frac{2}{11}$

2.3 Practice A

1. $\frac{3}{20}$ **2.** $\frac{4}{21}$ **3.** $\frac{14}{33}$ **4.** $\frac{5}{7}$

5. $\frac{1}{4}$ **6.** $\frac{4}{25}$ **7.** $\frac{8}{39}$ **8.** $10\frac{1}{2}$

9. $1\frac{2}{3}$ **10.** $\frac{16}{81}$ **11.** $\frac{1}{6}$ **12.** $1\frac{31}{33}$

13. The products of the numerators and denominators were added and not multiplied.

$\frac{3}{4} \times \frac{4}{9} = \frac{\overset{1}{\cancel{3}} \times \overset{1}{\cancel{4}}}{\underset{1}{\cancel{4}} \times \underset{3}{\cancel{9}}} = \frac{1}{3}$

14. $\frac{1}{5}$ **15.** $\frac{2}{3}$ **16.** $\frac{1}{3}$

17. $\frac{8}{15}$ **18.** $\frac{1}{3}$ hour **19.** $\frac{1}{15}$

20. a. $\frac{1}{10}$ of the class **b.** 3 students

2.3 Practice B

1. $\frac{5}{54}$ **2.** $\frac{7}{20}$ **3.** $\frac{4}{21}$ **4.** $\frac{11}{15}$

Answers

5. $\frac{4}{7}$ **6.** $\frac{1}{10}$ **7.** 15 **8.** $\frac{1}{9}$

9. 6

10. The denominators should have been multiplied.

$$\frac{5}{6} \times \frac{5}{6} = \frac{5 \times 5}{6 \times 6} = \frac{25}{36}$$

11. $\frac{1}{2}$ **12.** $\frac{3}{8}$ **13.** $\frac{3}{5}$ **14.** $1\frac{13}{40}$

15. $\frac{1}{10}$ **16.** $\frac{8}{343}$ **17.** $\frac{9}{49}$ **18.** 4 CDs

19. a. Find the area of the garden and the fraction of the area that is the rose section and multiply. Or, find the dimensions of the rose section and multiply.

b. 144 ft^2

2.3 Enrichment and Extension

1. Each number is $\frac{1}{8}$ the previous number; $\frac{1}{64}, \frac{1}{512}$

2. Each number is $\frac{1}{2}$ the previous number; $\frac{1}{32}, \frac{1}{64}$

3. Each number is $\frac{1}{4}$ the previous number; $\frac{3}{64}, \frac{1}{256}$

4. Each number is $\frac{2}{3}$ the previous number; $\frac{32}{27}, \frac{64}{81}$

5. Each number is $\frac{3}{2}$ the previous number; $\frac{27}{8}, \frac{81}{16}$

6. Each number is $\frac{5}{2}$ the previous number; $\frac{125}{16}, \frac{625}{32}$

7. Each number is $\frac{4}{3}$ the previous number; $\frac{256}{3}, \frac{1024}{9}$

8. Each number is $\frac{5}{4}$ the previous number; $\frac{625}{8}, \frac{3125}{32}$

9. a. pattern in Exercise 2 **b.** 32, 16, 8, 4, 2, 1

10. a. $1000 \times \frac{11}{10} \times \frac{11}{10}$ **b.** $1210

2.3 Puzzle Time

HE ALREADY HAD DRUMSTICKS

2.4 Start Thinking!

For use before Activity 2.4

Sample answer: A numerator of 10 and denominator of 1 form the fraction with the greatest value. A numerator of 1 and denominator of 10 form the fraction with the least value. In order to maximize the value, the numerator should be as large as possible and the denominator should be as small as possible. In order to minimize the value, the numerator should be as small as possible and the denominator should be as large as possible.

2.4 Warm Up

For use before Activity 2.4

1. $\frac{39}{10}$ **2.** $\frac{17}{7}$ **3.** $\frac{15}{2}$

4. $\frac{95}{8}$ **5.** $\frac{42}{5}$ **6.** $\frac{94}{15}$

2.4 Start Thinking!

For use before Lesson 2.4

Sample answer: You run around a $\frac{1}{4}$-mile track $3\frac{1}{2}$ times. How far have you run?

2.4 Warm Up

For use before Lesson 2.4

1. $\frac{21}{10}$, or $2\frac{1}{10}$ **2.** $\frac{4}{3}$, or $1\frac{1}{3}$ **3.** $\frac{11}{20}$

4. $\frac{25}{12}$, or $2\frac{1}{12}$ **5.** $\frac{9}{28}$ **6.** $\frac{17}{4}$, or $4\frac{1}{4}$

2.4 Practice A

1. $\frac{3}{5}$ **2.** 4 **3.** $3\frac{1}{2}$ **4.** 80

5. $8\frac{1}{4}$ **6.** 24 **7.** $1\frac{2}{3}$ **8.** $43\frac{1}{2}$

9. 1 **10.** $9\frac{55}{81}$ **11.** $21\frac{2}{3}$ **12.** $6\frac{2}{3}$

13. The mixed numbers must be changed to improper fractions before multiplying.

$$3\frac{7}{8} \times 6\frac{2}{5} = \frac{31}{8} \times \frac{32}{5} = \frac{31 \times \cancel{32}^{4}}{\cancel{8}_{1} \times 5} = \frac{124}{5} = 24\frac{4}{5}$$

14. $8\frac{4}{5}$ **15.** $27\frac{1}{2}$ **16.** $6\frac{7}{18}$ **17.** $14\frac{3}{8}$

Answers

18. $192\frac{1}{2}$ ft

19. $6\frac{1}{2}$ ft

20. $26\frac{2}{3}$ cases

21. $584\frac{3}{8}$ in.^{2}

2.4 Practice B

1. $\frac{11}{18}$

2. $\frac{3}{5}$

3. 1

4. $1\frac{1}{6}$

5. $1\frac{2}{3}$

6. $5\frac{3}{5}$

7. 232

8. 13

9. $37\frac{1}{2}$

10. $6\frac{1}{8} = \frac{49}{8}$, not $6 \times \frac{1}{8}$.

$$6\frac{1}{8} \times 8 = \frac{49}{8} \times \frac{8}{1} = \frac{49 \times \cancel{8}^{1}}{\cancel{8}_{1} \times 1} = 49$$

11. 4

12. $4\frac{2}{3}$

13. 8

14. $32\frac{4}{9}$

15. 9

16. $82\frac{1}{2}$

17. $68\frac{4}{7}$

18. $4\frac{1}{8}$ in.

19. $3\frac{6}{7}$ in.^{2}; *Sample answer:* Find the area of the large rectangle and subtract the small unshaded rectangle from it.

2.4 Enrichment and Extension

1. 85 ft^2

2. $87\frac{1}{2}$ ft^2

3. $93\frac{3}{4}$ ft^2

4. $81\frac{1}{8}$ ft^2

5. $87\frac{4}{9}$ ft^2

6. $99\frac{77}{100}$ ft^2

7. The design with the least shaded area requires the least amount of paint. The design in Exercise 4 has the least area, so you choose this design.

2.4 Puzzle Time

IT HAD TOO MANY DROP CALLS

2.5 Start Thinking!

For use before Activity 2.5

Sample answer: You can find the number of servings left in the box. The quotient of the number of cups left and the serving size is the number of servings left.

2.5 Warm Up

For use before Activity 2.5

1. $\frac{1}{20}$

2. $\frac{5}{28}$

3. $\frac{1}{24}$

4. $\frac{3}{4}$

5. 10

6. $\frac{12}{5}$, or $2\frac{2}{5}$

2.5 Start Thinking!

For use before Lesson 2.5

Sample answer: Think of a reciprocal as a person standing on his head.

2.5 Warm Up

For use before Lesson 2.5

1. $\frac{16}{5}$, or $3\frac{1}{5}$

2. $\frac{23}{7}$, or $3\frac{2}{7}$

3. $\frac{8}{3}$, or $2\frac{2}{3}$

4. $\frac{64}{3}$, or $21\frac{1}{3}$

5. $\frac{1}{12}$

6. $\frac{12}{5}$, or $2\frac{2}{5}$

2.5 Practice A

1. $\frac{8}{5}$

2. $\frac{1}{6}$

3. 3

4. $\frac{4}{7}$

5. $\frac{1}{2}$

6. $\frac{1}{8}$

7. $4\frac{1}{2}$

8. 35

9. $\frac{2}{3}$

10. $\frac{1}{27}$

11. $\frac{5}{9}$

12. $1\frac{7}{20}$

13. $\frac{1}{24}$

14. The reciprocal of 6 was not written.

$$\frac{3}{4} \div 6 = \frac{3}{4} \times \frac{1}{6} = \frac{3}{24} = \frac{1}{8}$$

15. $\frac{2}{3}$

16. 9

17. $\frac{2}{27}$

18. $\frac{1}{9}$

19. yes

20. yes

21. No, $\frac{2}{3} = \frac{12}{18}$. The reciprocal of $\frac{2}{3}$ and $\frac{12}{18}$ is $\frac{3}{2}$.

22. No. The reciprocal of $\frac{4}{7}$ is $\frac{7}{4}$. The reciprocal of $\frac{7}{8}$ is $\frac{8}{7}$.

23. $\frac{1}{6}$

24. 8 chairs

Answers

25. No, the reciprocal of $\frac{3}{4}$ is $\frac{4}{3}$. For any fraction with a numerator of 1, the reciprocal is a whole number.

2.5 Practice B

1. $\frac{12}{7}$ **2.** 10 **3.** $\frac{1}{15}$ **4.** $\frac{5}{21}$

5. $\frac{2}{5}$ **6.** $7\frac{1}{2}$ **7.** 3 **8.** $\frac{1}{45}$

9. $\frac{2}{3}$ **10.** $\frac{2}{75}$ **11.** $1\frac{2}{7}$ **12.** $31\frac{1}{2}$

13. $1\frac{1}{15}$

14. You do not find the reciprocal of the first fraction.
$\frac{3}{5} \div \frac{9}{10} = \frac{3}{5} \bullet \frac{10}{9} = \frac{2}{3}$

15. >; When you divide a number by a fraction less than 1, the quotient is larger than the number.

16. >; When you divide a number by a fraction less than 1, the quotient is larger than the number.

17. >; When you divide a fraction by a fraction less than 1, the quotient is larger than the fraction.

18. $\frac{1}{75}$ **19.** $10\frac{7}{9}$ **20.** $\frac{11}{32}$

21. 16 times larger

22. a. 900 pictures **b.** 225 pictures

2.5 Enrichment and Extension

1. $\frac{\boxed{1}}{2} \div \frac{3}{8} = \frac{\boxed{4}}{3}$ **2.** $\frac{\boxed{9}}{4} \div 2 = \frac{\boxed{9}}{8}$

3. $\frac{\boxed{3}}{2} \div \frac{1}{4} = \boxed{6}$ **4.** $\frac{5}{14} \div \frac{2}{\boxed{7}} = \frac{\boxed{5}}{4}$

5. $\frac{1}{\boxed{8}} \div \frac{3}{8} = \frac{1}{\boxed{3}}$ **6.** $\frac{7}{2} \div \boxed{7} = \frac{1}{\boxed{2}}$

7. $\frac{5}{9} \div \frac{2}{\boxed{3}} = \frac{\boxed{5}}{6}$ **8.** $\frac{1}{\boxed{4}} \div \frac{3}{2} = \frac{1}{\boxed{6}}$

9. 49-65-32-56

2.5 Puzzle Time

HE HAD A LOT OF TIME ON HIS HANDS

2.6 Start Thinking!

For use before Activity 2.6

No; *Sample answer:* $1\frac{4}{3}$ is not the reciprocal of $1\frac{3}{4}$. Amber forgot to convert the mixed number $1\frac{3}{4}$ to the improper fraction $\frac{7}{4}$ before attempting to find the reciprocal.

2.6 Warm Up

For use before Activity 2.6

1. $2\frac{2}{5}$ **2.** $5\frac{1}{4}$ **3.** $1\frac{7}{9}$

4. $4\frac{1}{2}$ **5.** $26\frac{2}{3}$ **6.** $6\frac{1}{9}$

2.6 Start Thinking!

For use before Lesson 2.6

Sample answer: You can divide the length of the string by the diameter of a bead to find the number of beads.

2.6 Warm Up

For use before Lesson 2.6

1. $\frac{14}{5}$, or $2\frac{4}{5}$ **2.** $\frac{35}{4}$, or $8\frac{3}{4}$ **3.** 5

4. $\frac{64}{27}$, or $2\frac{10}{27}$ **5.** $\frac{85}{126}$ **6.** $\frac{25}{6}$, or $4\frac{1}{6}$

2.6 Practice A

1. 10 **2.** $6\frac{1}{2}$ **3.** 8 **4.** $3\frac{3}{4}$

5. $1\frac{2}{5}$ **6.** $\frac{2}{9}$ **7.** $\frac{4}{57}$ **8.** 4

9. $\frac{4}{5}$ **10.** $\frac{5}{7}$ **11.** $4\frac{2}{7}$ **12.** $\frac{14}{15}$

13. The error was finding the reciprocal of both fractions, not just the divisor.

$8 \div 2\frac{3}{4} = 8 \div \frac{11}{4} = 8 \times \frac{4}{11} = \frac{32}{11} = 2\frac{10}{11}$

14. $\frac{9}{13}$ **15.** $1\frac{1}{3}$ **16.** $\frac{4}{5}$ **17.** $\frac{3}{4}$

18. 12 pieces **19.** $1\frac{4}{7}$ times **20.** 14 plots

21. Six loaves can be made and there is 4 cups of flour left over.

Answers

2.6 Practice B

1. $1\frac{5}{9}$ 2. $4\frac{10}{13}$ 3. $1\frac{11}{16}$ 4. $4\frac{1}{2}$

5. $1\frac{1}{6}$ 6. $9\frac{1}{3}$ 7. $\frac{1}{11}$ 8. $1\frac{11}{15}$

9. $4\frac{2}{7}$ 10. $\frac{1}{12}$ 11. $3\frac{1}{3}$ 12. $1\frac{2}{3}$

13. The order of the problem was incorrectly switched. The problem should be written as

$$8\frac{1}{3} \div 15 = \frac{25}{3} \times \frac{1}{15} = \frac{25}{45} = \frac{5}{9}$$

14. 2 15. $3\frac{6}{25}$ 16. $1\frac{43}{125}$ 17. 10

18. $45\frac{2}{5}$ 19. $12\frac{1}{3}$ 20. $2\frac{1}{4}$

21. 26 wires with $2\frac{1}{2}$ inches of wire left over.

22. $101\frac{1}{4}$ pounds; 6 bags; *Sample answer:* Divide 27,000 by 5000 and then multiply by $18\frac{3}{4}$ to find the number of pounds needed. Because the number of pounds needed is in $5\frac{2}{5}$ bags, you need 6 bags.

2.6 Enrichment and Extension

The values of the expressions from the top to the bottom of the board are:

$11\frac{7}{16}, 2\frac{9}{20}, 4\frac{19}{20}, 7\frac{23}{28}, \frac{175}{186}, \frac{19}{26}, 4\frac{2}{3}, 14\frac{5}{38}, 9\frac{5}{7}$

2.6 Puzzle Time

EXTRA HOLD BUG SPRAY

2.7 Start Thinking!

For use before Activity 2.7

Sample answer: It may be preferable to use a fraction in the following situations:

- If the numerator or denominator of the equivalent fraction will cancel easily $\left(\text{for example, } 0.125 \times 16 \text{ versus } \frac{1}{8} \times 16\right)$
- If the decimal is not a terminating decimal
- When the context of the situation commonly uses fractions instead of decimals, for example, customary measurements $\left(\frac{3}{16}\text{ inch}\right)$, time $\left(\frac{1}{2}\text{ hour}\right)$, cooking $\left(\frac{1}{3}\text{ cup}\right)$, distances $\left(\frac{1}{4}\text{ mile}\right)$, and sewing $\left(2\frac{1}{2}\text{ yards of fabric}\right)$

2.7 Warm Up

For use before Activity 2.7

1. $\frac{9}{25}$ 2. $\frac{1}{20}$ 3. $\frac{33}{200}$

4. $2\frac{21}{50}$ 5. $1\frac{9}{50}$ 6. $4\frac{111}{200}$

2.7 Start Thinking!

For use before Lesson 2.7

$\frac{4}{5}$; *Sample answer:* It is easier to add several decimals than fractions with unlike denominators. All you have to do is line up the decimal points and add instead of having to convert to like denominators, add, and simplify the result.

2.7 Warm Up

For use before Lesson 2.7

1. $\frac{1}{5}$ 2. $\frac{3}{10}$ 3. $\frac{19}{100}$

4. $\frac{1}{4}$ 5. $\frac{13}{20}$ 6. $\frac{31}{200}$

2.7 Practice A

1. $\frac{7}{10}$ 2. $\frac{2}{5}$ 3. $\frac{1}{5}$ 4. $\frac{1}{2}$

5. $\frac{4}{25}$ 6. $\frac{11}{25}$ 7. $\frac{7}{20}$ 8. $\frac{59}{100}$

Answers

9. $\frac{37}{50}$ 10. $\frac{27}{125}$ 11. $\frac{24}{25}$ 12. $\frac{78}{125}$

13. $7\frac{13}{20}$ 14. $4\frac{2}{25}$ 15. $2\frac{21}{25}$ 16. $8\frac{69}{200}$

17. Four is in the hundredths place, so 100 should be in the denominator.

$$0.04 = \frac{4}{100} = \frac{1}{25}$$

18. $\frac{9}{4}$ 19. $\frac{137}{25}$ 20. $\frac{29}{8}$

21. $0.10; $\frac{1}{10}$ 22. $0.06; $\frac{3}{50}$ 23. $0.80; $\frac{4}{5}$

24. $1.20; $1\frac{1}{5}$ 25. $\frac{8}{25}$ in. 26. $9\frac{4}{5}$ seconds

27. $3\frac{1}{4}$ 28. $\frac{9}{20}$ mi

29. $\frac{9}{25}$; 27 students

2.7 Practice B

1. $\frac{3}{10}$ 2. $\frac{4}{5}$ 3. $\frac{1}{4}$ 4. $\frac{3}{20}$

5. $\frac{9}{25}$ 6. $\frac{12}{25}$ 7. $\frac{18}{25}$ 8. $\frac{63}{100}$

9. $\frac{17}{100}$ 10. $\frac{9}{40}$ 11. $\frac{43}{50}$ 12. $\frac{54}{125}$

13. $5\frac{29}{100}$ 14. $6\frac{63}{125}$ 15. $2\frac{5}{8}$ 16. $7\frac{7}{125}$

17. Two is in the hundredths place, so 100 should be in the denominator.

$$0.42 = \frac{42}{100} = \frac{21}{50}$$

18. $\frac{79}{25}$ 19. $\frac{35}{8}$ 20. $\frac{2261}{250}$

21. $0.35, $\frac{7}{20}$ 22. $0.12, $\frac{3}{25}$ 23. $2.30, $2\frac{3}{10}$

24. $3.08, $3\frac{2}{25}$ 25. 0.55, $\frac{11}{20}$ 26. 0.72, $\frac{18}{25}$

27. 0.12, $\frac{3}{25}$ 28. 132 cards

29. a. $\frac{39}{125}$ b. 156 hits

2.7 Enrichment and Extension

1. $\frac{7}{125}$ 2. $\frac{3}{10}$

3. $\frac{33}{170}$ 4. $\frac{50}{7}$, or $7\frac{1}{7}$

5. $\frac{400}{31}$, or $12\frac{28}{31}$ 6. $\frac{71}{1300}$

7. $\frac{165}{119}$, or $1\frac{46}{119}$ 8. $\frac{112}{155}$

9. $\frac{11}{17}$ 10. $\frac{390}{71}$, or $5\frac{35}{71}$

2.7 Puzzle Time

TO VISIT AN ANT IN NEW YORK

2.8 Start Thinking!

For use before Activity 2.8

Sample answer: It may be preferable to use a decimal in the following situations:

- If you need to add or subtract numbers. Addition and subtraction are easier with decimals than with fractions. It is a one-step process, as opposed to with fractions, which involves finding a common denominator, writing each fraction with the common denominator, performing the operation, and then simplifying the answer.
- If you want to compare numbers, such as batting average or race times.
- When the context of the situation commonly uses decimals instead of fractions, for example, money, sports statistics, metric measurements, stock market data, and population statistics.

2.8 Warm Up

For use before Activity 2.8

1. 64 2. 27 3. 73

4. 21 5. 52 6. 23

2.8 Start Thinking!

For use before Lesson 2.8

No; *Sample answer:* 0.4555 terminates in the ten-thousandths place and 0.4555… and $0.4\overline{5}$ are equivalent repeating decimals in which the digit 5 repeats without end.

Answers

2.8 Warm Up

For use before Lesson 2.8

1. 0.24
2. 0.55
3. 0.4
4. 0.98
5. $0.08\overline{3}$
6. $0.6\overline{1}$

2.8 Practice A

1. 0.3
2. 0.4
3. 0.64
4. $0.41\overline{6}$
5. 0.475
6. 0.55
7. 0.875
8. $0.4\overline{3}$
9. 0.4375
10. $0.1\overline{5}$
11. 0.3375
12. 2.16
13. $1.8\overline{6}$
14. $1.0\overline{5}$
15. 4.94
16. $2.8\overline{3}$
17. The numerator and denominator were put in the wrong locations.

$$\begin{array}{r} 0.22\ldots \\ 9\overline{)2.00000} \\ \underline{18} \\ 20 \\ \underline{18} \\ 20 \end{array}$$

$$\frac{2}{9} = 0.\overline{2}$$

18. >
19. =
20. <
21. 5.875 in.
22. $\frac{8}{15}$, $0.5\overline{3}$
23. $1\frac{3}{8}$ lb, 1.375 lb
24. $22\frac{219}{400}$ m, 22.5475 m

2.8 Practice B

1. 0.1
2. 0.625
3. 0.84
4. $0.\overline{8}$
5. 0.85
6. 0.3875
7. $0.91\overline{6}$
8. $0.2\overline{7}$
9. 0.56
10. 0.325
11. $0.3\overline{285714}$
12. 3.04
13. $0.9\overline{6}$
14. 1.575
15. 2.24
16. $2.7\overline{3}$
17. Five should be divided into 4.0.

$$\begin{array}{r} 0.8 \\ 5\overline{)4.0} \\ \underline{4\,0} \\ 0 \end{array}$$

$$\frac{4}{5} = 0.8$$

18. >
19. =
20. <
21. $\frac{11}{15}$, $0.7\overline{3}$
22. $\frac{17}{40}$, 0.425
23. $\frac{71}{30}$, $2.3\overline{6}$
24. $4\frac{3}{4}$ in., 4.75 in.
25. a. 0.24, 0.325, 0.26, 0.06, 0.04, 0.075; ordered from least to greatest: 0.04, 0.06, 0.075, 0.24, 0.26, 0.325

    b. 171 students

2.8 Enrichment and Extension

1.

Fraction	Numerator	Decimal form	Repeating digit
$\frac{1}{9}$	1	$0.\overline{1}$	1
$\frac{4}{9}$	4	$0.\overline{4}$	4
$\frac{7}{9}$	7	$0.\overline{7}$	7

2.

Fraction	Numerator	Decimal form	Repeating digit
$\frac{1}{99}$	1	$0.\overline{01}$	01
$\frac{8}{99}$	8	$0.\overline{08}$	08
$\frac{17}{99}$	17	$0.\overline{17}$	17
$\frac{54}{99}$	54	$0.\overline{54}$	54

3. $\frac{2}{9}$
4. $\frac{5}{9}$
5. $\frac{8}{9}$
6. $\frac{5}{99}$
7. $\frac{34}{99}$
8. $\frac{98}{99}$
9. $\frac{1}{999}$
10. $\frac{125}{999}$
11. $\frac{305}{9999}$

2.8 Puzzle Time

IT WILL START TO PEEL

Answers

School-to-Work

1. Check students' recipes. **2.** *Sample answer:* 4

3. *Sample answer:* (using Garlic Pasta recipe)

$30 \bullet \frac{1}{4} = \frac{30}{4} = 7\frac{1}{2}$ cups olive oil

$30 \bullet 12 = 360$ cloves of garlic

$30 \bullet \frac{1}{3} = \frac{30}{3} = 10$ cups chopped bacon

$30 \bullet \frac{3}{2} = \frac{90}{2} = 45$ cups canned tomatoes

$30 \bullet 1 = 30$ lb penne pasta

4. *Sample answer:* (using Garlic Pasta recipe amounts in Exercise 3)

$\frac{1}{6} \bullet 7\frac{1}{2} = \frac{15}{12} = 1\frac{1}{4}$ cups; total of $8\frac{3}{4}$ cups olive oil

$\frac{1}{6} \bullet 360 = \frac{360}{6} = 60$ more cloves of garlic; total of 420 cloves of garlic

$\frac{1}{6} \bullet 10 = \frac{10}{6} = 1\frac{2}{3}$ cups; total $11\frac{2}{3}$ cups chopped bacon

$\frac{1}{6} \bullet 45 = \frac{45}{6} = 7\frac{1}{2}$ cups; total $52\frac{1}{2}$ cups canned tomatoes

$\frac{1}{6} \bullet 30 = \frac{30}{6} = 5$ lb; total 35 lb penne pasta

5. *Sample answer:* (using Garlic Pasta recipe)

$\frac{1}{2} \bullet \frac{1}{4} = \frac{1}{8}$ cup olive oil

$\frac{1}{2} \bullet 12 = 6$ cloves of garlic

$\frac{1}{2} \bullet \frac{1}{3} = \frac{1}{6}$ cup chopped bacon

$\frac{1}{2} \bullet \frac{3}{2} = \frac{3}{4}$ cup canned tomatoes

$\frac{1}{2} \bullet 1 = \frac{1}{2}$ lb penne pasta

Study Help

Available at *BigIdeasMath.com.*

Financial Literacy

1. 100; $\frac{1}{100}$; 0.01 **2.** 20; $\frac{1}{20}$; 0.05

3. 10; $\frac{1}{10}$; 0.10 **4.** 4; $\frac{1}{4}$; 0.25

5. *Sample answer:* To avoid getting a lot of change back; $\$5.25 - \$4.23 = \$1.02$ change.

6. *Sample answer:* Pay \$11.05 so that you get a \$5 bill and two pennies in change instead of three \$1 bills and 97 cents.

7. \$10.02; Sample explanation: They paid \$21 instead of \$20 so they could get one \$10 bill instead of one \$5 bill, four \$1 bills, and lots of change. It is easier to carry.

Technology Connection

1. $(3 + 7 \div 12)$ or $((3 \times 12 + 7) \div 12)$

2. $(5 \div 6)$

3. $(3 + 7 \div 12) \div (5 \div 6)$; 4.3

4. 3.125 **5.** 1.5 **6.** 14 **7.** 3.6

8. No; Kelly divides 6 by $\frac{2}{3}$, Jamie divides by 2 and then divides by 3 which is the same as dividing by 6, not by $\frac{2}{3}$. The correct answer is 9.

9. 5.25 **10.** 9.777778 **11.** 5.2

Life Connections

1. Check students' work.

a. *Sample answer:* I have to go to the library, mow the lawn, and attend the cookout. I should go to soccer practice and try to help at the shelter for a little while. I'd like to work on the float.

b. *Sample answer:* I can do all of the activities but I need to limit my time at the more optional activities (animal shelter and making the float).

c. *Sample answer:*

7:55–8:15	Walk to library
8:15–9:30	Work at the library
9:30–9:50	Walk back home
9:50–10:30	Mow the lawn
10:30–10:40	Walk to shelter
10:40–11:50	Volunteer at shelter
11:50–12:00	Walk home
12:00–12:30	Eat lunch
12:30–12:45	Walk to school
12:45–2:25	Work on float
2:25–2:30	Walk to soccer practice
2:30–5:00	Play soccer
5:00–5:20	Walk home
6:00	Leave for cookout

d. *Sample answer:* I will not be able to volunteer at the shelter for as long as I usually do. I may also be tired.

Stories in History

1. 

2. 

3. 

4. 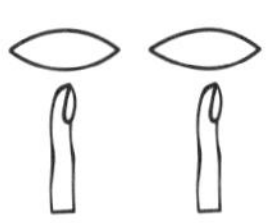

5. 

6. 

7. Check students' work. *Sample answer:* Yes, because all the numerators are one so you only have to multiply the denominators.

Cumulative Practice

1. $\frac{1}{2}$ 2. 5 3. 1

4. 7 5. 0 6. 3

7. 31 8. 0 9. 2

10. \$6; *Sample answer:* To the nearest dollar, the milk costs \$4 and the bread costs \$2; \$4 + \$2 = \$6.

11. $\frac{3}{4}$ 12. $\frac{5}{12}$ 13. 42 14. $\frac{3}{4}$

15. $1\frac{1}{2}$ 16. $\frac{33}{49}$ 17. $\frac{1}{2}$ 18. 20

19. $\frac{5}{6}$ 20. 28 21. $\frac{2}{45}$ 22. 1

23. $3\frac{1}{3}$ miles 24. $2\frac{19}{24}$ inches

25. $\frac{3}{8}$, 1, 1.75, 2.05, $\frac{9}{4}$ 26. $\frac{22}{5}$, 4.45, $4\frac{7}{10}$, 5, 5.18

27. $\frac{1}{3}$, 0.37, $\frac{1}{2}$, $\frac{2}{3}$, 0.68 28. 2.89, $2\frac{9}{10}$, 3, 3.01, $\frac{31}{10}$

Chapter 3

3.1 Start Thinking!

For use before Activity 3.1

Sample answer: If you are buying several items, you can round to the nearest dollar and find the sum to estimate the total cost.

3.1 Warm Up

For use before Activity 3.1

1. 16 2. 99 3. 8

4. 1 5. 7 6. 0

3.1 Start Thinking!

For use before Lesson 3.1

Sample answers are given.

10; 30
7; 28
4; 32
3; 27
29 ÷ 11

3.1 Warm Up

For use before Lesson 3.1

1. 12 2. 40 3. 60

4. 36 5. 15 6. 88

3.1 Practice A

1. 18 2. 40 3. 28 4. 30

5. 70 6. 15 7. 60 8. 3

9. 6 10. 24 11. 90 12. 6

13. 160 14. 120 15. 100 16. 3

17. 8 18. 11 19. 54 ft^2 20. 51 cm^2

21. 35 $in.^2$

22. 60 ÷ 10 = 6, rounding to a compatible number

23. 5 weeks 24. 32 liters

25. a. 1250 cm^2; too little b. 3750 cm^3

Answers

3.1 Practice B

1. 28 2. 45 3. 49 4. 15
5. 130 6. 24 7. 48 8. 3
9. 4 10. 75 11. 102 12. 4
13. 360 14. 350 15. 450 16. 6
17. 9 18. 9 19. $80\ m^2$ 20. $92\ ft^2$
21. $102\ cm^2$ 22. 81 23. 5
24. $40 \div 5 = 8$ or $40 \div 4 = 10$. Each number is rounded to nearby whole or compatible number.
25. $500, $1040
26. **a.** 200 fl oz **b.** 7 quarts

3.1 Enrichment and Extension

1. 1500 2. 30 3. 3
4. 2.75 5. 11 6. $6.\overline{6}$
7. 22.5 8. 5 9. 32
10. 28 11. HALF DOLLAR

3.1 Puzzle Time

IN A PETRIFIED FOREST

3.2 Start Thinking!

For use before Activity 3.2

Sample answer: It costs $1.70 to take the subway. You want to find the cost of a certain number of rides on the subway.

3.2 Warm Up

For use before Activity 3.2

1. 1.2 2. 0.36 3. 8.1
4. 1.23 5. 2.4 6. 0.6

3.2 Start Thinking!

For use before Lesson 3.2

Spider: 17.16 ft, 171.6 ft, 1716 ft, 17,160 ft

Three-toed sloth: 2.49 ft, 24.9 ft, 249 ft, 2490 ft

Garden snail: 0.44 ft, 4.4 ft, 44 ft, 440 ft

The teacher may want to ask students how many minutes 100 and 1000 seconds are or how many hours 10,000 seconds is. 100 seconds = 1 min 40 seconds, 1000 seconds = 16 minutes 40 seconds, 10,000 seconds = 2 hours 46 minutes 40 seconds

3.2 Warm Up

For use before Lesson 3.2

1. 7.8 2. 18.2 3. 14.44
4. 88.5 5. 9.84 6. 36.12

3.2 Practice A

1. 3.2 2. 5.4 3. 20.4 4. 19.2
5. 40.8 6. 4.76 7. 25.2 8. 43.4
9. 3.9 10. 11.88 11. 91.95 12. 22.08
13. 0.108 14. 0.09 15. 0.176 16. 0.279
17. 0.064 18. 0.0644
19. The answer should have 4 decimal places; 0.0032
20. 10.54 21. 0.47 22. 56,204 23. $17.38
24. 225.4 pounds 25. $49.95
26. between 4.08 feet and 4.5 feet

3.2 Practice B

1. 4.5 2. 13.6 3. 40.6 4. 36
5. 36.6 6. 7.2 7. 27.6 8. 37
9. 148 10. 59.64 11. 16.77 12. 57.4
13. 0.168 14. 0.0095 15. 0.245 16. 0.585
17. 0.216 18. 0.0548
19. A "1" was left out of the answer; 0.0168
20. 4512 21. 2.49 22. 278,950
23. 28.6 24. 60.44 25. 297.9 26. $6a + 1.5$
27. The expression $7.8 \bullet (6.9 + 3.5)$ is greater than the expression $7.8 + (6.9 \bullet 3.5)$. The largest number 7.8, is being multiplied in the first problem, but only added in the second problem.
28. **a.** In 2000, all of the coins cost less to make. In 2007, the quarter and dime cost less to make, but the nickel and penny cost more to make.
    b. quarter; 270 cents ($2.70), dime: 108.5 cents ($1.085), nickel: 318.5 cents ($3.185), penny: 42.5 cents ($0.425)

3.2 Enrichment and Extension

1. 10.5 mm 2. 7.5 mm
3. **a.** 300; 160 **b.** $43 **c.** 58.4 oz

Answers

4. a. You; $1.50 **b.** Your friend; 16.24 oz

3.2 Puzzle Time

HE RAN OVER THE GHOUL LINE

3.3 Start Thinking!

For use before Activity 3.3

Sample answer: The area of the rectangle that is 0.7 by 0.5 is 35 out of 100 total squares.

3.3 Warm Up

For use before Activity 3.3

1. 24 **2.** 21 **3.** 30

4. 81 **5.** 0 **6.** 45

3.3 Start Thinking!

For use before Lesson 3.3

Sample answer: The price per gallon (given to the thousandths' place) times the number of gallons (to the thousandths' place) is the total cost of the gas.

3.3 Warm Up

For use before Lesson 3.3

1. 0.12 **2.** 0.02 **3.** 0.00024

4. 0.000028 **5.** 0.0045 **6.** 0.0000006

3.3 Practice A

1. 20.54 **2.** 3.7086 **3.** 18.4080 **4.** 0.12

5. 0.045 **6.** 0.238 **7.** 0.73 **8.** 93.12

9. 1.1776 **10.** 2.7324 **11.** 0.05248 **12.** 9.51172

13. There needs to be one more decimal place in the answer: 0.000032

14. 22.5 **15.** 34.76 **16.** 0.686 **17.** 0.1225

18. 7.9 **19.** 20 **20.** $19.50

21. 16.15 pounds **22.** 65,767 mi^2

23. $3.05 **24.** 82.8 miles

3.3 Practice B

1. 0.15 **2.** 0.392 **3.** 0.000006

4. 1.896 **5.** 123.12 **6.** 1.5625

7. 7.6636 **8.** 0.11682 **9.** 11.438034

10. The number of decimal places used was just for one factor; 0.00000016

11. 21.6 **12.** 84.5 **13.** 1.156 **14.** 0.2448

15. 0.4 **16.** 15.8068

17. Multiply each number by 3: 56.7, 170.1, 510.3

18. Multiply each number by 5: 81.25, 406.25, 2031.25

19. Multiply each number by 0.02: 0.00000032, 0.0000000064, 0.000000000128

20. Multiply each number by 0.3: 0.0486, 0.01458, 0.004374

21. 0.07875 **22.** 0.304 **23.** 12.46875

24. $1.64 **25.** week 4

3.3 Enrichment and Extension

1. Yes;

$$\begin{array}{r} 3.92 \\ \times\ 7.18 \\ \hline 31\ 36 \\ 39\ 2\ \ \\ 2744\ \ \ \ \\ \hline 28.14\ 56 \end{array}$$

2. Yes; $0.392 \times 71.8 = 28.1456$,

$39.2 \times 0.718 = 28.1456$

3.
$$\begin{array}{r} 5.6 \\ \times\ 9.6 \\ \hline 33\mathbf{6} \\ 5\mathbf{0}4\ \ \\ \hline 5\mathbf{3}.76 \end{array}$$

4.
$$\begin{array}{r} 7.82 \\ \times\ \ 3.4 \\ \hline 3128 \\ 2346\ \ \\ \hline 26.588 \end{array}$$

5.
$$\begin{array}{r} 2.\mathbf{1}15 \\ \times\ \mathbf{8.34} \\ \hline 84\mathbf{60} \\ \mathbf{6}345\ \ \\ 1\mathbf{6}920\ \ \ \ \\ \hline 17.\mathbf{63}910 \end{array}$$

3.3 Puzzle Time

TO SERVE UP A GREAT ALIBI

3.4 Start Thinking!

For use before Activity 3.4

Sample answer: Multiplication and division are inverse operations. That is, multiplication and division "undo" each other. For each product, there is a related division. For example, $4 \times 5 = 20$ and $20 \div 5 = 4$.

3.4 Warm Up

For use before Activity 3.4

1–6. Check students' drawings.

3.4 Start Thinking!

For use before Lesson 3.4

Sample answer: If one person purchased the tickets for a group of people, divide the total cost of the tickets by the number of people to find the cost per ticket.

Answers

3.4 Warm Up

For use before Lesson 3.4

1. 6.1 **2.** 2.6 **3.** 3.8

4. 9.9 **5.** 5.7 **6.** 0.7

3.4 Practice A

1. 0.2 **2.** 0.53 **3.** 4.1 **4.** 0.32

5. 2.4 **6.** 4.9 **7.** 6.7 **8.** 1.205

9. 0.064 **10.** 2.7 **11.** 0.35 **12.** 1.74

13. 25.4 **14.** 2.05 **15.** 31.05

16. The dividend was not aligned properly. The answer should be 0.27.

17. 30.3 **18.** 5.31 **19.** 37.2 **20.** 10.7

21. 3.43 **22.** 14

23. about 145 stories **24.** \$0.35

25. \$32.10 **26.** 3.245 in.

27. 16-fluid ounce bottle; It costs 6 cents less per ounce.

3.4 Practice B

1. 0.8 **2.** 6.4 **3.** 6.5 **4.** 0.27

5. 9.7 **6.** 8.6 **7.** 20.8 **8.** 23.7

9. 0.028 **10.** 13.2 **11.** 8.1 **12.** 2.37

13. 2.077 **14.** 23.4 **15.** 4.56

16. The 6 was placed in the wrong location. The answer should be 3.06.

17. 32 **18.** 0.88 **19.** 34.4 **20.** 2.92

21. 3.8 **22.** 28.96 **23.** \$22.02

24. The 18-ounce box is the best buy because $\$3.96 \div 18 = \$0.22$ per ounce, which is the lowest per-ounce price.

25. \$12.76

3.4 Enrichment and Extension

1. $r = 4$ **2.** $a = 7$ **3.** $h = 9$

4. $y = 8$ **5.** $t = 5$ **6.** $b = 2$

7. $e = 3$ **8.** $n = 1$ **9.** $p = 6$

10. A PENNY **11.** A BREATH

3.4 Puzzle Time

SHE WAS PLANE CRAZY

3.5 Start Thinking!

For use before Activity 3.5

Sample answer: You can divide the number of dollars you want to exchange by the exchange rate to find the amount of the country's currency that you will have when you arrive. The total cost of the flight divided by the number of people you are traveling with gives the per person cost of the flight. The total distance of the flight divided by the length of time gives you the average speed of the airplane.

3.5 Warm Up

For use before Activity 3.5

1. 30 **2.** 50 **3.** 200

4. 2 **5.** 3, 7 **6.** 24

3.5 Start Thinking!

For use before Lesson 3.5

Sample answer:

1. You can multiply the quotient by the divisor and compare it to the dividend.
2. You can do the division a second time to make sure that you get the same quotient.
3. You can use a calculator to check.

3.5 Warm Up

For use before Lesson 3.5

1. 8 **2.** 17 **3.** 27

4. 9 **5.** 23 **6.** 12

3.5 Practice A

1. $231 \div 42$ **2.** $1287 \div 15$ **3.** $8760 \div 535$

4. 21 **5.** 16 **6.** 6 **7.** 27

8. 18.2 **9.** 3 **10.** 25.2 **11.** 14.7

12. 36.5 **13.** 3.4 **14.** 30 **15.** 410

16. 51,000 **17.** 14 **18.** 27

19. The dividend should be 2500. Both divisor and dividend should be multiplied by 100.

20. 36.57 **21.** 6476.67 **22.** 222.35

23. 10.20 **24.** 7.81 **25.** 2681.43

26. 750 people **27.** 300 sheets of paper

28. \$0.38 **29.** 10.3 meters **30.** 8400 songs

Answers

3.5 Practice B

1. 4 **2.** 430 **3.** 63 **4.** 20.5

5. 12.6 **6.** 17 **7.** 127 **8.** 657

9. 1505 **10.** 0.17 **11.** 300 **12.** 6.8

13. The decimal point did not move in the divisor;
$24\overline{)96{,}000}$

14. 12.7 **15.** 45.05 **16.** 13.31 **17.** 8.25

18. 67.37 **19.** 40.68 **20.** < **21.** >

22. > **23.** =

24. The 20.5-ounce can is the best buy. It is about 0.5 cents cheaper than the 14.5-ounce can.

25. a. 16 weeks

b. 4 weeks; *Sample answer:* Divide the difference in prices by $15.50.

3.5 Enrichment and Extension

1. $x = 3.2$ m **2.** $x = 4.6875$ m

3. $x = 1$ m **4.** $x = 3.2$ m

3.5 Puzzle Time

SHE WAS A BEE STUDENT

School-to-Work

1. a. needed dose ÷ dose in 1 tablet = number of tablets

b. 4 tablets

2. 1.5 tablets **3.** 2.5 tablets **4.** 0.5 tablet

5. 0.5 tsp **6.** 1.25 tsp **7.** 1.75 tsp

8. 0.75 cubic centimeter

Study Help

Available at *BigIdeasMath.com.*

Financial Literacy

1. *Sample answer:* A party for grandparents' anniversary, with extended family.

2. *Sample answer:* Our extended family and close friends, for about 40 people.

3–7. *Sample answer:* Corn tamales with tomato salsa, green salad, brownies.

40 ears of corn; $0.79 per ear	$32
1 bag of cornmeal; $2.39 per bag	$2
2 pounds of butter; $3.15 per pound	$6
sugar, baking powder, salt; on hand	n/a
10 heads of lettuce; $1.42 per head	$15
1 bag of carrots; $2.13	$2
8 lb of tomatoes; $3.14 per pound	$24
4 bunches of cilantro; $2.79 per bunch	$12
hot sauce; on hand	n/a
5 boxes brownie mix; $3.89 per box	$20
2 cartons of eggs; $4.19 per carton	$8
oil; on hand	n/a
13 6-packs of soda; $3.89 per 6-pack	$48
4 6-packs of seltzer; $3.89 per 6-pack	$16
3 pkgs. of plates; $4.20 per package	$12
1 pkg. of napkins; $4.49 per package	$5
Estimated total cost:	$202

8. *Sample response:* $48; I would spend half on flowers for the party and half on a gift for my grandparents.

Cumulative Practice

1. 0 **2.** 18 **3.** 2 **4.** 24

5. 74 **6.** 7 **7.** 93 **8.** 61

9. 40 **10.** 8 **11.** 200 **12.** 5

13. 49 **14.** 21 **15.** $65 **16.** $13

17. 65.8 **18.** 21.13 **19.** 15 **20.** 7.5

21. 4.26 **22.** 12.36 **23.** 64.828 **24.** 30.4856

25. 6.25 **26.** 2.5 **27.** 15.99 **28.** 36.027

29. 106.34932 **30.** 123.774 **31.** 5

32. 12.4 **33.** 0.1 **34.** 91.16

35. a. $10.50 **b.** $11.13 **c.** $3.87

36. a. 16-ounce: $0.21; 20-ounce: $0.22

b. the 16-ounce box; the cereal costs less per ounce in the smaller box than in the larger one.

Answers

Chapter 4

4.1 Start Thinking!

For use before Activity 4.1

Sample responses include: store sales, political polls, test scores, geography facts, taxes

4.1 Warm Up

For use before Activity 4.1

1. $\frac{60}{100}$ **2.** $\frac{25}{100}$ **3.** $\frac{60}{100}$

4. $\frac{66}{100}$ **5.** $\frac{135}{100}$ **6.** $\frac{540}{100}$

4.1 Start Thinking!

For use before Lesson 4.1

A, C, and D represent equivalent numbers (75%).

4.1 Warm Up

For use before Lesson 4.1

1. $\frac{3}{25}$ **2.** $\frac{1}{25}$ **3.** $\frac{39}{50}$

4. 50% **5.** 36% **6.** 95%

4.1 Practice A

1. $\frac{1}{5}$ **2.** $\frac{1}{4}$ **3.** $\frac{13}{20}$ **4.** $\frac{11}{25}$

5. $\frac{8}{25}$ **6.** $\frac{9}{100}$ **7.** $2\frac{1}{2}$ **8.** $1\frac{1}{5}$

9. $\frac{111}{250}$ **10.** $2\frac{11}{250}$ **11.** $\frac{1}{200}$ **12.** $\frac{3}{400}$

13. The denominator should be 100 instead of 10.

$$8\% = \frac{8}{100} = \frac{2}{25}$$

14. 24%; Multiply numerator and denominator by 4.

15. 15%; Multiply numerator and denominator by 5.

16. 70%; Multiply numerator and denominator by 10.

17. 75%; Multiply numerator and denominator by 25.

18. 18%; Multiply numerator and denominator by 2.

19. 48%; Multiply numerator and denominator by 4.

20. 62.5%; Rewrite as a decimal, then as a fraction with denominator of 100.

21. 31.25%; Rewrite as a decimal, then as a fraction with denominator of 100.

22. 40%; Multiply numerator and denominator by 20.

23. 22.5%; Rewrite as a decimal, then as a fraction with denominator of 100.

24. 28.75%; Rewrite as a decimal, then as a fraction with denominator of 100.

25. 65%; Multiply numerator and denominator by 5.

26. A step was skipped and the decimal was converted to the wrong number of hundredths.

$$\frac{39}{40} = 0.975 = \frac{975}{1000} = \frac{97.5}{100} = 97.5\%$$

27. $\frac{1}{4}$ **28.** $2\frac{3}{4}$ times heavier

29. 80% **30.** 92%

31. $\frac{3}{5}$; 60% **32.** $\frac{2}{5}$; 40% **33.** $\frac{1}{4}$; 25%

4.1 Practice B

1. $\frac{7}{20}$ **2.** $\frac{407}{500}$ **3.** $2\frac{1}{10}$ **4.** $\frac{1}{125}$

5. 74%; Multiply numerator and denominator by 2.

6. 84%; Multiply numerator and denominator by 4.

7. 18.75%; Rewrite as a decimal, then as a fraction with denominator of 100.

8. 47.5%; Rewrite as a decimal, then as a fraction with denominator of 100.

9. 325% **10.** 250% **11.** 180%

12. 514% **13.** 412% **14.** 655%

15. 212.5% **16.** 331.25% **17.** 55%

18. 25%; You sold $\frac{3}{4}$ of the cups and your friend sold $\frac{1}{4}$.

19. $1\frac{1}{20}$; The perspective in the figure makes the Florida flag look taller, but the U.S. flag is actually taller than the Florida flag.

Answers

20. the reticulated python; The anaconda is 118% as long as the Burmese python, but only 115% as long as the reticulated python, so the reticulated python must be longer than the Burmese python.

21. 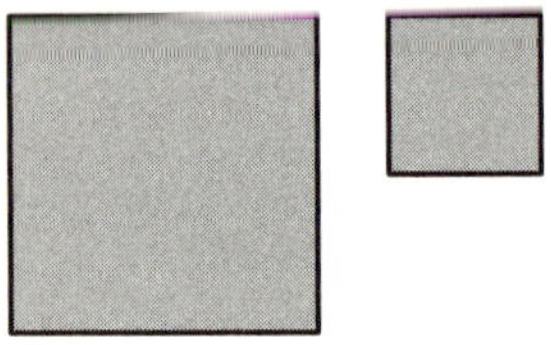

The area of the smaller square is $\frac{1}{4}$, or 25%, the area of the original square.

22. 12 years

4.1 Enrichment and Extension

1. $\frac{1}{4}$; 25%
2. $\frac{1}{2}$; 50%
3. $\frac{1}{8}$; 12.5%
4. $\frac{3}{16}$; 18.75%
5. $\frac{11}{16}$; 68.75%
6. $\frac{3}{4}$; 75%
7. $\frac{5}{16}$; 31.25%
8. $\frac{3}{8}$; 37.5%

4.1 Puzzle Time

LEAD ME TO YOUR READERS

4.2 Start Thinking!

For use before Activity 4.2

Sample answer: The percent has the decimal point two places to the right; No; Andrew only moved the decimal point one place to the right. It should be 98%.

4.2 Warm Up

For use before Activity 4.2

1. $\frac{3}{8}$
2. $\frac{1}{2}$
3. $\frac{1}{4}$
4. $\frac{1}{8}$
5. $\frac{5}{6}$
6. $\frac{3}{8}$

4.2 Start Thinking!

For use before Lesson 4.2

Sample answer: In basketball, free throw percentage, field goal percentage, and three point percentage are often given as decimals, which makes them easier to compare.

4.2 Warm Up

For use before Lesson 4.2

1. 0.19
2. 0.02
3. 0.075
4. 89%
5. 54%
6. 10%

4.2 Practice A

1. 0.81
2. 0.78
3. 0.05
4. 0.08
5. 0.4
6. 0.6
7. 0.237
8. 0.1675
9. 1.5
10. 2.1
11. 1.86
12. 4.16
13. 1.008
14. 0.0517
15. 0.004
16. 0.0004
17. The decimal point is moved to the right instead of the left.

    1.475% = 01.475% = 0.01475

18. 66%
19. 32%
20. 51%
21. 97%
22. 1%
23. 4%
24. 31.2%
25. 46.8%
26. 50%
27. 120%
28. 108%
29. 116%
30. 0.3%
31. 2.5%
32. 2.45%
33. 202.5%
34. The decimal point is moved from the wrong place.

    1.8 = 1.80 = 180%

35. 0.54
36. 15%
37. 88%
38. 0.86, 86%
39. 5 times
40. $\frac{21}{100}$; 0.21
41. $\frac{3}{4}$, 0.75
42. $\frac{16}{25}$, 0.64
43. $\frac{17}{20}$, 0.85

4.2 Practice B

1. 54%
2. 37%
3. 22.2%
4. 92.9%
5. 140%
6. 250%
7. 2000%
8. 0.5%
9. $\frac{17}{25}$, 0.68
10. $\frac{9}{100}$, 0.09
11. $\frac{11}{20}$, 0.55
12. $\frac{13}{50}$, 0.26
13. $\frac{53}{125}$, 0.424
14. $\frac{92}{125}$, 0.736
15. $\frac{5}{16}$, 0.3125
16. $\frac{893}{2000}$, 0.4465
17. 64%
18. a. 0.48, $\frac{12}{25}$; 0.28, $\frac{7}{25}$; 0.14, $\frac{7}{50}$

    b. 10%

Answers

c.

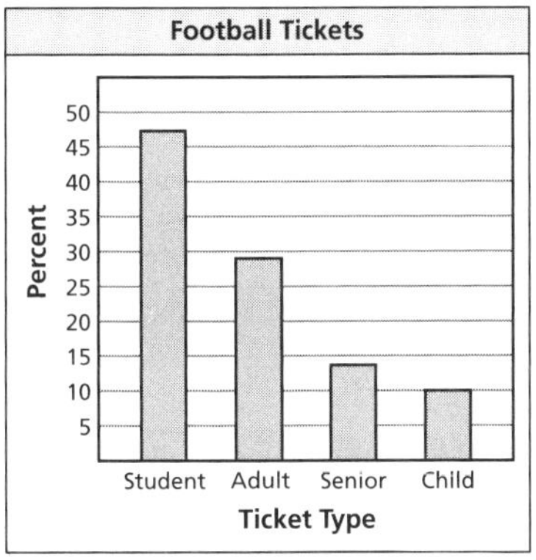

19. a. 33% **b.** 35% **c.** $1\frac{3}{4}$ times **d.** 27%, 0.27

20. a–d.

Distance	within 1 mi	1–2 mi	2–3 mi	Over 3 mi
Percent	50%	28%	10%	12%

4.2 Enrichment and Extension

1. 50%; 0.5 **2.** 50%; 0.5

3. a. 25%; 0.25 **b.** 75%; 0.75
c. 50%; 0.5 **d.** 50%; 0.5

4. a. 56.25%; 0.5625 **b.** 43.75%; 0.4375

5. a. 4 **b.** 9; 28.125%, 0.28125

4.2 Puzzle Time

BOTH ARE SUPPOSED TO BE DUE

4.3 Start Thinking!

For use before Activity 4.3

Sample answer: You can rewrite each fraction as a decimal and compare decimals or you can write each fraction using the same denominator and compare the numerators. $\frac{7}{12}$; methods may vary.

4.3 Warm Up

For use before Activity 4.3

1. 1.4, 1.44, 3.3, 4.1, 4.3

2. 0.5, 0.7, 1.5, 1.57, 1.75

3. 0.02, 0.04, 0.2, 0.24, 0.44

4. 0.06, 0.6, 6.06, 6.6, 6.63

5. 0.02, 0.17, 0.2, 0.72, 2.27

6. 0.18, 0.81, 1.8, 1.88, 8.1

4.3 Start Thinking!

For use before Lesson 4.3

Answers will vary. Check students' work.

4.3 Warm Up

For use before Lesson 4.3

1. 45% **2.** 0.4 **3.** $\frac{9}{17}$

4. 0.6 **5.** $\frac{2}{5}$ **6.** 240%

4.3 Practice A

1. $\frac{3}{4}$ **2.** 0.54

3. Neither: $0.21 = 21\%$ **4.** $\frac{2}{3}$

5. 49% **6.** 16%

7. Neither: $\frac{12}{25} = 48\%$ **8.** 12%

9. 1.2 **10.** $\frac{31}{50}$ **11.** $50\frac{1}{4}$ **12.** $\frac{1}{8}$

13. $\frac{68}{100} = 68\%$, not 0.68%.

So, $68\% > 0.7\%$ and $\frac{17}{25}$ is the greater number.

14. 63%, 0.64, $\frac{13}{20}$ **15.** $\frac{11}{25}$, 45%, 0.46

16. 0.12, $\frac{1}{8}$, 13%, 0.135

17. $\frac{7}{8}$, 90%, 0.925, 0.93, $\frac{15}{16}$

18. $3\frac{3}{5}$, 362%, 3.66, $3\frac{2}{3}$, 36

19. 0.27, 27.3%, $\frac{11}{40}$, 28%, 0.3

20. the punch that you made

21. You made a higher percent of baskets.

22. Yes; 62.5% is greater than 60%.

23. grade 7, grade 6, grade 8

Answers

4.3 Practice B

1. $\frac{1}{4}$ **2.** $\frac{5}{9}$ **3.** 3.2 **4.** 1

5. $0.3, \frac{8}{25}, 33\%, \frac{1}{3}, 33.6\%$

6. $210\%, \frac{43}{20}, 2.2, 2.\overline{2}$

7. B **8.** C **9.** A **10.** D

11. *Sample answer:* Write 31% and 32% as the decimals 0.31 and 0.32. Graph 0.31 and 0.32 on a number line. Then identify a decimal between them, such as 0.315.

12. Yes; 5% of a meter $= 0.05(100 \text{ cm}) = 5$ cm.

Because 6 centimeters is greater than 5 centimeters, 6 centimeters is greater than 5% of a meter.

13. No; 6% of 1 lb $= 0.06(16 \text{ oz}) = 0.96$ oz. Because 0.96 ounce is less than 1 ounce, 6% of a pound is less than an ounce.

14. 0.0004 of a day, 1% of an hour, $\frac{2}{3}$ of a minute

15. a. $0.45\%, 0.0082, \frac{1}{36}, \frac{1}{16}$

b. The population of Florida has grown faster than the population of the U.S.; *Sample answer:* People tend to move to Florida because it is a nice place to live.

c. *Sample answer:* No; Some day Florida will become too crowded to keep growing.

16. No;

$$0.000002\% = 0.00000002 = \frac{2}{100{,}000{,}000}$$
$$= \frac{20}{1{,}000{,}000{,}000}$$
$$= 20 \text{ parts per billion.}$$

Because this is greater than 10 parts per billion, it is not an allowable amount.

4.3 Enrichment and Extension

$1\%, \frac{11}{40}, \frac{17}{60}, 31\%, \frac{9}{25}, 0.3\overline{81}, \frac{31}{75}, 41.5\%, \frac{56}{99}, 0.\overline{75}, \frac{101}{120}, 85.25\%, 100\%$

4.3 Puzzle Time

IT WAS TRULY IN A JAM

4.4 Start Thinking!

For use before Activity 4.4

10; 15; 20; 25; $5n$

4.4 Warm Up

For use before Activity 4.4

1. 12 **2.** 60 **3.** 13

4. 30 **5.** 5 **6.** 9.2

4.4 Start Thinking!

For use before Lesson 4.4

Sample answer: The sales tax in Florida is 6% and the sales tax in Utah is 4.65%. If you buy the same shirt for \$30 in each state, how much more would you spend in Florida? Round to the nearest cent. Answer: \$0.41

4.4 Warm Up

For use before Lesson 4.4

1. 10 **2.** 35 **3.** 24

4. 4.6 **5.** 12 **6.** 10

4.4 Practice A

1. 20 **2.** 12 **3.** 21 **4.** 10

5. 32 **6.** 120 **7.** 21 **8.** 14

9. 52 **10.** 28 **11.** 123 **12.** 6

13. 31 **14.** 9 **15.** 2.7 **16.** 284

17. 4% is equal to 0.04, not 0.4;
$4\% \text{ of } 65 = 0.04 \times 65 = 2.6$

18. 36 **19.** 36 **20.** 15.6 **21.** 34

22. 165 **23.** 494 **24.** 22.88 **25.** 46.8

26. 54.6 **27.** 22 **28.** 7.54 **29.** 24.84

30. 6 sixth graders **31.** 14 min

32. \$8.75 **33.** 18.9 mi **34.** 0.7 lb

35. 10 ft **36.** 44 **37.** 45

38. 0.32 **39.** 2.16

4.4 Practice B

1. 0.9 **2.** 39 **3.** 63.18 **4.** 14.84

5. 120 **6.** 103.5 **7.** 1053.7 **8.** 149.5

9. 8.36 **10.** 9.36 **11.** 3 **12.** 17.4

13. $<$ **14.** = **15.** = **16.** $>$

17. 54 h **18.** 4224 ft **19.** 11.2 fl oz

20. a. C **b.** B **c.** B

21. \$15; 45% is equal to 3 times 15%, so the discount is 3 times \$5, or \$15.

22. *Sample answer:*

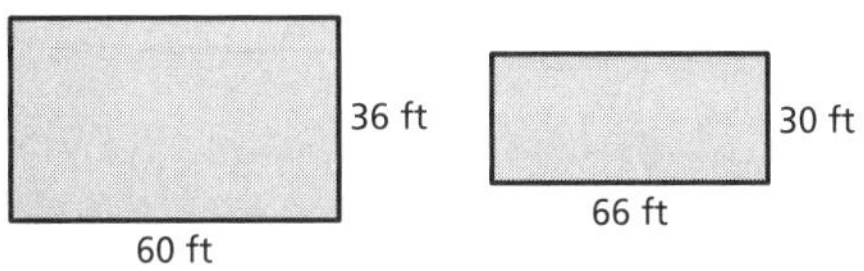

23. 68% **24.** \$17.28; no

4.4 Enrichment and Extension

1. Store A: \$26.22; Store B: \$23.80; Store B is the better buy because it is cheaper.

2. Store A: \$48.99; Store B: \$48.81; Store B is the better buy because it is cheaper.

3. Store A: \$191.25; Store B: \$191.25; The cost is the same at both stores.

4. a. \$546.52

b. No; you get the same price no matter which percent you calculate first.

4.4 Puzzle Time

PROBLEMS FILLED THE PAGES

4.5 Start Thinking!

For use before Activity 4.5

Estimates and methods may vary. *Sample answer:* about 30%

4.5 Warm Up

For use before Activity 4.5

1. 5 **2.** 7.2 **3.** 0.9

4. 27 **5.** 6 **6.** 10

4.5 Start Thinking!

For use before Lesson 4.5

Sample answer: In general an estimate would probably be sufficient in calculating the tip at a restaurant and comparing student survey results. If the basketball players' field goal percentages are close together, then an estimate is probably not sufficient.

4.5 Warm Up

For use before Lesson 4.5

Sample answers are given.

1. 60 **2.** 300 **3.** 4

4. 200 **5.** 45 **6.** 56

4.5 Practice A

1–25. Sample answers are given.

1. 60 **2.** 240 **3.** 30 **4.** 8

5. 80 **6.** 30 **7.** 20 **8.** 30

9. 10 **10.** 270 **11.** 140 **12.** 12

13. 60 **14.** 4 **15.** 10 **16.** 180

17. 168 songs **18.** 150 fish **19.** 45 pounds

20. a. 70 stores **b.** 28 stores **c.** 14 stores

21. a. 90 students **b.** 270 students **c.** 180 students

22. 24 **23.** 140 **24.** 180 **25.** 175

4.5 Practice B

1–24. Sample answers are given.

1. 240 **2.** 18 **3.** 5.6 **4.** 57

5. 25 **6.** 20 **7.** 80 **8.** 500

9. 160 **10.** 40 **11.** 450 **12.** 300

13. 48 **14.** 90 **15.** 98 **16.** 128

17. 150 **18.** 275 **19.** 175 **20.** 800

21. Sometimes: You are rounding one factor up and another factor down before multiplying. The result will sometimes be greater than the actual product and it will sometimes be less than the actual product.

22. Never: You are rounding both factors up before multiplying. The result will always be greater than the actual product.

23. a. 135 students

b. 90 students; The "Other" section is about the same size as the "Italian" section in the graph. So about 20% answered *Other* and $\frac{1}{5} \times 450 = 90$.

Answers

24. a. 60 people **b.** 40 people

c. 68%; 32% of the results are not shown. Any stations not shown were chosen by 8% or less. Because $32 \div 8 = 4$, there are at least 4 more stations. So, there are at least 8 FM station in all.

4.5 Enrichment and Extension

1. 0.5 **2.** 2 **3.** 3 **4.** 0.45

5. 2.4 **6.** 2 **7.** 1.75 **8.** 3

9. 1.8

10. Answer will vary. One method is:

For 2380, use the compatible number 2400.
Find 1% of 2400.

$$1\% \text{ of } 2400 = 0.01 \times 2400$$
$$= 24$$

$$0.3\overline{4}\% \approx 0.\overline{3}\% \text{ or } \frac{1}{3}\%.$$

$$\frac{1}{3} \text{ of } 24 = \frac{1}{3} \times 24 = 8$$

So, $0.3\overline{4}\%$ of 2380 is about 8.

4.5 Puzzle Time

BECAUSE WHEN FISH GET ON THE LINE THEY CAN NOT SEEM TO GET OFF

School-to-Work

1. 96.8% **2.** 3.2%

3. No; almost 97% of orders arrive on time.

4. 22 orders **5.** 20%

6. The company has 80% of the employees participating, so they are eligible for the discount.

7. *Sample answer:* What percent of employees use each type of health care plan? What percent have various numbers of vacation days? What percent are using extra education benefits? You might also calculate changes in the cost of any benefits.

Study Help

Available at *BigIdeasMath.com*.

Financial Literacy

1. *Sample answer:* Move the decimal point one place to the left.

2. *Sample answer:* $3.66

3. *Sample answer:* Multiply the 10% value by 2.

4. *Sample answer:* Divide the 10% value in half to estimate 5% of the bill. Then add the 10% value to the 5% value to estimate 15%

5. *Sample answer:* about $5.50 to $7.00

6. *Sample answer:* In Japan, tipping is not common; in Canada, tipping is similar to the U.S. but at slightly lower rates, such as 10–15% of a restaurant bill. Tipping is not common in many countries outside of the local tourist industry.

7. *Sample answer:* A large restaurant meal for a family birthday; To tip 20%, round the bill to the nearest $10 and find 10% of that value. Then double that value.

8. *Sample answer:* 6 tables an hour at about $75 per table would be $65 to $90 in tips.

9. *Sample answer:* $6 to $9

10. *Sample answer:* Round the night's total to the nearest $50 and take 1% of that, then double it for 2%.

Cumulative Practice

1. 0.375, 37.5%, $\frac{3}{8}$ **2.** 2.25, 225%, $2\frac{1}{4}$

3. 0.95, 95%, $\frac{19}{20}$ **4.** 0.1, 10%, $\frac{1}{10}$

5. 0.75, 75%, $\frac{3}{4}$ **6.** 3.5, 350%, $3\frac{1}{2}$

7. 0.06, 6%, $\frac{3}{50}$ **8.** 0.8, 80%, $\frac{4}{5}$

9. 1.15, 115%, $1\frac{3}{20}$ **10.** $\frac{3}{4}$

11. 0.02, 2.1%, $\frac{1}{5}$, 0.21, $\frac{3}{10}$

12. $\frac{11}{20}$, 56%, 0.567, $\frac{3}{5}$

13. 2.78, 280%, 2.87, $2\frac{7}{8}$

14. 0.97, 98%, $\frac{99}{100}$, 1, $\frac{110}{99}$

Answers

15. 2 **16.** $3\frac{7}{15}$ **17.** $1\frac{1}{2}$ **18.** $12\frac{1}{4}$

19. $\frac{11}{20}$ **20.** $\frac{2}{5}$ **21.** 60 **22.** 1.75

23. 648 **24.** 10 **25.** 2

26. 46.2 **27.** 3.5 gallons

28–33. Sample answers are given.

28. 4000 **29.** 15 **30.** 10

31. 800 **32.** 320 **33.** 15

34. 67,500 square miles

Chapter 5

5.1 Start Thinking!

For use before Activity 5.1

Sample answer: Mixing paint colors in a given ratio will ensure the same color no matter how much you need to make. For example, using 1 tablespoon of blue and 3 tablespoons of white will make the same light blue as using 4 cups of blue and 12 cups of white.

5.1 Warm Up

For use before Activity 5.1

1. $\frac{1}{3}$ **2.** $\frac{1}{4}$ **3.** $\frac{1}{2}$ **4.** $\frac{5}{7}$

5. $\frac{3}{5}$ **6.** $\frac{2}{3}$ **7.** $\frac{3}{8}$ **8.** $\frac{3}{5}$

5.1 Start Thinking!

For use before Lesson 5.1

$\frac{3}{5}$; *Sample answer:* The ratio of boys to girls in the class is $\frac{6}{7}$.

5.1 Warm Up

For use before Lesson 5.1

Sample answers are given.

1. $\frac{8}{14}, \frac{12}{21}$ **2.** $\frac{4}{30}, \frac{8}{60}$ **3.** $\frac{1}{2}, \frac{5}{10}$

4. $\frac{4}{5}, \frac{8}{10}$ **5.** $\frac{3}{4}, \frac{6}{8}$ **6.** $\frac{2}{3}, \frac{20}{30}$

5.1 Practice A

1. $\frac{2}{3}$, 2 to 3, 2 : 3; There are 2 flies for every 3 lizards.

2. $\frac{2}{5}$, 2 to 5, 2 : 5; There are 2 cars for every 5 trucks.

3. $\frac{9}{3}$, 9 to 3, 9 : 3 $\left(\text{or } \frac{3}{1}, 3 \text{ to } 1, 3 : 1\right)$; There are 9 notebooks for every 3 pencils.

4. $\frac{5}{6}$, 5 to 6, 5 : 6; There are 5 hamburgers for every 6 hot dogs.

5. $\frac{2}{3}$ **6.** $\frac{2}{1}$ **7.** 11 : 2 **8.** 3 : 10

9–16. Sample answers are given.

9. $\frac{1}{5}, \frac{4}{20}$ **10.** $\frac{1}{4}, \frac{2}{8}$ **11.** $\frac{3}{1}, \frac{9}{3}$ **12.** $\frac{3}{1}, \frac{6}{2}$

13. 2 : 12, 3 : 18 **14.** 6 to 8, 12 to 16

15. 6 to 2, 3 to 1 **16.** 5 : 12, 10 : 24

17. 1 to 32 **18.** 5 to 3

19. 1 to 5 **20.** 6 mi

21. a. 45 pencils **b.** 54 pencils
c. 93 pencils **d.** 183 pencils

22. 21 mi; Write the ratio of miles to days in simplest form, 3 to 1. Then write the equivalent ratio, 21 to 7, to find the number of miles you walk in 7 days.

5.1 Practice B

1. 7 to 3 **2.** 2 : 5 **3.** $\frac{11}{13}$ **4.** $\frac{9}{7}$

5. *Sample answer:* 6 to 5, 12 to 10

6. *Sample answer:* 4 to 10, 2 to 5

7. 5 to 10 **8.** 8 : 2 **9.** 15 to 5 **10.** 5 : 8

11. 3 pints; Because 16 fl oz = 1 pint, write the ratio as 1 to 6. Then write the equivalent ratio, 3 to 18.

12. 18 sixth graders; Write the ratio of sixth graders to sixth and seventh graders combined, 2 to 5. Then write the equivalent ratio, 18 to 45.

Answers

13. *Sample answer:* $a = 2$

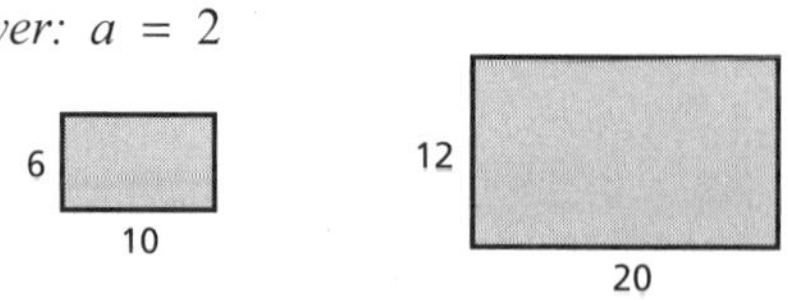

a. w to ℓ:	3 to 5	3 to 5
w to P:	3 to 16	3 to 16
ℓ to P:	5 to 16	5 to 16

b. Each ratio in the first rectangle is equivalent to the corresponding ratio in the second rectangle. This is true because each dimension in the second rectangle is $a = 2$ times as long as in the first rectangle, so the 2s divide out when you simplify the ratio for the second rectangle.

c. w to w: 1 to 2; ℓ to ℓ: 1 to 2; P to P: 1 to 2; A to A: 1 to 4

d. The ratio of the areas; Because area is a product of two dimensions, $a = 2$ is a factor twice in the area of the second rectangle: $A = 2\ell \bullet 2w = 4\ell w$. So, the ratio of the areas is $\ell w : 4\ell w = 1 : 4$.

14. **a.** The ratio 7 to 13 compares the cotton to the polyester, not to the total material blend.

b. 7 to 20

c. Find a ratio equivalent to 7 : 20 comparing the percent of cotton to a total material blend of 100%. 35% is cotton.

5.1 Enrichment and Extension

1. *Sample answer:* Rectangle B
2. *Sample answer:* 1.625 : 1
3. **a.** Each number is the sum of the two previous numbers.

   b. 144, 233, 377

   c. $\frac{1}{1}, \frac{2}{1}, \frac{3}{2}, \frac{5}{3}, \frac{8}{5}, \frac{13}{8}, \frac{21}{13}, \frac{34}{21}, \frac{55}{34}, \frac{89}{55}, \frac{144}{89}, \frac{233}{144}, \frac{377}{233}$

   d. $1 : 1, 1 : 2, 1.5 : 1, 1.\overline{6} : 1, 1.6 : 1, 1.625 : 1, 1.615 : 1, 1.619 : 1, 1.618 : 1, 1.6\overline{18} : 1, 1.618 : 1, 1.618 : 1, 1.618 : 1$

   e. 1.618 : 1
4. *Sample answers:* Credit card, driver's license, 8 x 5 photograph

5.1 Puzzle Time

SHE DIDN'T WANT TO WAKE UP ANY OF THE SLEEPING BAGS AROUND THE FIRE.

5.2 Start Thinking!

For use before Activity 5.2

Speed limits will vary; miles and hours; examples will vary. *Sample answer:* minutes of homework per day

5.2 Warm Up

For use before Activity 5.2

1. 60
2. 7
3. $\frac{1}{3}$
4. 12
5. 365
6. 52

5.2 Start Thinking!

For use before Lesson 5.2

Sample answer: Write a rate comparing the cost to the amount in the package (number of cups, number of ounces, or number of servings). The best buy has the lowest cost per amount.

5.2 Warm Up

For use before Lesson 5.2

1. $\frac{\$15}{1 \text{ hour}}$
2. $\frac{55 \text{ miles}}{1 \text{ hour}}$
3. $\frac{2 \text{ cups}}{1 \text{ serving}}$
4. $\frac{\$4.05}{1 \text{ pound}}$
5. $\frac{5 \text{ feet}}{1 \text{ second}}$
6. $\frac{4 \text{ degrees}}{1 \text{ hour}}$

5.2 Practice A

1. $\frac{50 \text{ candles}}{4 \text{ boxes}}$ or $\frac{25 \text{ candles}}{2 \text{ boxes}}$
2. $\frac{6 \text{ miles}}{7 \text{ minutes}}$
3. $\frac{\$25}{10 \text{ minutes}}$ or $\frac{\$5}{2 \text{ minutes}}$
4. $\frac{52 \text{ students}}{3 \text{ classrooms}}$
5. $\frac{5 \text{ pounds}}{3 \text{ months}}$
6. $\frac{4 \text{ days}}{\$350}$ or $\frac{2 \text{ days}}{\$175}$
7. $\frac{10 \text{ kilograms}}{9 \text{ cans}}$
8. $\frac{87 \text{ strokes}}{18 \text{ holes}}$ or $\frac{29 \text{ strokes}}{6 \text{ holes}}$
9. $\frac{\$11}{1 \text{ day}}$
10. $\frac{3 \text{ haircuts}}{1 \text{ hour}}$
11. $\frac{64 \text{ heartbeats}}{1 \text{ minute}}$
12. $\frac{5 \text{ liters}}{1 \text{ minute}}$

Answers

13. $\frac{2 \text{ cans}}{1 \text{ person}}$

14. $\frac{3 \text{ outs}}{1 \text{ inning}}$

15. $\frac{17 \text{ drops}}{1 \text{ minute}}$

16. $\frac{3 \text{ children}}{1 \text{ family}}$

17. equivalent

18. not equivalent

19. The error is that you cannot divide 5 minutes by 5 to get 1 hour. Instead you must multiply by 12.

$$\frac{70 \text{ gallons}}{5 \text{ minutes}} \overset{\times 12}{=} \frac{840 \text{ gallons}}{1 \text{ hour}} \quad (\times 12)$$

20. $\frac{45 \text{ text messages}}{1 \text{ hour}}$

21. $\frac{2 \text{ hours of study}}{1 \text{ day}}$

22. 20 minutes

23. $9.60

24. 3 sandwiches for $4.95

25. 20 ounces for $4.40

5.2 Practice B

1. $\frac{6 \text{ kittens}}{3 \text{ boxes}}, \frac{2 \text{ kittens}}{1 \text{ box}}$

2. $\frac{\$96}{16 \text{ hours}}, \frac{\$6}{1 \text{ hour}}$

3. $\frac{72 \text{ biscuits}}{9 \text{ batches}}, \frac{8 \text{ biscuits}}{1 \text{ batch}}$

4. $\frac{1800 \text{ revolutions}}{50 \text{ seconds}}, \frac{36 \text{ revolutions}}{1 \text{ second}}$

5. equivalent

6. not equivalent

7. $\frac{\$5.04}{28 \text{ ounces}}$

8. $\frac{\$0.24}{6 \text{ fluid ounces}}$

9. $\frac{\$0.85}{500 \text{ milliliters}}$

10. $\frac{\$9.60}{1 \text{ dozen}}$

11. 4 fluid ounces per hour

12. 7.3 kilometers per year

13. Grade 6; Grade 7 needs 22 more books to have the same rate.

14. a. about $\frac{107.8 \text{ miles}}{1 \text{ hour}}$

b. about $\frac{0.0093 \text{ hour}}{1 \text{ mile}}$

c. nearly two miles per minute; $\frac{107.8 \text{ miles}}{1 \text{ hour}}$ is about 1.8 miles per minute.

5.2 Enrichment and Extension

1. about \$0.097; $18 \times 32 = 576$ mi;
$18 \times \$3.09 = \$55.62$;
Unit cost $= \frac{\$55.62}{576} \approx \$0.097$

2. about \$0.097; $12 \times 30 = 360$ mi;
$12 \times \$3.29 = \$39.48$;
Unit cost $= \frac{\$39.48}{360} \approx \$0.11$

3. Because $\$0.097 < \$0.11$, your aunt is getting the better value for her vehicle.

4. His unit cost is $\frac{\$38.28}{360} \approx \$0.106$; So, it goes down slightly but is still greater than \$0.097.

5. No;
80% of $\$3.09 \approx \$2.47$; 80% of $\$3.39 \approx \$2.63$

Your aunt's unit cost is $\frac{\$44.46}{576} \approx \$0.077$.

Your uncle's unit cost is $\frac{\$31.56}{360} \approx \$0.088$.

Your aunt is still getting the better value.

6. a. 400,000,000 gallons
b. 250,000 gallons
c. 5000 gallons

5.2 Puzzle Time

BEHIND THE PLATE

5.3 Start Thinking!

For use before Activity 5.3

Sample answer: You are making bracelets to sell at a craft fair. You can make 4 bracelets per hour. How many bracelets can you make in 5 hours?

5.3 Warm Up

For use before Activity 5.3

1. 35 **2.** 30 **3.** 54

4. 2 **5.** 132 **6.** 1460

5.3 Start Thinking!

For use before Lesson 5.3

Answers will vary.

Answers

5.3 Warm Up

For use before Lesson 5.3

1. 25 feet **2.** 90 miles **3.** 7000 feet **4.** 1000 feet

5.3 Practice A

1. 60 ft **2.** 75 mi **3.** 168 m

4. 1008 ft **5.** 510 km **6.** 1144 in.

7. 66 mi/h **8.** 82 km/h **9.** 41 in./sec

10. 1211 ft/min **11.** 25 ft/min **12.** 20 yd/sec

13. 645 mi **14.** 72 mm **15.** 352 ft

16. 640 mi **17.** 5280 ft, or 1 mi

18. 136 mi/h **19.** \$7.50 **20.** \$6.45

21. 3 mi; *Sample answer:* It took you 20 minutes to jog 2 miles, so your unit rate was 10 minutes per mile. The rate $\frac{10 \text{ min}}{1 \text{ mi}}$ is equivalent to $\frac{30 \text{ min}}{3 \text{ mi}}$.

5.3 Practice B

1. 144 ft **2.** 55 mi/h **3.** 7 mi/h

4. 1284 km **5.** about 63 mi/h

6. 32 baseballs **7.** 60 min

8. a. 5520 ft

b. No, you won't meet in 20 minutes, because in 10 minutes, or half the time, you covered less than half the distance.

9. a. 60 minutes: Use the \$0.20 per minute option for a cost of \$12; 100 minutes: Use the \$6.99 plus an additional \$0.10 per minute option for a cost of \$16.99; 275 minutes: Use the \$30 for 400 minutes option for a cost of \$30.

b. Use the \$6.99 plus an additional \$0.10 per minute option because it is the best price for the typical number of minutes and gives you flexibility to pay less when you use fewer minutes.

5.3 Enrichment and Extension

1. a. 160 mi **b.** 2 h 40 min

2. a. 350 mi

b. Tallahassee to Jacksonville; 40 min

c. $58.\overline{3}$ mi/h

3. a. 6 h **b.** 5.5 h **c.** 60 mi/h

5.3 Puzzle Time

THE LETTERS U, V, W, X, Y, AND Z, BECAUSE THEY COME AFTER T

5.4 Start Thinking!

For use before Activity 5.4

Sample answer: Average means typical or ordinary. For example, you could be an average tennis player. Mathematically, average is a number used to describe the middle of a numerical set of data.

5.4 Warm Up

For use before Activity 5.4

1. Pena; 31 homeruns

2. 20% **3.** 50%

5.4 Start Thinking!

For use before Lesson 5.4

Answers will vary.

5.4 Warm Up

For use before Lesson 5.4

1. 2 **2.** 13 **3.** 3.5 **4.** 45.8

5.4 Practice A

1. 6 **2.** 12 **3.** 84

4. 58 **5.** 3 movies **6.** 2.2 students

7. 55 members **8.** 14 problems **9.** 64 lb

10. 203 text messages **11.** 53 mi/h

12. a. 8.7 bikes

b. 1 and 2 are outliers.

c. The outliers cause the mean to decrease. The daily mean without the two outliers is 10.5 bikes, which is more representative of the data.

13. 115 min

5.4 Practice B

1. 86 **2.** 275.375 **3.** 0.3

4. 62.425 **5.** 120 **6.** 32.5

7. Because the mean science test score is 94 and the mean math test score is 93, the mean for your science class is greater by 1 point.

Answers

8. a. The temperature of 103.5° is an outlier since it is more than 4° greater than any of the other temperatures.
   b. with outlier: 99.3°; without outlier: 98.6°; The outlier increases the mean, so eliminating the outlier makes the mean more representative of the temperatures.
   c. The student had a fever.

9. *Sample answer:* 4000 km; Round each radius to the nearest thousand, find the sum, then estimate the quotient of the result and 4; actual mean: 4566.75 km

10. a. The radius of Pluto, 1151 km, is an outlier because it is much smaller than any other values.
    b. with outlier: 36,646.8 km; without outlier: 45,520.75 km; The outlier decreases the mean.
    c. *Sample answer:* The average has more meaning without Pluto, because Pluto is a dwarf planet that is very much unlike the other four objects, which are planets and are all called "gas giants."

11. No; Because the boys' mean height is 60 inches, the sum of the heights of all 6 boys is $6 \times 60 = 360$ inches. So, the sum of the heights of the other 5 boys is $360 - 72 = 288$ inches. Their mean height is $288 \div 5 = 57.6$ inches. This is shorter than the girls' mean height of 59 inches.

5.4 Enrichment and Extension

1. Mrs. Rodriquez's class: 280 lb,
   Mr. Woo's class: 340 lb,
   Mrs. Smith's class: 342 lb;
   Mrs. Smith's class wins.

2. Mrs. Rodriquez's class: 10 lb/student,
   Mr. Woo's class: 14.16 lb/student,
   Mrs. Smith's class: 11.4 lb/student;
   Mr. Woo's class wins.

3. a. Mrs. Rodriquez's class: 520 lb,
   Mr. Woo's Class: 500 lb,
   Mrs. Smith's class: 472 lb
   b. The winner in Exercise 1 is now Mrs. Rodriquez's class. The winner in Exercise 2 is still Mr. Woo's class.

5.4 Puzzle Time

BIG TROUBLE

5.5 Start Thinking!

For use before Activity 5.5

Answers will vary. The mean is probably not a good representation because the teacher's age is most likely an outlier.

5.5 Warm Up

For use before Activity 5.5

1. 5
2. 24.4
3. 48
4. 31
5. 83.5
6. 16

5.5 Start Thinking!

For use before Lesson 5.5

Sample answer: Yes; the data set containing the values 1, 2, and 3 has no mode. Yes; the data set containing the values 1, 1, 2, 3, and 3, has two modes, 1 and 3.

5.5 Warm Up

For use before Lesson 5.5

1. 44; yes
2. 72; no, the mean is greater than all the values except 234. The median is a better average.
3. 13.5; yes
4. 5.875; yes

5.5 Practice A

1. median: 6; mode: none; range: 10
2. median: 7; mode: 2; range: 12
3. median: 12; mode: 12; range: 12
4. median: 18.5; mode: 22; range: 10
5. median: 58.5; mode: none; range: 20
6. median: 50; mode: none; range: 34
7. median: 93.5; mode: 89; range: 9
8. median: 57.5; mode: 60; range: 50
9. median: 160; modes: 160 and 210; range: 140
10. median: 7; modes: 5 and 9; range: 31
11. The ordered data set should include two values of 24. The correct ordered data set has a median of 25.5 and a mode of 24.
12. lab and pit bull
13. no mode

Answers

14. No, the data are not numerical values, so you cannot find a range of values.

15. mean: 10.9; median: 10.5; mode: 9; range: 10

16. mean: 6.6; median: 4.5; mode: 3; range: 15

5.5 Practice B

1. mean: 27; median: 28; mode: none; range: 11

2. mean: 21.2; median: 21; mode: 27; range: 23

3. mean: 0.4; median: 0.405; mode: none; range: 0.22

4. mean: 2.1; median: 2.25; mode: 2.4; range: 1.5

5. mean: $\frac{1}{2}$; median: $\frac{1}{2}$; mode: $\frac{1}{2}$; range: $\frac{2}{3}$

6. mean: $3\frac{1}{4}$; median: $3\frac{9}{16}$; mode: $4\frac{1}{8}$; range: $3\frac{1}{2}$

7. B

8. X

9. mean: about 9.8 min; median: 8 min; mode: none; range: 21 min

10. 2, 3, 2, 1, 3;

a. mean: 5; median: 5; modes: 4 and 6; range: 6

b. mean: 2.5; median: 2.5; modes: 2 and 3; range: 3

c. Each answer in part (a) is twice the corresponding answer in part (b) because there are twice as many free throw attempts as free throws made in each game.

11. There is more likely to be a mode for the home runs. *Sample answer:* Each set will have the same number of data. Because the range of numbers of home runs is likely to be much smaller than the range of number of hits, the chance that at least two of those numbers are the same is greater.

12. No; A difference in the weights of people is likely to be several pounds, but a difference in the weights of cars is likely to be several hundred of pounds.

5.5 Enrichment and Extension

1. a. The ages are 3, 10, 13, 42, and 42 years old.

b. No; At some point during the year, the 3-year old will turn 4 or one of the parents will turn 43. This will change the range at that point of the year. The only way the range would always stay the same is if the 3 year-old and both parents all share the same birthday.

2. a. The ages are 10, 12, 12, 16, 21, 45, and 45 years old.

b. *Sample answer:* Both parents are 45 years old and they have 12-year old twins.

3. a. $5\frac{1}{2}$ ft, 6 ft, $7\frac{1}{4}$ ft

b. 10.25 lb, 12.75 lb. 15.25 lb

c. It is likely that the eagles with bigger wingspans will weigh more.

Sample answer:

	Wingspan (feet)	Weight (pounds)
Bald Eagle 1	$5\frac{1}{2}$	10.25
Bald Eagle 2	6	12.75
Bald Eagle 3	$7\frac{1}{4}$	15.25

5.5 Puzzle Time

UMBRELLA

5.6 Start Thinking!

For use before Activity 5.6

Answers will vary.

5.6 Warm Up

For use before Activity 5.6

1.

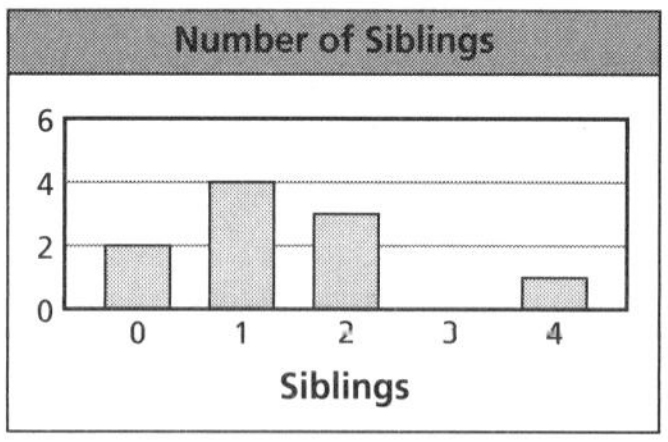

2.

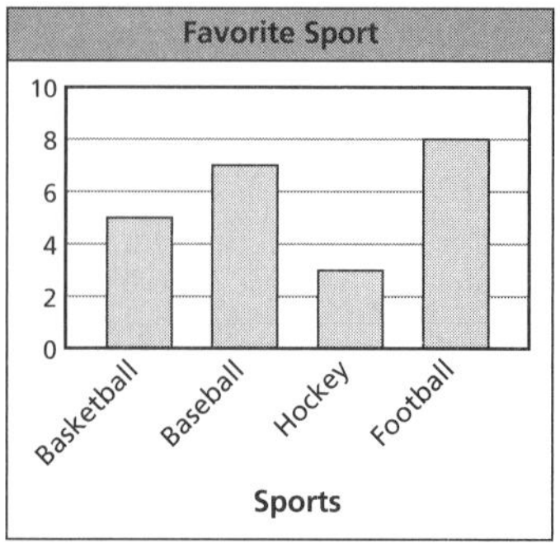

Answers

3. 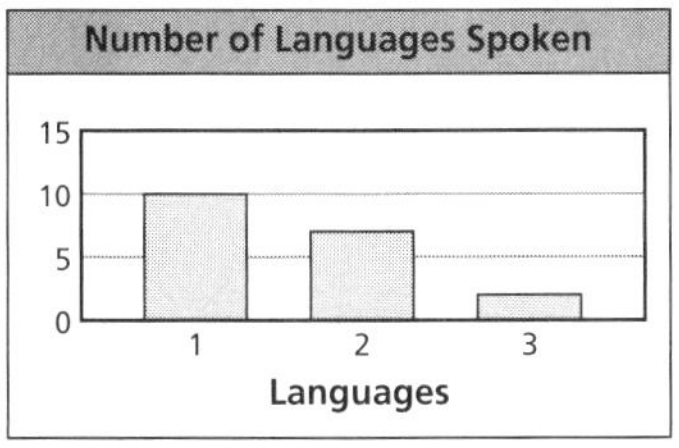

4. 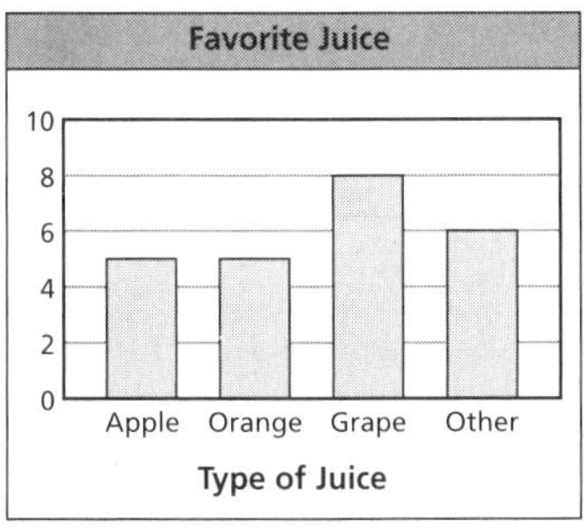

5.6 Start Thinking!

For use before Lesson 5.6

Answers will vary.

5.6 Warm Up

For use before Lesson 5.6

1.

Number	Tally	Total
4	|	1
5	||	2
6	||	2
7	||	2

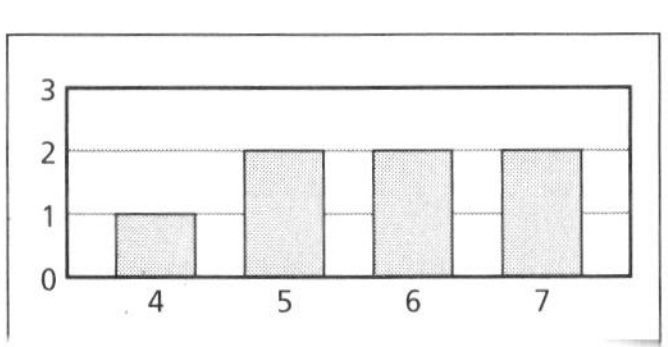

2.

Number	Tally	Total
0	||	2
1	|	1
2		0
3	||	2
4	|	1
5	|	1

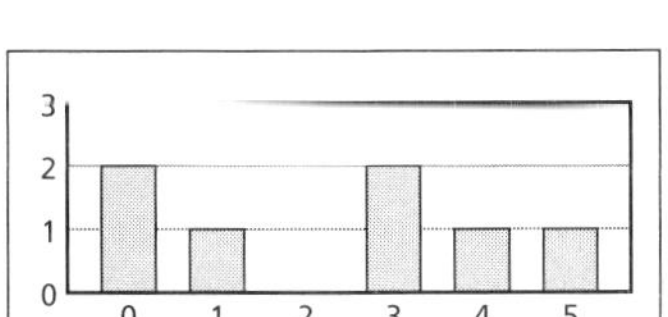

3.

Number	Tally	Total
5	||	2
10	||	2
15	||	2

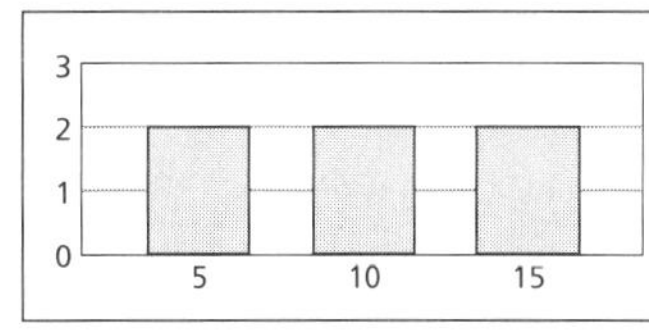

4.

Number	Tally	Total
1	||	2
2	|	1
3	|	1
4		0
5	|	1
6		0
7		0
8	|	1
9		0
10		0
11		0
12		0
13	|	1

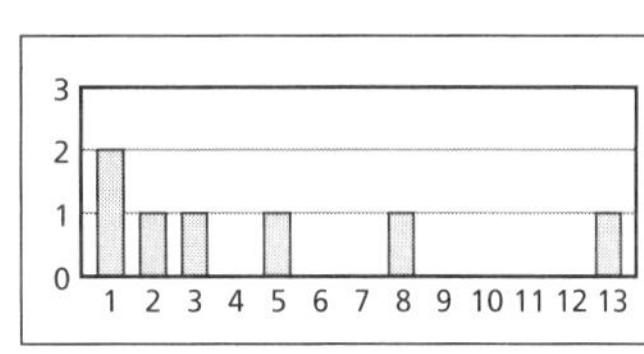

5.6 Practice A

1. mean: 7.75; median: 8.5; mode: 10; *Sample answer:* The median is probably best, because seven of the eight data values are between 6 and 11, and 8.5 is right in the middle of these.

2. mean: 52.75; median: 61.5; mode: 66; *Sample answer:* The mode or median is probably best. The mode is the same as three of the data. The median is in the middle of the data. The mean is quite low compared to most of the data.

3. mean: 14; median: 8; mode: 6; *Sample answer:* The mean is probably best. The median and mode are both close to the lower values but far away from the greater values.

4. mean: 77; median: 96; mode: 100; *Sample answer:* The mode or median is probably best. They are nearly the same and are both very close to most of the values. The mean is lower than most of the values.

5. *with outlier:* mean: 80, median: 92, mode: 99; *without outlier:* mean: 92, median: 93.5, mode: 99; The mean is most affected. It is much lower with the outlier. The median is slightly lower with the outlier. The mode is unchanged by the outlier.

6. *with outlier:* mean: 19, median: 12, modes: 10 and 12; *without outlier:* mean: 13, median: 12, modes: 10 and 12; The mean is the only measure affected by the outlier. It is higher with the outlier.

7. *with outlier:* mean: 62, median: 66, mode: 74; *without outlier:* mean: 69, median: 70, mode: 74; The outlier reduces the mean by 7 and the median by 4. The mode is not affected by the outlier.

8. *with outlier:* mean: 21, median: 22, mode: none; *without outlier:* mean: 13.5, median: 13, mode: none; The outlier increases the mean by 7.5 and increases the median by 9. There is no mode.

9. a. mean: 4.9 h; median: 2.5 h; mode: 1 h
   b. *Sample answer:* The median is best. The mean is greater than many of the data and the mode is less than many of the data.
   c. 25; It is much greater than the other values.
   d. *Sample answer:* The mean will probably decrease quite a bit, the median may decrease slightly, and the mode will not change.

10. a. the median b. the mean
    c. Multiply the number of books by the mean, 5 lb.

11. Answers should show the daily high and low temperature predictions for 7 to 10 days.
    a. Answers should show the mean, median, and mode(s) of the daily high temperatures and of the daily low temperatures.
    b. Answers should give the best measure for each set of data with an explanation showing why it is best.

5.6 Practice B

1. mean: 79; median: 78; mode: none; *Sample answer:* The mean and median are close in value and both represent the data well.

2. mean: 53.75; median: 56.5; mode: 65; *Sample answer:* The median is probably best, because the mode is the highest data value and the mean is less than most of the data values.

3. mean: 92; median: 75; mode: 73; *Sample answer:* The mean is probably best. The median and mode are both close to the lower values but far away from the greater values.

4. mean: 24.4; median: 20.45; mode: none; *Sample answer:* The median is probably best, because the mean is greater than all but one of the data values.

5. *with outlier:* mean: 64, median: 71, mode: 72; *without outlier:* mean: $70\frac{1}{3}$, median: 71.5, mode: 72; The mean is most affected. It is much lower with the outlier. The median is slightly lower with the outlier. The mode is unchanged by the outlier.

6. *with outlier:* mean: 84.375, median: 77.5, mode: 85; *without outlier:* mean: 75, median: 70, mode: 85; The mean and median are both greater with the outlier. The mode is unchanged by the outlier.

7. a. mean: $12.98; median: $9.95; mode: $9.95
   b. *Sample answer:* Either the median or the mode represent the typical price well, because four of the five gifts cost around $10 and the median and mode are both $9.95.
   c. The mean represents each child's share of the cost exactly, because they are sharing the total cost equally.

8. 44 yd; *Sample answer:* Because the mean is 30, the quotient of the sum of the data values and 5 must be 30, so the sum of the data values must be 150. The fifth punt has to be 44 yards for the sum to be 150.

9. a. The mode is 20 feet.
   b. The players took the shortest shots that would earn them 3 points by the 3-point rule.
   c. Both are less than 20 feet. Because the mode is the distance of the longest shots, the mean of all the shots must be a shorter distance. There are 13 shots, so the median is the seventh shot when the distances are written in order. There are 5 shots on the 3-point line, so the seventh shot is less than 20 feet.

5.6 Enrichment and Extension

1. *Sample answer:* Hawks: 6'3"; Tigers: 6'1"; Panthers: 6'8"

2. Hawks: 6'3"; Tigers: 6'2"; Panthers: 6'7"

3. Hawks: 6'2"; Tigers: 5'11" and 6'0"; Panthers: 6'8"; The mode(s) and means of each data set are close to each other.

Answers

4. Hawks: 1'4"; Tigers: 1'1"; Panthers: 1'4"; The Hawks and Panthers have the same range. No; The heights of the players on the Hawks appear to be more evenly spread out compared to the Panthers.

5. The heights of the players on the Hawks appear to be evenly spread out with most of the heights between 6'1" and 6'5".

   The heights of the players on the Tigers appear to be toward the lower end of the height scale with most of the heights clustered around 6'0".

   The heights of the players on the Panthers appear to be toward the upper end of the height scale with most of the heights greater than 6'6".

6. Yes; The gaps occur between the shortest players and tallest players on the teams.

   Gaps can be caused by outliers.

5.6 Puzzle Time

A SOLE

School-to-Work

Sample answer: Encourage students to consider survey questions with answers that are easy to group. For example "yes/no/maybe" rather than open-ended questions.

Talk about price points—what price is too high? What types of people buy the product, and how would you reach them? (For example, almost everyone would stop at a snack bar, but only certain groups go to a store selling team baseball caps.)

Study Help

Available at *BigIdeasMath.com*.

Financial Literacy

1–2. *Sample answer:* Dried pasta, unit is the pound

manicotti: \$1.89, 8 oz; \$3.78/lb

whole wheat spaghetti: \$2.29, lb; \$2.29/lb

tri-color rotini: \$1.39, 12 oz; \$1.85/lb

fettuccine: \$1.39, 16 oz; \$1.39/lb

macaroni 1: \$1.09, 1 lb; \$1.09/lb

egg noodles: \$1.89, 1 lb; \$1.89/lb

whole wheat lasagna: \$2.49, 8 oz; \$4.98/lb

spaghetti: \$1.39, 16 oz; \$1.39/lb

rotini: \$1.39, 1 lb; \$1.39/lb

macaroni 2: \$1.39, 1 lb; \$1.39/lb

3. *Sample answer:* highest whole wheat lasagna, lowest macaroni 1; I guessed the macaroni was least expensive, but I didn't realize how expensive the whole wheat lasagna was until I found the unit rate, because the quantity in the box was much less than most boxes.

4. **a.** *Sample answer:* To make comparison shopping easier for consumers.

   b. *Sample answer:* Sale items do not have the unit price recalculated, so you need to be able to determine if the savings are worthwhile.

5. Store A: $(\$3.99 \times 2) \div (13 \text{ oz} \times 3) \approx \$0.20/\text{oz}$;

   Store B: $(\$2.99 - \$0.50) \div (9 \text{ oz}) \approx \$0.28/\text{oz}$;

   Store A is the better buy.

Technology Connection

1. http://academic.reed.edu/math/faculty.html and http://people.reed.edu/~davidp/111/; Reed College; http://academic.reed.edu/math/faculty.html, There is no ~ signifying a personal directory.

2. www.k12.hi.us; "HI" is the state abbreviation for Hawaii, so the school is in Hawaii. A homework policy is posted by a "konawahs."

3. www.ed.gov; You would find editor picks for homework resources available to students.

4. An organization is noted by org: www.nctm.org; a company is noted by com, so the company URL is www.bigideaslearning.com.

5. www.k12.hi.us/~konawahs and http://people.reed.edu/~davidp/111; *Sample answer:* Go to the main site and search for information about the person. For example, working backwards to the main website, you can discover who "davidp" is at http://academic.reed.edu/math/faculty.

6. Check students' work.

Cumulative Practice

1–16. Sample answers are given.

1. $\frac{1}{3}, \frac{2}{6}$ **2.** $\frac{1}{10}, \frac{2}{20}$ **3.** $\frac{2}{10}, \frac{3}{15}$ **4.** $\frac{1}{4}, \frac{2}{8}$

5. $\frac{3}{1}, \frac{6}{2}$ **6.** $\frac{6}{16}, \frac{9}{24}$ **7.** $\frac{1}{2}, \frac{2}{4}$ **8.** $\frac{14}{12}, \frac{21}{18}$

9. $\frac{18}{50}, \frac{27}{75}$ **10.** $\frac{2}{3}, \frac{4}{6}$ **11.** $\frac{7}{4}, \frac{14}{8}$ **12.** $\frac{7}{15}, \frac{14}{30}$

Answers

13. $\frac{3}{7}, \frac{6}{14}$ 14. $\frac{4}{3}, \frac{8}{6}$ 15. $\frac{4}{5}, \frac{8}{10}$ 16. $\frac{3}{4}, \frac{6}{8}$

17. 24 strawberries
18. 16 miles per minute
19. $1.25 per pound
20. 29 words per minute
21. 7 fish per aquarium
22. $0.15 per gram
23. 33 chairs per room
24. 30 bottles for $13.20
25. 585 m
26. 13 in./min
27. 330 mm
28. $32
29. 15
30. 22
31. 2
32. 59
33. 32
34. 53
35. mean: 8; median: 7.5; no mode; *Sample answer:* Either the mean or the median represent the data because the values are approximately equal.
36. mean: 62; median: 70; mode: 70; *Sample answer:* Either the median or mode represent the data well; the mean is lower than most data values.
37. mean: 9; median: 10.5; mode: 11; *Sample answer:* The median represents the data well. The mean is lower than most data values and the mode is the greatest value.
38. mean: 32.0; median: 30.2, no mode; *Sample answer:* The median represents the data well. The mean is greater than most data values.

Chapter 6

6.1 Start Thinking!

For use before Activity 6.1

Check students' piems.

6.1 Warm Up

For use before Activity 6.1

1. 24 m
2. 40 cm
3. 36 yd
4. 120 ft
5. 16 in.
6. 60 mm

6.1 Start Thinking!

For use before Lesson 6.1

Sample answer: The circumference of a tree trunk is easier to measure. Unless the tree is cut down, you do not have access to the circular cross section in order to measure the diameter. The diameter of a quarter is easier to measure. You only need a ruler to measure across the quarter. In general, it is usually easier to measure the diameter of a circular object. The circumference is easier to measure only when you cannot access the diameter to measure it.

6.1 Warm Up

For use before Lesson 6.1

1. about 37.68 in.
2. about 12.56 ft
3. about 50.24 cm
4. about 62.8 ft
5. about 18.84 mm
6. about 3.14 in.

6.1 Practice A

1. 30 ft
2. 4 m
3. 32 mm
4. 5 cm
5. 12 in.
6. 3.5 yd
7. about 53.38 m
8. about 18.84 ft
9. about 88 in.
10. about 56.54 mm
11. about 20.56 in.
12. about 25.7 yd
13. The circumference calculation used the radius of the circle instead of using the diameter.

$$C = \pi d$$
$$\approx 3.14 \bullet 4$$
$$= 12.56 \text{ m}$$

14. 76 mm
15. $131.56
16. about 5.14 ft

6.1 Practice B

1. about 81.64 m
2. about 37.68 ft
3. about 264 in.
4. about 71.96 yd
5. about 38.55 ft
6. about 28.27 cm
7.

Circle	*A*	*B*	*C*	*D*
Radius	2.5 ft	1 ft	32 ft	3.5 ft
Diameter	60 in.	24 in.	768 in.	84 in.

8. about 9.42 in.
9. yes; The diameter of each of the small circles in diagram B is 2 feet. Because there are 5 circles along each side of the square, the length of each side is $5 \bullet 2 = 10$ feet.
10. about 10 times

6.1 Enrichment and Extension

1. 25.12 in., 50.24 in.
2. 43.96 ft, 87.92 ft
3. 47.1 cm, 94.2 cm
4. The circumference is multiplied by 2.
5. The circumference is multiplied by *n*.

Answers

6. 12.85 mm, 6.425 mm **7.** 5.14 in., 2.57 in.

8. 51.4 m, 25.7 m

9. The perimeter is multiplied by $\frac{1}{2}$.

10. The perimeter is multiplied by n.

6.1 Puzzle Time

BECAUSE HE MIGHT GET A HOLE IN ONE.

6.2 Start Thinking!

For use before Activity 6.2

Sample answer: To find the perimeter of an irregular shape, you can add all of the different straight sections (or circular sections). To find a distance on a map, assuming that the route traveled is not along a straight line, you must add all of the straight sections to estimate the total distance.

6.2 Warm Up

For use before Activity 6.2

1. 16 cm **2.** 17 ft **3.** 52 in.

4. 25 m **5.** about 56.52 in. **6.** about 62.8 ft

6.2 Start Thinking!

For use before Lesson 6.2

Sample answer: Yes; it is made up of a rectangle and two semicircles. Some examples: windows (rectangle and semicircle), some company and organization logos, a first place ribbon (circle and rectangles), some flowers

6.2 Warm Up

For use before Lesson 6.2

1. 25 in. **2.** 24 in.

3. 15 in. **4.** about 13.71 in.

5. about 20.5 in. **6.** about 22.28 in.

6.2 Practice A

1. 24 in. **2.** 28 in. **3.** about 21.42 in.

4. 34 yd **5.** 24 in. **6.** 60 mm

7. 31 m **8.** 24 ft **9.** 64 cm

10. about 23.42 ft **11.** \$3685.90

6.2 Practice B

1. 22 in. **2.** 28 in.

3. 22 in. **4.** 27 m

5. about 140.76 cm **6.** 28 in.

7. The perimeter calculation included the circumference of the circle instead of the perimeter of the semicircle.

$$\text{Perimeter} \approx 2 + 7 + 2 + 10.99$$
$$\approx 21.99 \text{ ft}$$

8. about \$211.31

9. *Sample answer:*

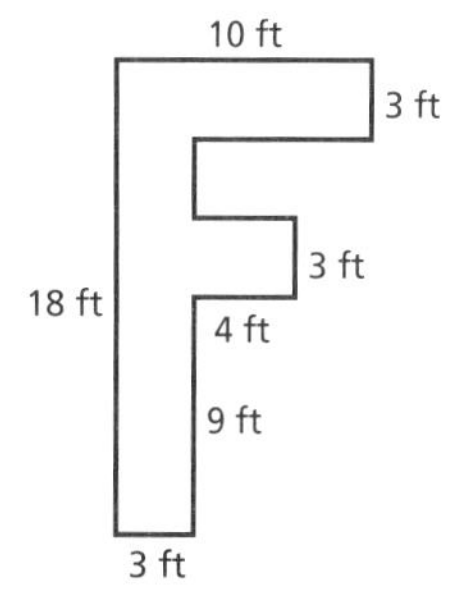

Yes; for example, if the 11 ft by 3 ft rectangle on the bottom of the F is shortened to be a 6 ft by 3 ft rectangle, the perimeter decreases by 10 ft.

6.2 Enrichment and Extension

1. about 25.12 in. **2.** about 18.84 m

3. about 34.26 ft **4.** about 30.26 yd

5. about 55.4 cm **6.** about 50.24 mm

6.2 Puzzle Time

LIBRARY

6.3 Start Thinking!

For use before Activity 6.3

Sample answer: **1.** You can divide it into several triangles each with a vertex at the center of the circle. **2.** You can put it on grid paper and estimate the number of squares that are covered. **3.** You can divide it into tall, narrow rectangles.

6.3 Warm Up

For use before Activity 6.3

1. 63 in.^{2} **2.** 14 m^2 **3.** 66 yd^2

4. 50 cm^2 **5.** 22.4 m^2 **6.** 1 mm^2

6.3 Start Thinking!

For use before Lesson 6.3

Sample answer: It is easier to use $\frac{22}{7}$ if the radius is a multiple of 7. It is easier to use 3.14 in all other instances. Check students' word problems.

Answers

6.3 Warm Up

For use before Lesson 6.3

1. about 50.24 in.^{2} **2.** about 154 ft^2

3. about 3.14 cm^2

6.3 Practice A

1. about 78.5 m^2 **2.** about 7850 mm^2

3. about 1256 in.^{2} **4.** about 38.5 ft^2

5. about 1386 mm^2 **6.** about 3850 cm^2

7. about 50.24 ft^2 **8.** about 15,400 cm^2

9. about 113.04 in.^{2} **10.** about 28.26 in.^{2}

11. $\frac{99}{224}$ in.^{2} **12.** about 153.86 in.^{2}

6.3 Practice B

1. about 962.5 in.^{2} **2.** about 19.625 m^2

3. about 7234.56 mm^2 **4.** about 157 in.^{2}

5. about 100.48 m^2 **6.** about 693 ft^2

7. about 2464 ft^2 **8.** about 28.26 ft^2

9. a. $A = \pi r^2$

b. $A = \pi\left(\frac{r}{2}\right)^2$

c. The circle's area is one fourth of the area of the circle whose radius is twice as large; The area formula for the smaller circle can be rewritten as $A = \pi\left(\frac{r}{2}\right)^2 = \frac{1}{4}\pi r^2$; this is one fourth of the area of the larger circle.

10. The radius is 2 units; Set the circumference equal to the area to get $2\pi r = \pi r^2$. So $\pi \bullet r \bullet 2 = \pi \bullet r \bullet r$; the value of r must be 2.

6.3 Enrichment and Extension

1. 1.5625 **2.** 2.25 **3.** 16% **4.** $\frac{1}{9}$

5. a. $3.\overline{5}\%$

b.

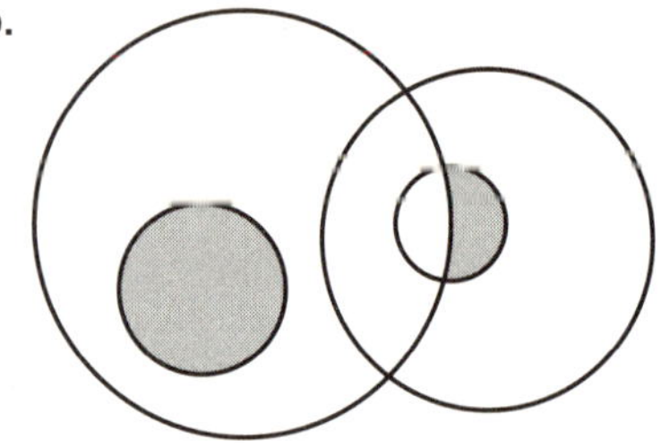

13,816 mi^2

c.

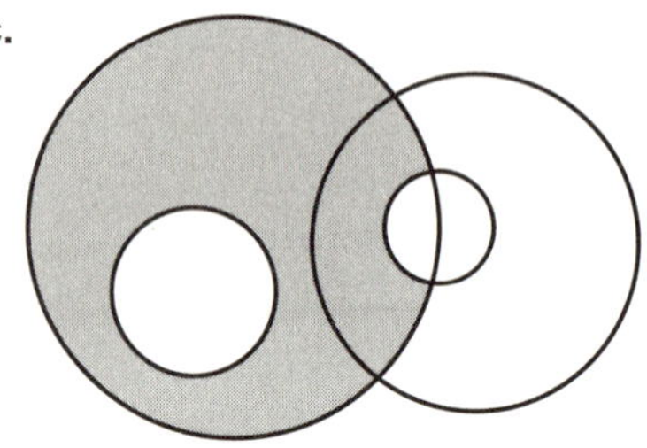

56,834 mi^2

6.3 Puzzle Time

HAMWORK

6.4 Start Thinking!

For use before Activity 6.4

Answers will vary. *Sample answer:* If you are planning on painting a room, you will need to calculate the area of each section of wall to figure out how much paint to buy. The windows and doorways make each wall a composite figure.

6.4 Warm Up

For use before Activity 6.4

1. 100 ft^2 **2.** 256 in.^{2}

3. 150 m^2 **4.** 30 cm^2

5. about 314 ft^2 **6.** about 7850 yd^2

6.4 Start Thinking!

For use before Lesson 6.4

Check students' drawings and calculations.

6.4 Warm Up

For use before Lesson 6.4

1. 21 in.^{2} **2.** 20 in.^{2}

3. 12 in.^{2} **4.** about 12.53 in.^{2}

5. 18 in.^{2} **6.** about 17.72 in.^{2}

Answers

6.4 Practice A

1. 30 m^2 2. 33 m^2 3. 16 m^2
4. 48 in.^2 5. about 178.5 mm^2
6. 402 ft^2 7. 210 cm^2
8. 60 yd^2 9. 40 m^2
10. Perimeter: about 37.85 ft; Area: about 69.625 ft^2
11. a. 144 in.^2 b. 72 in.^2 c. $172.80

6.4 Practice B

1. Perimeter: about 23.42 ft; Area: about 38.13 ft^2
2. Perimeter: 18 mm; Area: 16 mm^2
3. Perimeter: 26.6 cm; Area: 42.6 cm^2
4. 27.98 in.^2 5. about 4.55 m^2
6. 6.63 yd^2
7. The area calculation included the area of the circle instead of the area of the semicircle.

$$\begin{aligned}\text{Shaded area} &\approx (4 \bullet 4) - \left(\frac{3.14 \bullet 2^2}{2}\right)\\ &\approx 16 - 6.28\\ &\approx 9.72 \text{ ft}^2\end{aligned}$$

8. a. 39 in. b. 54 in.^2
   c. No. The dimensions of the logo are 10.5 in. by 12 in., but the dimensions of the notebook cover are 8.5 in. by 11 in.

6.4 Enrichment and Extension

1. I: 78 m^2 2. P: about 50.7 m^2
3. M: 50 m^2 4. S: about 75.36 m^2
5. C: about 21.98 m^2 6. T: 82.775 m^2
7. O: about 58.875 m^2 8. E: 98.935 m^2
9. O: about 37.68 m^2 10. COMPOSITE

6.4 Puzzle Time

DO WE WALK OR HOP ON A DOG?

School-to-Work

1. 64 feet of fence 2. 42 feet of edging
3. about 228 square feet
4. about 252 square feet of flower beds; 63 cubic yards of mulch
5. Check students' work.
6. Check students' work.

Study Help

Available at *BigIdeasMath.com*.

Cumulative Practice

1. 12 2. 56 3. 8.68 4. 72
5. 75 6. $\frac{9}{5}$ 7. 156 in. 8. 92 yd^2
9. about 37.68 mm 10. 238 ft^2
11. about 452.16 cm^2 12. 92 km
13. 225 yd^2 14. about 58 mm
15. about 2688 m^2 16. $4\frac{3}{8} \text{ in.}^2$
17. about 10.28 cm
18. about 2.512 in.; about 0.5024 in.^2

Chapter 7

7.1 Start Thinking!

For use before Activity 7.1

Check students' problems; *Sample answer:* Advantages are that you need to understand the problem in order to rewrite it and the language is probably simpler and easier to understand. Disadvantages are that you may introduce errors by copying numbers incorrectly or not include all of the necessary information.

7.1 Warm Up

For use before Activity 7.1

1. 60 mi 2. 80 in.^2 3. 2 mi

7.1 Start Thinking!

For use before Lesson 7.1

Answers will vary.

7.1 Warm Up

For use before Lesson 7.1

Answers will vary. Sample answers are given.

1. How far do you walk for 30 minutes at 15 minutes per mile?; 2 mi

Answers

2. What is $\frac{2}{3}$ of 99 pitches?; 66 pitches

7.1 Practice A

1. *Sample answer:* You type at a steady rate of 20 words per minute for 48 minutes to enter your report on a computer. How many words long is your report? 960 words

2. *Sample answer:* A park is 6 miles from your home. You and your friend ride your bikes at a steady rate for one-half hour to get home from the park. What was your riding rate? 12 miles per hour

3. $4 = 16 \div n$ or $4 = \frac{16}{n}$

4. $k + 5 = 11$

5. $x - 2 = 7$

6. $7a = 42$

7. $6 = \frac{1}{3}s$

8. $13 + t = 17$

9. $b - 5 = 16$

10. $30 = 6y$

11. $z \div 7 = 4$ or $\frac{z}{7} = 4$

12. $m + 5 = 8$

13. The number that should be multiplied by x is 8, not 24; $24 = 8x$

14. $n - 217 = 36$

15. $4n = 72$

16. $0.6h = 27$

17. $4070 + 2710 + p = 8180$

7.1 Practice B

1. $17 = 41 - n$

2. $168 \div x = 14$ or $\frac{168}{x} = 14$

3. $17s = 153$

4. $b + 37 = 112$

5. $23 = t \div 61$

6. $114 = g + 37$

7. $a - 46 = 33$

8. $\frac{1}{5}d = 22$

9. $2\pi x = 36\pi$

10. $108 = \frac{1}{2}(12)x$

11. $10w = 520(0.25)$

12. $2(25) + 12(10) + 23 + p = 200$

13. $24 = \frac{1}{2}h + \frac{1}{2}\left(\frac{1}{2}h\right)$

14. $6(27) + 8m = 626$

15. a. Because 3 to 10 fibers are used for each thread, it isn't possible to write an equation to find the exact total length of the fibers.

    b. *Sample answer:* Using 6 fibers, an equation is $300 = \ell \div 6$.

7.1 Enrichment and Extension

1. $n^2 + 7 = 23$

2. $10 - p^2 = 9$

3. $x - y = 8$

4. $x^2 + y^2 = 4$

5. Maria

6. The variable x is not squared. The only number being squared is 2.

7. *Sample answer:* A number x plus 2 quantity squared is equal to 9.

8. *Sample answer:* Three minus a number x quantity squared is equal to 4.

7.1 Puzzle Time

IT SEEMS NO ONE CAN EAT JUST ONE POTATO SHIP.

7.2 Start Thinking!

For use before Activity 7.2

Sample answer: You can represent the situation with an equation. Assume that your scale reads 133 pounds for your combined weight and 100 pounds for your weight alone. Then you would solve the equation $100 + w = 133$ to find the weight w of your pet.

7.2 Warm Up

For use before Activity 7.2

1. 5
2. 20
3. 15
4. 8
5. 0
6. 14

7.2 Start Thinking!

For use before Lesson 7.2

Addition and subtraction are inverse operations. Multiplication and division are inverse operations. Inverse actions are:

1. Button
2. Descend
3. Withdraw
4. Turn on

Examples will vary.

Answers

7.2 Warm Up

For use before Lesson 7.2

1. $n = 3$ **2.** $r = 4$ **3.** $p = 3$

4. $t = 6$ **5.** $d = 7$ **6.** $x = 4$

7.2 Practice A

1. yes **2.** yes **3.** no **4.** no

5. $x = 14$ **6.** $y = 12$ **7.** $q = 17$

8. $f = 36$ **9.** $s = 2$ **10.** $r = 4.9$

11. $v = 1.6$ **12.** $n = \frac{1}{3}$ **13.** $g = \frac{5}{12}$

14. You must subtract 13 from both sides of the equation.

$$\begin{aligned} 13 + m &= 56 \\ -13 \quad & \quad -13 \\ m &= 43 \end{aligned}$$

15. You must add 15 to both sides of the equation.

$$\begin{aligned} 27 &= n - 15 \\ +15 \quad & \quad +15 \\ 42 &= n \end{aligned}$$

16. $20 = 8 + y$; $y = 12$ **17.** $x + 12 = 15$; $x = 3$

18. $g - 4 = 9$; $g = 13$ **19.** $w - 10 = 3$; $w = 13$

20. Let x be the height of the desk in inches; $x + 11 = 41$; $x = 30$; 30 in.

21. Let h be the average height in inches of the Northern Mockingbird; $h + 20 = 30$; $h = 10$; 10 in.

22. Let p be the price of the cell phone before the rebate; $p - 50 = 149$; $p = 199$; \$199.

7.2 Practice B

1. yes **2.** no **3.** yes **4.** yes

5. $a - 13.4 = 2.6$; $a = 16$

6. $h - 27 = 3.5$; $h = 30.5$

7. $46 = 2.5 + z$; $z = 43.5$

8. $b + 4.7 = 10.9$; $b = 6.2$

9. $x = 208$ **10.** $j = 103$ **11.** $a = \frac{1}{5}$

12. $n = 28$ **13.** $v = 59$ **14.** $c = 47$

15. $y = 61$ **16.** $e = 84$ **17.** $\ell = 5$

18. $x + 13 + 5 = 30$; $x = 12$; 12 in.

19. $x + 10 + 15 + 10 = 43$; $x = 8$; 8 m

20. $x + 5 + 5 = 16$; $x = 6$ ft

21. a. $m - 1.56 = 2.9$, $m = 3.75$, \$3.75;
$e + 2.36 = 3.75$, $e = 1.39$, \$1.39;
$c - 3.20 = 1.39$, $c = 4.59$, \$4.59

b. \$0.08

22. $c - 10 - 5.80 = 33.19$; $c = 48.99$; \$48.99

7.2 Enrichment and Extension

1. yes **2.** no **3.** yes **4.** no

5. no **6.** no **7.** yes **8.** yes

9. *Sample answer:* $x + 3 = 13$ and $x + 7 = 17$

10. *Sample answer:* $x - 2 = 5$ and $x - 4 = 3$

11. $8 + 10 + 9 + 3 + x = 40$; $x = 10$

12. $2.5 + 1\frac{3}{4} + 3\frac{1}{4} + 2 + x = 12$; $x = 2.5$

7.2 Puzzle Time

MICE KRISPIES

7.3 Start Thinking!

For use before Activity 7.3

Sample answer: The cost of four movie tickets is \$28. How much does each ticket cost?

7.3 Warm Up

For use before Activity 7.3

1. $x = 3$ **2.** $y = 8$ **3.** $z = 24$

4. $a = 28$ **5.** $b = 13$ **6.** $c = 16$

7.3 Start Thinking!

For use before Lesson 7.3

Sample answer: If you know the distance and time it takes, you can plug in the distance d and time t into the equation $d = rt$ and solve for r. If you want to find the rate in miles per hour, be sure to use distance in miles and time in hours.

Answers

7.3 Warm Up

For use before Lesson 7.3

1. $x = 41$ **2.** $w = 2$ **3.** $b = 11$

4. $v = 42$ **5.** $p = 60$ **6.** $k = 104$

7.3 Practice A

1. $x = 18$ **2.** $t = 16$ **3.** $w = 80$

4. $s = 42$ **5.** $a = 3$ **6.** $d = 5$

7. $m = 3$ **8.** $g = 2$ **9.** $y = 8$

10. $n = 21$ **11.** $v = 44$ **12.** $c = 35$

13. $b = 11$ **14.** $h = 150$ **15.** $k = 2.5$

16. $r = 30$

17. Both sides should be divided by 9.

$$\frac{x}{9} = 3$$
$$9 \bullet \frac{x}{9} = 9 \bullet 3$$
$$x = 27$$

18. Both sides should be divided by 4.

$$4 \bullet z = 32$$
$$\frac{4 \bullet z}{4} = \frac{32}{4}$$
$$z = 8$$

19. $36 = 4s$; 9 students

20. $384 = 12w$; 32 weeks

21. $\frac{c}{8} = 352$; 2816 calories

22. $\frac{s}{8} = 7.25$; 58 points

23. $36w = 576$; 16 cm; the area is given in square centimeters and the length is given in centimeters, so the width must be in centimeters also.

7.3 Practice B

1. $a = 13$ **2.** $s = 140$ **3.** $p = 4$

4. $g = 108$ **5.** $v = 108$ **6.** $x = 6.5$

7. $c = 61.56$ **8.** $b = 9.5$ **9.** $x = 7$ units

10. $x = 14$ units **11.** $x = 6$ units

12. Use the circumference formula, $c = \pi d$, to write an equation, then solve for d. The diameter is about 36 feet.

13. *Sample answer:* The price of gas is $3.299 per gallon; $\frac{p}{15} = 3.299$; $p = 49.485$; it costs $48.49 to fill the empty tank.

14. 20% **15.** 16 cups

16. $2.70; *Sample answer:* Write and solve an equation to find the weight w of each biscuit: $25w = 0.75$; $w = 0.03$ lb. Multiply to find the weight of 40 biscuits; $40(0.03) = 1.2$ lb. Then multiply to find the price of 1.2 pounds of biscuits: $1.2(2.25) = \$2.70$.

7.3 Enrichment and Extension

1. $\frac{7}{9}$ **2.** $\frac{2}{3}$ **3.** $\frac{4}{11}$

4. $\frac{17}{33}$ **5.** $\frac{43}{99}$ **6.** $\frac{1}{11}$

7. $4\frac{89}{99}$ **8.** $\frac{245}{333}$ **9.** $\frac{29}{999}$

10. Let $x = 0.2\overline{3}$.

$$10x = 2.333\ldots$$
$$-x = 0.2333\ldots$$
$$9x = 2.1$$
$$90x = 21$$
$$\frac{90x}{90} = \frac{21}{90}$$
$$x = \frac{7}{30}$$

7.3 Puzzle Time

IF THIS HEAVY RAIN KEEPS UP MY NAME WILL CHANGE AND I WILL BE CALLED MUD.

7.4 Start Thinking!

For use before Activity 7.4

Sample answer: The result is always the chosen number. In the steps, there are 2 pairs of inverse operations. Multiplying and dividing by 2 are inverses. At first glance, adding 3 and subtracting 6 do not seem to be inverses, but because your result after Step 2 is $2(x + 3) = 2x + 6$, subtracting 6 will undo the addition of 3 (multiplied by 2).

Answers

7.4 Warm Up

For use before Activity 7.4

1. Subtract 3. **2.** Add 5. **3.** Divide by 6.

4. Multiply by 2. **5.** Subtract 4.57. **6.** Divide by 2.5.

7.4 Start Thinking!

For use before Lesson 7.4

Sample answer: At online photo websites, there is usually a flat shipping rate and a cost per print. So, if you have a certain amount of money that you are willing to spend, and know the shipping rate and cost per print, you can solve a two-step equation to find the number of prints you can buy. If the shipping rate is \$4 and the cost per print is \$0.10, and you want to spend \$10, then you can solve the equation $10 = 0.1x + 4$ to find the number x of prints you can buy.

7.4 Warm Up

For use before Lesson 7.4

1. $x = 56$ **2.** $a = 18$ **3.** $y = 20$

4. $p = 7$ **5.** $t = 3$ **6.** $w = 11$

7.4 Practice A

1. $j = 2$ **2.** $b = 27$ **3.** $a = 4$

4. $m = 12$ **5.** $z = 14$ **6.** $r = 5$

7. $t = 5$ **8.** $f = 49$ **9.** $n = 140$

10. $z = 9$ **11.** $j = 3$ **12.** $e = 8$

13. $x = 54$ **14.** $s = 45$ **15.** $y = 3$

16. You must add 5 to each side before multiplying.

$$\frac{d}{2} - 5 = 8$$
$$\frac{d}{2} = 13$$
$$d = 26$$

17. You must add 4 to each side of the equation.

$$16 = 4g - 4$$
$$20 = 4g$$
$$5 = g$$

18. Let x be the number of swimming classes; $12x + 10 = 82$; $x = 6$; 6 classes

19. Let x be the number of days you rent the surfboard and wetsuit; $10x + 15x = 75$; $x = 3$; 3 days

20. Let x be the per person discount; $10x + 530 = 650$; $x = 12$; \$12

21. a. When you solve $4x + 15 = 47$ and $5x + 7 = 47$ for x, you get the same result: $x = 8$.

b. 8 marbles

7.4 Practice B

1. $m = 13$ **2.** $c = 90$ **3.** $x = 8$

4. $y = 9$ **5.** $n = 330$ **6.** $z = 88$

7. $w = 280$ **8.** $k = 9$ **9.** $f = 10$

10. $b = 48$ **11.** $t = 39$ **12.** $r = 32$

13. $y = 7$ **14.** $p = 6$ **15.** $b = 45$

16. $x = 5$ in. **17.** $x = 4.5$ m **18.** $x = 2$ ft

19. Let x be the weight of each cookie box; $45x + 52x = 776$; $x = 8$; 8 oz.

20. Let x be the maximum amount each person can spend on a fishing pole; $5x + 100 = 350$; $x = 50$; \$50.

21. a. 85 points **b.** less than

c. Yes; this will happen if you receive a score of 92 points on the seventh test.

7.4 Enrichment and Extension

1. $x + x + 1 = 11$; 5, 6

2. $x + x + 1 = 39$; 19, 20

3. $x + x + 1 = 141$; 70, 71

4. $x + x + 1 = 689$; 344, 345

5. Add a third number, $x + 2$, to the equation. So, the equation becomes $x + x + 1 + x + 2 = 15$ The numbers are 4, 5 and 6.

6. $x + x + 1 + x + 2 = 36$; 11, 12, 13

7. $x + x + 2 = 50$; 24, 26

7.4 Puzzle Time

HE WAS FEELING CRUMMY.

7.5 Start Thinking!

For use before Activity 7.5

Sample answer: If you know the circumference, you can solve the equation $C = \pi d$ for d to find the diameter.

Answers

7.5 Warm Up

For use before Activity 7.5

1. 144 $in.^2$ **2.** 36 ft^2

3. 114 mm^2 **4.** 36 cm^2

5. 210 m^2 **6.** about 153.86 yd^2

7.5 Start Thinking!

For use before Lesson 7.5

Sample answer: No. Solving the equation $19x = 655$, gives you a width of 35 inches for the first television. Solving $2(19) + 2y = 90$ gives you a width of 26 inches for the second television.

7.5 Warm Up

For use before Lesson 7.5

1. $4x = 24$; $x = 6$ mm **2.** $\frac{1}{2}(5x) = 5$; $x = 2$ in.

3. $2x + 8 = 18$; $x = 5$ ft

4. $3.14x = 314$; $x = 100$ m

7.5 Practice A

1. $15 = \frac{1}{2} \bullet x \bullet 6$; $x = 5$ in.

2. $44 = \frac{22}{7}x$; $x = 14$ ft

3. $46 = 2(8) + 2x$; $x = 15$ yd

4. $60 = 12x$; $x = 5$ cm **5.** $54 = 6x$; $x = 9$ km

6. $24 = 6 + 4 + 9 + x$; $x = 5$ m

7. $56 = 4x$; $x = 14$ m

8. $50 = \frac{1}{2} \bullet x \bullet (12 + 8)$; $x = 5$ in.

9. $25.12 = 2\pi x$; $x = 4$ ft

10. The circumference formula in terms of the diameter should have been used rather than the circumference formula in terms of the radius.

$$157 = 3.14x$$
$$\frac{157}{3.14} = \frac{3.14x}{3.14}$$
$$50 = x$$

The diameter is 50 cm.

11. $x = 32.7$ cm **12.** 41 in.

7.5 Practice B

1. $308 = \frac{1}{2}(22)x$; $x = 28$ m

2. $62.8 = 2(3.14)x$; $x = 10$ ft

3. $496 = \frac{1}{2} \bullet x \bullet (25 + 37)$; $x = 16$ in.

4. $118 = x + x + 40$; $x = 39$ mm

5. $285 = 15x$; $x = 19$ cm

6. $228.5 = x + x + x + 0.5(3.14x)$; $x = 50$ ft

7. about 1.27 in. **8.** about $9\frac{5}{8}$ ft^2

9. about 56.52 ft^2 **10.** 24 ft

7.5 Enrichment and Extension

1. $r = 3$ in. **2.** $r = 7$ cm **3.** $d = 30$ ft

4. 7.5 mm **5.** 10.5 yd

7.5 Puzzle Time

ENVELOPE

7.6 Start Thinking!

For use before Activity 7.6

Sample answer: Volume, measured in cubic units (like $in.^3$, m^3, or ft^3) is the amount of space an object takes up. You may want to know the volume of a swimming pool to know how much water will fill it.

7.6 Warm Up

For use before Activity 7.6

1. $x = 5$ **2.** $x = \frac{1}{2}$ **3.** $x = 9$

4. $x = 12$ **5.** $x = 18$ **6.** $x = 1.5$

7.6 Start Thinking!

For use before Lesson 7.6

A. no **B.** yes **C.** yes

D. no **E.** yes

Sample answer: the amount of water in a bathtub

7.6 Warm Up

For use before Lesson 7.6

1. $15w = 30$; $w = 2$ m

Answers

2. $160w = 5600$; $w = 35$ ft

3. $16\ell = 64$; $\ell = 4$ in.

4. $50h = 100$; $h = 2$ cm

7.6 Practice A

1. 12
2. 16
3. 48
4. 160 m^3
5. 75 cm^3
6. 480 yd^3
7. 11 in.
8. 16 m
9. 9 yd
10. 75 cm
11. 1 ft
12. 280 ft^3
13. 16 cm^3
14. The volume is multiplied by 8.

7.6 Practice B

1. 3.75 ft^3
2. 84,525 cm^3
3. 2 km
4. 4.8 ft
5. 0.75 mi
6. 2.5 mm
7. $1\frac{1}{3}$ in.
8. 5 cm
9. 75 mm
10. 1 $in.^3$
11. 0.268 $in.^3$
12. The volume is multiplied by 8.

7.6 Enrichment and Extension

1. 12 in.
2. 6 in.
3. 10 in. by 10 in. by 10 in.
4. Before: 10 in.; After: 11.25 in.

7.6 Puzzle Time

YOU GO AHEAD I WILL JUST HANG AROUND.

School-to-Work

1. $x + 2 = 6$; 4 milliamperes
2. 10.5 amperes
3. 332.5 feet
4. 58.5 feet
5. $\frac{3}{4}x = 12$; 16 pieces
6. $3\frac{1}{4}$ inches
7. 2 ohms
8. 780 meters

Study Help

Available at *BigIdeasMath.com*.

Financial Literacy

1. *Sample answer:* refrigerator: 0.7 kW, computer: 0.28 kW, light: 0.2 kW

2. *Sample answer:* refrigerator: 8 h/day, computer: 12 h/day, light: 3 h/day

3. *Sample answer:* refrigerator: 5.6 kW-h/day, 168 kW-h/month, 2016 kW-h/year; computer: 3.36 kW-h/day, 100.8 kW-h/month, 1209.6 kW-h/year; light: 0.6 kW-h/day, 18 kW-h/month, 216 kW-h/year

4. *Sample answer:* If electricity is \$0.13/kW-h, then refrigerator costs \$21.84 per month and \$262.08 per year; the computer costs \$13.10 per month and \$157.20 per year; the light costs \$2.34 per month and \$28.08 per year.

5. *Sample answer:* \$111.84 per year

6. *Sample answer:* The cost equals the power times the hours used times the cost per kW-h. Write the formula. Simplify by multiplying. Then solve the equation by dividing both sides by 0.42.

$$\begin{aligned}(3.5\text{ kW})(\text{hours})(\$0.12/\text{kW-h}) &= \$13.44\\ 0.42(\text{hours}) &= 13.44\\ \text{hours} &= 32\end{aligned}$$

Technology Connection

1. 47
2. 1.28
3. 2618
4. 5668
5. 1593
6. 938
7. 81
8. 3.7
9. 305.2
10. 5.135

Cumulative Practice

1. 5
2. 1
3. 30
4. a. $\frac{d}{6}$ b. \$9.75
5. $n = 51$
6. $d = 10$
7. $w = 6$
8. $s = 12$
9. $j = 7$
10. $a = 1$
11. $u = 4$
12. $x = 7.5$
13. $y = 0.9$
14. $q = 11.25$
15. $c = 90$
16. $z = 48$
17. Let x be the cost of each plant; $6x + 18 = 39$; $x = 3.5$; \$3.50.
18. $w = 4$ in.
19. $y = 7$ cm
20. $b = 12$ m
21. $h = 6$ ft
22. 8 in.

Answers

Chapter 8

8.1 Start Thinking!

For use before Activity 8.1

Equality means equal, in- means not; *Sample answer:* inaccurate, inappropriate, inattentive, incombustible, incompatible, incomplete, inconsiderate, inconsistent, incorrect, incurable, indefinite, independent, indestructible, indirect, inedible, ineffective, inexpensive, inflammable, informal, insane, insignificant, invisible

8.1 Warm Up

For use before Activity 8.1

1. $<$ **2.** $>$ **3.** $<$

4. $>$ **5.** $=$ **6.** $<$

8.1 Start Thinking!

For use before Lesson 8.1

Sample answer: A speed limit is the fastest that you are allowed to drive; in a school zone, the speed that you are allowed to drive can be written using the inequality $s \leq 15$.

8.1 Warm Up

For use before Lesson 8.1

1. B **2.** C **3.** A **4.** D

8.1 Practice A

1. $2 > v$ **2.** $h \leq 15$ **3.** $p < \frac{1}{2}$

4. $12 < n$ **5.** $\frac{3}{5} \geq a$ **6.** $b \leq 17$

7. $6 + x \geq 12$ **8.** $k - 7 > 10$

9. solution **10.** not a solution **11.** not a solution

12. not a solution **13.** solution **14.** solution

15. 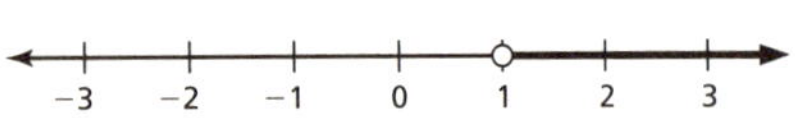

16. 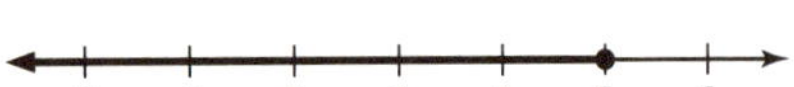

17. 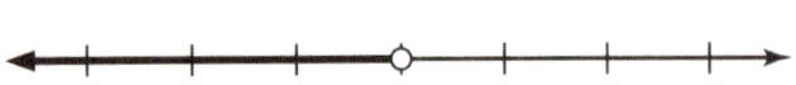

18. 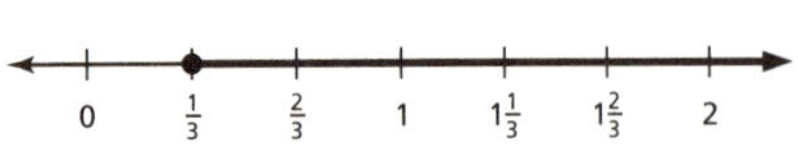

19. 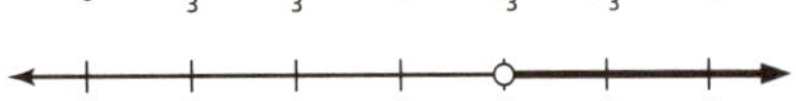

20. 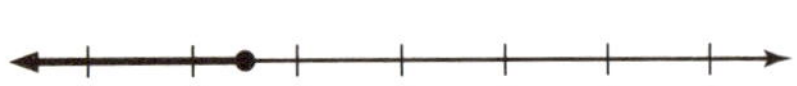

21. 

22. 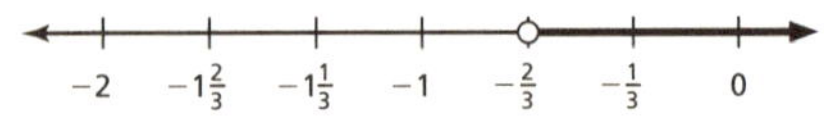

23. 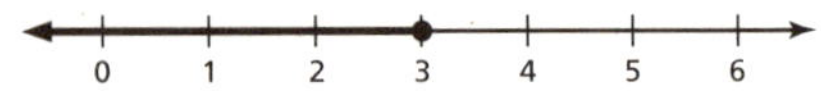

24. Let p be the number of people that the lifeboat can carry; $p \leq 24$.

25. a. $16u \leq 50$

b. no; *Sample answer:* 4 is not a solution of $16u \leq 50$.

26. a. Let w represent the number of pounds of potatoes the box can hold; $w \leq 25$;

b. yes

c. *Sample answer:* any number greater than 25

8.1 Practice B

1. $3.5b \geq 21$ **2.** $y \div 9 > 4$ **3.** $h - \frac{1}{4} \leq 0$

4. $w + 2.56 \geq 10.24$ **5.** $6c < 12$

6. solution **7.** not a solution **8.** not a solution

9. solution **10.** not a solution **11.** solution

12. 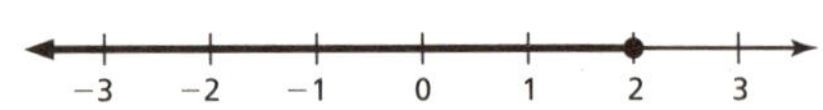

13. 

14. 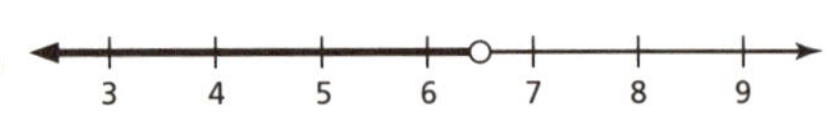

15. 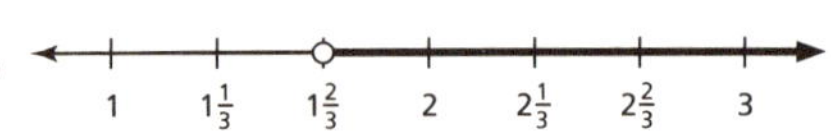

16. 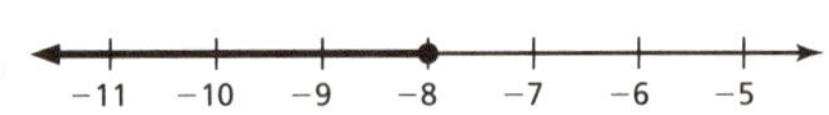

17. 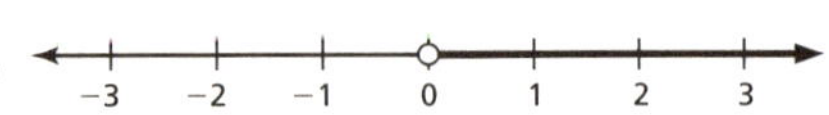

18. 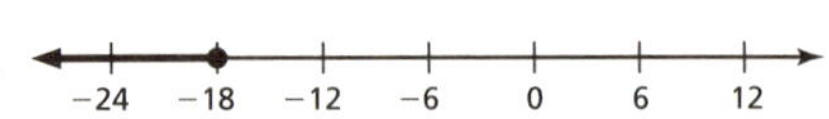

19. 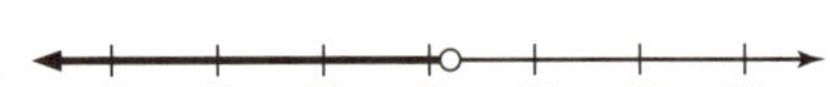

20. 

21. Let h be the height of a vehicle that can travel on the highway; $h < 7.75$;

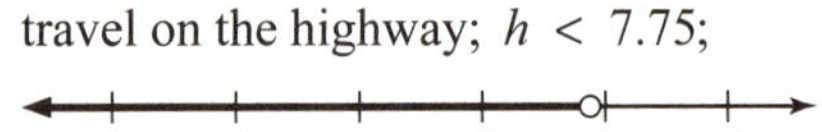

22. a. Scrambler: $h \geq 48$; Busy Boats: $h \leq 48$

b. 48 inches

23. numbers that are greater than 2 and less than or equal to 8

Answers

8.1 Enrichment and Extension

1. 1, 2, 3, 4, 5 **2.** 3, 4, 5, 6, 7, 8

3. 4, 5, 6, 7, 8, 9, 10 **4.** −1, 0, 1, 2

5. D **6.** C **7.** A **8.** B

9. 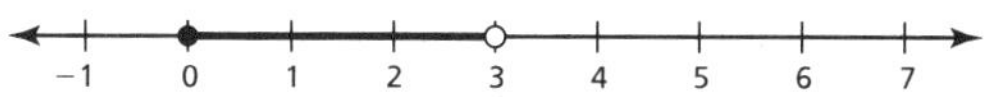

10. 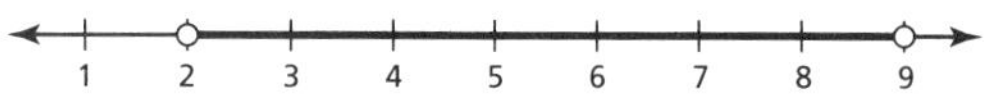

11. 

12. 

13. 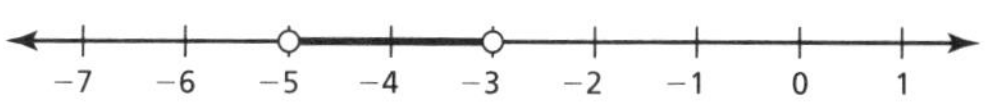

14. 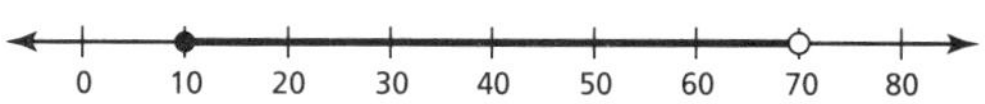

15. 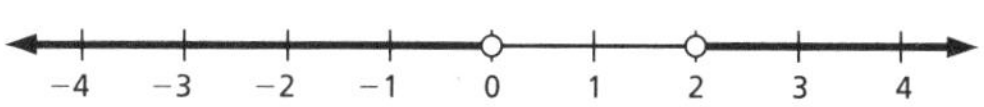

No; the two inequalities that make up the compound inequality do not share a common solution.

8.1 Puzzle Time

COTTAGE CHEESE

8.2 Start Thinking!

For use before Activity 8.2

Sample answer: You want to have more than 15 games in your collection. You already have 5 games. How many more do you need?

8.2 Warm Up

For use before Activity 8.2

1. 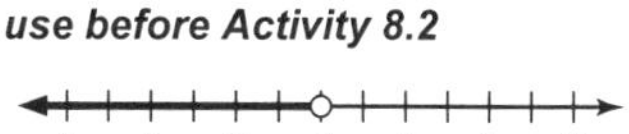

2. 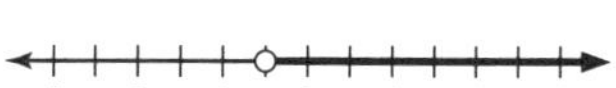

3. 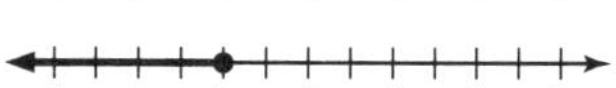

4. 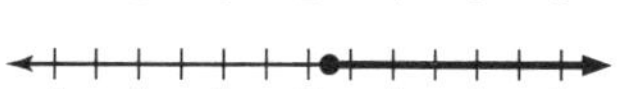

5. 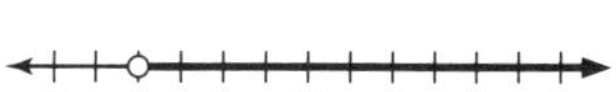

6. 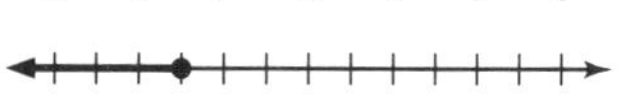

8.2 Start Thinking!

For use before Lesson 8.2

$x + 2 \leq 22; x \leq 20$

8.2 Warm Up

For use before Lesson 8.2

1. $x < 4$;

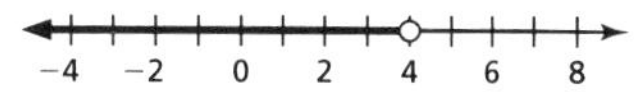

2. $h > 9$;

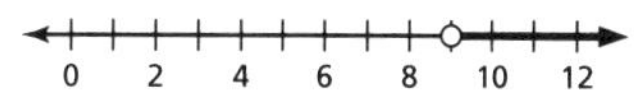

3. $y \leq 5$;

4. $p \geq 15$;

5. $k \leq 11$;

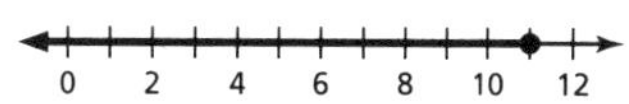

6. $b > 6$;

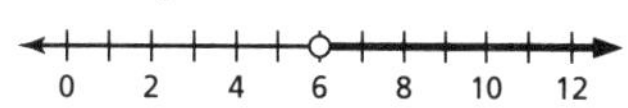

8.2 Practice A

1. $n \geq 11$

2. $v \leq 4$

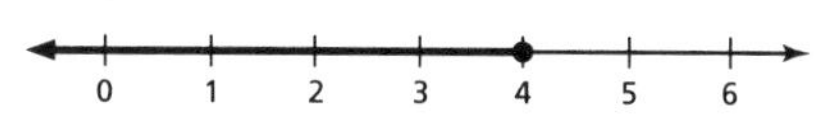

3. $p < 1$

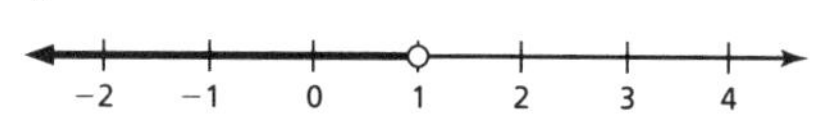

4. $x > 11$

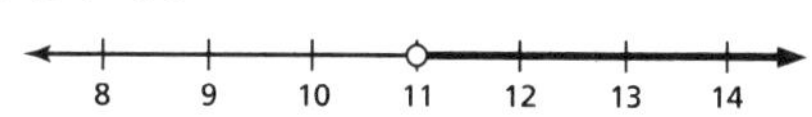

5. $k > 5$

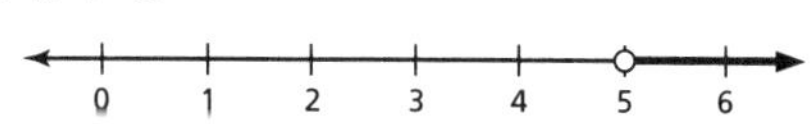

6. $m \geq 1$

7. $h \leq 20$

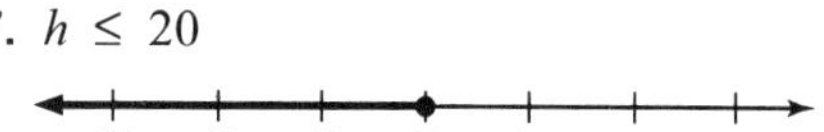

8. $b < 2$

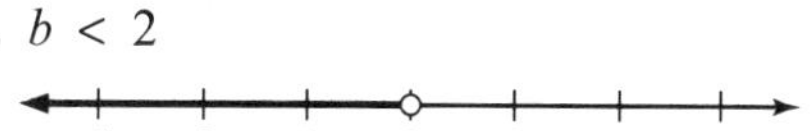

9. $w \leq 100$

10. $a \geq 9$

11. $r > 2$

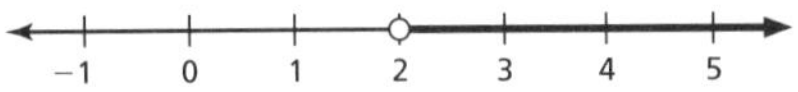

12. $y < 70$

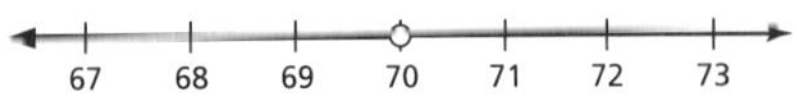

13–16. Choice of variables may vary.

13. $d + 6 \le 10$; $d \le 4$ **14.** $x - 4 > 3$; $x > 7$

15. $0.6 \ge n - 2.4$; $n \le 3$

16. $z + 14 \ge 18$; $z \ge 4$

17. The inequality sign should not change; $7 > g$ or $g < 7$.

18. Five should be subtracted from both sides of the equation.

$$\begin{aligned} x + 5 &\ge 11 \\ \underline{-5} & \quad \underline{-5} \\ x &\ge 6 \end{aligned}$$

19. Let d be the amount you can spend on the cold drink: $d + 6.75 \le 10$; $d \le 3.25$; You can spend up to \$3.25.

20. Let n be the number of days for your books to arrive; $n + 9 \ge 28$; $n \ge 19$; at least 19 days.

21. Let p be the number of people who can attend the play; $p + 185 \le 480$; $p \le 295$; at most 295 people.

8.2 Practice B

1. $x \ge 3$

2. $c < 12$

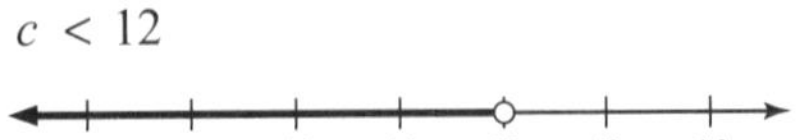

3. $q > 11$

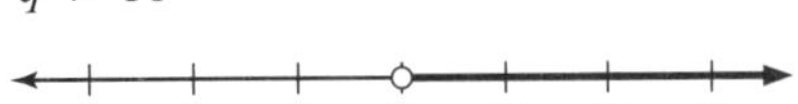

4. $t \le 31.6$

5. $z > 13\frac{3}{8}$

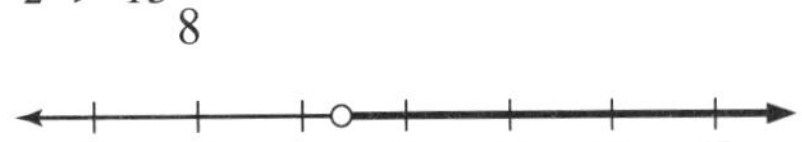

6. $g \le \frac{3}{5}$

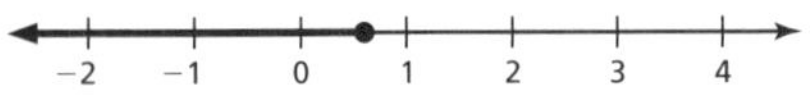

7. $k \ge 0$

8. $x < 5\frac{2}{3}$

9. $u \le \frac{1}{6}$

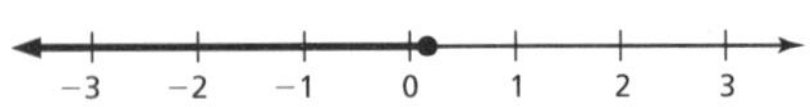

10. $y < 132$

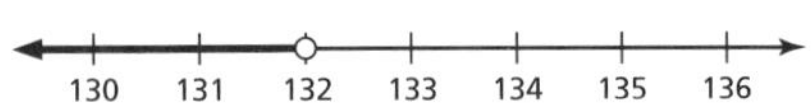

11. $d \le \frac{3}{8}$

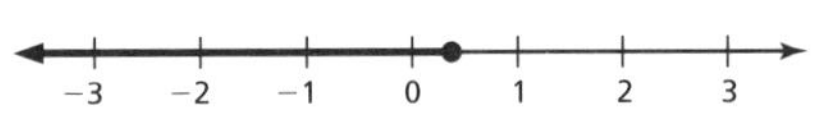

12. $h > 4.4$

13–17. Choice of variables may vary.

13. $k - 4.5 > 8$; $k > 12.5$

14. $\frac{3}{4} + x \le 2$; $x \le 1\frac{1}{4}$

15. $m - 9.1 < 4.6$; $m < 13.7$

16. $n - \frac{7}{12} \ge 3\frac{1}{2}$; $n \ge 4\frac{1}{12}$

17. $5\frac{1}{4} > y + 2\frac{1}{10}$; $y < 3\frac{3}{20}$

18. $a < 1$

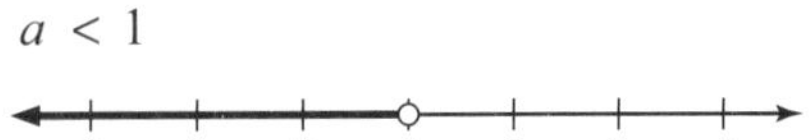

19. $n > 5.5$

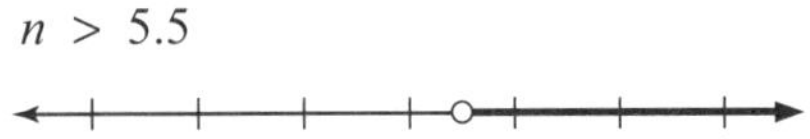

20. $v \ge 26.6$

Answers

21. Let d be the number of degrees the temperature must rise; $135 + d \geq 165$; $d \geq 30$; at least 30 degrees.

22. Let n be the number of additional phone numbers; $n + 21 \leq 50$; $n \leq 29$; up to 29 additional phone numbers.

23. 1.1

8.2 Enrichment and Extension

1. $3 < x < 6$

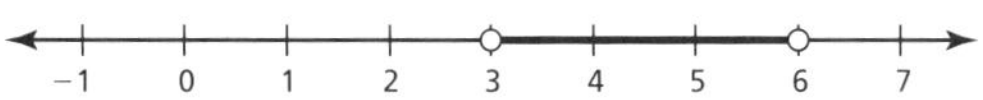

2. $0 \leq x \leq 4$

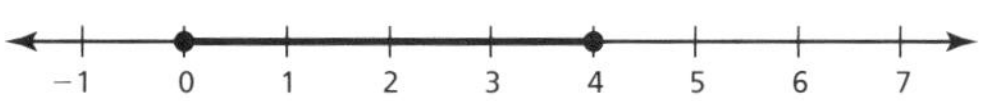

3. $7 < x \leq 11$

4. $4 \leq x < 8$

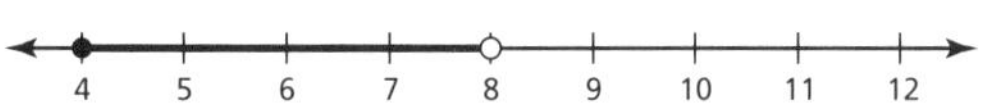

5. $x < 4$ or $x > 6$

6. $x \leq 3$ or $x \geq 5$

7. $x \geq 7$ or $x < 0$

8. $x < 3$ or $x \geq 4$

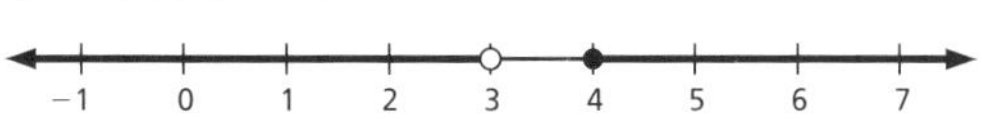

9. The solution is the entire number line.

8.2 Puzzle Time

I'M SORRY SIR SAID THE ATTENDANT
THE MOON IS FULL RIGHT NOW

8.3 Start Thinking!

For use before Activity 8.3

Sample answer: Each has $5x$ on the left-hand side and 20 on the right-hand side of the equation or inequality. To solve $5x = 20$, divide each side by 5. Yes, you can use the same method to solve inequalities. Dividing each side by 5, $5x < 20$ yields the solution $x < 4$ and $5x > 20$ yields the solution $x > 4$.

8.3 Warm Up

For use before Activity 8.3

1. $x < 2$ 2. $y > 24$ 3. $r \leq 13$

4. $c \geq 13$ 5. $s < 5$ 6. $x \leq 5$

8.3 Start Thinking!

For use before Lesson 8.3

Sample answer: This week, Shelley hopes to spend less than \$3.50 on text messages. If each text message costs \$0.20, what are the possible numbers of text messages Shelley should send this week?

8.3 Warm Up

For use before Lesson 8.3

1. $x > 8$;

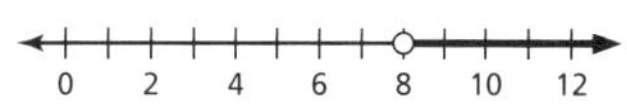

2. $t \geq 2$;

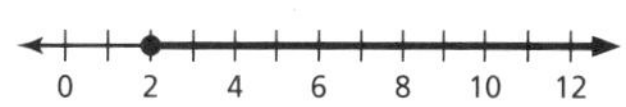

3. $n \leq 9$;

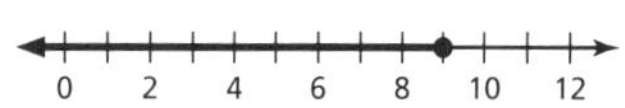

4. $f < 3$;

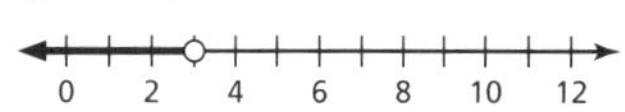

5. $b > 6$;

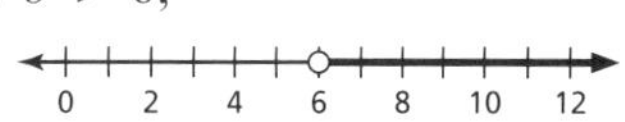

6. $y \leq 0$;

8.3 Practice A

1. $b > 18$

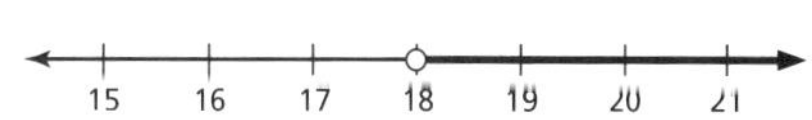

2. $w < 8$

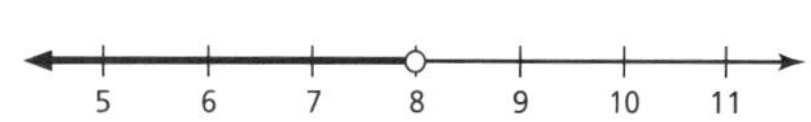

3. $p \leq 10$

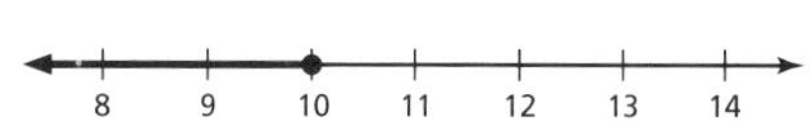

4. $n \geq 36$

5. $x > 20$

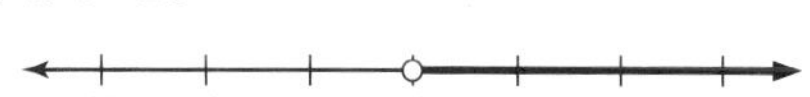

Answers

6. $a < 5$

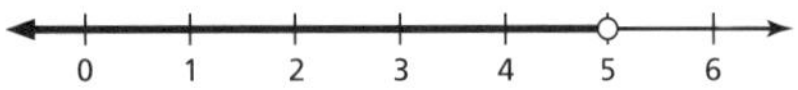

7. $g \le 84$

8. $k \ge 153$

9. $h \le 0$

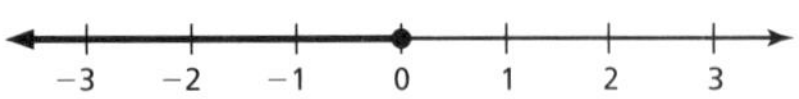

10. $v \ge 25$

11. $t > 3$

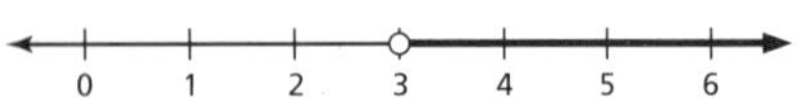

12. $m < 35$

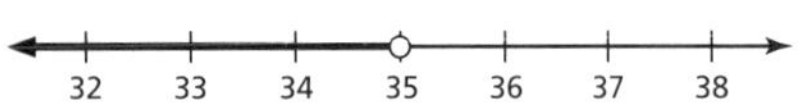

13. $y \ge 8$

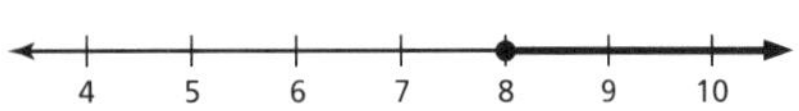

14. $r \le 60$

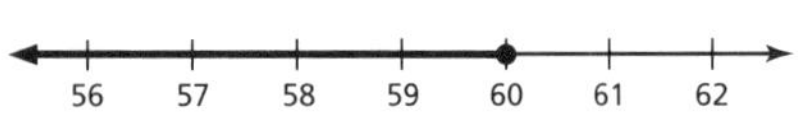

15. $c < 78$

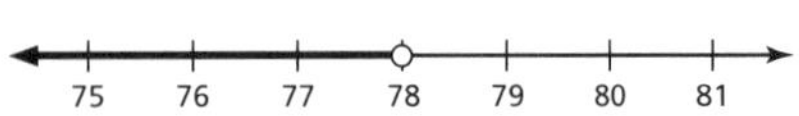

16. $s > 7$

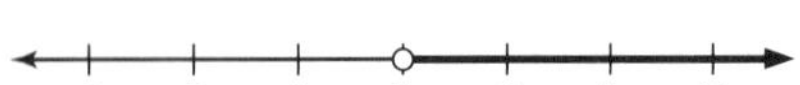

17. $\frac{q}{14} \ge 4$; $q \ge 56$ **18.** $21u \le 126$; $u \le 6$

19. $z \div 7 \ge 12$; $z \ge 84$ **20.** $a \div 22 > 5$; $a > 110$

21. $13g \le 117$; $g \le 9$

22. You must divide both sides by 5.

$$5x < 10$$
$$\frac{5x}{5} < \frac{10}{5}$$
$$x < 2$$

23. You must divide both sides by 2, not subtract 2.

$$2n \ge 28$$
$$\frac{2n}{2} \ge \frac{28}{2}$$
$$n \ge 14$$

24. Let b be the number of bags of soil the wheelbarrow can carry; $20b \le 300$; $b \le 15$; at most 15 bags of soil.

25. Let ℓ be the length of the table; $3\ell > 24$; $\ell \ge 8$; at least 8 feet.

26. Let s be the number of steps; $8s \ge 10(12)$; $s \ge 15$; at least 15 steps.

8.3 Practice B

1. $m \le 8$

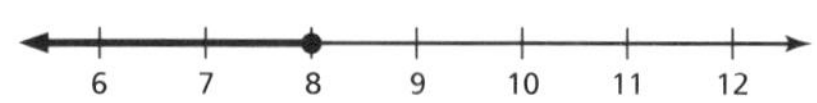

2. $v \ge 5$

3. $b > \frac{1}{4}$

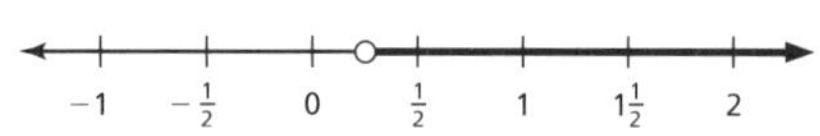

4. $x < 7\frac{1}{2}$

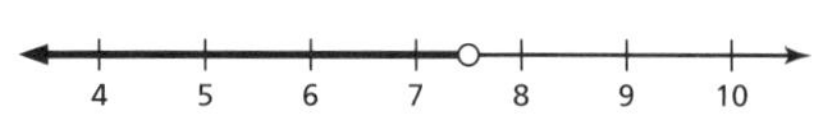

5. $h \ge 2.5$

6. $u \le 126$

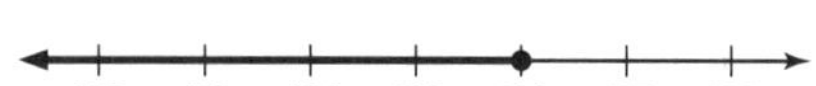

7. $r < 12.4$

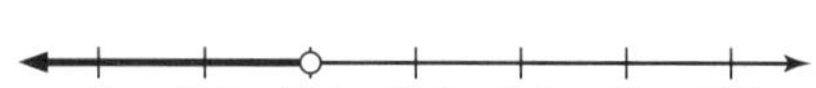

8. $a > 2.4$

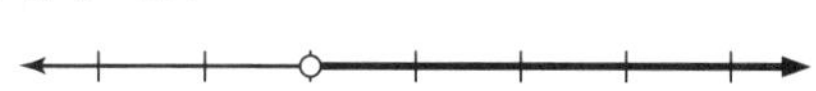

9. $p > 9$

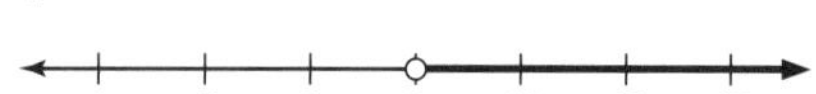

Answers

10. $k < \frac{7}{10}$

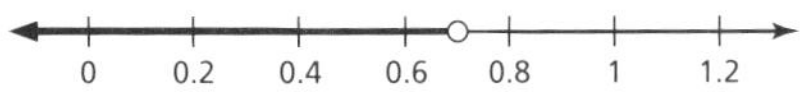

11. $y \leq 40.5$

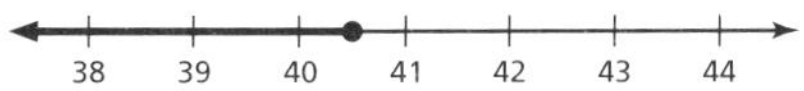

12. $g \geq 0.5$

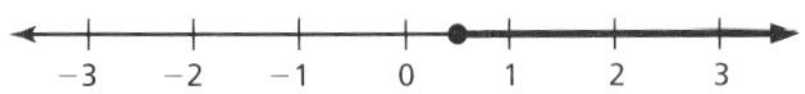

13. $n < 7\frac{1}{2}$

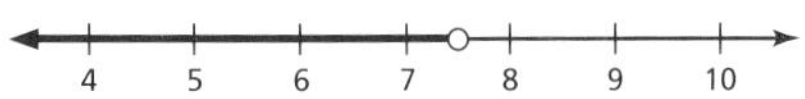

14. $c > 8$

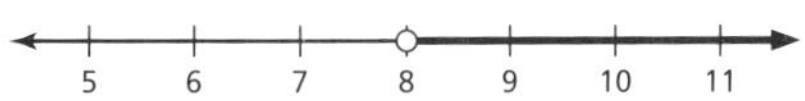

15. $s \geq 2.5$

16. $w \leq 3.6$

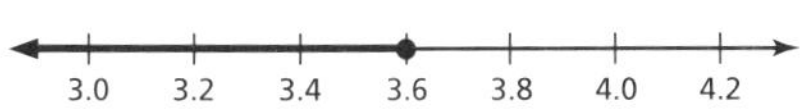

17. $10q \geq 2.01$; $q \geq 0.201$

18. $z \div 18 \leq \frac{2}{3}$; $z \leq 12$

19. $\frac{d}{8} \leq 3.43$; $d \leq 27.44$

20. $5t > 12$; $t > 2.4$

21. $b \div 3 < \frac{1}{12}$; $b < \frac{1}{4}$

22.

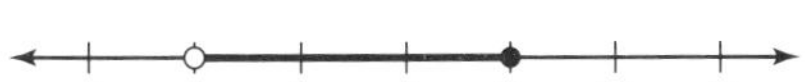

23.

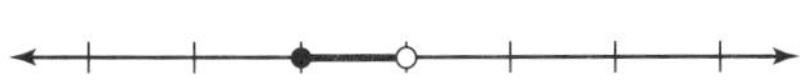

24.

25.

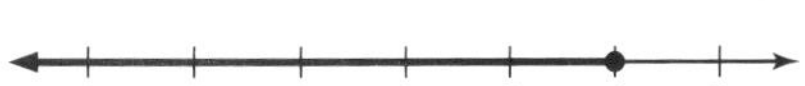

26. Let x be each person's share of the rental cost each month; $3x \geq 1200$; $x \geq 400$; at least \$400 per month.

27. Let m be the number of minutes required to print a 380-page report; $80m \geq 380$; $m \geq 4.75$; at least 4.75 minutes.

28. Let r be the number of rest stops in 200 miles of highway; $25r \geq 200$; $r \geq 8$; at least 8 rest stops.

8.3 Enrichment and Extension

1. $2 < x < 6$

2. $3 \leq x \leq 5$

3. $12 \leq x < 24$

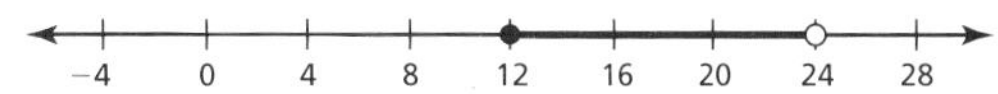

4. $6 < x \leq 8$

5. $x > 5$ or $x < 4$

6. $x \leq 3$ or $x \geq 6$

7. $x > 10$ or $x \leq 4$

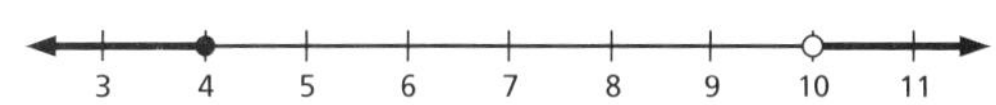

8. $x < 2$ or $x \geq 3$

9. The graphs for Exercises 1–4 share common solutions. The graphs for Exercises 5–8 do not.

8.3 Puzzle Time

ONE PAW AND MAW

8.4 Start Thinking!

For use before Activity 8.4

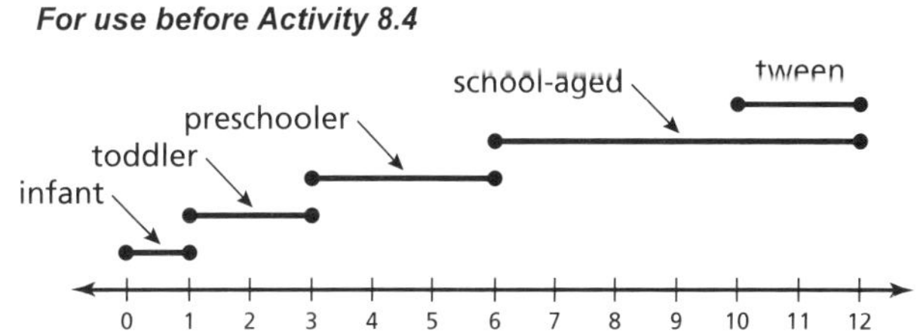

yes; no

8.4 Warm Up

For use before Activity 8.4

1. no
2. yes
3. yes
4. no
5. yes
6. yes

Answers

8.4 Start Thinking!

For use before Lesson 8.4

The length must be less than or equal to 15 units; Real-world situations will vary. *Sample answer:* Tom has 50 feet of fencing for his vegetable garden. If the width of the garden is 10 feet, what are the possible lengths for the garden?

8.4 Warm Up

For use before Lesson 8.4

1. $x \leq 20$;

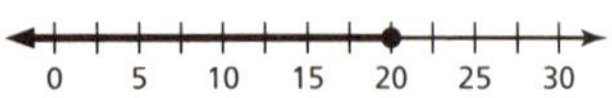

2. $x > 1.5$;

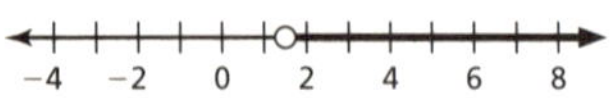

3. $x < 6$;

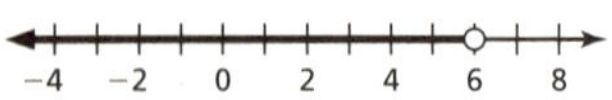

4. $x \geq 7$;

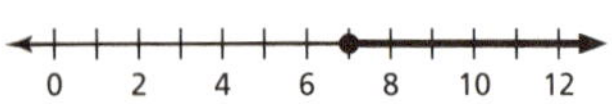

5. $x \leq \frac{1}{2}$;

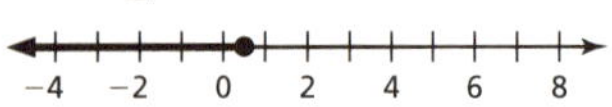

6. $x > 0.8$;

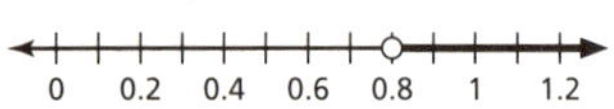

8.4 Practice A

1. $s \leq 2$

2. $k < 6$

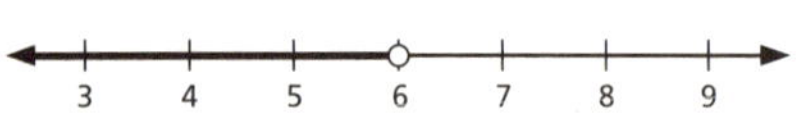

3. $n > 39$

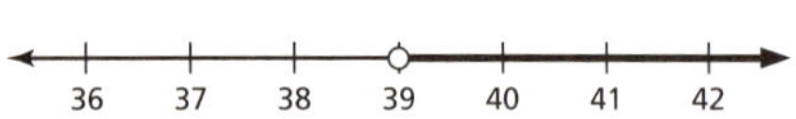

4. $x \geq 21$

5. $g > 7$

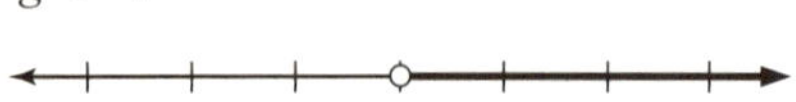

6. $b < 15$

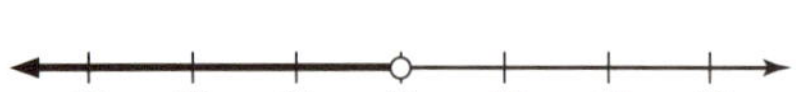

7. $p \geq 6$

8. $c \leq 40$

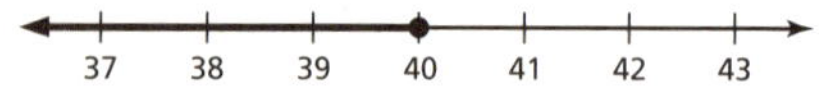

9. $v > 72$

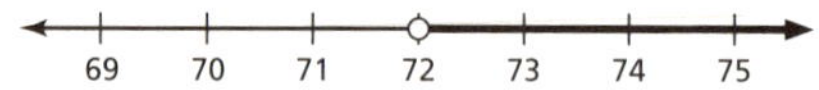

10. $h < 0$

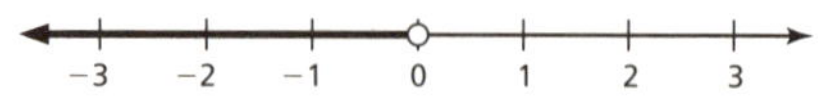

11. $y \leq 60$

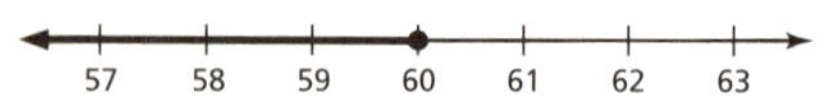

12. $a \geq 5$

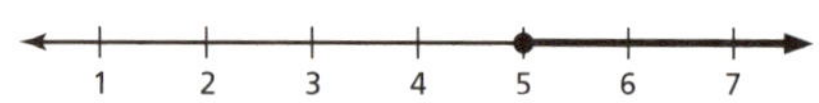

13. You must add 9 to each side of the inequality.

$$3q - 9 \geq 18$$
$$3q \geq 27$$
$$q \geq 9$$

14. You must subtract 14 from each side of the equation.

$$7d + 14 < 98$$
$$7d < 84$$
$$d < 12$$

15. Let f be the number of files that can be packed in the box; $50f + 380 \leq 1440$; $f \leq 21.2$; at most 21 files.

16. Let s be the number of songs you can store; $6s + 550 \geq 5800$; $s \geq 875$; at least 875 songs.

17. Let m be the amount of money you must have in your account; $m - 2(6.50) \geq 50$; $m \geq 63$; at least \$63.

18. Let d be the number of days she must practice; $1.5d + 10 \geq 40$; $d \geq 20$; at least 20 days.

8.4 Practice B

1. $d < 3$

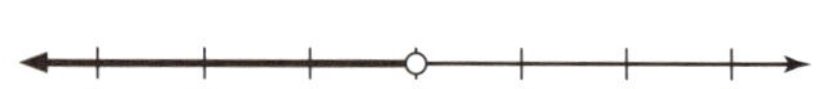

2. $r \leq 25$

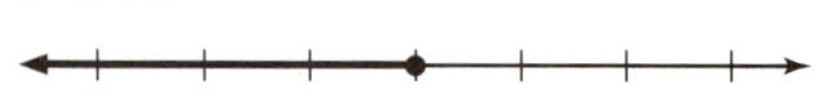

Answers

3. $a > 5$

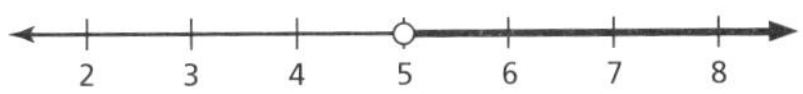

4. $k > 1.5$

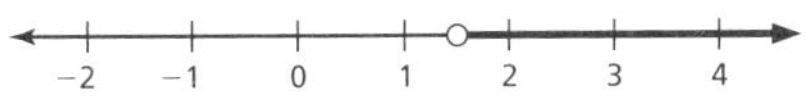

5. $u \leq 0.6$

6. $n \geq 141.12$

7. $z > 0$

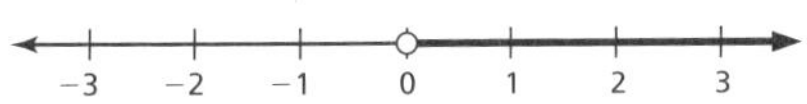

8. $x < 140$

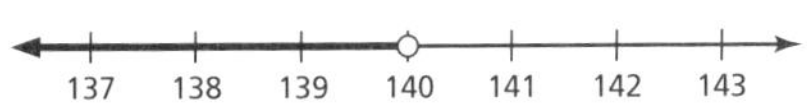

9. $r \leq 7$

10. $c > 47$

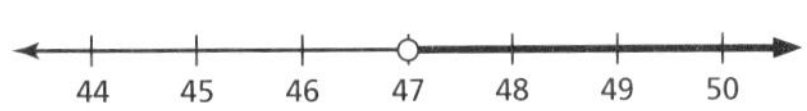

11. $m < 2.6$

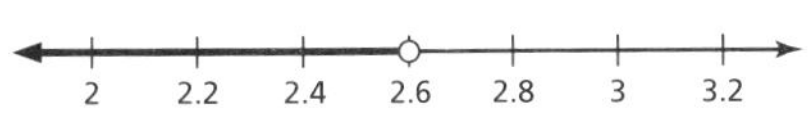

12. $y \geq \frac{3}{4}$

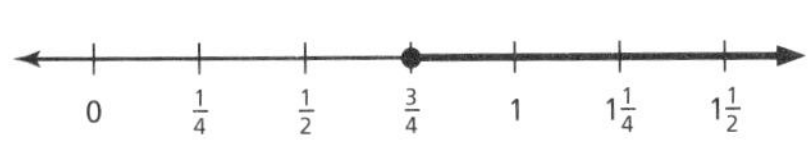

13. $18x + 15 \bullet 25 \leq 600$; $x \leq 12.5$

14. $20x + \frac{1}{2} \bullet \pi \bullet 10^2 \leq 600$; $x \leq 22.15$

15. $\frac{1}{2} \bullet 20 \bullet x + 20 \bullet 22 \leq 600$; $x \leq 16$

16. Let t be the number of tickets you buy; $0.5t + 3 < 15$; $t < 24$; less than 24 tickets; 23 tickets.

17. Let s be your score on the sixth level; $\frac{250 + 190 + 380 + 420 + 310 + s}{6} \geq 350$; $s \geq 550$; at least 550.

18. Let r be your average speed for the remainder of the trip; $62 \bullet 4 + r \bullet 5 \geq 500$; $r \geq 50.4$; at least 50.4 miles per hour.

19. Let n be the number of T-shirts you can buy; $15n + 14.75 \leq 60$; $n \leq 3.01\overline{6}$; up to 3 T-shirts.

8.4 Enrichment and Extension

1. $x < \frac{1}{2}$

2. $x \geq 8$

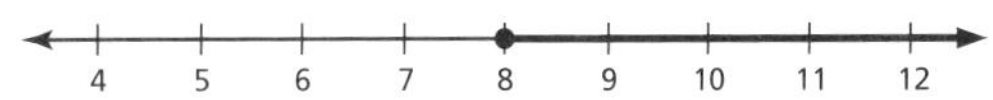

3. $x \leq 14$

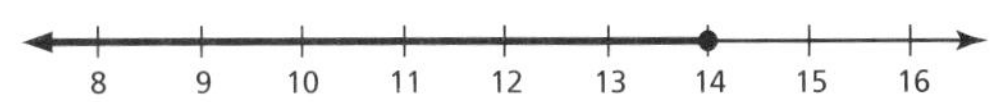

4. $x > 7$

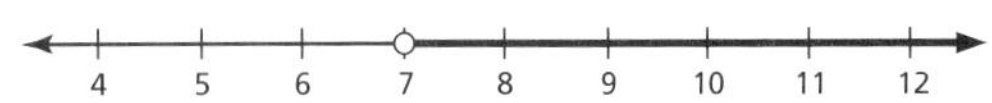

5. $x \geq \frac{5}{2}$

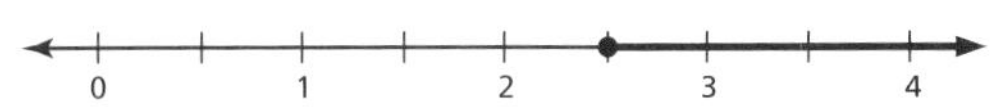

6. $x \leq 11$

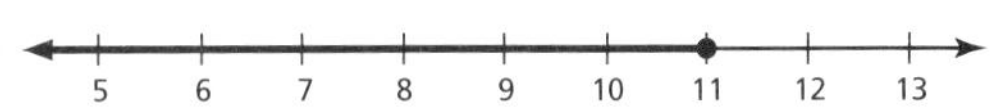

7. $x > 2$

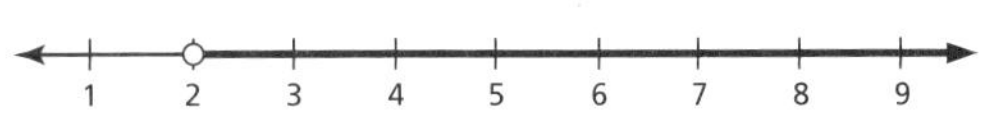

8. $x \geq 1$

9. $x < 0$

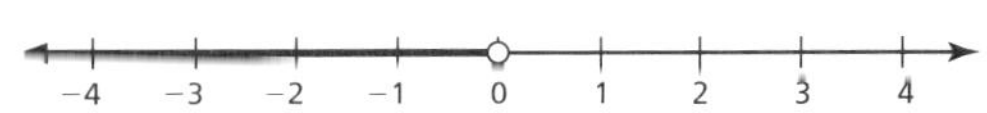

10. $\frac{85 + 91 + x}{3} \geq 90$; $x \geq 94$; You need to get at least a 94.

11. $\frac{30 + 20 + x}{3} \leq 20$; $x \leq 10$; You can wait at most 10 minutes.

12. $5(3x + 2) > 130$; $x > 8$; The value of x must be greater than 8.

8.4 Puzzle Time

TO TRAIN HIM

Answers

School-to-Work

1. **a.** $w \leq 9$ **b.** this medication
2. The puppy will grow to be a medium to large dog.
3. **a.** $T \geq 86, T \leq 93$

   b. *Sample answer:* 90 degrees

   c. *Sample answer:* 82 degrees, too cold
4. **a.** $s \geq 1.6f$

   b. Yes, it grew 73%, which is more than 60%.
5. $65 + x \leq 130$; $x \leq 65$

   The sugar glider should be weaned within 65 days of coming in.

Study Help

Available at *BigIdeasMath.com.*

Cumulative Practice

1. $9 > h$
2. $v - 6 = 12$
3. $8 + k$
4. $b \geq 13$
5. $24x$
6. $w \div 40 > 3$
7. $4 - a$
8. $4 + c < 15$

9–13. Samples answers are given.

9. A number c divided by 6
10. The difference of a number m and 9 is 12.
11. The product of 4 and a number u is at least 28.
12. $x \geq 2$; A number x is greater than or equal to 2.
13. $y < 8$; A number y is less than 8.
14. $p + 135 \leq 500$
15. 
16. 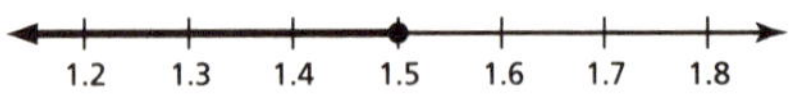
17. 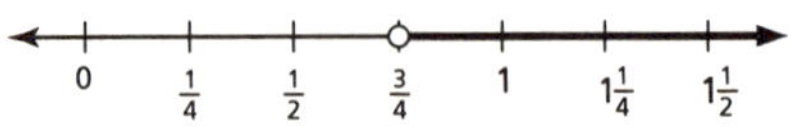
18. $x \leq 10$

    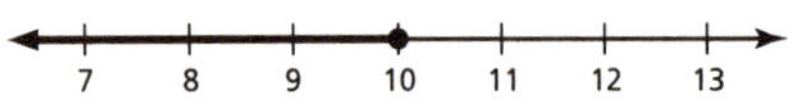
19. $y = 8$
20. $b = 12$
21. $p \geq 5$

22. $u > 1$

    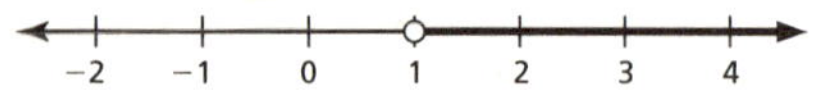
23. $c \geq 6$

24. $z = 3$
25. $m = 1.75$
26. $k \geq 192$

27. $v < 18$

    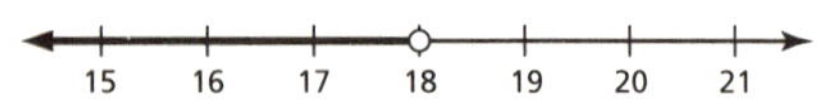
28. Let b be the number of books you can pack; $3b + 5 < 50$; $b < 15$; fewer than 15 books.

Chapter 9

9.1 Start Thinking!

For use before Activity 9.1

Answers will vary depending on the cost of milk. For example, if one milk costs \$0.25, then the entries in the mapping diagram would be \$0.25, \$0.50, \$0.75, and \$1.00. A mapping diagram maps one value to another.

9.1 Warm Up

For use before Activity 9.1

1. 36 m^3
2. 800 ft^3
3. 256 in.^3
4. 1400 cm^3

9.1 Start Thinking!

For use before Lesson 9.1

The output is 15. Observe students playing the Guess the Function Game.

9.1 Warm Up

For use before Lesson 9.1

1. Add 4; missing entries are 8, 9, and 10.
2. Multiply by 5; missing entries are 20, 25, and 30.

Answers

9.1 Practice A

1. As each input increases by 1, the output increases by 3.

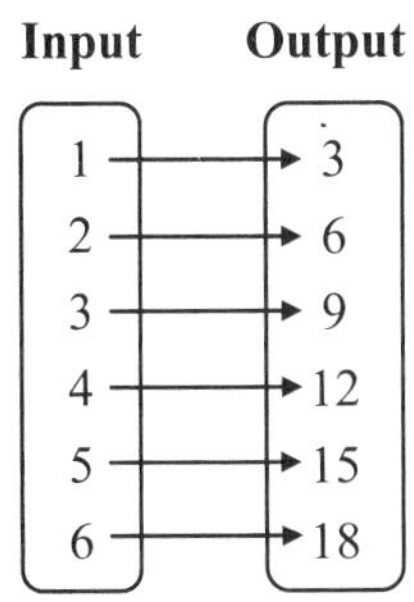

2. As each input increases by 1, the output increases by 2.

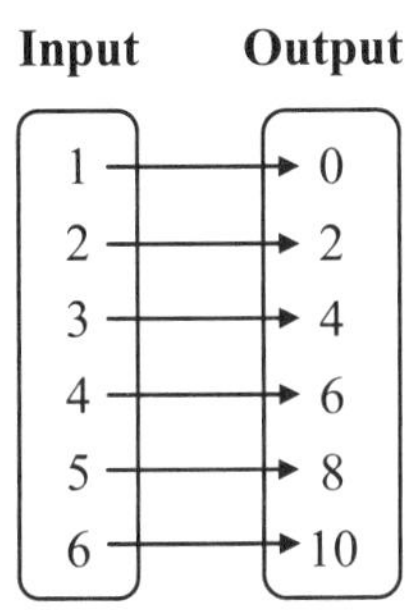

3. (0, 1), (2, 2), (4, 3), (6, 4)

4. (1, 9), (4, 7), (7, 5), (10, 3)

5. 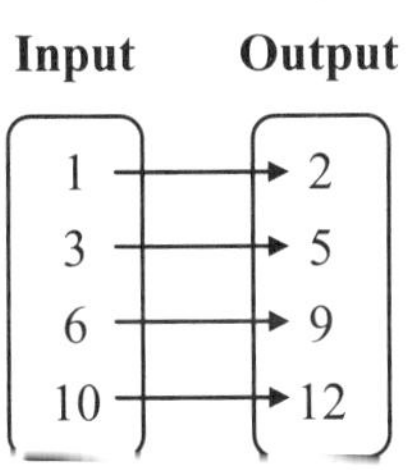

6. 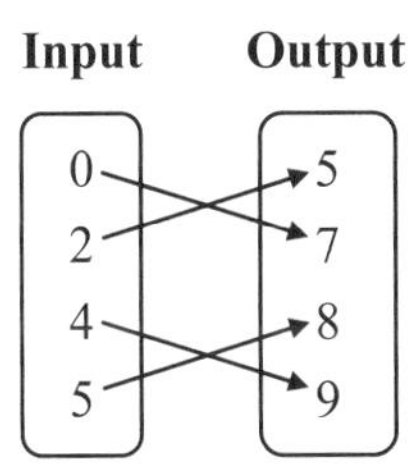

7. a. 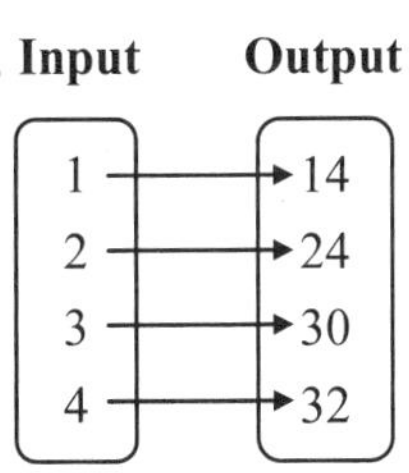

   b. For each additional ticket purchased, the price per ticket decreases $2.

   c. $30

   d. The cost is the same, you should invite two more friends and get 5 tickets rather than 3 tickets.

   e. If the pattern continued, the cost for 8 tickets would be $0.

9.1 Practice B

1. (1, 2), (5, 6), (9, 4), (13, 8)

2. (3, 1), (4, 0), (5, 1), (6, 0)

3. 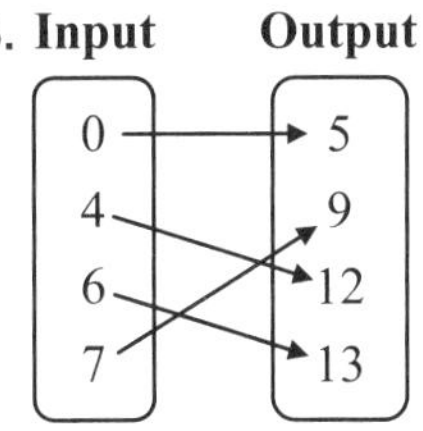

4. 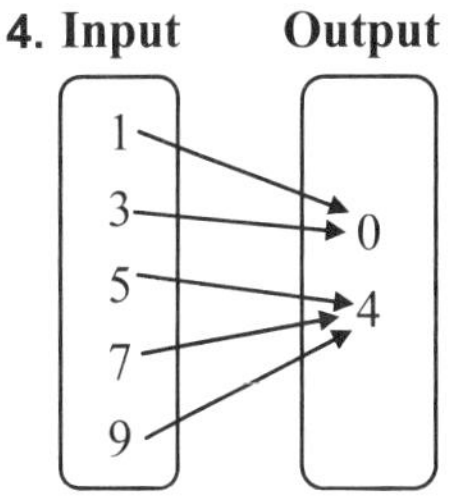

5. 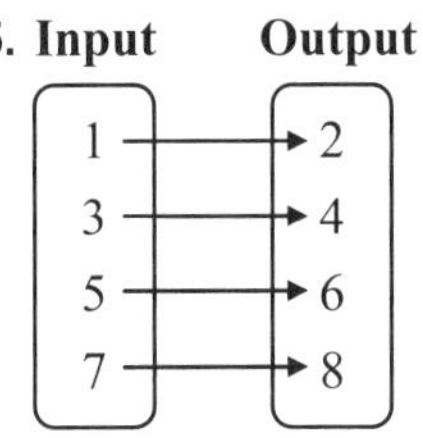

   As each input increases by 2, the output increases by 2 (or each output is 1 more than the input).

6. 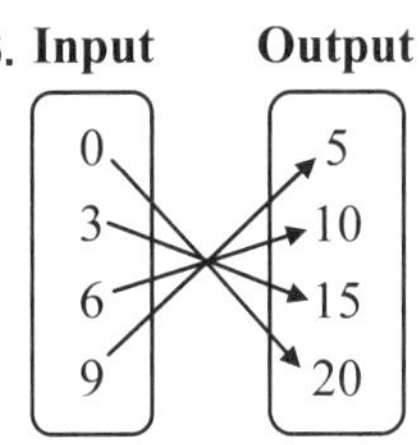

   As each input increases by 3, the output decreases by 5.

7. a. $3.75, $4.00, $4.25, $4.50

   b. 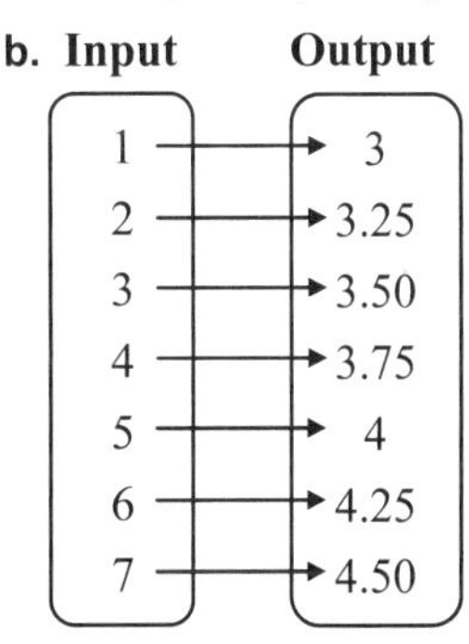

   c. (1, 3), (2, 3.25), (3, 3.50), (4, 3.75), (5, 4), (6, 4.25), (7, 4.50)

   d. 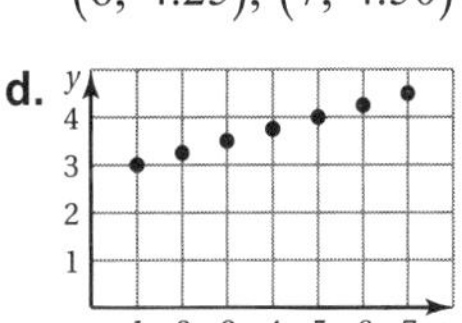

   e. The cost is $3 for the first minute and $0.25 for every additional minute.

Answers

9.1 Enrichment and Extension

1. Function; Each input is paired with exactly one output.

2. Function; Each input is paired with exactly one output.

3. Not a function; The input 1 has two outputs, 2 and 4.

4. Not a function; The input 3 has two outputs, 1 and 5.

5. a. *Sample answer:*

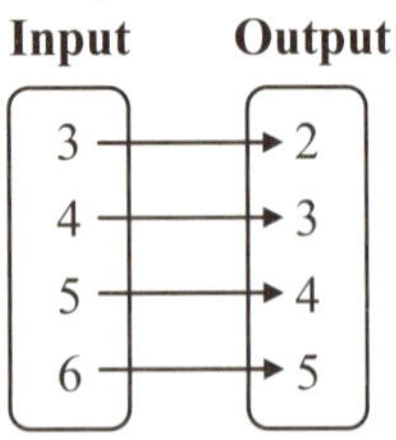

b. *Sample answer:*

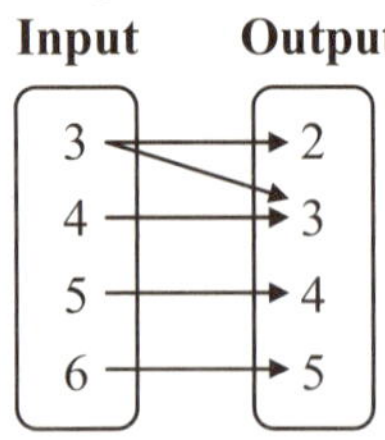

6. a. *Sample answer:*

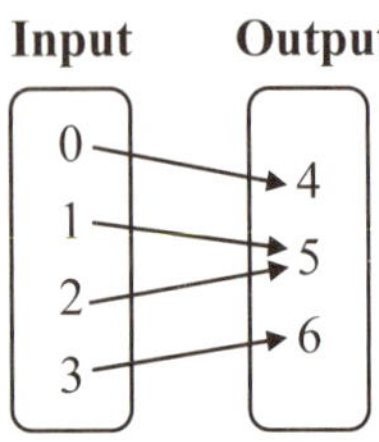

b. *Sample answer:*

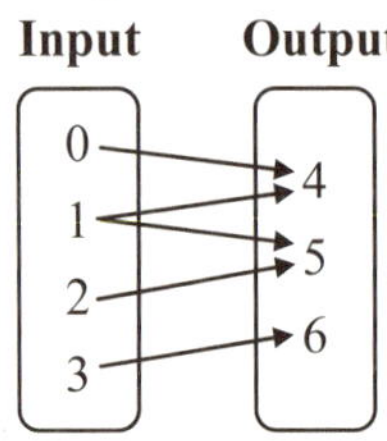

7. a. *Sample answer:* (3, 3)

b. *Sample answer:* (2, 3)

8. a. *Sample answer:* (7, 10)

b. *Sample answer:* (4, 1)

9. a. *Sample answer:* (0, 0)

b. *Sample answer:* (5, 0)

9.1 Puzzle Time

TULIPS

9.2 Start Thinking!

For use before Activity 9.2

$y = 7x$; If the length is 7 meters and the width is x meters, then the area y is the product of the length and width.

9.2 Warm Up

For use before Activity 9.2

1. 20 in. **2.** 15 ft^2

3. about 75.36 ft **4.** 150 cm^2

5. 32 m **6.** about 78.5 $in.^2$

9.2 Start Thinking!

For use before Lesson 9.2

Sample answer: The cost y of buying x apples at $0.59 per apple can be represented by the equation $y = 0.59x$. The variable x represents the number of apples and y represents the total cost.

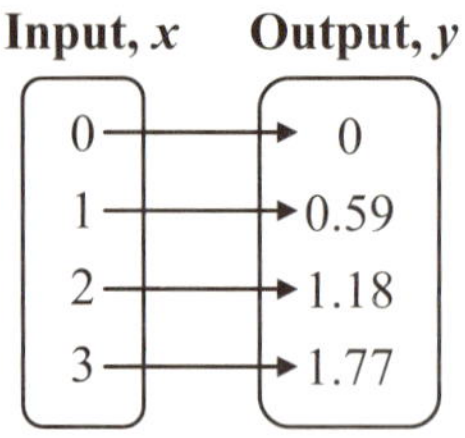

9.2 Warm Up

For use before Lesson 9.2

1. $y = 10x$ **2.** $y = x + 16$

3. $y = \frac{x}{2}$ **4.** $y = x - 10$

9.2 Practice A

1. $y = x + 10$ **2.** $y = 3x$

3. $y = x - 8$ **4.** $y = 2x$

5. $y = 4$ **6.** $y = 24$

7. $y = 39$ **8.** $y = 6$

9. solution **10.** not a solution

11. not a solution **12.** solution

13. a. $d = 6h$ **b.** $d = 12$ miles

14. a. $s = a - 4$ **b.** $s = \$3.50$ **c.** $a = \$6$

9.2 Practice B

1. $y = x + 4$ **2.** $y = \frac{x}{2}$ **3.** $y = 5x$

4. $y = x - 2$ **5.** $y = 2$ **6.** $y = 1$

7. not a solution **8.** solution

9. a. $d = 70h$ **b.** $d = 245$ miles

10. $x = 4$ **11.** $x = 10$

12. a. $P = 1.50b - 90$ **b.** $b = 60$ bags

13. $A = \left(\frac{P}{4}\right)^2$

Answers

9.2 Enrichment and Extension

1. $f(x) = 2x$
2. $f(x) = x - 1$
3. $f(x) = 4x + 7$
4. $f(x) = \frac{3}{4}x + 5$
5. $f(0) = 0,\ f(2) = 16,\ f(4) = 32$
6. $f(0) = 7,\ f(2) = 9,\ f(4) = 11$
7. $f(0) = 5,\ f(2) = 3,\ f(4) = 1$
8. $f(0) = 8,\ f(2) = 14,\ f(4) = 20$
9. $f(0) = 9,\ f(2) = 5,\ f(4) = 1$
10. $f(0) = 0,\ f(2) = 1,\ f(4) = 2$
11. $f(0) = 6,\ f(2) = \frac{13}{2}$, or $6\frac{1}{2}$, $f(4) = 7$
12. $f(0) = 3,\ f(2) = 6.2,\ f(4) = 9.4$
13. $f(0) = 1,\ f(2) = 5,\ f(4) = 17$
14. *Sample answer:* $f(x) = 2$

9.2 Puzzle Time

FROM SECOND TO THIRD BECAUSE THERE IS A SHORTSTOP IN THE MIDDLE

9.3 Start Thinking!

For use before Activity 9.3

Answers will vary; tables help you organize information, which can help you analyze the information.

9.3 Warm Up

For use before Activity 9.3

1. 3
2. 0.2
3. 3
4. 17.5
5. 12
6. 30

9.3 Start Thinking!

For use before Lesson 9.3

Sample answer: An ATM; the input is your ATM card and the output is money.

9.3 Warm Up

For use before Lesson 9.3

1.

Input, x	10	11	12	13
Output, y	7	8	9	10

2.

Input, x	1	2	3	4
Output, y	8	16	24	32

3.

Input, x	0	1	2	3
Output, y	6	7	8	9

4.

Input, x	0	2	4	6
Output, y	0	1	2	3

9.3 Practice A

1.

Input, x	1	2	3	4
Output, y	4	5	6	7

2.

Input, x	1	3	5	7
Output, y	5	15	25	35

3. $y = x + 6$

Input, x	0	1	2	3
Output, y	6	7	8	9

4. $y = 3x$

Input, x	0	2	4	6
Output, y	0	6	12	18

5. $y = x + 5$
6. $y = 4x$

7.

Input, x	0	2	4	5	7	9
Output, y	1	5	9	11	15	19

8. a.

Number of jobs completed in a week, x	1	4	8	10	12
Money earned, y	8	17	29	35	41

b. For each output, multiply the input by 3, then add 5.

c. $y = 3x + 5$

9.3 Practice B

1.

Input, x	1	3	5	7
Output, y	8	10	12	14

2.

Input, x	0	2	4	6
Output, y	0	20	40	60

Answers

3. $y = \frac{x}{4}$ **4.** $y = x - 3$

5.

Input, x	0	1	3	4	5	8
Output, y	3	8	18	23	28	43

6. a.

Eastern Time Zone, x	3	5	6	8	10	11
Central Time Zone, y	2	4	5	7	9	10

b. $y = x - 1$

c.

Eastern Time Zone, x	3	5	6	8	10	11
Mountain Time Zone, y	1	3	4	6	8	9

d. $y = x - 2$ **e.** 11 A.M.

9.3 Enrichment and Extension

1. *Sample answer:* 5 turns

2.

Input, x	1	2	3	4
Output, y	1	2	3	4

Input, x	1	2	3	4
Output, y	4	8	12	16

Input, x	1	2	3	4
Output, y	3	6	9	12

Input, x	1	2	3	4
Output, y	2	3	4	5

Input, x	1	2	3	4
Output, y	1	3	5	7

Input, x	1	2	3	4
Output, y	3	5	7	9

Input, x	1	2	3	4
Output, y	3	4	5	6

Input, x	1	2	3	4
Output, y	1	4	7	10

Input, x	1	2	3	4
Output, y	5	4	3	2

Input, x	1	2	3	4
Output, y	1	4	9	16

Input, x	1	2	3	4
Output, y	6	10	14	18

Input, x	1	2	3	4
Output, y	2	4	6	8

Input, x	1	2	3	4
Output, y	1	5	9	13

Avoid spaces whose functions have only even number outputs. So, avoid the 2nd, 11th, and 12th spaces.

3. Choose 3, then 4, then 1

9.3 Puzzle Time

A FENCE

9.4 Start Thinking!

For use before Activity 9.4

The table gives the Celsius-Fahrenheit temperature conversions for 4 different temperatures. The graph represents the table and in addition shows all of the conversion values in between. It is generally easier to use a graph for predictions. Predictions may vary, but the answer should be close to 24° Celsius.

9.4 Warm Up

For use before Activity 9.4

1. (4, 1) **2.** (1, 3)

9.4 Start Thinking!

For use before Lesson 9.4

Answers will vary. Check students' tables and graphs.

9.4 Warm Up

For use before Lesson 9.4

1.

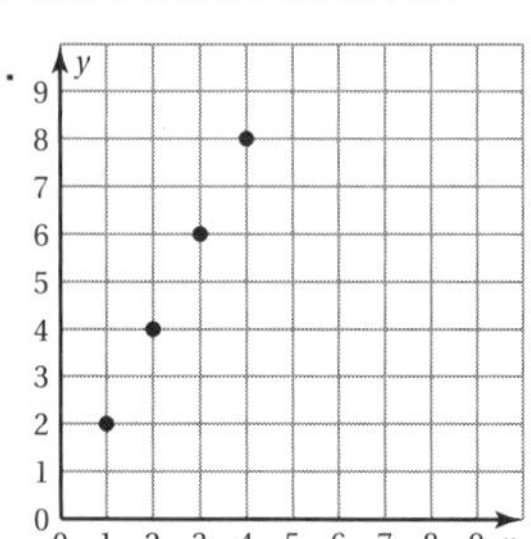

2.

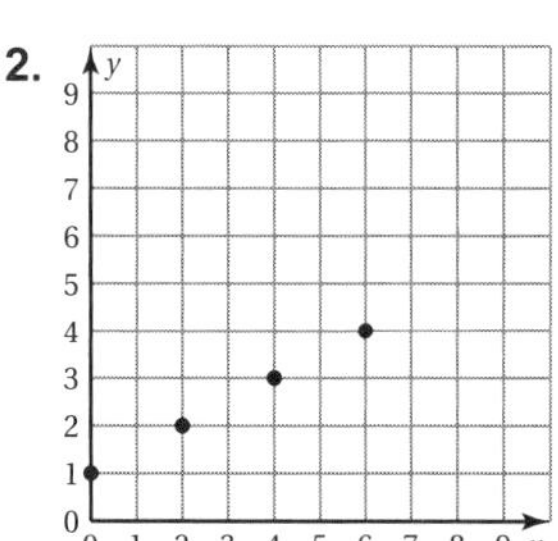

Answers

3.

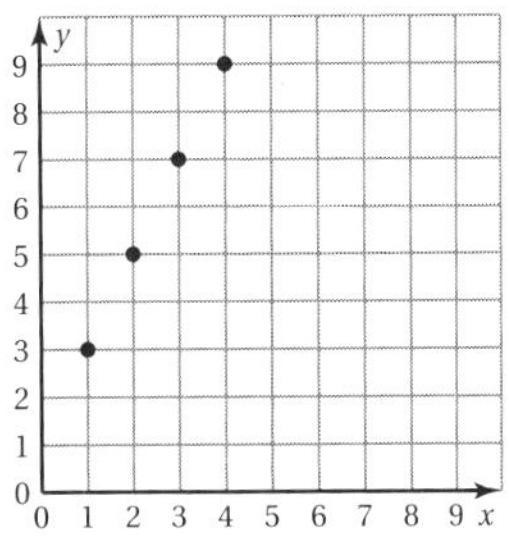

9.4 Practice A

1.

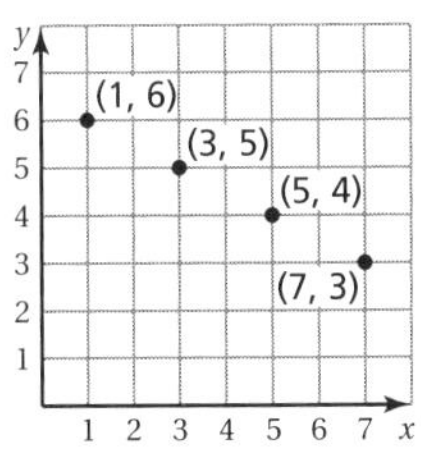

2.

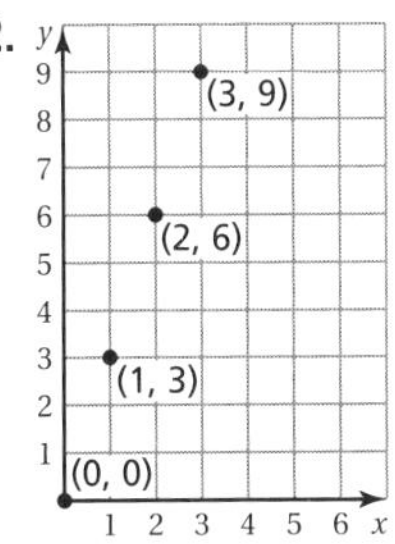

3.

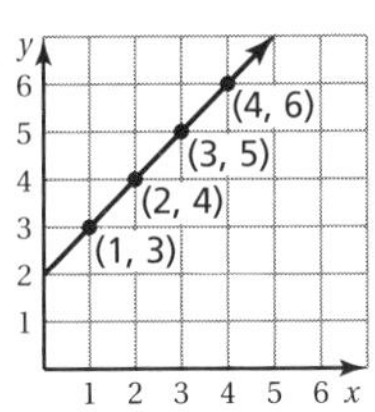

4.

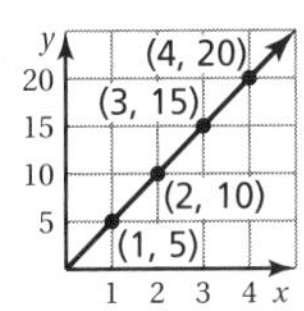

5.

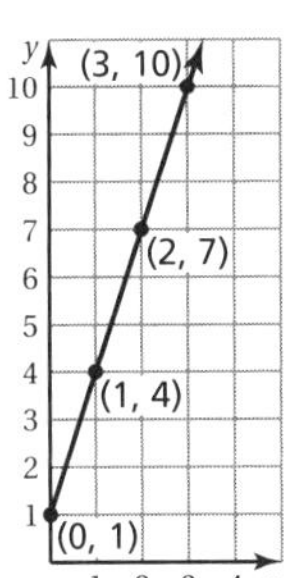

6.

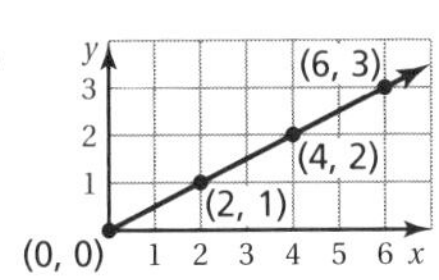

7.

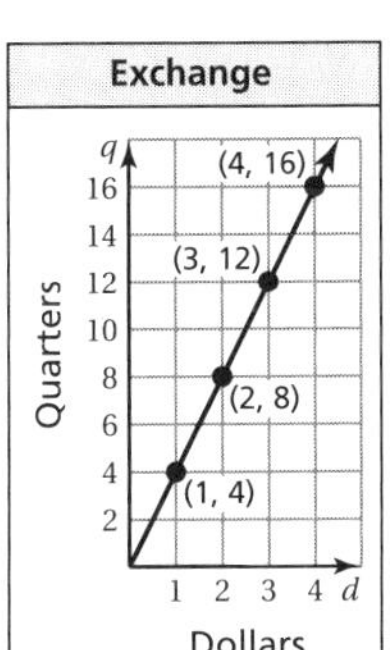

8. B

9. A

10. a. $d = 0.30p$

b.

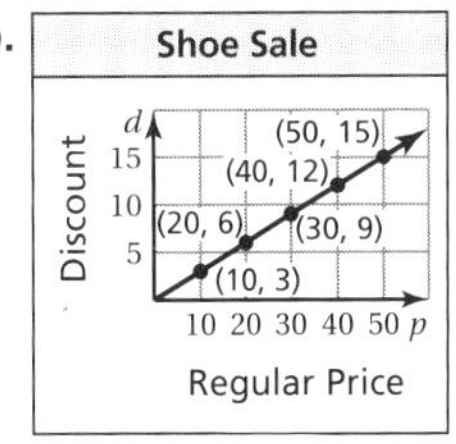

c. \$7.50; \$17.50

d. \$15

9.4 Practice B

1.

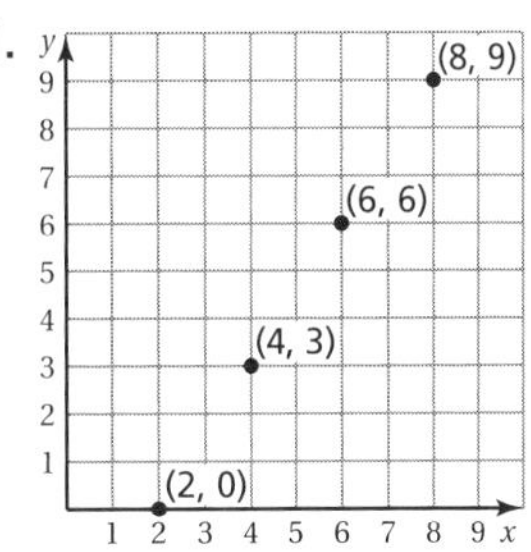

2.

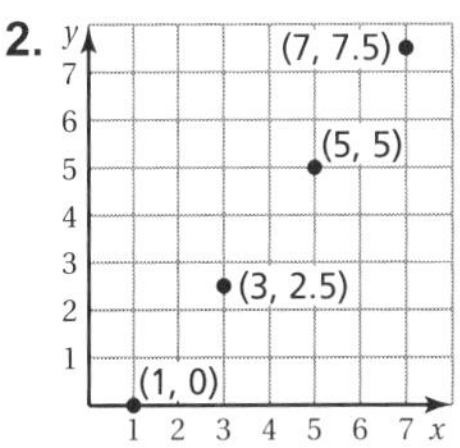

3.

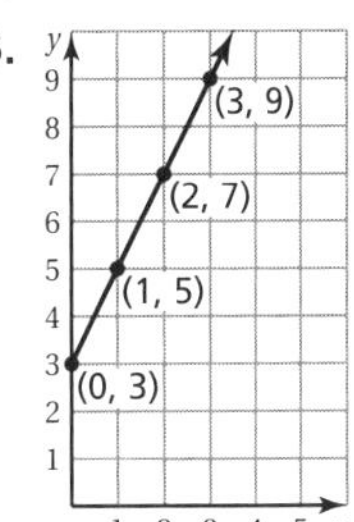

4.

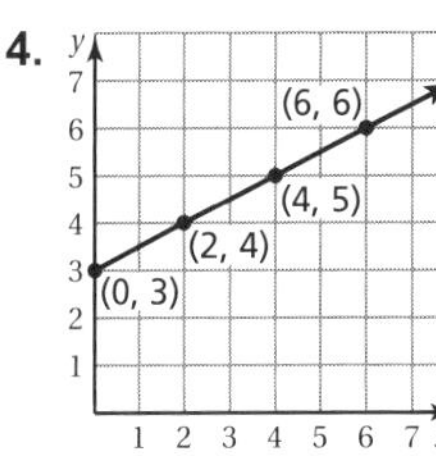

5.

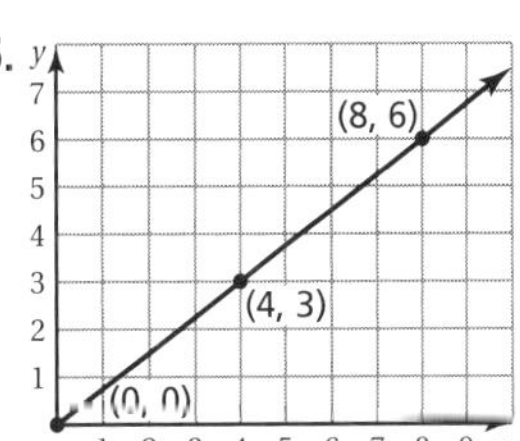

6.

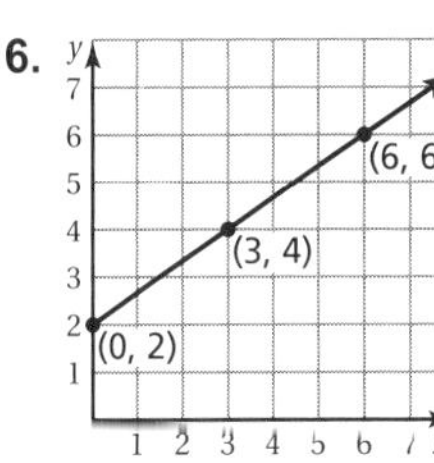

7.

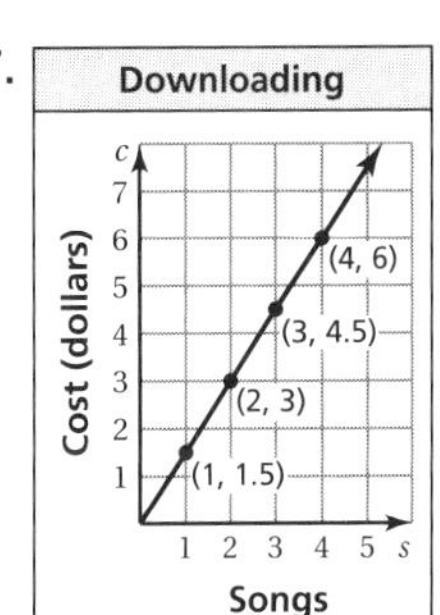

Answers

8. a.

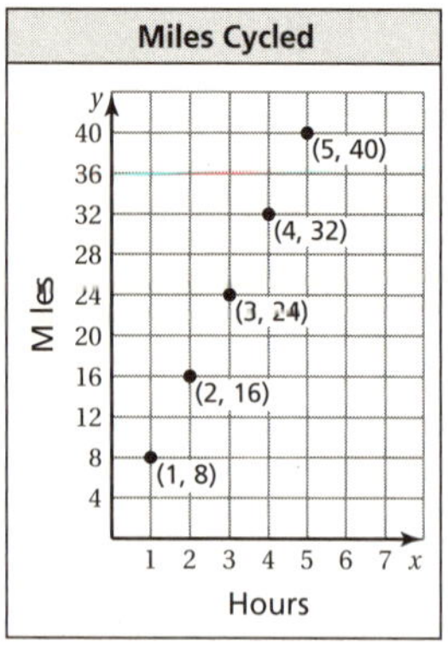

b. $y = 8x$ **c.** 28 mi **d.** 7.5 hours

9. $y = 10$

10. The solver plotted the points from the input-output table incorrectly.

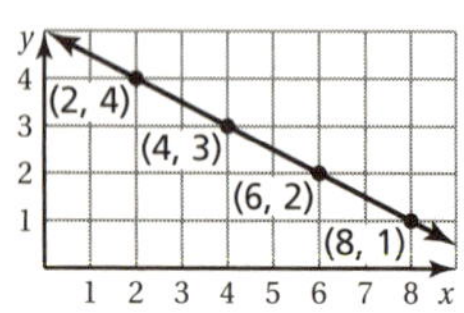

9.4 Enrichment and Extension

1. Point B; The point is on both graphs, so it is a solution of both equations.

2. (4, 5)

3. (1, 2)

4.

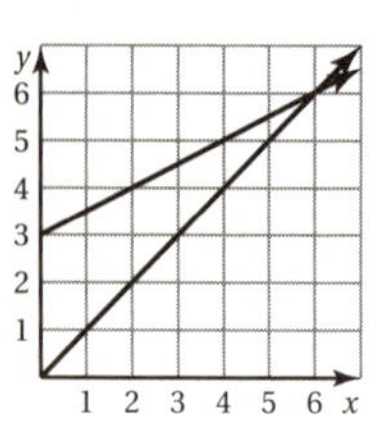

(6, 6)

5.

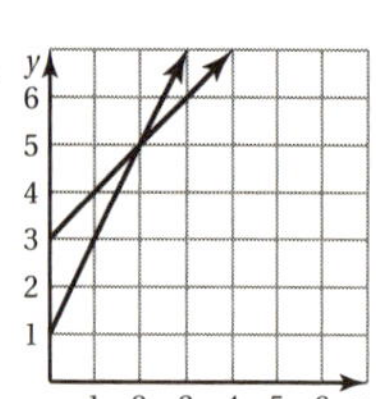

(2, 5)

6.

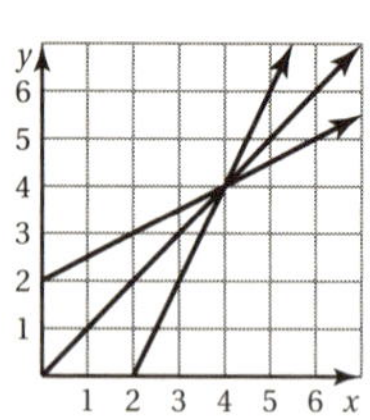

(4, 4)

7.

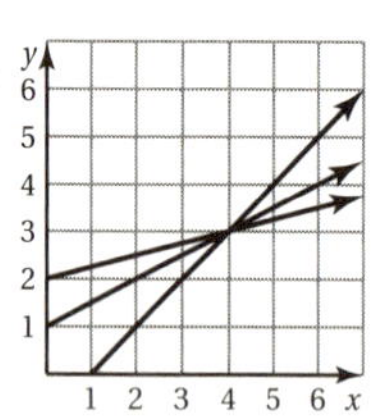

(4, 3)

9.4 Puzzle Time

TO SEE IF THERE WAS ANY CHANGE IN THE WEATHER

9.5 Start Thinking!

For use before Activity 9.5

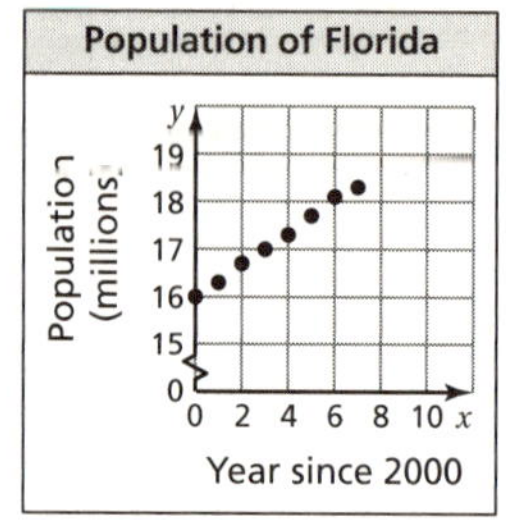

The data appears to follow a line that is going up to the right. Check students' predictions and current Florida population.

9.5 Warm Up

For use before Activity 9.5

1.

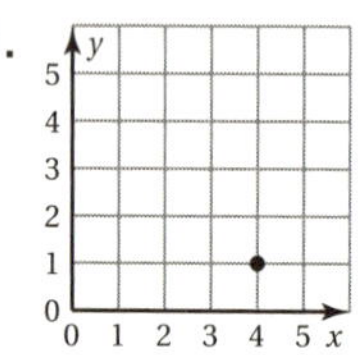

2.

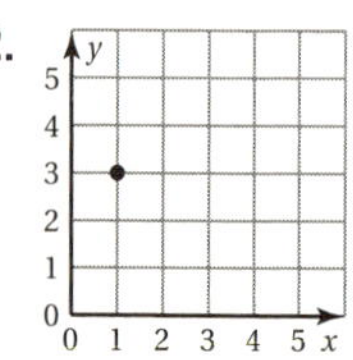

3.

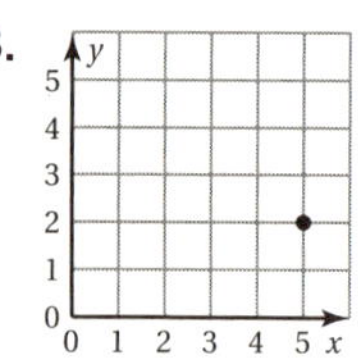

4.

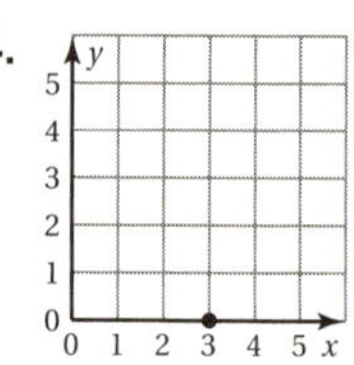

5.

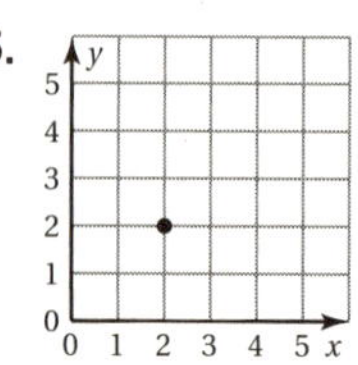

6.

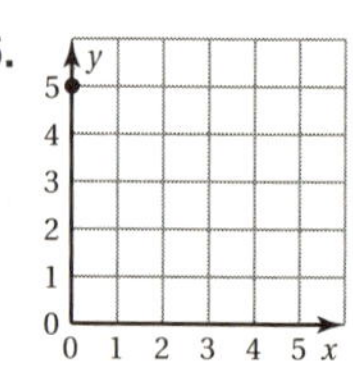

9.5 Start Thinking!

For use before Lesson 9.5

Sample answer: Linear has the word "line" in it. A linear function probably looks like it follows a line. Check students' graphs.

9.5 Warm Up

For use before Lesson 9.5

1.

Months, m	1	2	3	4
Cost, C	50	100	150	200

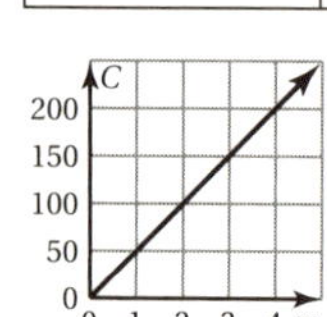

$C = 50m$

Answers

2.

Height, h	1	2	3	4
Area, A	3	6	9	12

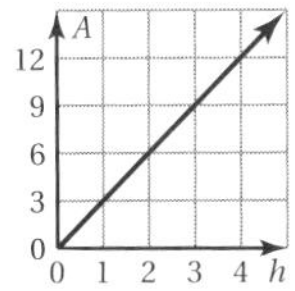

$A = 3h$

9.5 Practice A

1.

Tickets, t	1	2	3	4
Cost, c	7	14	21	28

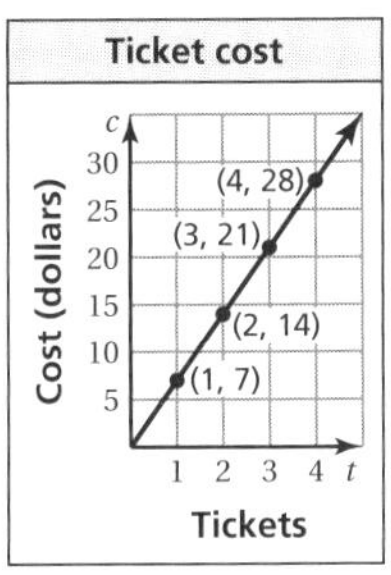

$c = 7t$

2. no; the graph is not a straight line.
3. yes; the graph is a straight line.
4. yes; the graph of the ordered pairs lie on a straight line.
5. no; the graph of the ordered pairs do not lie on a straight line.
6. **a.** you **b.** 5 hours **c.** 6 hours
   d. no; Every hour that you both drive, you drive 10 miles farther than your friend.
   e. 1 hour

9.5 Practice B

1.

Bracelets, b	1	2	3	4
Cost, c	5.50	11	16.50	22

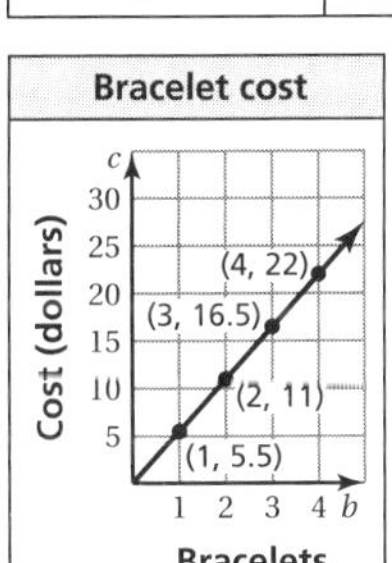

$c = 5.50b$

2. yes; the graph is a straight line.
3. no; the graph is not a straight line.
4. no; the graph of the ordered pairs do not lie on a straight line.
5. yes; the graph of the ordered pairs lie on a straight line.
6. Check graphs; $y = 4x$; *Sample answer:* For every value of x except 0, the corresponding y-value for $y = 4x$ is greater than the corresponding y-value for $y = \frac{1}{4}x$.
7. Check graphs; $y = 3x + 1$; *Sample answer:* For every value of x, the corresponding y-value for $y = 3x + 1$ is greater than the corresponding y-value for $y = x$.
8.

x	0	1	2	3	4	5
y	3	3.5	4	4.5	5	5.5

Differences: 0.5, 0.5, 0.5, 0.5; all the differences are the same, 0.5.

9.5 Enrichment and Extension

1.

Input, x	0	1	2	3
Output, y	0	1	4	9

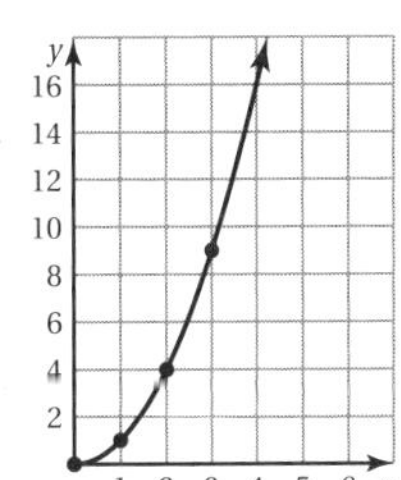

2.

Input, x	0	1	2	3
Output, y	1	2	5	10

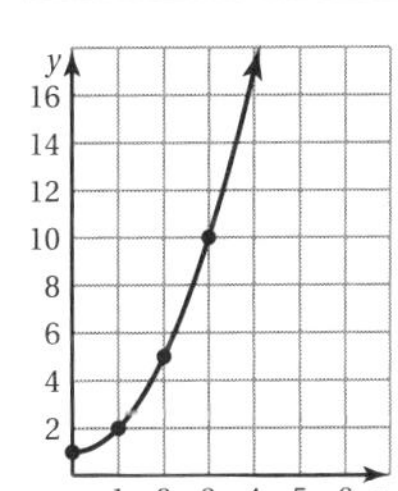

Answers

3.

Input, x	0	1	2	3
Output, y	4	5	8	13

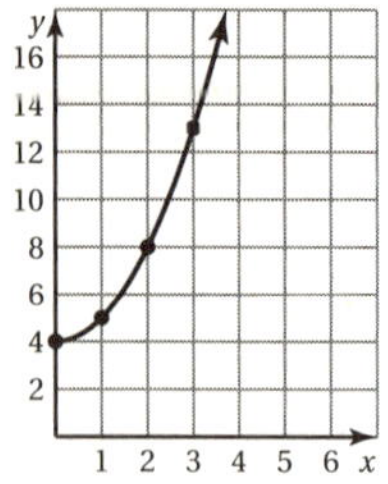

4.

Input, x	0	1	2	3
Output, y	8	9	12	17

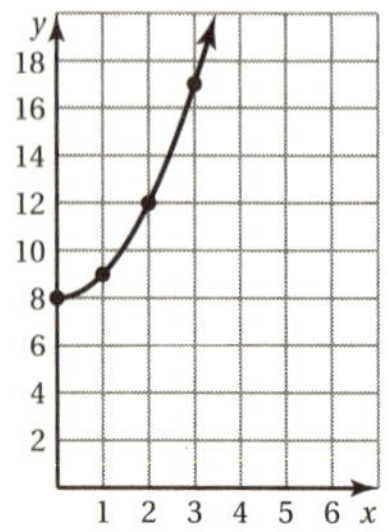

5. All of the graphs have the same shape and are not linear. The graphs are shifted up and cross the y-axis at different points.

6. 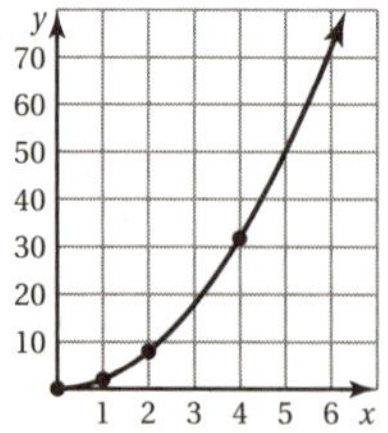

7. 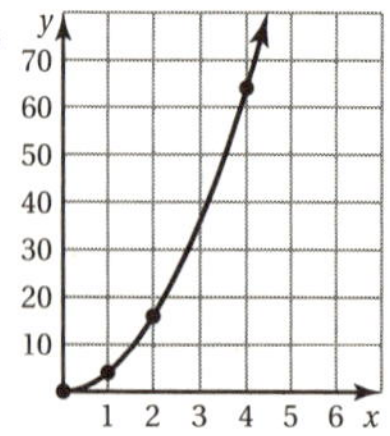

8. 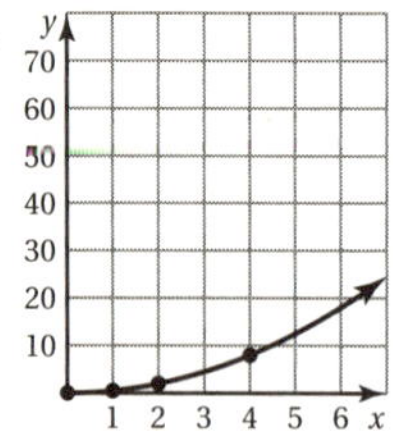

9. 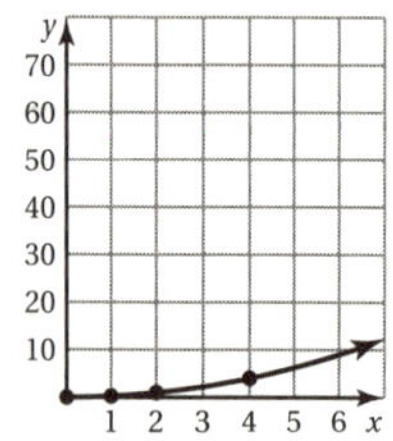

10. As a increases, the graph gets steeper. As a gets closer to 0, the graph appears to flatten out.

11. The function $y = x^2$ appears to increase more rapidly.

12. The function $y = 2^x$ increases more rapidly. The functions share common points at $(2, 4)$ and $(4, 16)$.

13. Using the input-output table, you can see that the values of y are getting farther and farther apart for the functions as x increases.

9.5 Puzzle Time

A CHEETAH

School-to-Work

1. $y = 18 + 0.5x$ **2.** 14 ft

3. 25 ft **4.** 10 ft, 14 ft

5.

Input	0	1	2	3	4	5	6
Output	0	1.5	3.0	4.5	6.0	7.5	9.0

6. $y = 12 - 1.5x$ **7.** $y = 18x + 1$

8. \$1540 **9.** \$1500

Study Help

Available at *BigIdeasMath.com.*

Financial Literacy

1. In each case a number, such as employees or sales, changes depending on changes in an input.

2. Employees increase when sales increase, sales increase when the weather is good, business increases when word of mouth is good.

3. *Sample answer:* Profits are a function of sales; when sales go up, profits go up.

4. down

5. a. 40 discs **b.** \$5

c. When they give the discs away for free, people will want 60 discs.

d. When the price is \$30.

6. Demand should rise. The graph will shift upwards—at a given price, more people will want to buy the disc. When it is free $(x = 0)$ more people will want a copy; when it is very expensive (e.g. $x = 30$) a few people may still buy it.

Cumulative Practice

1–8. 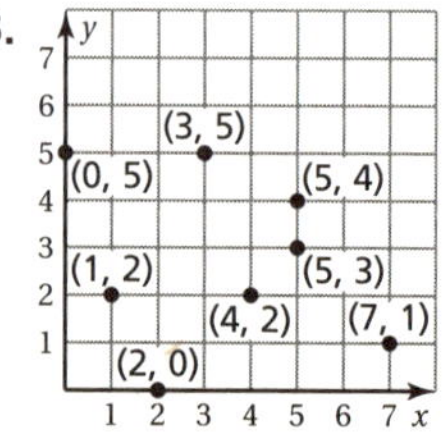

9. No; the graph is not a straight line.

10. Yes; the graph is a straight line.

11. Yes; the graph is a straight line.

Answers

12. $y = \dfrac{12}{x}$

Input, x	1	2	3	4
Output, y	12	6	4	3

13. $y = x + 15$

Input, x	1	2	3	4
Output, y	16	17	18	19

14.
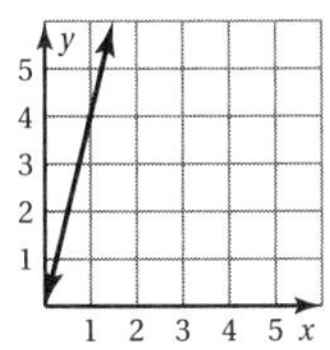

15.
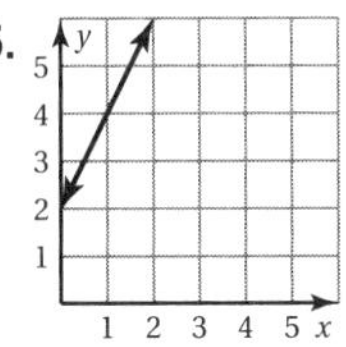

16.
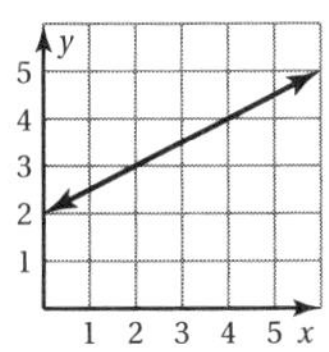

17. a. $d = 0.6p$

b.
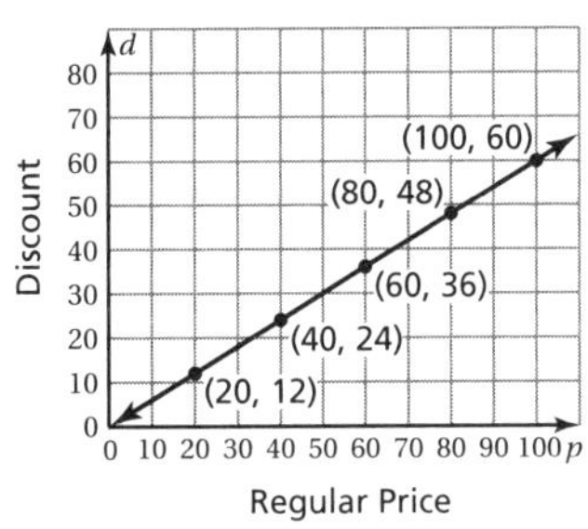

c. 12

18. a.
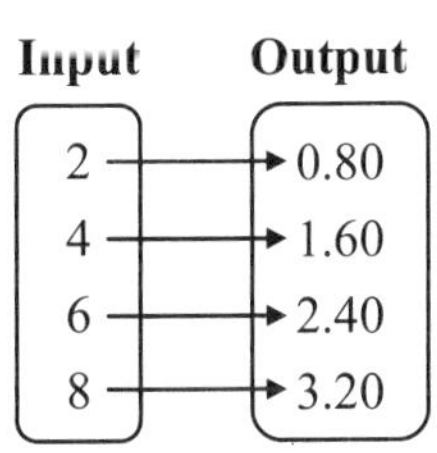

b. For every pound of aluminum cans that you recycle, you receive \$0.40.

Appendix B

B.1 Start Thinking!

For use before Activity B.1

Answers will vary. Check students' number lines.

B.1 Warm Up

For use before Activity B.1

1.
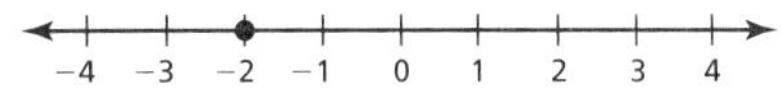

2.

3.

4.
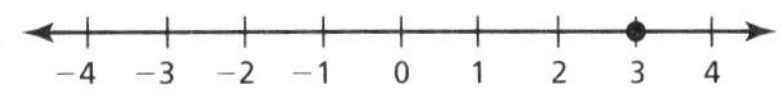

5.

6.
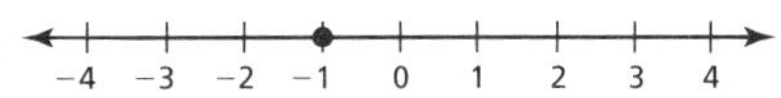

B.1 Start Thinking!

For use before Lesson B.1

Answers will vary. Check to see that 6 A.M. corresponds to −6 on the number line.

B.1 Warm Up

For use before Lesson B.1

1. 57;

2. −420;

3. 128;

4. −300;
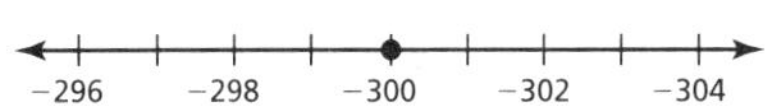

B.1 Practice A

1. 10;
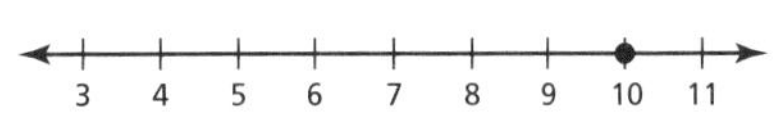

2. −1;
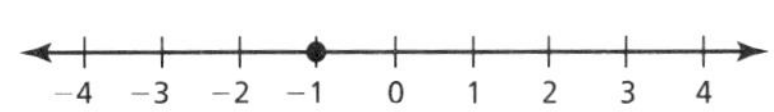

3. 22;
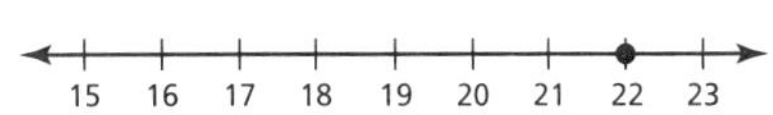

4. −3;
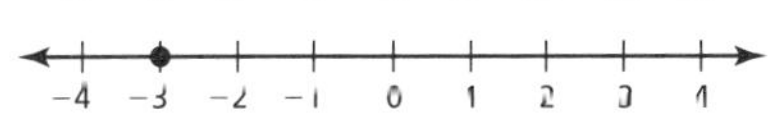

5. −16;
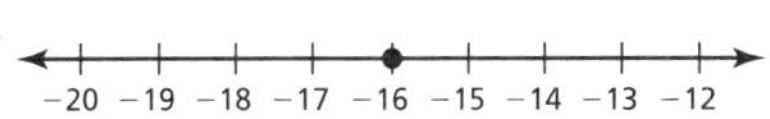

6. 2;
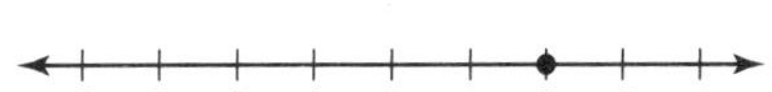

7. 11 **8.** 9 **9.** 8 **10.** −2

11. −3 **12.** 17 **13.** 34 **14.** −12

15. 6 **16.** 7 **17.** 10 **18.** 11

19. 45 **20.** 21 **21.** 32 **22.** 61

23. 1 **24.** 4

25. your donations: −8; friend's donations: 8

Answers

26. a. eagle: 200; jellyfish: -210

b. $|200| = 200; |-210| = 210$

c. the jellyfish

27. a. -1 **b.** -16

28. a. $26, -28$

b. the second character; because $|-28| > |26|$

B.1 Practice B

1. 4; 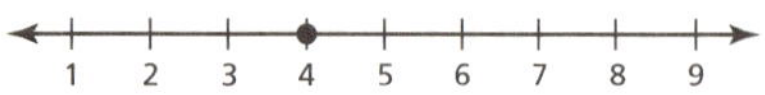

2. -7; 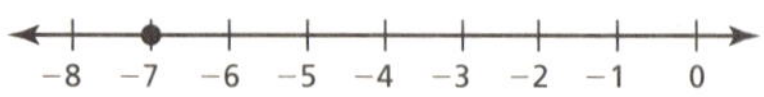

3. 32; 

4. -6; 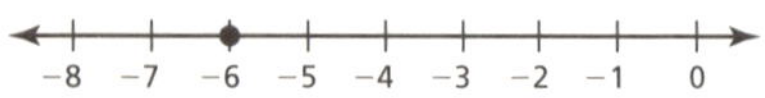

5. -3; 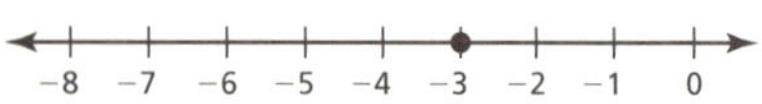

6. 142; 

7. $>$ **8.** $<$ **9.** $=$ **10.** $>$

11. $-3, 0, |1|, 2, |-3|, 4, 5$

12. $-8, -7, -6, |-6|, |-7|, |8|$

13. $-11, 0, |-9|, |-10|, |11|, 12$

14. $-30, -22, |-21|, |24|, 28, |32|$

15. 0 **16.** -3 **17.** -16 **18.** -9

19. *Sample answer:* -24

20. a. *Sample answer:* -20

b. *Sample answer:* -10

c. *Sample answer:*

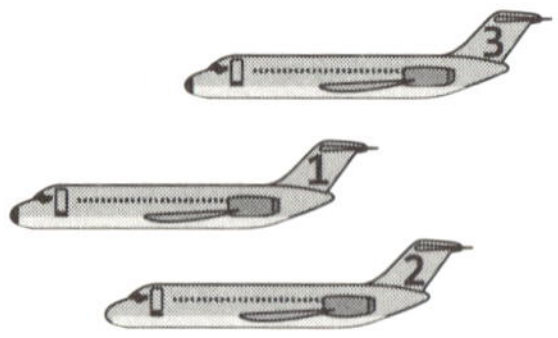

21. a. 7:00 A.M.

b.

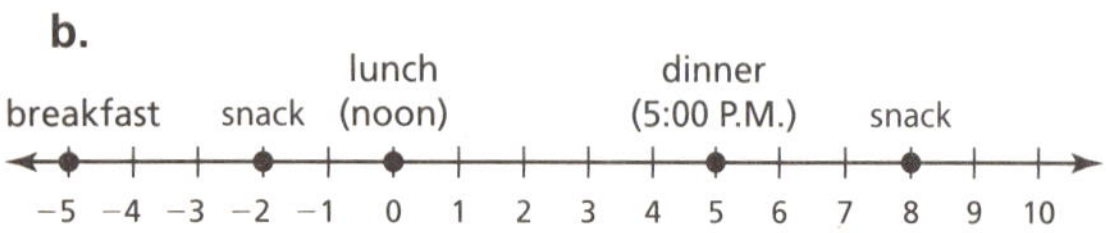

c. 10:00 A.M.

B.1 Enrichment and Extension

1. sometimes **2.** always **3.** never

4. sometimes **5.** The Bahamas

6. no integers **7.** all integers

8. all negative integers **9.** all integers

10. no integers **11.** Jamaica

B.1 Puzzle Time

BECAUSE THEY ALWAYS HAVE THEIR HANDS RIGHT IN FRONT OF THEIR FACES

B.2 Start Thinking!

For use before Activity B.2

Answers will vary. Check students' work; 17°F

B.2 Warm Up

For use before Activity B.2

1. 5 **2.** 1 **3.** 0

4. -5 **5.** -1 **6.** 6

B.2 Start Thinking!

For use before Lesson B.2

Check students' problems and drawings and make sure that the parking levels are labeled $0, -1, -2,$ and -3. *Sample answer:* A man gets on the elevator at the tenth floor and goes to the bottom floor of the parking garage. How many floors did he travel?

B.2 Warm Up

For use before Lesson B.2

1. 1 **2.** -6 **3.** -2

4. -7 **5.** -6 **6.** -3

B.2 Practice A

1. 2 **2.** -4 **3.** -1 **4.** -3

5. -1 **6.** 3 **7.** -4 **8.** -11

9. -1 **10.** -5 **11.** -15 **12.** -4

13. -5 **14.** 8 **15.** 2 **16.** -16

17. -11 **18.** 5

19. The arrow subtracting 4 starts at 0 instead of 2;

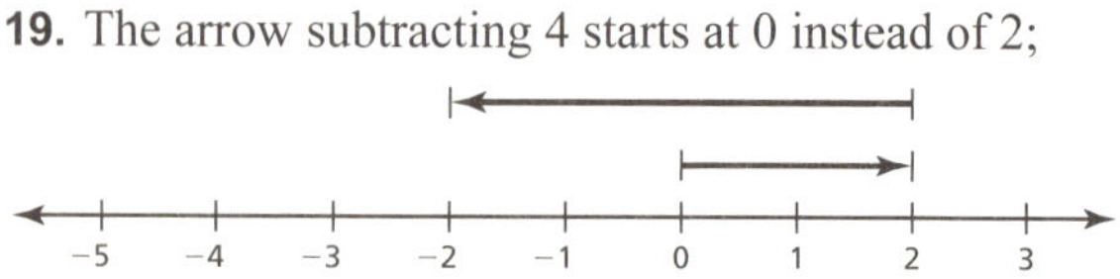

$2 - 4 = -2$

Answers

20. a. Boston: $-7°C$, Chicago: $-10°C$, Miami: $15°C$, New York: $-2°C$, Tampa: $6°C$

b. $-10°C$, $-7°C$, $-2°C$, $6°C$, $15°C$; Chicago

21. a. -1

b. Because $2 - 3 = -1$, the result of two steps forward and three steps back is negative, as you said.

B.2 Practice B

1. 2	**2.** -2	**3.** -5	**4.** -9
5. -3	**6.** -6	**7.** 3	**8.** 4
9. -13	**10.** -21	**11.** -15	**12.** -12
13. 6	**14.** -3	**15.** -5	**16.** -1
17. -4	**18.** -13	**19.** -8	**20.** -6
21. 3	**22.** 11	**23.** 11	**24.** -3

25. a. 0; -2; 1; 0; 1; 1 **b.** 1

c. 3 shots; The golfer's score against par is $+1$ after 8 holes, so a score of -1 is needed on hole 9. Because par is 4 shots, the golfer needs 3 shots.

26. a. The dolphin is at the surface.

b. 7 and 11; 7 is a leap, because after this change, the dolphin is 3 feet above the water; 11 is a leap, because after this change, the dolphin is 4 feet above the water.

27. a. all integers greater than 12

b. the integers from 1 to 11

c. 12

B.2 Enrichment and Extension

1. true

2. false; A negative number plus a positive number can be negative, such as $-3 + 1 = -2$.

3. false; A positive number minus a positive number can be negative, such as $4 - 5 = -1$.

B.2 Puzzle Time

TO A RETAIL STORE

B.3 Start Thinking!

For use before Activity B.3

Sample answer: You buy $2\frac{1}{2}$ pounds of apples to make apple pies. A football team loses $2\frac{1}{2}$ yards on a play.

B.3 Warm Up

For use before Activity B.3

1. 2

2. *Sample answer:* 0, 1, 2, 3, or 4

3. *Sample answer:* -6 or -5

4. *Sample answer:* -7 or -6

5. -1

6. *Sample answer:* 1, 2, 3, or 4

B.3 Start Thinking!

For use before Lesson B.3

Answers will vary. Oversee the activity and make sure that the students have lined up correctly.

B.3 Warm Up

For use before Lesson B.3

1. *Sample answer:* $\frac{1}{4}$ **2.** *Sample answer:* $-1\frac{3}{8}$

3. *Sample answer:* $-\frac{3}{4}$

B.3 Practice A

1. *Sample answer:* 1 **2.** *Sample answer:* $-\frac{1}{2}$

3. *Sample answer:* $-1\frac{1}{2}$ **4.** *Sample answer:* 1

5. $-\frac{1}{5}$, The number is farther right on the number line.

6. $3\frac{2}{9}$, The number is farther right on the number line.

7. $-\frac{7}{4}$, The number is farther right on the number line.

8. $\frac{4}{5}$, The number is farther right on the number line.

9. $-\frac{6}{7}$, The number is farther right on the number line.

Answers

10. $-2\frac{2}{3}$, The number is farther right on the number line.

11. $-\frac{7}{10}$, The number is farther right on the number line.

12. $4\frac{5}{6}$, The number is farther right on the number line.

13. $-\frac{3}{10}$ **14.** $\frac{7}{12}$ **15.** $-1\frac{2}{3}$ **16.** $-\frac{1}{4}$

17. first

18. a.

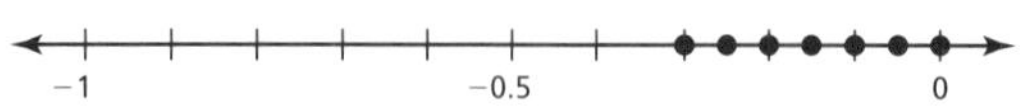

b. 2 min

B.3 Practice B

1. *Sample answer:* $4\frac{2}{3}$ **2.** *Sample answer:* $-3\frac{1}{5}$

3. *Sample answer:* $-\frac{3}{4}$ **4.** *Sample answer:* $-10\frac{1}{6}$

5. $-\frac{3}{8}$, The number is farther right on the number line.

6. $-\frac{2}{9}$, The number is farther right on the number line.

7. $\frac{3}{4}$, The number is farther right on the number line.

8. $-\frac{13}{4}$, The number is farther right on the number line.

9. $-1\frac{3}{10}$, The number is farther right on the number line.

10. $-\frac{3}{10}$, The number is farther right on the number line.

11. $-\frac{11}{5}$, The number is farther right on the number line.

12. $-1\frac{1}{2}$, The number is farther right on the number line.

13. $-\frac{2}{3}$ **14.** $-\frac{3}{4}$ **15.** $-5\frac{3}{7}$ **16.** $-\frac{4}{25}$

17. first

18.

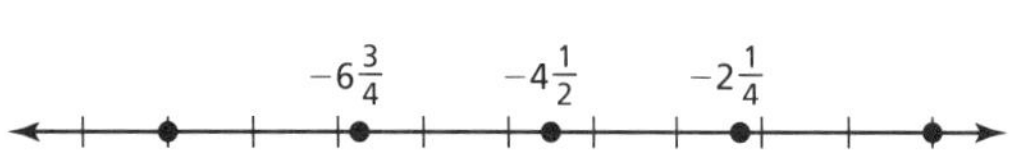

On the third day, if the pool is at 0 on the first day.

B.3 Enrichment and Extension

1–10. *Sample answer:*

In Exercise 1, Player A may write down $\frac{1}{2}$ and player B may write down $\frac{4}{3}$.

Player B takes the cards because $\frac{4}{3} > \frac{1}{2}$.

B.3 Puzzle Time

ONLY ONE BECAUSE AFTER THAT THE JAR ISN'T EMPTY ANYMORE

B.4 Start Thinking!

For use before Activity B.4

Sample answer: Elevation is the height above or below sea level. So, elevations that are below sea level can be written as negative decimals. For example, 1.2 miles below sea level is an elevation of −1.2 miles.

B.4 Warm Up

For use before Activity B.4

1. $\frac{1}{8}$ **2.** $-\frac{2}{3}$ **3.** $-\frac{15}{8}$

4. $-1\frac{1}{5}$ **5.** $2\frac{1}{2}$ **6.** $-3\frac{1}{2}$

B.4 Start Thinking!

For use before Lesson B.4

Sample answer: 1. You can line up their decimal points and compare digits from left to right. 2. You can graph them on a number line and compare. You can compare negative decimals using a number line. If you are comparing two negative decimals using the first method, the negative decimal with the lesser digit is the greater number. Consider the following example:

−1.43

−1.52

In the above example, $4 < 5$. That means that $-1.43 > -1.52$.

B.4 Warm Up

For use before Lesson B.4

1. *Sample answer:* $-2\frac{3}{4}$ **2.** *Sample answer:* $-\frac{1}{2}$

3. *Sample answer:* $1\frac{1}{3}$ **4.** *Sample answer:* $1\frac{7}{8}$

B.4 Practice A

1. *Sample answer:* −2.2 **2.** *Sample answer:* −0.75

Answers

3. *Sample answer:* -0.3 4. *Sample answer:* 1.6

5. 2.1, The number is farther right on the number line.

6. 4.8, The number is farther right on the number line.

7. -0.08, The number is farther right on the number line.

8. -3.08, The number is farther right on the number line.

9. 5.05, The number is farther right on the number line.

10. -4.82, The number is farther right on the number line.

11. -3.5, The number is farther right on the number line.

12. -4.70, The number is farther right on the number line.

13. yes 14. no 15. no 16. yes

17. March; February

18. -2.83 km, -2.70 km, -2.48 km, -2.15 km

19. They compared the absolute values of the two numbers instead of using a number line.

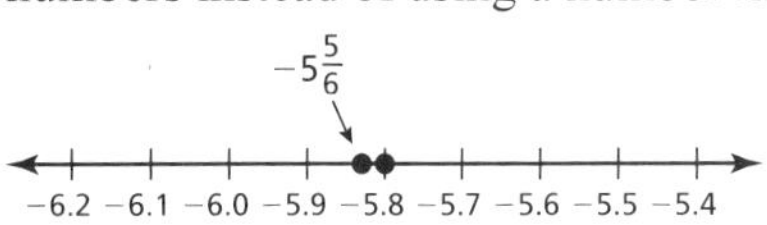

-5.8 is greater than $-5\frac{5}{6}$ because it is to the right of $-5\frac{5}{6}$ on the number line.

20. Tuesday 21. dock

B.4 Practice B

1. -0.7 2. -0.4 3. 0.1 4. 0.8

5. 5.38, The number is farther right on the number line.

6. -4.9, The number is farther right on the number line.

7. -0.11, The number is farther right on the number line.

8. -2.05, The number is farther right on the number line.

9. 0.7, The number is farther right on the number line.

10. -4.6, The number is farther right on the number line.

11. -0.1, The number is farther right on the number line.

12. $-3\frac{3}{8}$, The number is farther right on the number line.

13. no 14. yes 15. no 16. yes

17. a. 0 seconds b. 4 seconds
 c. Between 2 and 3 seconds, the speed is 0 when the ball is changing direction.

18. a. 3 seconds
 b. 6 seconds; Yes, the ball is falling down the cliff and going lower and lower.

B.4 Enrichment and Extension

1–12. *Sample answer:*
In Round 1, Player A may roll 2, 5, 6 and write the decimal -5.26. Player B may roll 1, 4, 6 and write the decimal -4.61. Player B wins the round because -4.61 is closer to -4.5.

B.4 Puzzle Time

LUNCH IS ON ME

B.5 Start Thinking!

For use before Activity B.5

The position where the rulers come together is 0. All of the numbers on the right hand ruler are positive. All of the numbers on the left hand ruler are negative. For the addition expressions, start at the first number and move to the right by the second number of centimeters. For the subtraction expressions, start at the first number and move to the left by the second number of centimeters.

$$-2\frac{1}{4} + 4 = 1\frac{3}{4}$$

$$-3\frac{1}{2} + 3 = -\frac{1}{2}$$

$$3\frac{1}{12} - 4 = -\frac{11}{12}$$

$$-6\frac{3}{4} - 2 = -8\frac{3}{4}$$

B.5 Warm Up

For use before Activity B.5

1. 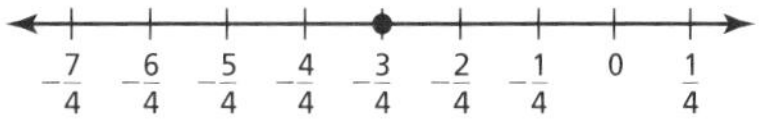

2. 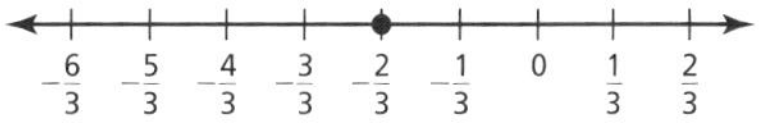

3. 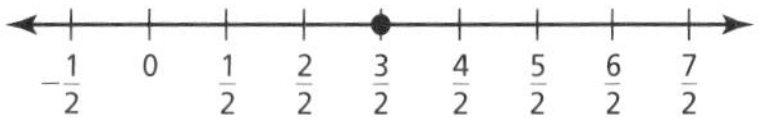

4. 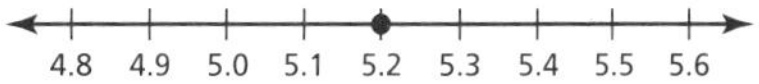

5. 

6. 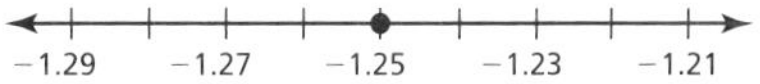

Answers

B.5 Start Thinking!

For use before Lesson B.5

Sample answer: A fish is swimming 5.5 meters below sea level and ascends 3.1 meters. What is the fish's current position?; −24 meters

B.5 Warm Up

For use before Lesson B.5

1. 3.8 **2.** $-1\frac{1}{2}$ **3.** −1.75

4. −2 **5.** −2.5 **6.** $1\frac{1}{4}$

B.5 Practice A

1. $\frac{2}{3}$ **2.** $5\frac{2}{5}$ **3.** $-1\frac{1}{4}$ **4.** 1

5. $-5\frac{2}{5}$ **6.** $\frac{1}{2}$ **7.** $-4\frac{3}{4}$ **8.** $-\frac{1}{5}$

9. 7.1 **10.** 3.5 **11.** −5.9 **12.** 0.25

13. −2.2 **14.** −0.4 **15.** −2.7 **16.** 1.6

17. −1 **18.** −2 **19.** $-2\frac{5}{6}$ **20.** −1.5

21. −8.6 **22.** 2.7

23. a. −1.7, or 1.7 meters below the top

b. −5 + 1.4 = −3.6; −5 represents descending 5 meters from the edge, +1.4 represents ascending 1.4 meters from that point

c. −18.5 m, 18.5 meters below the edge of the cliff

24. $5\frac{1}{2}$ yard line **25.** $5\frac{2}{3}$ feet below water

26. 20 + 12.5 − 13.85 = 18.65, she has $18.65

B.5 Practice B

1. $-\frac{1}{3}$ **2.** $-\frac{3}{4}$ **3.** $-1\frac{3}{5}$ **4.** $-2\frac{5}{6}$

5. $-1\frac{4}{5}$ **6.** $\frac{1}{2}$ **7.** $-2\frac{3}{8}$ **8.** $-\frac{1}{8}$

9. −1.8 **10.** −0.8 **11.** −4.8 **12.** −2.2

13. −0.02 **14.** −0.76 **15.** −6.2 **16.** 1.1

17. $\frac{1}{7}$ **18.** $\frac{1}{3}$ **19.** $-1\frac{3}{4}$ **20.** −5.7

21. −1.7 **22.** −11.4

23. a. 0.25 feet above the edge; 10.5 feet below the edge

b. 29.25 feet deep

24. The second arrow should go left for subtraction; −2.3 − 1 = −3.3

25. −6.5 + 1.5 + 1.5 = −3.5; Your younger brother owes you $3.50.

B.5 Enrichment and Extension

1. $-\boxed{9} + \frac{6}{5} = -7\frac{\boxed{4}}{5}$ **2.** $-\frac{3}{8} + \frac{3}{\boxed{4}} = \frac{3}{\boxed{8}}$

3. $-5\frac{2}{\boxed{3}} + 8\frac{\boxed{3}}{5} = 2\frac{14}{15}$

4. $\frac{7}{9} - \boxed{7} = -6\frac{\boxed{2}}{9}$ **5.** $1\frac{\boxed{2}}{3} - 2\frac{\boxed{5}}{9} = -\frac{8}{9}$

6. $-\frac{5}{\boxed{6}} - 2\frac{1}{15} = -2\frac{\boxed{9}}{10}$

7. $-\boxed{6}.2 - 11.\boxed{2} = -17.4$

8. $-9.\boxed{8} + \boxed{9}.8 = 0$ **9.** $-\boxed{3} + 7.\boxed{2} = 4.2$

10. $-\boxed{1}.8 + 7.4\boxed{1} = 5.61$

11. $-1\boxed{1}.7 - 9.\boxed{9} = -21.6$

12. $\boxed{8} - 23.\boxed{5} = -15.5$

13. $\boxed{1}\,\boxed{5}\,\boxed{1}\,\boxed{3}$

B.5 Puzzle Time

BETWEEN YOU AND ME IT SEEMS THAT SOMETHING IS STARTING TO SMELL

B.6 Start Thinking!

For use before Activity B.6

Check students' graphs. The new graphs include the other quadrants. The new part of the graph represents negative temperatures. For example, −4°F corresponds to −20°C.

B.6 Warm Up

For use before Activity B.6

1–6.

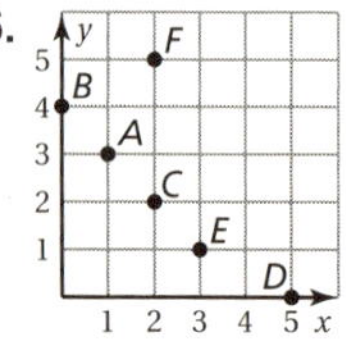

Answers

B.6 Start Thinking!

For use before Lesson B.6

Check students' work.

B.6 Warm Up

For use before Lesson B.6

1–6.

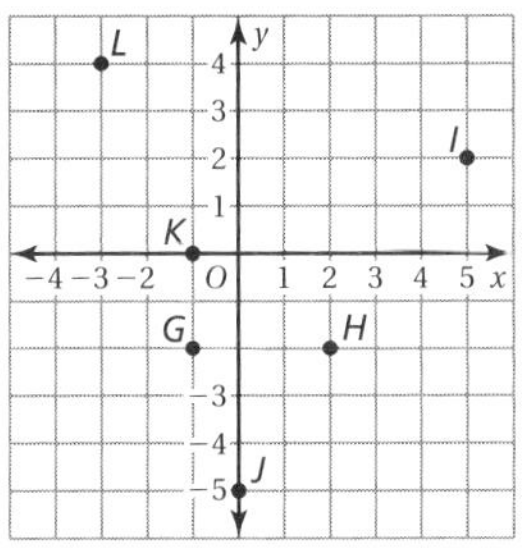

1. Quadrant III **2.** Quadrant IV **3.** Quadrant I

4. y-axis **5.** x-axis **6.** Quadrant II

B.6 Practice A

1. $(4, 0)$ **2.** $(2, 2)$ **3.** $(0, 1)$

4. $(-1, 4)$ **5.** $(-3, 2)$ **6.** $(-2, 0)$

7. $(-4, -4)$ **8.** $(-3, -2)$ **9.** $(0, -4)$

10. $(1, -3)$ **11.** $(2, -2)$ **12.** $(3, -4)$

13–24.

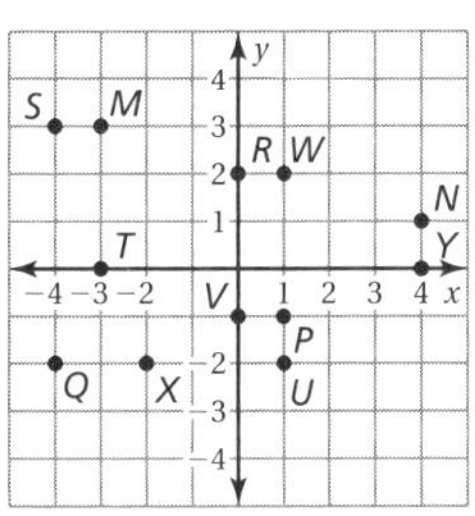

13. Quadrant II **14.** Quadrant I **15.** Quadrant IV

16. Quadrant III **17.** on the y-axis **18.** Quadrant II

19. on the x-axis **20.** Quadrant IV **21.** on the y-axis

22. Quadrant I **23.** Quadrant III **24.** on the x-axis

25. Moving left is negative. The correct answer is $(-6, -3)$.

26. They lie on a straight line.

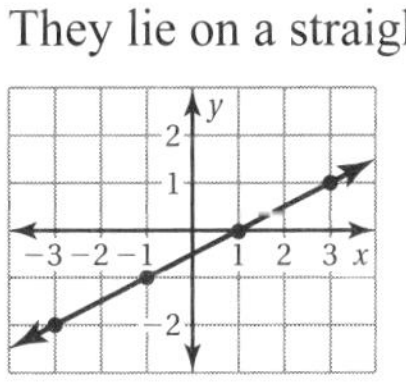

27. $(3, -2)$ **28.** $7\frac{1}{2}$ square units

B.6 Practice B

1. $(3, 3)$ **2.** $(2, 4)$ **3.** $(0, 3)$

4. $(-4, 4)$ **5.** $(-2, 1)$ **6.** $(-4, 0)$

7. $(-3, -1)$ **8.** $(-3, -3)$ **9.** $(0, -3)$

10. $(1, -4)$ **11.** $(3, -2)$ **12.** $(4, 0)$

13–24.

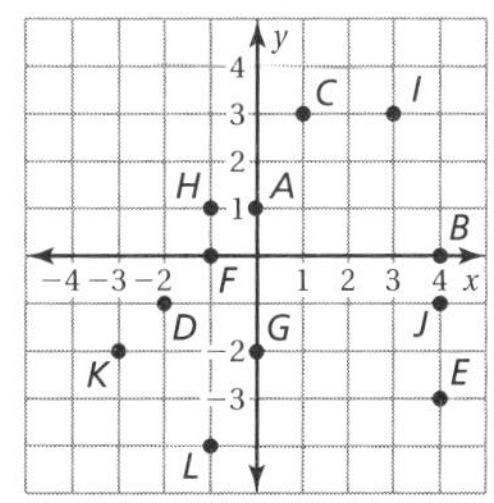

13. on the y-axis **14.** on the x-axis **15.** Quadrant I

16. Quadrant III **17.** Quadrant IV **18.** on the x-axis

19. on the y-axis **20.** Quadrant II **21.** Quadrant I

22. Quadrant IV **23.** Quadrant III **24.** Quadrant III

25. a. The vertical distance is 2 units, so go down 2 units from $(2, -1)$: the y-coordinate is -3. The horizontal distance is 5 units, so go right 5 from $(-3, -3)$: the x-coordinate is 2. The missing vertex is $(2, -3)$.

b. 14 units **c.** 10 square units

26. Move each vertex 1 unit left and 2 units up.

B.6 Enrichment and Extension

1–9.

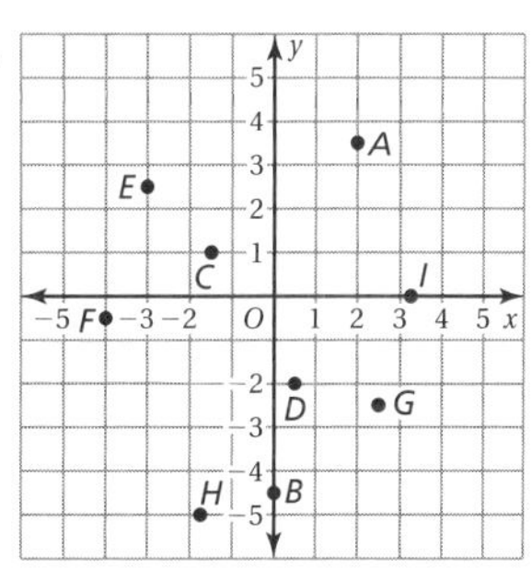

1. Point A is in Quadrant I.

2. Point B is on the y-axis.

3. Point C is in Quadrant II.

4. Point D is in Quadrant IV.

5. Point E is in Quadrant II.

Answers

6. Point F is in Quadrant III.

7. Point G is in Quadrant IV.

8. Point H is in Quadrant III.

9. Point I is on the x-axis.

10. *Sample answer:* $(-2, 4.25)$

11. *Sample answer:* $(4.5, 4.5)$

12. *Sample answer:* $(-3, -2.25)$

13. *Sample answer:* $(1, -1.75)$

14. Answers may vary. **15.** $(-2.5, -2)$

B.6 Puzzle Time

COOKIE SHEETS

B.7 Start Thinking!

For use before Activity B.7

Sample answer: The time spent walking and distance walked must both be positive. Yes, real-life graphs can have negative values—temperature and elevation are two examples.

B.7 Warm Up

For use before Activity B.7

1. 1 **2.** −1 **3.** −4

4. −3 **5.** 4 **6.** −5

B.7 Start Thinking!

For use before Lesson B.7

Monitor students playing the game.

B.7 Warm Up

For use before Lesson B.7

1.

x	−2	−1	0	1	2
y	8	7	6	5	4

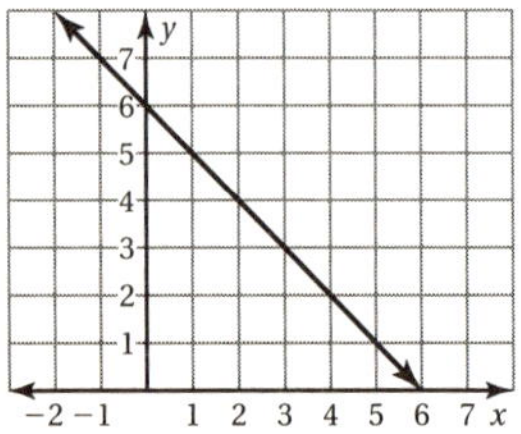

2.

x	−2	−1	0	1	2
y	0	−1	−2	−3	−4

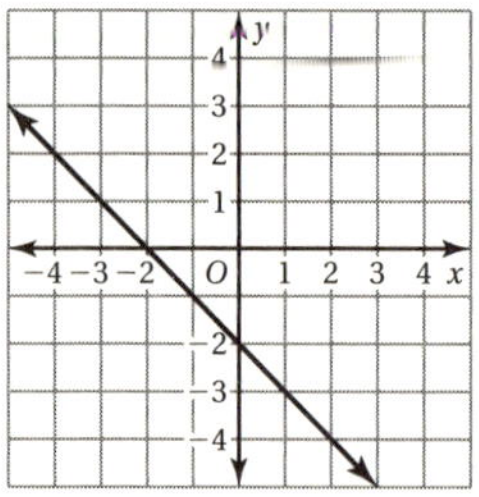

3.

x	−2	−1	0	1	2
y	2	1	0	−1	−2

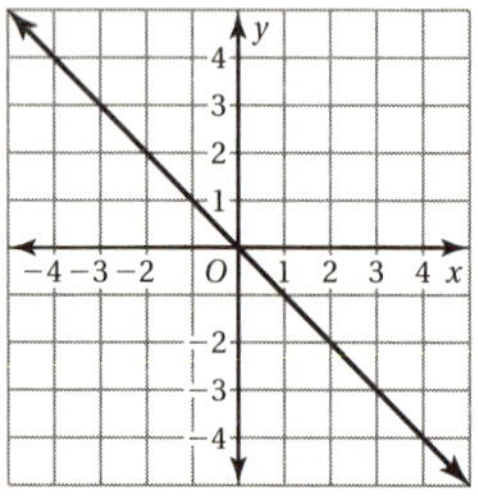

4.

x	−2	−1	0	1	2
y	−1	0	1	2	3

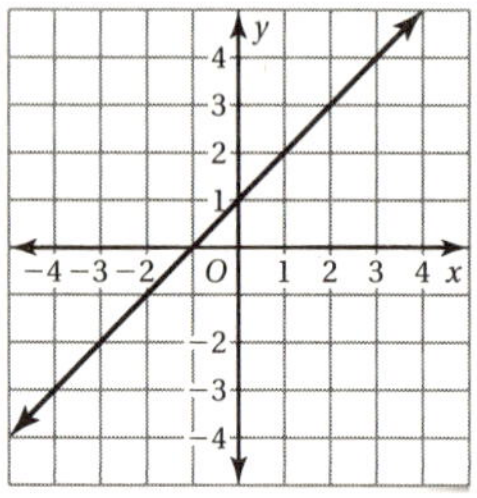

5.

x	−2	−1	0	1	2
y	−4	−2	0	2	4

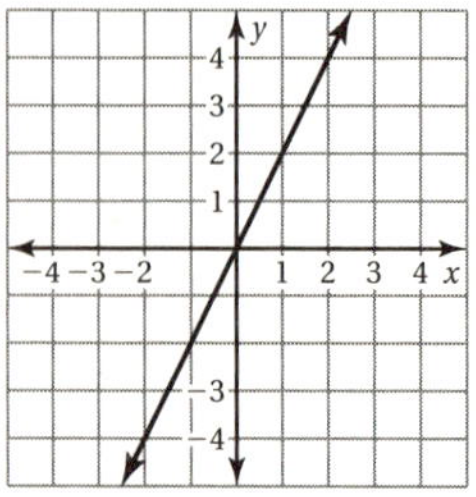

Answers

6.

x	−2	−1	0	1	2
y	2	3	4	5	6

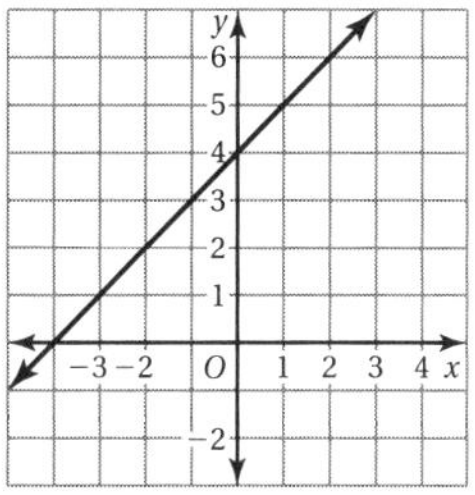

B.7 Practice A

1. The points lie on a line.

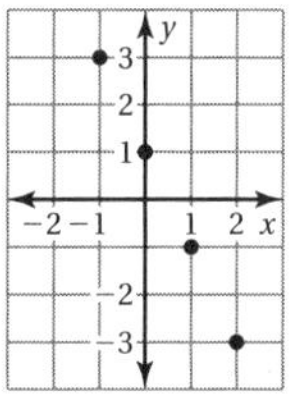

2. The points lie on a line.

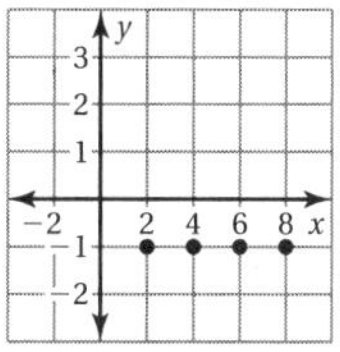

3. The points lie on a line.

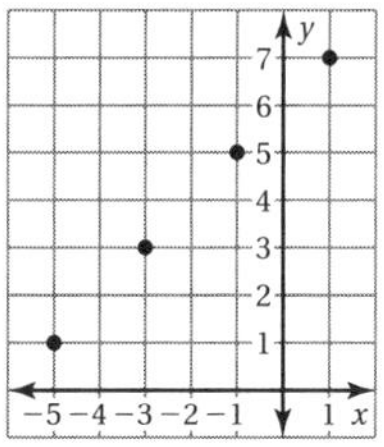

4. The points do not lie on a line.

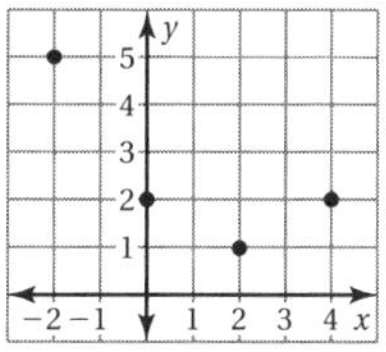

5.

Input, x	−4	−2	0	2	4
Output, y	−8	−4	0	4	8

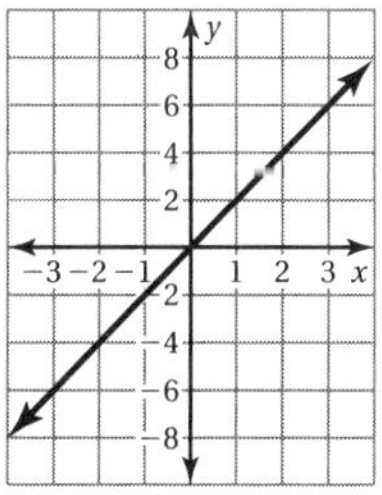

6.

Input, x	−4	−2	0	2	4
Output, y	−4	−2	0	2	4

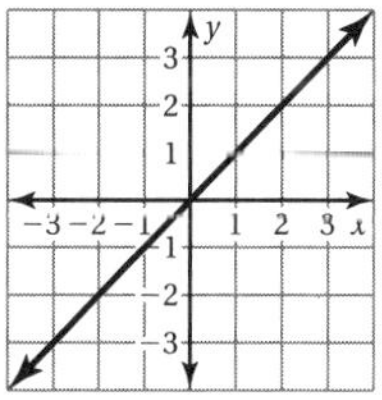

7.

Input, x	−4	−2	0	2	4
Output, y	−7	−3	1	5	9

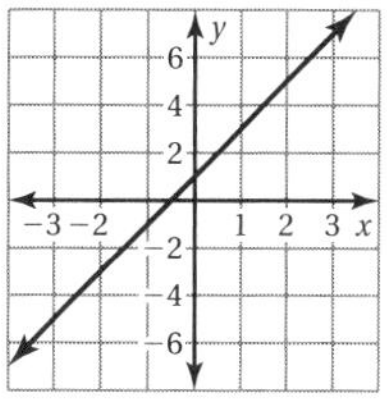

8.

Input, x	−4	−2	0	2	4
Output, y	−3	−1	1	3	5

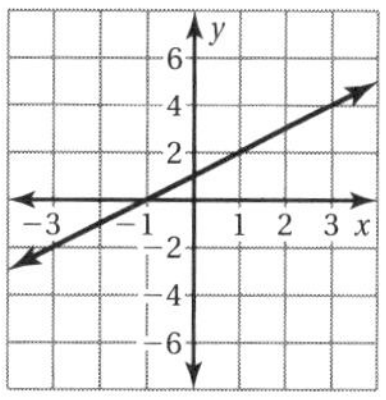

9.

Input, x	−4	−2	0	2	4
Output, y	−11	−5	1	7	13

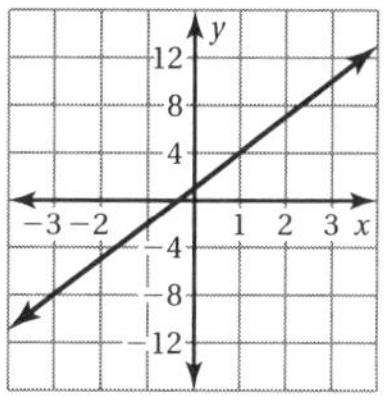

10.

Input, x	−4	−2	0	2	4
Output, y	2	0	−2	−4	−6

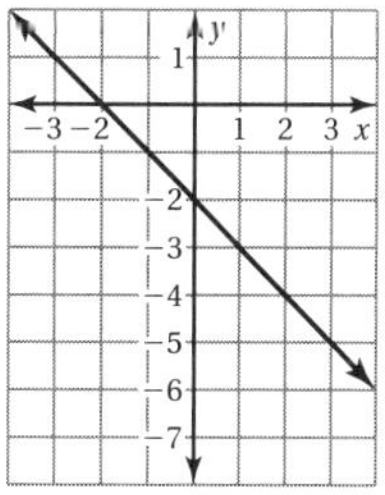

Answers

11.

Input, x	−4	−2	0	2	4
Output, y	−24	−14	−4	6	16

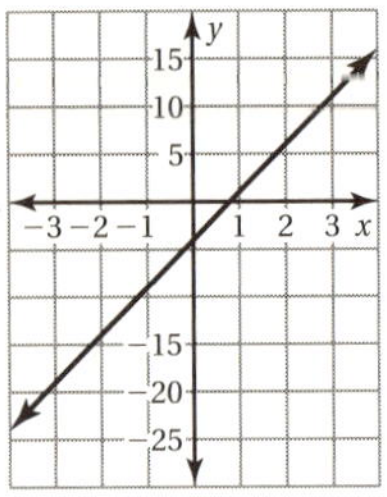

12.

Input, x	−4	−2	0	2	4
Output, y	−4	0	4	8	12

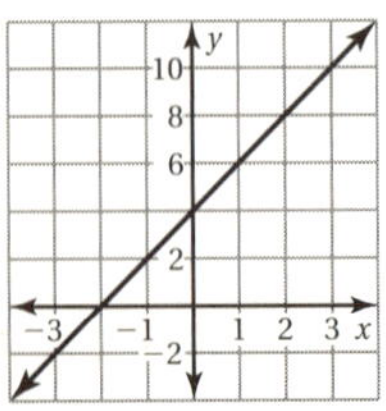

13. B

14. A

15. C

16. a. $C = V \div 10$ or $C = 0.1V$

b. The current would decrease by 2 amperes. Every decrease of 10 volts corresponds to a decrease of 1 ampere.

c. 6 amperes

17. $T = 4 - 2h$

Input, h	0	1	2	3	4	5
Output, t	4	2	0	−2	−4	−6

B.7 Practice B

1.

Input, x	−3	−1	0	1	3
Output, y	−9	−5	−3	−1	3

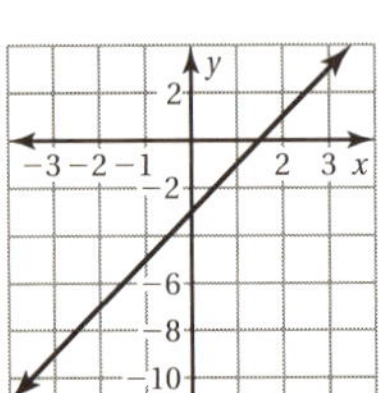

2.

Input, x	−3	−1	0	1	3
Output, y	8	6	5	4	2

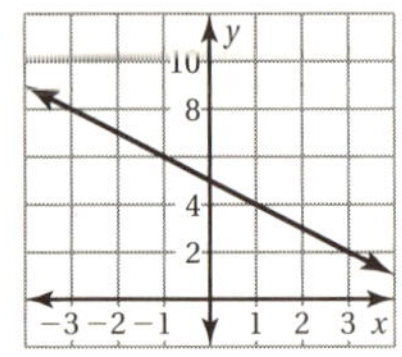

3.

Input, x	−3	−1	0	1	3
Output, y	−6	0	3	6	12

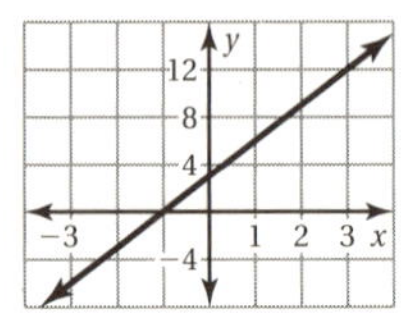

4.

Input, x	−3	−1	0	1	3
Output, y	2	0	−1	−2	−4

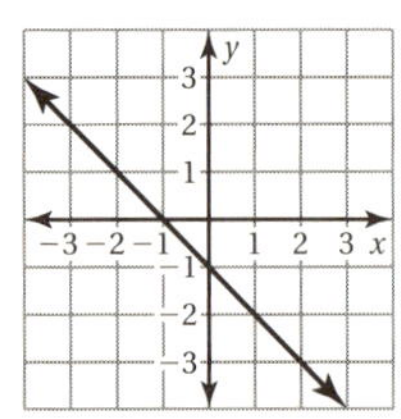

5.

Input, x	−3	−1	0	1	3
Output, y	1	1	1	1	1

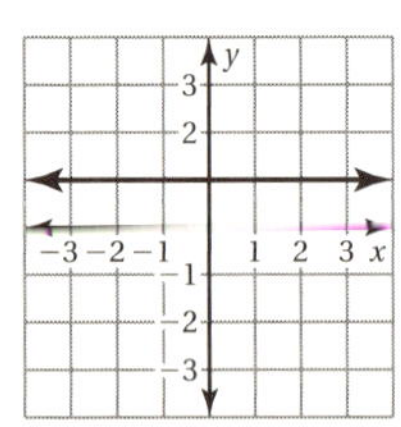

6.

Input, x	−3	−1	0	1	3
Output, y	−22	−14	−10	−6	2

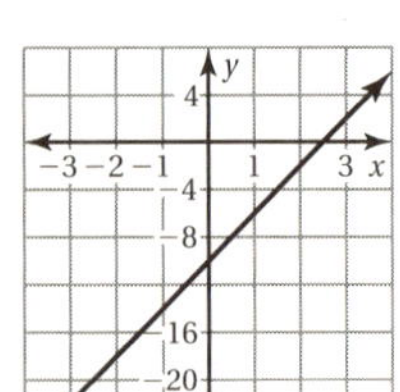

Answers

7.

Input, x	−3	−1	0	1	3
Output, y	0	−2	−3	−4	−6

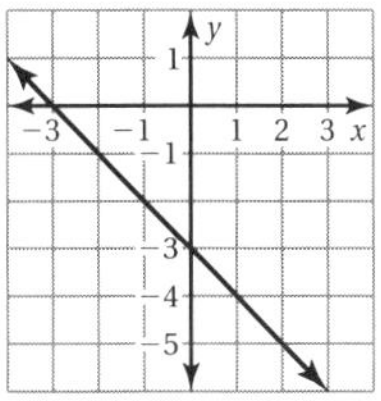

8.

Input, x	−3	−1	0	1	3
Output, y	−2	0	1	2	4

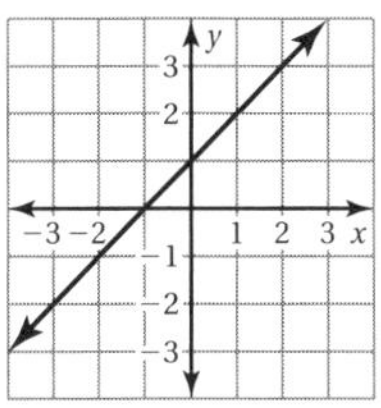

9. B **10.** C **11.** A

12. $y = -x$ **13.** $y = 5x$

14. a. Cooking time increases by 1 hour.

b. 3.5 hours; Because the data form a straight line and 12 is halfway between 10 and 14, the answer is halfway between 3 and 4, or 3.5.

15 a. $d = -30 + 2t$

b. −10 meters

c. After 15 minutes

B.7 Enrichment and Extension

1. a.

x	−2	−1	0	1	2
y	2.5	2.5	2.5	2.5	2.5

b.

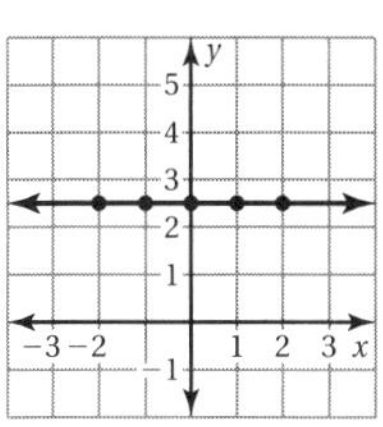

The graph of $y = 2.5$ is a horizontal line.

2. a.

x	−3	−3	−3	−3	3
y	−2	−1	0	1	2

b.

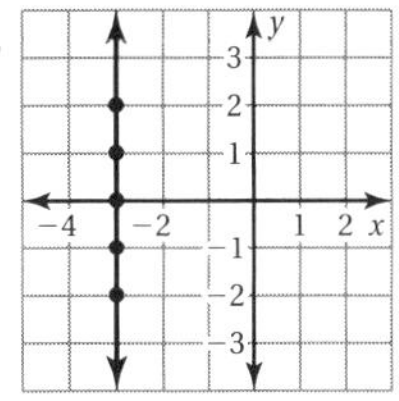

The graph of $x = -3$ is a vertical line.

3. False; the graph of $y = a$ is a horizontal line.

4. False; the graph of $x = b$ is a vertical line.

5. true **6.** true **7.** true

8. False; $x = b$ is not a function.

B.7 Puzzle Time

NEW JERSEY

School-to-Work

1. −89, −8, 20, 35, 45, 60, 75, 80

2. 80 **3.** −89 **4.** \$268

5. 972; paid out more

6. \$760 to Seafood Market, \$95 to Wholesale Baking

7. \$535 **8.** Merry Farm

Study Help

Available at *BigIdeasMath.com*.

Financial Literacy

1.

Input	1	2	3	4
Output	\$600	\$1500	0	−\$1375

Input	5	6
Output	−\$1000	−\$125

2.

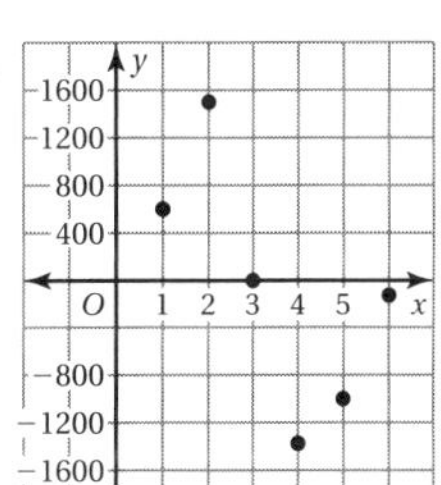

3. In Sept, Oct; out Dec, Jan, Feb

Answers

4.

Input	1	2	3	4
Output	\$600	\$2100	\$2100	\$725

Input	5	6
Output	−\$275	−\$400

5. 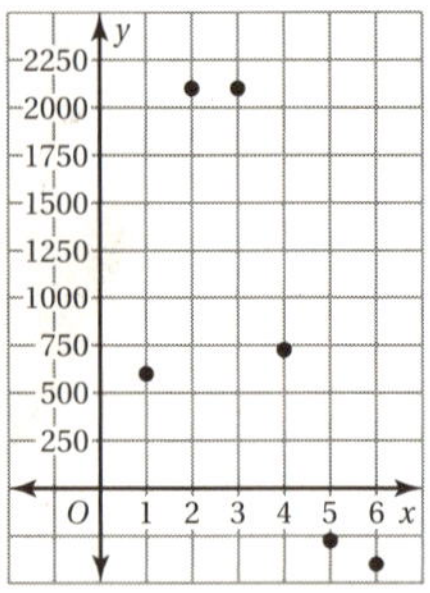

6. In month 5, January

7. \$400; at the beginning of January

8. \$900; 180 tickets

Technology Connection

1. −4.4 **2.** 5 **3.** −92.8

4. −3.1 **5.** −1.2 **6.** −7.9

7. −0.18 **8.** −32 **9.** −9

10. −9.9 **11.** −9 **12.** −0.19

Life Connections

1. 9 yd; no **2.** 12 yd; yes **3.** 8 yd; no

4. 6 yd; no **5.** 21 yd; yes **6.** 37 yd; yes

7. increased; 24 yards **8.** Team B's 10-yard line

9. Team B's 31-yard line **10.** Team B's 23-yard line

11. Team A's 38-yard line **12.** Team A's 20-yard line

Cumulative Practice

1. 1 **2.** $\frac{1}{4}$ **3.** 4 **4.** $\frac{3}{8}$

5. −19 **6.** 15.7 **7.** $\frac{4}{5}$ **8.** 25

9. −3 **10.** −1.77 **11.** −11 **12.** $\frac{5}{16}$

13. 0 **14.** −8 **15.** 1 **16.** 3

17. −1 **18.** 4

19. 40 + (−30) + 20; 30 feet

20. −250 + (−100); 350 feet

21. 4; 4 is to the right of −9 on a number line.

22. −2.05; −2.5 = −2.50, and −2.05 is to the right of −2.50 on a number line.

23. $-\frac{1}{6}$; $-\frac{1}{3} = -\frac{2}{6}$, and $-\frac{1}{6}$ is to the right of $-\frac{2}{6}$ on a number line.

24. 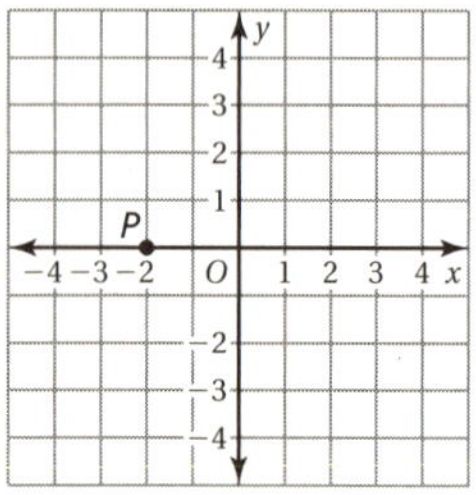

The point is on the *x*-axis.

25. 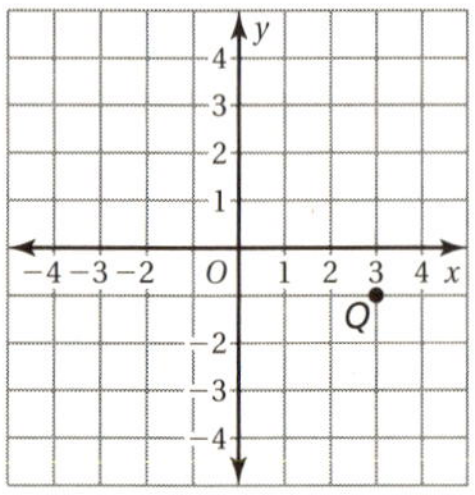

The point is in Quadrant IV.

26. 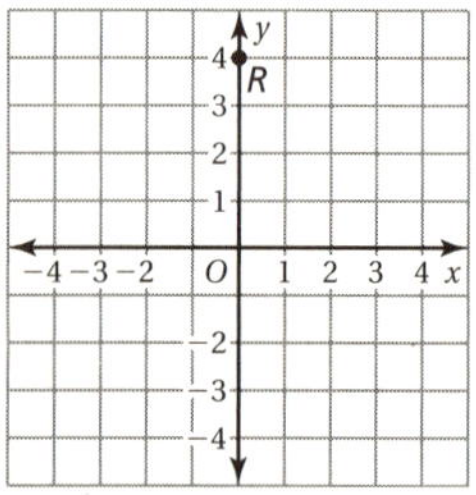

The point is on the *y*-axis.

27.

Input, *x*	−2	−1	0	1	2
Output, *y*	3	2	1	0	−1

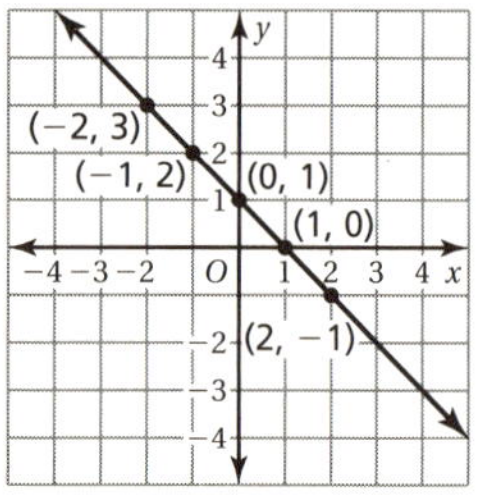

28. −32 meters; 3 meters greater